VEGETABLES I

HANDBOOK OF PLANT BREEDING

Editors-in-Chief:

JAIME PROHENS, *Universidad Politecnica de Valencia, Valencia, Spain*
FERNANDO NUEZ, *Universidad Politecnica de Valencia, Valencia, Spain*
MARCELO J. CARENA, *North Dakota State University, Fargo, ND, USA*

Volume 1
Vegetables I: Asteraceae, Brassicaceae, Chenopodicaceae, and Cucurbitaceae
Edited by Jaime Prohens and Fernando Nuez

Volume 2
Vegetables II: Fabaceae, Liliaceae, Solanaceae and Umbelliferae
Edited by Jaime Prohens and Fernando Nuez

VEGETABLES I

Asteraceae, Brassicaceae, Chenopodicaceae, and Cucurbitaceae

Edited by

Jaime Prohens

Universidad Politecnica de Valencia
Valencia, Spain

and

Fernando Nuez

Universidad Politecnica de Valencia
Valencia, Spain

Springer

Jaime Prohens
COMAV-UPV
Universidad Politecnica de Valencia
14 Camino de Vera
Valencia 46022
Spain
jprohens@btc.upv.es

Fernando Nuez
COMAV-UPV
Universidad Politécnica de Valencia
14 Camino de Vera
Valencia 46022
Spain
fnuez@btc.upv.es

ISBN: 978-0-387-72291-7 e-ISBN: 978-0-387-30443-4

Library of Congress Control Number: 2007936360

Cover illustration: A lettuce seed production field near Fresno, California, U.S.A. (courtesy of Y. Peng)

Printed on acid-free paper.

9 8 7 6 5 4 3 2 1

springer.com

Contents

Preface ...... **vii**

Contributors ...... **ix**

Family Asteraceae (=Compositae)

1. Chicory and Endive **3**
Margherita Lucchin, Serena Varotto, Gianni Barcaccia, and Paolo Parrini

2. Globe Artichoke and Cardoon **49**
Sergio Lanteri and Ezio Portis

3. Lettuce **75**
Beiquan Mou

Family Brassicaceae (=Cruciferae)

4. Cabbage and Kale **119**
Amando Ordás and M. Elena Cartea

5. Cauliflower and Broccoli **151**
Ferdinando Branca

Family Chenopodiaceae

6. Spinach **189**
Teddy E. Morelock and James C. Correll

7. Table Beet 219
Irwin L. Goldman and John P. Navazio

Family Cucurbitaceae

8. Cucumber 241
Jack E. Staub, Matthew D. Robbins, and Todd C. Wehner

9. Melon 283
Michel Pitrat

10. Pumpkin and Winter Squash 317
María Ferriol and Belén Picó

11. Summer Squash 351
Harry S. Paris

12. Watermelon 381
Todd C. Wehner

Index 419

Preface

The production and consumption of vegetables has expanded dramatically in the last years, with a global growth in the production of more than 50% in the last decade, a rate of increase that is much higher than for other plant commodities. Vegetables constitute an important part of a varied and healthy diet and provide significant amounts of vitamins, antioxidants and other substances that prevent diseases and contribute to an improvement in the quality of life. In consequence, it is expected that in the coming years, vegetable crops production will continue its expansion.

Improved varieties have had a main role in the increases in yield and quality of vegetable crops. In this respect, the vegetables seed market is very dynamic and competitive, and predominant varieties are quickly replaced by new varieties. Therefore, updated information on the state of the art of the genetic improvement of specific crops is of interest to vegetable crops breeders, researchers and scholars. During the last years an immense quantity of new knowledge on the genetic diversity of vegetables and the utilization of genetic resources, breeding methods and techniques, and on the development and utilization of modern biotechnologies in vegetables crop breeding has accumulated, and there is a need of a major reference work that synthesizes this information. This is our objective.

The diversity of vegetable crops is appalling, with hundreds of species being (or having been) grown. However, among this plethora of crops, there are some which are prominent, and for which there has been a greater development in the breeding science and development of varieties. In consequence, we have produced two volumes devoted to 20 of these most important vegetable crops. These crops belong to eight different botanical families. Because in many cases crops from the same botanical family share many reproductive, physiological, and agronomic features, as well as similar breeding techniques, we have decide to group them by this taxonomic category. In this respect, this first volume includes 12 chapters that deal with vegetables that belong to four families: Asteraceae or Compositae (chicory and endive, globe artichoke and cardoon, and lettuce), Brassicaceae or Cruciferae (cabbage, and cauliflower and broccoli), Chenopodiaceae (spinach and sugar beet) and Cucurbitaceae (cucumber, melon, pumpkin and winter squash, summer squash, and watermelon).

Chapters have been written by outstanding breeders with wide experience in the crop treated. Each chapter includes information on the origin and domestication, varietal groups, genetic resources, major breeding achievements and current goals of breeding, breeding methods and techniques, integration of the new biotechnologies in the breeding programmes, and the production of seed of specific crops.

The completion of this book would not have been possible without the contributions of the many authors, who have devoted much time to the task of writing the chapters. We also want to thank the staff of Springer, in particular Jinnie Kim and Shoshana Sternlicht, who have made possible to produce a high quality book in a very short time span. We are also indebted to many colleagues for useful suggestions that have contributed to improve this book.

Valencia, Spain

Jaime Prohens
Fernando Nuez

Contributors

Gianni Barcaccia
Dipartamento di Agronomia Ambientale e Produzioni Vegetali, University of Padova, 35122 Padova, Italy

Ferdinando Branca
Dipartimento di OrtoFloroArboricultura e Tecnologie Agroalimentari (DOFATA), Università di Catania, Via Valdisavoia 5, 95123 Catania, Italy

M. Elena Cartea
Misión Biológica de Galicia, Spanish National Research Council, P.O. Box 28, 36080 Pontevedra, Spain

James C. Correll
Department of Plant Pathology, University of Arkansas, Fayetteville, AR 72701, USA

María Ferriol
Institute of Mediterranean Agroforestry, Universidad Politécnica de Valencia, Camino de Vera 14, 46022 Valencia, Spain

Irwin L. Goldman
Department of Horticulture, University of Wisconsin, 1575 Linden Drive, Madison, WI 53706, USA

Sergio Lanteri
Di.Va.P.R.A., Plant Genetics and Breeding, University of Torino, Via Leonardo da Vinci 44, 10095 Grugliasco, Italy

Margherita Lucchin
Dipartamento di Agronomia Ambientale e Produzioni Vegetali, University of Padova, 35122 Padova, Italy

Beiquan Mou
United States Department of Agriculture, Agricultural Research Service, 1636 E. Alisal St., Salinas, 93905 CA, USA

Teddy E. Morelock
Department of Horticulture, University of Arkansas, Fayetteville, AR 72701, USA

John P. Navazio
Abundant Life Seeds, P.O. Box 157, Saginaw, OR 97472, USA

Amando Ordás
Misión Biológica de Galicia, Spanish National Research Council, P.O. Box 28, 36080 Pontevedra, Spain

Harry S. Paris
Department of Vegetable Crops, Newe Ya'ar Research Center, Agricultural Research Organization, P.O. Box 1021, Ramat Yishay 30095, Israel

Paolo Parrini
Dipartamento di Agronomia Ambientale e Produzioni Vegetali, University of Padova, 35122 Padova, Italy

Belén Picó
Institute of Conservation and Improvement of Agrodiversity, Universidad Politécnica de Valencia, Camino de Vera 14, 46022 Valencia, Spain

Michel Pitrat
Unité de Génétique et Amélioration des Fruits et Légumes, Station d'Amélioration des Plantes Maraîchères, Institut National de la Recherche Agronomique, Domaine St. Maurice, B.P. 94, 84143 Montfavet, France

Ezio Portis
Di.Va.P.R.A., Plant Genetics and Breeding, University of Torino, Via Leonardo da Vinci 44, 10095 Grugliasco, Italy

Matthew D. Robbins
USDA, ARS, Department of Horticulture, University of Wisconsin, 1575 Linden Drive, Madison, WI 53706, USA

Jack E. Staub
USDA, ARS, Department of Horticulture, University of Wisconsin, 1575 Linden Drive, Madison, WI 53706, USA

Serena Varotto
Dipartamento di Agronomia Ambientale e Produzioni Vegetali, University of Padova, 35122 Padova, Italy

Todd C. Wehner
Department of Horticultural Science, North Carolina State University, Raleigh, NC 27695-7609, USA

Family Asteraceae (=Compositae)

Chicory and Endive

Margherita Lucchin[1], Serena Varotto[1], Gianni Barcaccia[1], and Paolo Parrini[1]

[1] University of Padova, Dipartimento di Agronomia Ambientale e Produzioni Vegetali, paolo.parrini@unipd.it

1 Introduction

Chicory and Endive are common names that correctly indicate two different species. Their conversational use may nevertheless bring about some misunderstanding as they do not only refer to a series of different leafy vegetables but, more extensively, to substantially different crops from which many different products are obtained.

As leafy vegetables, chicory and endive are much less used than lettuce or cabbages, but they are anyway among the most known and popular horticultural products in the world and, although with great differences in cultural practices and type of utilization, they are diffused in almost every country and are included in the diet of most western as well as eastern populations. Mainly known as important components for fresh salads, they are also often cooked and differently prepared according to traditions and alimentary habits.

Chicory and endive are two traditional European horticultural crops and, although they cannot be considered as autochthonous, their evolution as vegetable crops has taken place in continental Europe where they have gradually differentiated in a variety of cultivated types.

Actually, the name "Endive" only indicates a leafy vegetable crop whose cultivated material usually refers to two groups of cultivars: the "Escarole Group" and the "Curled Endive Group". On the other side, the term "Chicory" indicates at least two kinds of crops: a leafy vegetable, very differentiated according to several cultural types, and a root crop whose industrial utilization seems at present mainly addressed to inulin extraction or, on a more limited scale, to the production of a coffee substitute. Both these types of "root chicory", have the same origin as they have been derived from the so called "Magdeburg chicory", the ancient root chicory known and traditionally used in some European countries as a coffee substitute since the end of 16th century and that gained outstanding importance with the continental block at the time of Napoleon. Also a very important leafy vegetable, the so called "Witloof chicory" or "Belgian endive", perhaps the most known among the leafy chicories, has to be considered a derivative of the Magdeburg chicory as it seems commonly accepted that its first well known pale yellowish sprouts have casually been obtained by a Belgian farmer who, around 1870, had observed and harvested them from a stock of roots piled up in autumn and left apart during the cold season, waiting to be dried, grounded, and toasted.

Lacking comprehensive, homogeneous, sufficiently detailed, and univocal data on horticultural productions and trade, it is impossible to give reliable figures on the diffusion and economic importance of the two cultures in Europe, where chicory and endive are mostly grown.

In the most recent statistics concerning the European market (EU Market Survey 2004 for fresh fruit and vegetables) too, chicory and endive are often confused under the general voice "salads", or considered together with lettuce which is by far the most important leafy vegetable at both European and world-wide scale. The situation is not very different if one considers, as a source of reliable information, the statistics of each single country. On the basis of accessible data it is however possible to figure out that Belgium, France, Italy, and Netherlands are the almost exclusive producers of chicory and endive. These two crops do not give a great contribution to each country's total agricultural income, but they are very important at local level, as they characterize the agriculture of limited areas where from 80 to 90% of the country's production is concentrated. This is the case of France, where 86% of the more than 15.000 ha of Witloof chicory grown in the country are localized in four northern departments, or of Italy, where the north eastern region accounts for 87% of the national acreage and 84% of the national production of that particular type of red or variegated chicory known as "Radicchio". The Escarole and Curly endive types, in both France and Italy, are much less concentrated and may extend far south where they are usually grown. The same occurs with other chicory types like the "Chicory of Catalogne". By the way, regarding the possible confusion between the two terms, it may be interesting to note that the Witloof chicory is officially registered by French statistics as "Endive", while the Escarole and the Curly endive are identified as "Chicorèe".

Furthermore, it is perhaps worth noting that chicory and endive are not only important for the local economies, but they may have significance at an international trade scale too. Altogether, the US imports of chicory in 2002 have been equal to 5996 Mt for a value of $ 8.193.000. About a half of these amounts, both in quantity and value, are represented by Witloof chicory, whose imports from Belgium and Netherlands sum up to more than 90% of the whole figure. Thus, although still on a regional scale, chicory and endive have their own place among more known and used vegetables and may represent a significant source of income for farmers in areas where they have been traditionally present.

Within this frame, two observations may be added. The first concerns the marked decrease of US not qualified chicory imports from Europe, in particular from Belgium, Netherlands, and Italy (2622 Mt in1996, 536 Mt in 2002), and the increase of imports from Central and Southern America (1046 Mt in 1996, 2522 Mt in 2002). Compared to the stable or lightly increasing figures recorded during the same period for Witloof chicory (between 2000 and 2400 Mt) this trend seems to indicate that Witloof has taken an advantage thanks to the quality and standardization of the marketable product. The second observation regards the "Radicchio" which is now considered with more and more attention both in Europe and in the US, as well as in other overseas countries, where its cultivation has started some years ago and seems to have an increasing evaluation as its red or variegated leaves are particularly appreciated as a component of prepared salads.

Finally, the recognised value of some compounds present in chicory's roots and leaves may enhance its appreciation beyond the horticultural use and has to be underlined to completely figure out the potentiality of the plant and the possible breeding goals. From this point of view, the industrial use of chicory for inulin production deserves particular attention. In Belgium, the acreage dedicated to this utilization has been constantly increasing during the last ten years and has passed from 11.700 ha in 1997 to 15.700 ha in 2005.

2 Taxonomy and Origin

Both Chicory (*Cichorium intybus* L.) and Endive (*Cichorium endivia* L.) belong to the family *Asteraceae*, a very large family with about 23.000 species subdivided in 1535 genera grouped in three subfamilies: *Barnadesioideae*, *Cichorioideae*, and *Asteroideae* (Bremer et al., 1994). The tribe *Lactuceae*, in the subfamily *Cichorioideae*, includes the genus *Cichorium* within which different species are recognized according to the source. Tutin et al. (1976), referring to the European flora, describe the three species *C. spinosum*, *C. intybus*, and *C. endivia* and subdivide this last one in subsp. *endivia* (cultivated), and subsp. *divaricatum* (wild). Pignatti (1982), taking into account the Italian flora, refers to the three wild species *C. spinosum*, *C. intybus*, with the var. *glabratum* (Presl) Fiori, and *C. pumilum*, maintaining *C. endivia* as a cultivated species only. In a revision of the genus made by Bedarff in 1985 (cited by Kiers, 2000) and partially published by Wagenitz and Bedarff (1989), seven species were described on the basis of morphological characters, and C. *endivia* and C. *intybus* were further divided in two subspecies (*C. endivia* subsp. *endivia* and C. *endivia* subsp. *divaricatum*; *C. intybus* subsp. *intybus* and *C. intybus* subsp. *glabratum*). This classification does not agree with the one given by the Royal Botanical Garden which, in the Flora Europea section, only referring to *C. intybus*, indicates three subspecies: subsp. *foliosum* (Hegi) Janch., subsp. *glabratum* (C. Presl) Arcang., and subsp. *sativum* (Bisch.) Janch.

Kiers et al. (2000), integrating morphological characters with molecular observations, describe the two cultivated and most known species *C. intybus* and *C. endivia* and the two wild species *C. spinosum* and *C. pumilum*. Moreover, two additional species, never observed in Europe, are added, *C. calvum* and *C. bottae*, the former endemic to the dry and hot environments of Middle East and South Western Asia and the latter from Yemen and Saudi Arabia. More recently, Conti et al. (2005), in their study of the Italian flora, recognized three species in the genus: *C. endivia*, with the two subspecies *endivia* Hegi and *pumilum* (Jacq) Cout., *C. intybus*, with the two subspecies *glabratum* (C. Presl) Arcang. and *intybus*, and C. *spinosum*.

Since the early nineties, when the analysis of DNA fragments became more and more familiar to taxonomists, several studies have allowed the task to explore, and possibly clarify, the relationships among the two cultivated species - *C. intybus* and *C. endivia* - and their wild relatives. Vermeulen et al. (1994), using mitochondrial RFLPs, suggest that *C. spinosum* may be considered an ecotype of *C. intybus* rather than a separate species. Gemeinholzer and Bachmann (2005), with other and more sensible molecular methods (ITS, AFLP, SSR), were unable to discriminate between these two species which, on the contrary, could be clearly delimited with two diagnostic and one overlapping morphological character. On the basis of chloroplast DNA RFLPs and chloroplast DNA and nuclear rDNA sequence analysis (Kiers et al.,1999) or using AFLP markers (Kiers et al., 2000) it has been confirmed that *C. intybus* is closely related to *C. spinosum*, while *C. endivia*, *C. pumilum*, and *C. calvum* show a close molecular resemblance among each other and are fairly well separated from the first two. The sixth species, *C. bottae*, has to be considered a sister species.

Besides morphological and molecular resemblances or diversities, a distinction among these six species can be made on the basis of their life cycle and reproductive system. Thus, two groups may be established: on one side *C. intybus*, *C. spinosum*, and *C. bottae*, perennials that are characterized by a strong self-incompatibility system, on the other *C. endivia*, *C. pumilum*, and *C. calvum*, annual and self-compatible species. Within this frame, the names of the recognized botanical varieties do not appear, although it is from them that the various cultivated types have originated.

The origin and differentiation of the genus is concordantly located in South-Eastern Europe, the Eastern Mediterranean basin and the South Western Asia. Within this large centre, *C. intybus* and *C. endivia* partially share their area of origin which, for *C. intybus*, tends to be located in the southern Balkan peninsula and northern Middle East and, for C. *endivia*, is claimed to be the whole Middle East with an extension to the northern Arabic peninsula. From there, they firstly migrated in the whole Mediterranean basin and, on the other side, toward Southern and Eastern Asia where they seem to have found different areas of diffusion and adoption as horticultural crops.

At present, *C. intybus* is mainly grown all over continental Europe, in South Western Asia, and on limited areas in Northern America, South Africa, and Australia. *C. endivia*, other than in continental Europe, perhaps in accordance with its more southern origin, is grown in Central and Southern America and all along the Mediterranean coast of the African continent. In Asia, it seems to have found particularly favourable conditions in the eastern part of the continent, on an area which includes South Eastern China, Korea, and the eastern part of Inner Mongolia.

Probably known by the Egyptians and used as food and/or medicinal plants by ancient Greeks and the Romans, in Europe both species gradually underwent a process of naturalization and, as said before, although they cannot be considered as autochthonous species, they became part of the natural and agricultural European flora. Thus, *C. endivia*, traditionally indicated only as a cultivated species, may be found in the spontaneous flora of at least some Italian regions (Conti et al., 2005). *Cichorium intybus* covers, as wild, a great portion of the whole European continent and traditionally enters into the diet of local populations as an important ingredient of typical local dishes. This might be the consequence and, at the same time, the cause of the great differentiation among a number of types which, mostly within *C. intybus*, have originated an always increasing number of cultivar groups, types, populations which, altogether, make the horticultural landscape of the genus *Cichorium* particularly rich and interesting from a historic, cultural, agronomic, commercial, and scientific point of view.

Several extensive lists of *Cichorium* species, subspecies, botanical varieties, and cultivar groups, have been published and are present in accessible internet sites, where scientific and technical news often mix up with commercial promotion, forming a mass of information not always easy to be interpreted.

The most exhaustive seems to be the list given by Mansfeld's World Database of Agricultural and Horticultural Crops (IPK Gatersleben 2002) which includes 49 entries. Many of them are synonyms and often refer to differences among commercial types rather than to taxonomic distinctiveness. Almost as large is the list at the web site of Melbourne University (2003) - M. H. Porcher mantainer – where 29 entries of species and synonyms are given together with a large, although not always accurate, picture of the cultivated types. More restricted, although still large, is the list given by GRIN (Germplasm Resources Information Network - USDA) where cultivar groups do not appear, but synonym subspecies and botanical varieties are nevertheless taken into consideration.

Altogether, at least six cultivar groups, mainly differentiated on the basis of their use, are recognizable (Kiers et al., 1999, 2000; Kiers, 2000; Van Stallen et al., 2001).

Aiming to schematise in a readable manner the whole of these information, a synopsis is proposed in table 1 where a correspondence between taxonomy, cultivar group and most frequent and known utilization has been attempted.

3 Biological Features

Although they strongly resemble each other on the basis of morphological characters, *C. intybus* and *C. endivia* have always been considered as two different species. For an accurate morphological description see Kiers et al. (1999). Here a very synthetic picture is proposed where attention is mainly brought on the life cycle and the breeding system, i.e. on characters and features particularly concerned with breeding and, as such, proper of the cultivated types rather than of wild or naturalized species. From this point of view it has to be stressed that a description of the behaviour of the two species is strictly dependent on the environmental and cultural conditions and, in particular, on the latitude which one refers to. Since both species are grown under very different situations, it is worth to underline that we will consider what occurs with direct sowing or transplanting in open field at a latitude of about 45° N, as this is the average latitude of North Eastern Italy where both species are usually grown using different cultural techniques: in greenhouse, under temporary covers, or in the open field.

Table 1. Chicory and Endive: European species of *Cichorium*, cultivar groups and use.

Taxonomic position	Cultivar group	Use
C. endivia		
subsp. *endivia*	wild [a]	
var. *latifolium*	Endive	salads
var. *crispum*	Crispum	salads
subsp. *pumilum*	wild [a]	
C. intybus		
subsp. *intybus*	wild [b]	
var. *foliosum*	Witloof chicory	cooked/salads
	Pain de sucre	cooked/salads
	Radicchio	salads
	Catalogne	cooked
var. *sativum*	Root chicory (roasted)	coffee substitute
	Root chicory (industrial)	inulin extraction
	Root chicory	cooked
subsp. *glabratum*	wild [b]	
C. spinosum	wild [b]	

[a]Exotic, naturalized in Europe, and [b]autochthonous (Conti et al. 2005).

Both species have a tap root which in *C. endivia* is subdivided in parallel branches and may deepen in the soil down to over 1 m (Tesi, 1965), while in *C. intybus* is larger and unique. Particularly large is in the industrial types, whose commercial product is the root, or in the horticultural types whose commercial product is the bunch of leaves obtained through "forcing", as it happens with the "Witloof chicory" or with the late type of "Radicchio di Treviso".

C. *endivia* has to be considered an annual, in as much as, independently of the moment of sowing or transplanting, between May and August, if temperature is sufficiently high, the plant forms an enlarged rosette of leaves which are characterized by a wide midrib and an extended flat (var. *latifolium*) or crisp (var *crispum*) lamina and, immediately thereafter, develops a flowering stalk.

C. intybus is a biennial or, in the wild, a perennial species. An early sowing or transplanting in spring, under long days, although with differences according to the cultivar group, brings about an almost generalized flowering. If sowing or transplanting are delayed to the month of July, the plant forms a rather loosen rosette, or a fairly compact "head", which remains in the field until the following spring when, between May and June, the central bud develops in a stem bearing, as in *C. endivia,* blue "flowers" (rarely white or mauve).

On the flowering stalk many clusters of 4-6 sessile "flowers" (2-4, rarely 8, in *C. intybus*) are inserted in axillary position, or single "flowers" are brought at the end of peduncles 10-20 cm long (4-7, rarely up to 13, in *C. intybus*). The "flower" is actually an inflorescence (*capitulum*) which is typical of the whole family and is a cluster of 15-25 single hermaphrodite flowers, borne on a receptacle and protected by an involucre. Each single flower has a gamopetalous and ligulate corolla, and bears five filamentous stamens fused by their anthers to form a column surrounding a pistil with a bifid stigma.

At flowering, the style elongates, the stigma is pushed up through the small channel made by the anthers, the two halves of the stigma separate and assume a rather pronounced spiral form that may bring the inner receptive surface, completely free from pollen, to touch the outer surface of the pistil which, extruding from the staminal column, has remained densely covered with pollen grains. Thus, in both species, independently of the intervention of external agents, self-pollination is possible. This does not mean that both species are self-fertile. As we will see later, while *C. endivia* is self-fertile, *C. intybus* is characterized by a strong sporophytic incompatibility system which inhibits self-fertilization.

4 Cultivar Groups

In table 1 nine cultivar groups are listed, mainly according to the product they give and its use. Two of these refer to root crops for industrial utilization, coffee substitute or inulin extraction, and thus they are outside the strict horticultural field. All the others, although after different culinary transformations, are directly used, as leaves or roots, as fresh or cooked foods.

Although, as it has been said, chicory and endive are cultivated all over Europe and tend to expand towards always new horticultural areas, most of these groups are well known and extensively adopted, as horticultural crops, at a local scale only and, as such, their description in the scientific and technical literature is often incomplete or inaccurate. This is particularly true when the crop is largely differentiated and has generated subgroups among which it may be really difficult to find differences and affinities. We are aware that, particularly for this kind of horticultural crops, diversity is often the most efficient tool for commercial success, both for the seedman and the farmer; thus, it may be unwise, in a continuously moving breeding world, to establish a rigid frame within which everything has to find its place. Nevertheless, it seems advisable to have an account, as complete as possible, of the material we are talking about. This is the reason why, giving attention to the most distinctive traits and with the help of some pictures, we feel stimulated to attempt a short description of the most frequently cultivated plant material. Doing this, we do not intend, in any way, to cover the whole landscape of the chicory and endive types, cultivars, local populations, and farmer's selections grown in Europe and outside: we only want to propose a tool, although far from being exhaustive and of general satisfaction, usable for classifying at least the major part of the cultivated material according to objective criteria rather than to commercial perceptions.

C. endivia subsp. *endivia* var. *latifolium* (2n = 18) (Fig 1A)
English: Escarole, Batavian endive, Broad leaved endive;
French: Scarole, Chicorée blanche;
Italian: Scarola;
German: Endivie-Eskariol;
Spanish: Escarola.

C. endivia subsp. *endivia* var. *crispum* (2n = 18) (Fig.1B)
English: Curled endive;
French: Chicorée frisée;
Italian: Indivia riccia;
German: Krause endivie;
Spanish : Escarola crespa.

The name “endive” correctly pertains to the cultivated material belonging to the two botanical varieties mentioned above. Their main use, alone or in mixtures, is in fresh salads for which their yellowish or pale green leaves are particularly appreciated. Cultivation techniques are much the same for the two types: they are typical spring-summer crops and are scarcely tolerant to low temperatures. Any out of season cultivation, both delayed or anticipated, although possible, needs artificial protection. The commercial value of the final product mainly depends on the ratio between the bunch of etiolated leaves which form the “heart” and the whole of the plant. In ancient times this ratio was traditionally increased by closing the rosette with a rubber band during the last period of permanence in the field. Recently bred cultivars, tendentiously self blanching, form a more or less closed rosette of leaves, so that the etiolated “heart” is naturally obtained directly in the field.

C. intybus subsp. *intybus* (2n = 18)

Two main groups can be recognized within this subspecies to which all the cultivated types of chicory belong: the first, which refers to the var. *foliosum*, traditionally includes all the cultivar groups whose commercial products are the leaves, while the second regards the var. *sativum* and comprises all the types whose commercial product, either destined to industrial transformation or direct human consumption, is the root.

It might be argued that if it is true, as commonly accepted, that Witloof chicory has been firstly obtained from roots of Magdeburg, then, strictly speaking, it should be grouped under the var. *sativum*, together with all the other root chicories. Nevertheless, the most recent scientific literature refers to Witloof chicory as a type belonging to the var. *foliosum* (Koch et al., 1997; Van Stallen et al., 2001; Van Stallen et al., 2003; Van Stallen, 2003; de Proft et al., 2003; Van Stallen et al., 2005) and as such it is considered here together with the cultivar groups Pain de sucre, Radicchio, and Chicory of Catalogne.

Witloof chicory (Fig. 1C)

English: Witloof;
French: Chicorée de Bruxelles, Chicorée witloof;
Italian: Cicoria witloof, Cicoria di Bruxelles, Cicoria belga;
German: Zichorienzalat;
Spanish: Endibia, Achicoria de Bruselas.

The productive cycle of Witloof chicory may be divided in two distinct phases. The first is aimed to obtain well developed and uniform roots which, in the second one, were traditionally forced under a soil coverage, ending up with the production of the well known firm etiolated heads (*chicons*) formed by leaves tightly grown. At present, due to the adoption of hydroponic culture techniques, a year round production is possible. In this evolution a role has been played by the development of specific hybrids which, thanks to both their targeted selection and uniformity, gradually replaced the original populations and the old farmer's selections, further reducing the narrow genetic basis of the crop (de Proft, l.c.). In spite of this, genetically differentiated populations might still be sporadically traceable and used. Besides to physiological or parasitic disturbances which may alter the overall productivity and the product's marketability, breeders pay attention to quality traits which may increase its commercial attractiveness. Among these, the pale yellowish colour of the leaf blade and an absolutely colourless midrib are most important, although much attention is also paid to the head's closeness and firmness. Intrinsic quality has been considered as well and the almost complete disappearance of the bitter taste from the commercial product has to be retained as one of the reasons for its generalized acceptance outside the area of origin.

More recently, attempts of innovation in the appearance of the commercial product have been made and, taking advantage from the within species variability and the interfertility among all the cultivar groups, new looking red or reddish leafed cultivars have been put on the market.

Pain de sucre (Fig. 1D)

English: Sugarloaf chicory, Tall heading chicory;
French: Chicorée pain de sucre ;
Italian: Pan di zucchero ;
German: Zichorien;
Spanish: Achicoria.

It is perhaps one of the most ancient results of selection from wild populations. The plant's appearance at maturity is more like Romaine lettuce or Chinese cabbage as it has very large leaves enveloping one over the other to form a large, firm, tightly closed head, yellowish green in colour, weighing up to 1.5-2.0 Kg. Its cultivation is not very widespread but, as it happens with other types of chicory, it may give a connotation to the horticulture of some restricted areas like, for instance, Southern France, North Western Italy and Southern Switzerland. Despite the name, it has maintained quite an accentuated bitter flavour which renders this vegetable perhaps more adapted to be cooked rather than to enter as a component in crude salads. Open pollinated populations as well as some hybrid varieties are available on the seed market.

Fig. 1. Varietal groups of endive and chicory: A escarole (photo ISI Sementi), B curled endive (photo ISI Sementi), C Witloof chicory, D pain de sucre, E radicchio Red of Chioggia (photo Veneto Agricoltura), F radicchio Early Red of Treviso (photo Veneto Agricoltura), G radicchio Late Red of Treviso (photo Veneto Agricoltura) , H radicchio Red of Verona (photo Veneto Agricoltura), I radicchio Variegated of Castelfranco (photo Veneto Agricoltura), L asparagus chicory (Catalogna) , M root chicory.

Radicchio

This Italian common name has been adopted by all the most internationally used languages and indicates a very differentiated group of chicories, with red or variegated leaves, traditionally cultivated in North Eastern Italy.

There is no documented history about the origin of coloured chicory in Italy. All the red types of Radicchio now being cultivated seem to derive from red-leaved individuals firstly introduced in XV century. According to Bianchedi (1961) the cultivation of red chicory goes back to the first half of XVI century. For sure, the original type has to be identified with the "Rosso di Treviso" which has been for long the only cultivated radicchio in the Venetian territories. Later on, possibly from spontaneous or controlled crosses between red leaved individuals and plants of *C. endivia*, the types with red spotted or variegated leaves have been originated. After spreading out to the nearest territories, the original type underwent an accentuated selection according to very different criteria suggested by each farmer's personal preference, but at least partially due, or depending on, the various environmental situations met by the crop. Thus, in the area of Verona, from the original "Rosso di Treviso" a small winter hardy type forming a rosette of deep-red coloured leaves has been initially selected; from this, the most recent populations of "Rosso di Verona" have been obtained around 1960. During the second half of the last century a further selection from the original Treviso type has been made, thus originating a long leaved and early maturing population with self closing plants. Differently, in the area of Chioggia, a traditional horticultural area established since ever on the sandy soils extending southward of this small sea sided town just south of Venice, a variegated type, able to form rather conic, firm, and tightly closed heads while in the field, had been originally selected around 1930. From this, a large leaved red type with an accentuated and white midrib and characteristic ball-shaped heads has been initially selected about twenty years later and an almost completely light-yellowish type of very limited cultivation has been obtained toward the end of the last century.

As a result, at least five grown types, named according to their province or town of origin, may be distinguished, at present, within this cultivar group:

Rosso di Chioggia = Red of Chioggia (Fig. 1E)
Rosso di Treviso Precoce = Early Red of Treviso (Fig. 1F)
Rosso di Treviso Tardivo = Late Red of Treviso (Fig. 1G)
Rosso di Verona = Red of Verona (Fig. 1H)
Variegato di Castelfranco = Variegated of Castelfranco (Fig. 1I)

Rosso di Chioggia (Fig. 1E). This is by far the most widely grown among the various types of Radicchio and the one which presents the highest within-type differentiation as far as the availability of cultivars able to guarantee an almost complete year round production. As a matter of fact, it has shown a great adaptability to very different environmental situations all around the world, becoming the most grown type of Radicchio outside Italy and, thanks to this fact, the most known at international level. Independently of the sowing time, it grows in the open field and only early cultivations, able to give a product in the months of May and June, need protection during the first part of the cycle. Its massive production is concentrated between September and the end of February of the following year. Its main features are, first of all, the ball shaped and very firm heads which, at harvest, may reach a weight of 500 g once the outside green leaves have been eliminated. Other distinctive traits are the deep red colour of the leaf lamina and the extension of the midrib which must be associated with its absolute whiteness. Although the seed industry has since long become interested to this crop and named commercial varieties have been adopted, the great majority of the farmers in the typical area of production still use seed of their own populations which they maintain through a yearly conservative selection and an on-farm seed production. Quite often this seed is sold and bought through private transactions, outside the official seed market, both inside and outside the typical area, while commercial seed is mainly used outside Italy. The majority of the commercial varieties are open pollinated populations derived through selection from the original genetic pool. In recent years so called hybrid varieties have been put on the market and are favourably adopted mainly for out of season productions.

Rosso di Treviso Precoce (Fig. 1F). It is characterized by having upright long leaves with a large and thick midrib sustaining a rather expanded deep-red coloured lamina. During the vegetative period, as the plant grows, the newly developed leaves do not expand in an open rosette, but tighten more and more to form closed and firm heads. It is sown or transplanted in the field from July to mid August and harvested in September through December. At harvest, the outer green leaves and the major part of the tap root are taken away in order to leave the inside red heart ready for the market. Although it is one of the most recent selections, it is becoming more and more known outside its initially limited area of production and, thanks to a cultivation technique very similar to the one of "Rosso di Chioggia", it is on the way to follow the same trend of expansion. As a consequence, the seed industry has been looking with increasing attention to this type of radicchio and, at present, besides some open pollinated commercial populations, one hybrid variety has been put on the market. Since an out of season cultivation has started to be adopted, an increasing need of genetically improved material is foreseeable. Anyway, at present, most farmers are using their own populations derived from the original genetic pool and maintained through yearly mass selection.

Rosso di Treviso Tardivo (Fig. 1G). It is the most ancient type of Radicchio grown in Italy and can be considered the legitimate ancestor of all the others. It is a typical winter crop in as much as it is sown or transplanted in the field from July to mid August and may be harvested in October through February. The plant grows with long, deep-green, basically upright leaves which form a loose rosette and whose both midrib and lamina assume an always more accentuated reddish colour as temperature lowers. At harvesting, the entire plants are dig out, stocked with all their leaves and roots, and maintained at low temperature (around 0°C) as long as possible. According to the market's request, plants are forced placing them under a black cover, with their roots in running water at 10-12°C.

After 10-18 days, according to the air temperature, the forcing period is concluded: plants are cleaned off, the outer leaves and a great portion of the tap root are eliminated leaving a bunch of bright-red coloured leaves with a white large midrib and a rather reduced lamina. As it seems clear enough, this crop has at least some features in common with Witloof chicory with which it shares the shape of the leaves, the growth habit, the large tap root, and the forcing process in order to obtain the commercial product. It is grown in a very restricted area and, together with the "Variegato di Castelfranco" and the "Rosso di Treviso Precoce" is one of the three radicchios recognized since the late '90s with the PGI (Protected Geographical Indication) mark. Its peculiar aspect and really superior culinary quality make this Italian Radicchio the most appreciated one. Its market price, particularly at Christmas, may reach as much as twice or three times the price of any other Radicchio. In spite of this, no named commercial variety is on the seed market except for selected open pollinated populations. As a matter of fact, its cultivation is very limited and the whole productive procedure is rather complicated and much less standardized than the one adopted for Witloof chicory. So, unless it reaches a comparable degree of popularity, it seems difficult that the seed industry would invest on this very peculiar crop. Almost the total present production relies on farmers' populations whose history may go back for generations and which are maintained through yearly mass selection.

Rosso di Verona (Fig. 1H). The first populations of this type of Radicchio, as we know it now, were obtained about fifty years ago. With respect to previous populations, the present ones have plants with much larger heads which may resemble those of "Rosso di Chioggia". In comparison to these, besides being smaller on the average, they are more egg-shaped and formed by less expanded leaves with a brighter red lamina and a large and thick midrib from which less evident and intersecting veins depart. Its cultivation is much like that of "Rosso di Chioggia" and "Rosso di Treviso Precoce". Sowing dates are from July to mid August, while harvesting starts at mid October and goes on until the end of February. During the vegetative period the plant develops a rosette of pale-green leaves which gradually close and tighten up to form a very firm head. It is a typical winter crop whose popularity and area of cultivation is increasing both in Italy, where it is expanding to more southern regions, and outside the country. The reason of this is much the same as for the early Rosso di Treviso: its cultivation can be standardized quite easily, there is no need for forcing, thanks to its attractiveness the product is well accepted by the market, and the consumer recognizes to it a better culinary quality in comparison to other Radicchio types. In spite of this increasing popularity, the seed market is rather poor and the available commercial varieties are selected open pollinated populations. The development of the first hybrid variety seems not too far anyway. The most frequently used seed is thus from the farmer's populations selected during the last decades and maintained through mass selection. It seems worth noting that in developing these populations, a procedure implying crosses of the initial small leaved "Rosso di Verona" with the larger headed type "Rosso di Chioggia" may have been adopted.

Variegato di Castelfranco (Fig. 1I). Together with the "Rosso di Treviso Tardivo", is the second most traditional type of Radicchio grown in North Eastern Italy. Its morphological traits make it easily distinguishable from any other type. Directly sown or transplanted between July and mid August, plants form a large rosette of more or less indented brilliant green leaves with a very extended red spotted or variegated lamina sustained by a not too evident white midrib. When developing, the inside leaves wrap up and tighten to form a closed but not too firm cone-shaped self blanching head. At harvest, the external green leaves are removed and the internal ones are open to form a bunch of pale yellowish red spotted leaves which looks very much like a flower. As a matter of fact, this Radicchio is also known as the "Rose of Castelfranco" and is one of the most appreciated components of fresh salads during the cold season The selected populations grown at present are all self blanching thus making the cultivation of the "Variegato di Castelfranco" comparable to the one described for the other types of Radicchio with the exception of the "Rosso di Treviso Tardivo" with which has for long been sharing the final forcing process. As far as the availability of commercial seed, the situation is much the same as for the majority of the other radicchios. Selected populations are available on the market, but the great part of the crop is planted with seed of farmers populations selected and maintained through mass selection by each farmer.

Catalogna (Fig. 1L)

English: Asparagus chicory; Large leafed chicory;
French: Chicorée asperge d'Italie;
Italian : Cicoria Catalogna;
German: Katalanische endivie.

Has a recognised Italian origin but, other than in Italy, it is grown in Southern France, Spain, and, generally speaking, in all the Mediterranean countries, as well as outside Europe where the environmental conditions are sufficiently mild and favourable. The plant has tall deep green coloured upright leaves, with indented or continuous lamina and a marked midrib, growing in a tuft without forming a head. Its rather bitter taste is particularly appreciated. It is usually cooked, but young sprouts which origin from inside the plant may be used as fresh salad (*puntarelle*). Many cultivars are known whose names often refer to the place they come from or to the culinary utilization, thus generating synonyms used at commercial level to indicate the same product. It is rather frequent to find these names included in lists or catalogues, as if they were belonging to different cultivar groups, together with other types of different chicories, thus generating a little bit of confusion.

Root chicory (Fig. 1M). This cultivar group has to be referred to *C. intybus* subsp. *intybus* var *sativum* and includes both horticultural crops, grown for direct consumption as cooked food, and industrial crops whose destination is the preparation of a coffee substitute or inulin extraction. As far as the first utilization is concerned, the crop has no great diffusion and may have significance at local level only. All the grown types, differing in name according to the place of origin, have been derived from the ancient Magdeburg chicory. In selected varieties, roots tend to be cylindrical, with smooth surface, few hairy roots, and the central part reduced at a minimum. Their bitter taste, which renders them particularly appreciated by the connoisseurs, has been strongly reduced in comparison to that of the original types. Selected open pollinated varieties are available on the seed market, but local populations maintained by farmers through mass selection are adopted in most cases.

5 Genetics and Breeding

5.1 Reproductive Systems and Population Genetics

In any breeding program, the breeding schemes that can be adopted as well as the variety types that can be constituted depend on the reproductive barriers and mating systems of plants, and hence on the genetic structure of populations.

The genetic structure of natural populations of cultivated *Cichorium* species cannot be referred to a unique model as there are basic differences between *C. endivia* and *C. intybus* in their reproductive system.

C. endivia (2n = 18) is a self-pollinated species with less than 1% of spontaneous crosses (Rick, 1953), whose populations are composed of a mixture of pure lines, genetically related but reproductively independent from each other. Thus, genetic as well as phenotypic variation are principally detectable among lines, due to the presence, within natural populations, of fixed genotypes, mainly homozygous for different alleles. Spontaneous hybridization is however possible to some extent, depending on environmental factors and germplasm sources. Commercial endive varieties are usually represented by pure lines obtained through repeated selfing of a number of plants selected from original genetically variable populations or of hybrid individuals stemmed from crosses between superior parental lines chosen for complementary morphological and commercial traits. The close autogamy of endive limits the choice among breeding strategies to mass selection, individual selection, pedigree breeding and back-crossing (Ryder, 1998). The first strategy enables to constitute multiple line varieties, whereas the other ones lead to single pure lines: from here comes the uniformity of the commercial product. Production of F_1 hybrids, which is usually more appropriate for cross-pollinated species, has been developed in other related self-pollinated leafy vegetables, such as lettuce, but it has yet to be exploited in endive. Although many commercial cultivars are on the seed market and represent the great majority of the cultivated material, local populations are still grown and, according to the area of cultivation, may give rise to different and very specific local productions.

C. intybus (2n = 18) is a strictly allogamous species for which selfing is strongly hampered by an efficient incompatibility system that prevents inbreeding and promots out-breeding.

The original populations of C. *intybus*, as far as their genetic structure is concerned, could be considered as natural since, independently of their historic background, the production of both Witloof and Radicchio has for long relied on populations, maintained by farmers for their own use, on which very little selection, if any, might have been applied according to personal criteria. All these populations, obtained by mass selection and maintained through the intercrossing of selected parents, have to be considered highly heterozygous and genetically heterogeneous whose behaviour and level of adaptation to different environments and/or cultural conditions depend on the frequency of favourable genes or gene combinations. As the interest for the edible product grew, farmers' selection criteria became more and more attentive to the consumer's request and most of them elaborated their own ideotype. This brought about a great deal of genetic and morphological differentiation which has been entirely preserved until organized breeding programs have been established, firstly by public institution and, in more recent times, by private firms. As it happens for most cross breeding species, in *C. intybus* detectable heterosis effects are present and hybridisation between selected genotypes give uniform and heterotic progenies; the constitution of F_1 hybrid varieties is thus feasible. Nowadays the situation is quite different between Witloof and Radicchio. Since some years, F_1 Witloof hybrids, released by private seed companies, are on the seed market and the crop is at present mainly based on them, thus determining the complete disappearance of most farmer's populations. In Radicchio, although with some differences among the various types, the major part of the crop is still based on farmer's populations which are yearly selected and maintained and whose seed is usually reutilized on farm but may also be sold through private and not officially registered transactions. These populations are very well distinguishable among types, but they are often recognizable within type as well, on the basis of morphological and physiological characters and agronomic performances, although, at the same time, they present an acceptable phenotypic uniformity among individuals. Regarding their genetic variation, as estimated by genetic analysis performed through the application of appropriately chosen molecular markers, it is a common observation, also applicable to Witloof chicory too (Kiers et al., 2000; Van Stallen et al., 2001), that the major part of the genetic variation takes place within populations, while a minor portion is attributable to among population differences (Barcaccia et al., 2003).

Since this is the plant material which, in recent years, has been representing, and still represents, the starting point for the constitution of new commercial varieties, it seems reasonable to state that, if preserved from extinction, it is an invaluable source of genetic diversity on which chicory breeding may rely for long in the future. It is however expected that F_1 hybrid varieties will be bred and adopted with increasing frequency for Radicchio too. This is particularly true for the types which take an advantage from the uniformity of the marketed product as this is often the key for the customers appreciation.

5.2 Breeding Achievements, Methods and Goals

Productive as well as qualitative traits are main objectives in chicory and endive breeding programs. General and common goals in breeding new varieties mainly concern i) single plant size, weight and yield; ii) resistance to biotic (fungal diseases and insects) and abiotic stresses; iii) adaptation to a specific climatic or agronomic environment; iv) uniformity of crop maturity; v) good market acceptance regarding extrinsic (color, shape, uniformity) and intrinsic (taste and texture) traits.

C. endivia is a minor crop compared to chicory. Most of the breeding work on this species has been done by private seed companies and, as a consequence, there is not much literature on it. The breeding approach adopted with endive is much the same as with any other autogamous species: selection of superior genotypes a) from genetically variable farmer's populations according to a pure line selection scheme and/or b) from segregating populations derived from crosses between previously selected superior parents, following a pedigree procedure. The first approach resulted in a series of varieties on which the crop almost completely relied at least during the whole first half of the last century. These varieties were often heterogeneous at morphological level and had to be considered as mixtures of more or less homozygous genotypes. Thus, although far from being genetically homogeneous, these varieties could be well differentiated from each other for leaf and head size and shape (Tesi, 1974) and for long represented the basis for the successive selection work which led to improved pure line varieties. Because of the almost complete disappearance of the original farmers' populations and the presence of good commercial varieties, at present breeders largely prefer the second strategy, based on selection within segregating populations, with which a number of favorable traits already present in the parents can be combined in superior genotypes. When selection for the favorable parental traits is made after each selfing, four to six generations might be enough to have genotypes with a sufficient level of fixation to be transferred in field trials and tested in comparison to the existing commercial varieties. It seems obvious that the first generations might be grown in dense stands and are mainly aimed to test the segregating material for traits usually under simple genetic control, like the resistance ones, which allow the elimination of large numbers of individuals. Selection for morphological traits, more strictly linked to productivity and marketable quality, should be made in later generations grown in the field. This last part of the work might cover a variable number of generations according to the urgency to release new varieties. When one or more pure lines are considered superior to the existing varieties, the field testing may be extended to different climatic and agronomic situations in order to define the limits or the general adaptability of the selected material. Thereafter, the initial quantities of commercial seed might be produced and, after inscription in the Register of Commercial Varieties, put on the seed market. Sometimes, although morphologically uniform, a new variety is released while still containing some genetic variation which might let further divergent selection, thus originating a number of rather similar sister varieties. Analogously, a variety might have had a long commercial life and thus, through mutation and repeated multiplication, might have been accumulating genetic changes which, by mean of selection, may produce variants of the original variety. This is a practice much used by seed companies to develop new varieties from publicly developed landmark varieties.

When a variety is outstanding in most respects but lacks a specific trait, particularly when this is under the control of one gene as often occurs with resistance traits, the backcrossing approach can be successfully used for transferring that gene from another variety, a landrace or a wild type, where it might be present, to the otherwise superior line or variety. Once known as a "surgical" method of breeding in as much as, if correctly applied, it preserves the genetic structure and the agronomic performance of the variety to be improved while introducing into it the desired trait, the backcross method has nowadays been efficiently integrated with more sophisticated procedures which make use of the molecular tools in Marker Assisted Selection (MAS) programs. Although very efficient and largely adopted in many important field crops, this approach needs however a molecular knowledge of the species which is not too difficult to reach but which is still lacking as far as *C. endivia* is concerned.

C. intybus is by far more important than *C. endivia* and has a much more ancient breeding history which goes back to at least eighty years ago, when the first varieties have been bred and sold on the seed market.

As far as Witloof is concerned, the main traits evaluated during selection programs are related to morpho-phenological, agronomic, and organoleptic characteristics. Important features are the time of cultivation, class of earliness, thickness and length of the main root, leaf shape and color, adaptation to local environments, disease resistances, taste and bitterness of edible parts.

As already stated, roots are harvested at the end of the growing season and stored under low temperature until they are forced to produce the leafy vegetable.

The production and quality of heads in Witloof chicory changes during the same season, and between production seasons, depending upon climatic factors, cultural practices, time of harvest, storage and forcing conditions, all of which influence the final yield and quality. However, there is no need to underline that the performance of a cultivar is strictly dependent upon its genetic value which, in turn, may be tightly linked to its genetic structure and thus to the strategy adopted for its constitution.

Traditionally, varieties of chicory were developed by mass selection in order to obtain uniform populations characterized by valuable production and acceptable commercial head size and shape. Newly released varieties are mainly synthetics produced by intercrossing a number of phenotypically superior plants, selected on the basis of morpho-phenological and commercial traits. More rarely, plants are also evaluated genotypically by means of progeny tests. Synthetics have a rather large genetic base and are represented by a heterogeneous mixture of highly heterozygous genotypes sharing a common gene pool.

In recent years, methods for the constitution of F_1 hybrids have been developed by private breeders and seed companies. Details on the procedure for the constitution of such hybrids are not available in the current literature and it may be presumed that each company has developed its own protocol, mainly in accordance to the genetic material it has at disposal and to the possibility of applying a more or less efficient control on the F_1 hybrid seed production phase. As a matter of fact, the strong self-incompatibility system, which hinders obtaining highly homozygous parents, and the absence of a male sterility factor within the species or in sexually compatible species, make it difficult to propose a F_1 seed production scheme and, most of all, to consider these newly commercial varieties as true F_1 hybrids.

Many so called Witloof hybrids are now in production and are highly appreciated by growers for their performances and especially for their uniformity. Some types are bred exclusively for tray production and others for either tray or soil-planted production. Cultivars are tailored for a specific season of production. Early varieties have longer core lengths than late ones and this is used as a criterion of selection for early-, mid- and late-season varieties (Huyskes, 1963). Varieties designed for early-season forcing require a shorter period of vernalization, store less well, and grow rapidly during forcing. Varieties designed for late-season forcing are harvested later, store better, require more chilling, grow relatively slowly during forcing, and produce more compact heads (Ryder, 1998). Other breeding objectives include tight closure of the tops of chicons, tolerance to internal browning, resistance to premature flowering, reduced bitterness and good presentation quality with regard to head color, shape and size.

Chicory breeding materials of the radicchio group are usually represented by local populations known to possess a high variation and adaptation to the natural and anthropological environment where they have originated and are still cultivated (Barcaccia et al., 2003). As said, these populations are maintained by farmers through phenotypic selection according to their own criteria; sometimes, controlled hybridizations among different types is exploited in order to obtain recombinant genotypes showing superior agronomic and commercial traits. The breeding programs under way at present by local breeders and regional institutions aim: i) to isolate, within the best local selections, individuals amenable to be used as parents for the constitution of synthetic varieties and, although not easily feasible, ii) to select inbred lines suitable for the production of commercial F_1 hybrids. These breeding procedures could be greatly assisted by the use of molecular markers that allow discarding of molecular off-types, to better exploit parental genetic polymorphisms for synthetics and to identify the most genetically distant inbreds as parental lines for hybrids.

For the constitution of synthetics in all radicchio types, a widely used scheme includes the selection in the field of about 100 mother plants responding to a prefixed breeder's ideotype. These plants should share similar morpho-phenological traits, and productive and qualitative properties. Local varieties, ecotypes and especially populations from mass selection represent an excellent breeding material to start with the selection of mother plants. In addition to the morpho-phenological, agronomic and commercial evaluation, a characterization based on molecular markers could be undertaken in order to choose and retain plants of the original population that associate a good genetic uniformity to a superior phenotypic value. Roots of the selected mother plants should be preserved during the winter until the next spring, when they will be transplanted in the field under an isolation cage. The seed from plant intercrossing is the basic stock from which commercial seed lots can be obtained and sold to farmers. It is generally accepted that seed multiplication should be performed in the radicchio cultivation areas.

The constitution of experimental hybrids has been attempted also for radicchio types (Lucchin et al., 2007). Progenies of 28 cross combinations performed between partial inbred lines were recently evaluated for agronomic traits and production components. Most hybrid progenies proved to be morphologically uniform and showed heterotic vigour, although results varied according to the parental genotypes and the radicchio type. Heterosis in terms of plant vigour was particularly evident for most hybrids of Castelfranco and Early Red of Treviso types, whereas it was less pronounced for Chioggia and Late Red of Treviso (Fig. 2). It is worth mentioning that most hybrid combinations showing heterotic vigour revealed root and leaf traits of generally low agronomic and commercial value. Five to six generations of selfing in a species like chicory, characterized by an efficient self-incompatibility system, most likely led to a selection for self-compatibility rather than for agronomic and commercial traits.

The application of molecular tools to chicory breeding, as it has been done with other crops where they have been extensively and routinely utilized, may be of great help to overcome scientific as well as practical problems, and this can improve the efficiency of the possible breeding procedures.

As previously seen, molecular markers can be applied to the analysis of genetic variation at different levels: between cultivar groups within species, among types within cultivar groups, among individuals within type. The kind and, most of all, the deepness and, hence, the usefulness of the information obtainable from such an analysis can differ according to the purpose it is made for. Phylogeny studies, distinctiveness of populations and their interrelationships, traceability of the commercial product are only few examples of the most frequent uses. On this basis, for instance, the possibility of discriminating the five different types of radicchio grown in Veneto (Italy) was established (Barcaccia et al., 2003). More refined applications concern the construction of genetic maps, the location of specific loci affecting important commercial traits and the possibility to quantify their contribution in the total explained variation of quantitative characters. The location of useful genes should make the identification of superior genotypes easier, thus implementing any marker-assisted selection program, be it aimed to the selection of parental genotypes of a synthetic or of an F_1 hybrid. In the backcrossing strategy, the process to recover the genome of the recurrent parent could be improved by selection for molecular markers tightly linked to the target gene in order to reduce linkage drag.

Molecular markers could also be useful to test other important varietal attributes, such as plant relationships, population uniformity and distinctiveness of synthetics, and to assess the homozygosity of inbred lines and diversity among inbred lines in order to maximize heterosis in the constitution of hybrids. Moreover, the efficiency of selection procedure can be tested. Multilocus PCR-derived markers were exploited by Lucchin et al. (2003) to evaluate the effects of phenotypic selection. Two generations of phenotypic selection for earliness on breeding populations of radicchio showed that the first selection cycle produced an increase of the genetic uniformity within each population, whereas the second cycle was ineffective. Moreover, no genetic differentiation among generations within types was observed.

5.3 Breeding for Industrial Applications

Both chicory and endive represent valuable crops not only as vegetables with positive effects on human health, but also in terms of industrial application for the production of fructans such as inulin and oligofructose. Fructans of chicory have been recently shown to be dietary fiber, while its phenolic components as well as the flavonoids contained in endive, are believed to act as natural antioxidants. As a consequence, these types of vegetable food should be a significant part of the diet.

Fig. 2. Field plots with radicchio Early Red of Treviso hybrids: A and B inbred parents, C and D hybrid progenies.

Let's consider chicory as source of dietary fibers. Inulin and oligofructose are fructans extracted on a commercial basis from the chicory roots and are known as polymers widely used in several industrial applications. They share the basic common characteristics of dietary fibers, that is, saccharides of plant origin, resistant to digestion and absorption in the small intestine, and fermenting in the colon to produce short-chain fatty acids that are absorbed and metabolized in various parts of the body (Flamm et al., 2001). In the last few years they have received increasing interest because of their positive effects on health. At present, fructans are mostly supplied by chicory, which is only grown and processed in Netherlands, France, Belgium and Northern Italy. It would therefore be an attractive concept to expand its cultivation to the southern European countries, although water shortage and high temperatures may hinder its growth and yield. Monti et al. (2005) have recently demonstrated that growth, fructan yield, and quality of chicory are related to photosynthetic capacity, harvest time, and water regime.

Only a few studies on the yield potential of root chicory have been undertaken (Frese et al., 1991). Occasionally, this crop species was discussed as a potential industrial crop, but until recently inefficient processing techniques have hampered the large-scale industrial exploitation of inulin-containing plant species. Meanwhile, biotechnological innovations have considerably improved the inulin processing and under these new conditions the reassessment of root chicory as an industrial crop can prove rewarding. Today, the economic importance of root chicory cultivation is restricted to the traditional production areas (North France and Belgium) and to very specific products (mainly coffee surrogates, little isoglucose).

Root chicory currently yields less utilizable carbohydrates than modern sugar beet varieties and other crop species. As an industrial crop it will have to compete, for instance, on the isoglucose market with starch crops such as potatoes, corn, and wheat as well as with another inulin containing species, the Jerusalem artichoke. The competitive position of root chicory might be enhanced if the total sugar yield is increased by breeding. About 50 years ago root chicory yielded as much total sugars as sugar beet varieties (Nuding, 1935; Oltmann et al., 1984) but only little effort has been directed towards root chicory breeding since that time. Hence, it can be assumed that intensified selection will considerably improve the performance of this neglected crop. Although root chicory is believed to have a great potential as industrial crop, this species seems to have a limited genetic basis which might hamper breeding progress. Frese and Dambroth (1987) have investigated several accessions in order to assess the yield potential of chicory and to ascertain whether leaf chicory might be suitable to broaden the genetic basis of root chicory breeding. Overall results suggested that leaf chicory may not contribute to a rapid breeding progress in root chicory within a short period.

Chicory is also known as source of fructose, the sweetest naturally occurring sugar. Fructose is a simple sugar (monosaccharide) found in many foods and one of the three most important blood sugars along with glucose and galactose. Fructose is often recommended for, and consumed by, people with diabetes mellitus or hypoglycemia. Chicory and other root vegetables, such as beets and sweet potatoes, contain fructose, usually in combination with sucrose and glucose. Fructose yield components of chicory were studied by Coppen's d'Eegkenbrugge et al. (1989) using more than 2.500 roots from several cultivars analyzed individually for root weight, root shape, total fructose, total glucose, potassium and amino-N contents.

Fructose yield was shown to be highly correlated with root weight and slightly with fructose content, both within and between varieties.

Phenolic components of chicory proved also to be potent antioxidants. Heavily red colored, red spotted and fully green varieties of *C. intybus* were investigated for their phenolic content and antioxidant activity. All of the studied chicories are characterized by the presence of a large amount of hydroxybenzoic and hydroxycinnamic acids, whereas the red color is due to cyanidin glycosides. The presence of these phenolics in red chicories confers to them an exceptionally high peroxyl radical scavenging activity in terms of both capacity and efficiency, particularly in their early stage of growth, and makes this popular and low-cost vegetable foods comparable or superior to many foods having well-known antioxidant properties such as red wine, blueberry and tomato (Rossetto et al., 2004).

Flavonoids composition and content in endive varieties and their biological activity were investigated by Du Pont et al. (2000). A diet containing a high proportion of fruit and vegetables has been advocated as one of the best ways to reduce the incidence of chronic disease in the Western world (Block, 1992). The protective effect, which this type of diet confers, is believed to be due to the antioxidant and other activities of the flavonoids, a class of compounds found at significant levels in many fruits and vegetables (Hertog et al., 1993). The antioxidant activity of some of the subclasses of the flavonoids, such as the flavonols and anthocyanins, has been reported to be greater than that of either vitamin C or E (Rice-Evans et al., 1995). It is important to further understand the mode of action of flavonols as biologically active components of the human diet. The flavonols, although present only in the form of a multiplicity of conjugates, have been demonstrated to be very important for human health. Du Pont et al. (2000) reported the composition and content of the flavonoid conjugates, which are based on the four aglycons (quercetin, kaempferol, luteolin, and cyanidin) found in commercial varieties of lettuce and endive. Processing and storage of the vegetables have also been shown to alter both their levels and compositions.

5.4 Breeding for Resistance Traits

Several diseases on endive and chicory have been reported (Ryder, 1998). These crops are affected by many of the same diseases and disorders as lettuce, a species much more cultivated and studied than chicory and endive. The diseases of endive include: sclerotinia drop, downy mildew, damping off, anthracnose, grey mould, bottom rot and other fungus-incited diseases, several bacterial diseases (soft drop and leaf spot), Bidens mottle and lettuce mosaic and tipburn. Moreover, a number of diseases have been specifically reported for chicory. Schoofs and De Langhe (1988) described fungal diseases incited by *Helicobasidium brebissonii* (violet root rot), *Phytophtora erythroseptica* (black or brown root rot), *Puccinia cichorii* (chicory rust) and *Verticillium albo-atrum* (verticillium wilt). The main fungal and bacterial diseases of chicory and endive are summarized in Table 2.

Very little breeding for disease resistance has been reported for endive and chicory. Zitter and Guzman (1977) tested several endive varieties and germplasm materials for reaction to the most common virus-mediated diseases and all were susceptible. Provvidenti et al. (1979) reported on resistance to turnip mosaic in chicory and proposed transferring the resistance to endive.

As far as insect pests are concerned, many aphids and lepidoptera attack chicory and endive plants. The maggots of two dipterous flies, *Ophiomyia* spp. and *Napomyza* spp., are common pests of Witloof chicory: they burrow in the petiole near the crown and, during forcing, feed on emerging *chicon* leaves. No germplasm source of resistance is known.

Table 2. Main fungal and bacterial diseases of chicory and endive.

Diseases	Causing agents
Alternaria	*Alternaria cichorii* Nattrass
Anthracnose	*Michrodochium panattoniana* Berl.
Black rot	*Phytophthora erythroseptica* Pethyb.
Bottom rot	*Rhizoctonia solani* Kühn
Damping off	*Pythium tracheiphilum* Matta
Downy mildew	*Bremia lactucae* Regel
Grey mould	*Botrytis cinerea* Pers.
Leaf spot	*Cercospora longissima* Sacc.
Powdery mildew	*Erysiphe cichoracearum* DC
Rust	*Puccinia cichorii* (DC) Bell.
Sclerotinia drop	*Sclerotinia minor* Jagger and *S. sclerotiorum* (Lib.) DeBary
Verticillium wilt	*Verticillium dahliae* Kleb
Violet root rot	*Helicobasidium brebissonii* (Tul.) Pat.
Leaf spot	*Pseudomonas cichorii* (Swingle) Stapp.
Soft rot	*Erwinia carotovora* (Jones) Holland

Plant resistance to many types of pathogens and pests can be achieved by the presence of disease resistance (R) genes. The nucleotide binding site-leucine rich repeat (NBS-LRR) class of R-genes is the most commonly isolated class of R-genes and makes up a super-family, which is often arranged in the genome as large multi-gene clusters. The NBS domain of these genes can be targeted by PCR using degenerate primers. Plocik et al. (2004) studied the recent evolution of NBS sequences within the *Asteraceae* and extended the comparison to the *Arabidopsis thaliana* genome. Using multiple sets of primers, NBS fragments were amplified from genomic DNA of three species, namely *Helianthus annuus*, *Lactuca sativa* and *Cichorium intybus*. This analysis suggests that *Asteraceae* species share distinct families of R-genes, composed of genes related to both coiled-coil (CC) and toll-interleukin-receptor homology (TIR) domain containing NBS-LRR R-genes. Among the most closely related species (lettuce and chicory), a striking similarity of CC subfamily composition was identified, while sunflower showed less similarity in structure. Overall data revealed that NBS families in the *Asteraceae* are ancient, but also that gene duplication and gene loss events occur and change the composition of these gene subfamilies over time.

The discovery of mutations conferring resistance or tolerance to specific pests or pesticides is another aspect to take into account. In particular, the finding of herbicide-resistant mutants could be crucial for breeding new varieties and for improving crop yields. Weed control in chicory and endive cultivations is primarily based on the use of selective herbicides. This is largely successful, even though related species of *Asteraceae* are difficult to control. Chlorosulphuron is a highly effective, broad spectrum, weed control agent, with low toxicity and requiring only small amounts for control (Ryder, 1998). A single gene for resistance has been identified in Witloof chicory, which may allow to use the material on chicory plantings (Lavigne et al., 1994).

The ecological concern for use of herbicide-resistant chicory was investigated by Lavigne et al. (1995). They compared two near-isogenic lines for various traits in field plots and no significant difference was found between them for vegetative and reproductive traits. This suggested that the mutation causing resistance does not have a deleterious effect on plant fitness. Consequently, if the resistance to herbicide can be transferred from cultivated chicory to wild chicory plants, there would be no selection against the latter and they would survive despite the herbicide treatment (Ryder, 1998).

5.5 Genetic Aspects of Interest for Breeding

C. intybus is a self-incompatible (SI) species. Self or self-related pollen are recognized by cells of the pistil and rejected after pollination, either immediately, on the stigma surface, or later on, during pollen tube growth in the transmitting tissue of the style. Genes located at the S-locus encode the male (pollen) and female (pistil) recognition determinants of SI. All the SI systems identified so far fall into two distinct groups: Gametophytic Self-Incompatibility (GSI) in which the incompatibility phenotype of the pollen is determined by its haploid genome; Sporophytic Self-Incompatibility (SSI) where pollen phenotype is determined by its diploid parent plant as the products of the pollen incompatibility genes are deposited on the pollen coating during microgametogenesis. Both systems are controlled by the S-locus, a single locus with multiple alleles, determining multiple haplotypes. Whereas studies of GSI have obtained genetic and molecular evidences from a variety of different families, investigations of SSI have focused mainly on species of *Brassicaceae*, particularly cultivated *Brassica* species. In *Brassica* spp. both the female and male determinant of SI have been identified.

The existence of a SSI system in Chicory was demonstrated in two different studies, analysing the progenies of two crosses between inbred lines of Witloof chicory (Eenink, 1981) and crossing a wild-type chicory plant with a "Rosso di Chioggia" genotype (Varotto et al., 1995). In both papers, a diallel cross among the F_1 plants and the F_1 backcrosses with both parents were used to analyse the progenies for incompatibility reaction. The mean number of achenes per flowerhead was chosen as criterion for evaluating the compatibility or incompatibility of a cross. The results indicated that in chicory a one locus sporophytic self-incompatibility system is present and pointed out the existence of dominance and co-dominance relationships between S-alleles in pollen and styles. In *Brassica*, it has been shown that dominant interactions between S-haplotypes can act independently in pollen and stigma, determining extremely complex patterns of incompatibility among individuals

(Hiscock and McInnis, 2003). This seems to be the case also in chicory, since only one difference in alleles activity in pollen and stigma can explain the seed set observed in certain cross-combinations (Varotto et al., 1995).

Different levels of self- and cross-incompatibility are found when chicory genotypes and commercial cultivars are crossed (Castaño et al., 1997; Varotto et al., 1995). It has been suggested that self-compatibility could be due to the presence of heterozygous genotypes bearing S-alleles, whose relations of dominance in stigma and pollen are not linear, and of homozygous genotypes with weak S-alleles (d'Eeckenbrugge, 1990). Moreover, recessive S-haplotypes can reach high frequencies in chicory populations because their effects are masked by dominant S-haplotypes. It has to be underlined that despite the SSI mechanism, when self-pollinated, chicory plants may produce 1 or 2 fertile seeds per flowerhead; considering the high number of flowerheads produced by a single plant during its flowering period (8-10 weeks), the seed set can be significative.

Rejection of incompatible pollen on the chichory stigma is very rapid in the case of self-pollinations since self-pollen does not adhere to papilla cells. In incompatible crosses, more frequently papilla cells permit the development of the pollen tubes that burst and do not reach the transmitting tissue of the styles. The rejection process in chicory differs from that of *Brassica* spp, since in *Brassica* the SI response is manifested within few minutes by the inhibition of pollen hydration, pollen germination or pollen invasion of the stigma epidermis (Nasrallah, 2000).

Little data are available on the molecular mechanisms operating in chicory or in other species belonging to *Asteraceae*. In Radicchio molecular investigations sought to identify orthologues of SRK and SLG, the female components of *Brassica* S-locus. Anyway, the identification of SRK-like genes in chicory showed that their transcripts were not expressed exclusively in stigmas, thus indicating that they were also unlikely candidates for stigma S genes (Varotto, unpublished results). Recently, analyses of stigmatic proteins in *Senecio squalidus* have identified a family of polymorfic basic proteins associated with specific S genotypes. These proteins bear no resemblance to the S-locus product of *Brassica* spp. (Hiscock and McInnis, 2003). It will be interesting to discover whether the *Asteraceae* possess their own system of self-incompatibility at molecular level, since further information is needed for the manipulation of SI system in chicory breeding.

An important role in chicory breeding could be played by male sterility in hybrid seed production, particularly since SI of parent lines is inadequate for reliable hybrid production. Male sterile mutants which cannot produce fertile pollen or functional anthers allow the exploitation of heterosis in F_1 hybrid varieties of many agricultural and horticultural crops. Two kinds of male sterility can be observed in plants: nuclear and cytoplasmic male sterility (CMS). Nuclear male sterility is based solely on recessive mutations which affect different functions in nuclear genes, while CMS is maternally inherited and mainly due to mutations in mitochondrial gene expression. Moreover, in CMS genotypes male fertility can be restored by nuclear-encoded fertility restorer (Rf) genes. In several species CMS has been used to produce female parental lines and used for the production of hybrid seeds. In absence of CMS system, male floral organs must be removed mechanically or other devices must be adopted.

In chicory, the presence of a naturally occurring CMS system has not been reported, but approaches to genetically engineering male sterility were used in Magdeburg, Witloof and Chioggia genotypes.

In a first approach, genic male sterile (S) lines of chicory were produced by expressing the ribonuclease genes Rnase T1 from *Aspergillus oryzae* or "barnase" from *Bacillus amyloliquefaciens* under the control of a tapetum specific promoter originally isolated from tobacco (TA-29). Restorer (R) lines for the "barnase" S lines were obtained by expressing the gene coding for "barstar", the intracellular inhibitor of "barnase" under control of the same promoter (Reynaerts et al., 1993). The development of inbred lines and male-sterile lines provided a reliable pollination control and allowed a new hybrid seed production system which has been patented as SeedLink™.

Protoplast symmetric fusion between chicory and the CMS line of sunflower PET-1, allowed the regeneration of interspecific hybrid plants. PET-1 CMS in sunflower was identified in an interspecific cross between *Helianthus petiolaris* and *Helianthus annuus*. This CMS is associated with the expression of the mitochondrial gene ORF522, which encodes a 15-kD polypeptide. The ORF522 gene was originated by a recombination event at the 3' of atp1 gene and its protein is detectable in flowers of CMS but not of restored lines (Horn et al., 1991; Moneger et al., 1994). The hybrid plants obtained after somatic symmetric fusion were cybrids and showed mtDNA rearrangements, indicating that symmetric fusion had the tendency to maintain the chicory mitochondrial genome. Three different kinds of sterility were observed: a) plant with anthers lacking dehiscence without, or with non-viable, pollen, b) complete absence of the anthers, c) absence of both anthers and styles or the presence of reduced styles. One cybrid plant was used for the production of hybrids whose yields were equal to or higher that those of traditional varieties (Rambaud et al., 1993; Rambaud et al.,1997).

In a subsequent work, three different CMS chicory cybrids that originated from three different fusion events were characterized at molecular level and backcrossed to Witloof chicory, in order to transfer the male sterile cytoplasms from an industrial chicory nuclear environment to a Witloof context. The expression analysis, carried out to investigate whether orf522 was expressed in the cybrids, showed that one line did not express the sequence, while the other two cybrid lines showed a weak expression of the gene. These observations, along with the restoration of fertility obtained when the cybrids were backcrossed to the same pollinator, led the authors to conclude that orf522 is not associated to the CMS observed in the chicory cybrids and to suggest that the cybrids represented a novel CMS, different from the CMS of sunflower (Dubreucq et al., 1999).

Protoplast asymmetric fusion was used to produce male sterile somatic hybrids between a "Rosso di Chioggia" genotype and a PET-1 sunflower CMS line. In these experiments mesophyll chicory protoplasts, inactivated with iodoacetic acid, were fused with hypocotyl sunflower protoplasts irradiated with γ rays. Regenerated cybrids presented mitochondrial re-arrangments, at anthesis produced fewer pollen grains which could not germinate either *in vitro* or *in situ*, and produced seeds when free-pollination occurred. In some cybrids orf522 was detected in the mitocondrial genome (Varotto et al., 2001). Cybrid plants obtained by protoplast asymmetric somatic fusion between an inbred (S_6) "Variegato di Castelfranco" line and a PET-1 sunflower CMS line are currently under investigation (Varotto et al., unpublished results).

The results obtained so far using interspecific protoplast fusion experiments in order to produce CMS chicory plants seem to indicate that, after fusion, male sterile cybrids can be regenerated. It appears evident that mitochondrial genome re-arrangements led to the creation of novel CMS instead of tranferring the desidered trait from PET-1 CMS sunflower lines.

6 Integration of Biotechnologies in Chicory Breeding Programs

6.1 QTL Mapping and Marker-Assisted Selection

Productivity and quality have always been the main focus for chicory production, in addition to uniformity and stability of the varieties. However, little is known about the genetics of the yield traits and quality characteristics in this species. Quantitative trait loci (QTL) mapping of commercially important traits can broaden our knowledge of the chicory genome and be of help in breeding. Moreover, detecting and mapping QTLs seem to be crucial for the potential application of marker-assisted selection (MAS) strategies.

The RAPD technology was efficiently used for the construction of a basic genetic linkage map for chicory including 129 marker loci (Van Stallen et al., 2003), a prerequisite for QTL analysis. QTL mapping within the *Asteraceae* family was previously described for *Microseris* spp. (Bachmann and Hombergen, 1997) and lettuce (Johnson et al., 2000). More recently, Van Stallen et al. (2005) investigated the inheritance patterns and the genetic basis of quality characteristics of the pith in chicory by means of QTL analysis using RAPD markers. Several QTLs were mapped in an F_2 population of chicory derived from the F_1 of a cross between two phenotypically antagonist inbred lines. Each F_2 individual was selfed and plant characteristics were measured in F_3 populations. Although variation in pith characteristics was largely environmentally influenced, several QTLs for the length of pith, browning of the pith, hollow pith and apple pith were detected. Interactions between QTLs were also found.

The detected QTLs confirmed the different forcing suitability of the early and late parental inbred lines, and strengthened the potential feasibility of marker-assisted breeding for this trait. Nevertheless, if it is true that QTLs were found in many linkage groups for all characteristics studied, it is also true that most of these QTLs explained only a small proportion of the total variance observed. This implies that most measured traits were strongly influenced by environmental factors. Many QTLs were found in at least two of the three forcing periods, especially for pith length and brown colouring, validating their reliability. It is commonly known that browning of the pith is very much influenced phenotypically and it has been shown that Ca-mobility is an important factor in this phenomenon (Van Melckebeke, 1993). Root characteristics are also important in pith browning. Shorter and smaller roots will lead to chicons with lesser browning of the pith (Van Stallen et al., 1999;

Lapage, 2002). Although environmental conditions can strongly influence these characteristics, Van Stallen et al. (2005) stated that there is genotype influence in the expression of browning of the pith. QTLs for pith abnormalities were often mapped at similar positions on the chicory genetic linkage map, whereas QTLs located elsewhere in the genome can have an interfering effect, as demonstrated between QTLs for pith length, and between QTLs for browning of the pith and QTLs for hollow pith.

The genetic bases of yield traits and taste characteristics have been recently investigated in chicory by Van Stallen et al. (2005). Production components were measured on F_3 populations and QTLs were mapped on the genetic linkage map previously constructed in chicory (Van Stallen et al., 2005). QTLs were also found for bitterness and sweetness in the fresh non-cooked chicon.

In most cases, the amount of variation explained by the QTLs detected was rather small. The amount of variation explained is an approximation of the genotypic effect on that specific trait in the population studied. There are several causes for the overall limited amount of variation explained by the observed QTLs. First of all, a large portion of the variation may be due to environmental factors influencing a given trait and hence to its low heritability. Yield and taste characteristics of the chicon are, for example, influenced by nitrogen and magnesium fertilization in the field, soil structure and forcing conditions and period (Savonet, 1961; Peters and Van Amerongen, 1997). Moreover, scoring of the quantitative traits was done on F_3 populations in which markers at the loci that were heterozygous in the F_2 plant necessarily segregate. Thus, variation can also be explained by different genotypes at the loci being investigated. Finally, other factors influencing the amount of variation explained by a QTL include the accuracy of the linkage map and the coverage of the mapped genome (Jansen et al., 1995).

For most plant characteristics, a large part of the total variation was explained by the QTLs detected in the first forcing period. In the winter and late forcing period, less variation was explained. This can be due to the effect of the extra parameter, such as conservation of the roots, since bacteriological and fungal contamination can occur and influence the forcing results.

Production and quality of chicons will depend upon the roots. Problems in forcing associated with an inferior root quality cannot be completely solved by modifying forcing conditions. The weight, length and number of leaves in the field are influenced by plant density, nutritional value of the field, climate and cultivar (see Van Stallen et al., 2005). A clear relationship between the characteristic weight of the longest leaf in the field and the chicon weight was observed in all three forcing periods. Not much is published about the association between field characteristics and subsequent forcing performance. Although production characteristics are strongly influenced by several environmental factors, almost 46% of the total variance for chicon gross weight was explained by the QTLs detected (Table 3).

The QTLs for chicon gross weight detected on linkage groups 1 and 5 seem the most promising for future use since they were detected in all three forcing periods and the amount of variation explained was more than 10%. For pith length, the QTL on linkage group 7 deserves more attention: this QTL was also observed in all three forcing periods and the amount of variation explained was, with amounts exceeding

20%, high compared with the other QTLs detected. For chicon diameter, no QTLs explaining more than 10% of the variation were detected. The QTL on linkage group 5 however was detected in all three forcing periods. At first sight, no markers associated with QTLs useful for marker-assisted breeding were found for the taste characteristics of the chicon. However, bitterness and sweetness properties proved to be partly genotypically determined. Interestingly, the smaller the chicon, the sweeter it tasted and vice versa.

Table 3. Summary of the QTLs for production components, longest leaf weight, chicon length and diameter, and taste characteristics with indication of the percentage of variance explained (range) in the three forcing periods (from Van Stallen et al., 2005).

QTL	Linkage group	Position (cM)	Variance (%)
Production components			
Chicon gross weight	1	57.4-90.4	5.2-13.9
	3	0.0-23.5	8.9-13.0
	5	30.3-82.5	8.0-17.1
Pith length	7	42.0-54.0	7.4-22.3
Chicon diameter	1, 3, 5, 6, 7	-	13.0-30.0[a]
No. of leaves (field)	3	18.5-35-3	n.a.
	6	45.9-61.7	n.a.
Longest leaf weight	2, 5, 6, 7	-	n.a.
Taste characteristics			
Bitterness	2	0.0-14.0	n.a.
	8	24.0-47.0	n.a.
Sweetness	2	31.0-39.0	n.a.

[a]Overall variance expressed by all detected QTLs; n.a. not available.

Theoretically, it should be possible to make use of the QTLs identified so far and introduce marker-assisted selection into Chicory breeding programs. However, more precise localization of the QTLs and their linked DNA markers need to be established before attempting any large-scale analysis of other crosses. QTL analysis was performed on the basis of the linkage map obtained using RAPD markers. Because QTL mapping is very sensitive to errors in marker placement and between marker distance, exact localization of QTLs on linkage groups and better estimation of explained variation by these QTLs are only possible when using codominant markers, such as SSR or SNP markers.

Optimization of the genetic linkage map with the addition of codominant markers and more precise localization of the QTLs detected need to be achieved in order to select DNA markers linked as closely as possible to the QTLs of interest. The usefulness of these markers in marker-assisted breeding needs subsequently to be tested in other crosses, with other cultivars being used.

6.2 Gene Flow and Genome Polymorphism in *C. intybus* and *C. endivia* Species and Varieties

A few genetic studies using molecular markers have been carried out on *Cichorium* spp. mainly to characterize commercial varieties and experimental materials (Bellamy et al., 1995, Koch and Jung, 1997; Van Stallen et al., 2000; Barcaccia et al., 2003), to evaluate the genetic homogeneity and purity, respectively, of inbreds and hybrids (Bellamy et al. 1996), and to investigate phylogenetic relationships between cultivars and cultivar groups of *C. intybus* and other species, both cultivated and wild, belonging to the same genus (Vermeulen et al., 1994; Kiers et al., 2000; Van Stallen et al., 2001). AFLP and RAPD markers were also used to construct the only two genetic linkage maps of *C. intybus* reported so far in the literature (De Simone et al. 1997; Van Stallen et al., 2003).

As mentioned earlier, Kiers et al. (2000) exploited AFLP markers and multivariate statistics to establish the genetic relationships among the species and cultivar groups of *C. intybus* and *C. endivia*. At the species level, the results correspond with previously obtained phylogenetic relationships in that *C. bottae* is the most divergent species, and *C. intybus* and *C. spinosum*, as well as *C. endivia*, *C. pumilum* and *C. calvum* are clustered in distinct subgroups. Based on the congruence between phylogenetic and genetic analyses, unique markers were expected for all species, exception made for *C. bottae*. The analysis of *C. intybus* materials resembled the species analysis in terms of grouping cultivars according to cultivar groups. In contrast to *C. intybus*, the cultivar series of *C. endivia* did not form distinct groups, which would reflect that crosses have been made among the various cultivar groups.

Molecular markers were used to distinguish between several *C. intybus* genotypes, comprising four white Witloof inbred lines, three red Witloof experimental inbred lines and a number of experimental hybrids (Bellamy et al., 1996). In particular, RAPD fingerprints allowed the identification of all the materials analyzed. All the discriminant polymorphisms were confirmed both on individual heads and young seedlings for each genotype. This information was applied and proved to be useful also to evaluate the genetic uniformity of inbred lines and the genetic purity of F_1 hybrid seed samples (Bellamy et al., 1996).

Molecular marker-based investigations were also aimed to evaluate the genetic relationships among the five types of Radicchio and to set up a molecular reference system that would allow a precise identification of the different cultivated types. The five major cultivated types of Radicchio were investigated by PCR-derived markers (Barcaccia et al., 2003). The experimental material was represented by two outbred populations (one of "Variegato di Castelfranco" (CF) and one of "Rosso di Verona" (VR)) and by eight partial inbred lines (three of early "Rosso di Treviso" (TVP), three of late "Rosso di Treviso" (TVT) and two of "Rosso di Chioggia" (CH)). The different types were well distinguished from one another if analyzed by means of bulks using AFLP markers at the population level, while they were not if analyzed at the individual level using RAPD, Inter-SSR and AP-PCR markers. The genetic variation was shown to be much higher within types than between types (Fig. 3). This result suggests that, in each Radicchio type, populations produced by breeders through controlled intercrossing or repeated selfing conserved their gene pools well separated over the years.

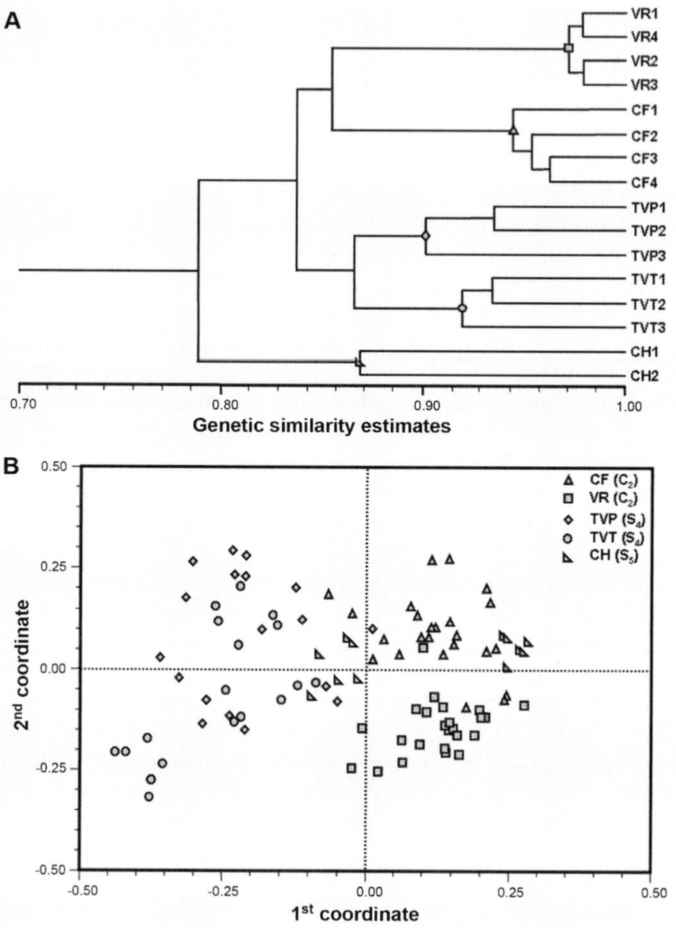

Fig. 3. Results of the molecular characterization of radicchio types: (A) UPGMA dendrogram of 16 bulked DNA samples obtained from the genetic similarity matrix based on AFLP markers; (B) Centroids of 96 individuals belonging to the five radicchio types constructed by the first two coordinates according to RAPD, AP-PCR, and I-SSR markers and based on the Dice's similarity coefficients (modified from Barcaccia et al., 2003).

On the basis of the reproductive system of *C. intybus*, such a finding may be explained taking into account i) the SSI system of the species that limits both selfing and intercrossing between plants with an identical phenotype at the multi-allelic S-locus, thus allowing a certain amount of heterozygousity to be maintained even in inbred populations; ii), the selection criteria of mother plants applied each year by each farmer to maintain his own population probably limited contamination between types, preserving the phenotypic identity of each type. At the same time, seed multiplication carried out over the years produced a clear genetic differentiation between types within the species.

The setting up of a molecular reference system seems to be feasible and suitable both for the precise identification of the single types of radicchio and for the evaluation of the extent of natural hybridization that can occur between different types. Diagnostic molecular markers, along with morphological and phenological descriptors, will be useful for the certification of typical local products of radicchio and for the recognition of a Protected Geographical Indication (PGI) mark.

6.3 *In vitro* Approaches

Plant cell and tissue culture is a fundamental tool in both basic and applied strategies of chicory improvement. In many breeding procedures, the possibility of maintaining and/or multiplying plants through vegetative propagation, independently of the adopted technique, may allow the genotypic evaluation of selected individuals through the observation of their clonal, half-sib, or full-sib progenies and, consequently, the adoption of more efficient selection approaches (Wricke and Weber, 1986). This is particularly true when, as it happens with resistance traits, the reaction is genotype-specific and clonal evaluation may be of notable significance.

In France the first *in vitro* cultures of *C. intybus* were initiated by Gautheret (1941). An extensive list of papers dealing with *in vitro* culture mainly focussed on Witloof chicory and prior to 1988 can be found in Schoofs and de Langhe (1988), where an exhaustive introduction to *in vitro* approaches are presented, comprising methods of sterilization, media composition and different culture techniques.

Currently, the applications of the *in vitro* technology go well beyond micropropagation: they group all the *in vitro* approaches that are relevant or possible for chicory and include all the tools which, during the years, have shown up to be essential in the study of plant biology.

6.3.1 *In vitro* Organogenesis

Explants of roots, leaves and stems of chicory can form calli and differentiate buds, generally at the surface of the explants. Appropriate culture medium, containing low doses of citokinine and lower doses of auxin can be used for *in vitro* direct organogenesis, limiting the proliferation of callus. In fact, Caffaro et al. (1982) investigating on nuclear cytology of callus and on chromosome number variation in *in vitro* regenerated shoots and roots, demonstrated amitosis in developing primary callus and chromosome number mosaicism in almost all apical meristems of plantlets regenerated from primary callus. Explants of young leaves are routinely used in our lab to propagate the genotypes of interest. They are induced to differentiate buds, usually from wounds or veins, without callus proliferation. Somaclonal variation is usually avoided using low concentrations of hormones (0.1mg/l citokinine and 0.01mg/l auxine) and suitable periods of culture, with subculture every 15-20 days.

Young buds are generally transplanted in small pots and transferred to the greenhouse, where they produce the root (Varotto, 1995). Several factors are known to influence the production of shoots in chicory, such as the age of the explant, carbohydrates, and ethylene (Lefebvre and Sadki, 1989; Lefebvre et al., 1992). It has been shown that long days promote a faster and more abundant production of buds and that an increase of light intensity and exogenous sucrose supply enhance the amounts of newly formed buds (Badila et al., 1991).

The proliferation of callus from different explants can be induced in chicory using high doses of auxin (2,4D) and equal or lower doses of citokinines (6BAP). Somaclonal variation has given a thermosensitive male sterile plant (Dubois et al., 1987) and selected lines resistant to sulfonylurea (Millecamps, 1989) and glyphosate (Sellin, 1991; Sellin et al., 1992)

Culture of zygotic embryos *in vitro* has been shown to be a suitable technique to accelerate the breeding process in biennial cultivated "Rosso di Chioggia " chicory. Plantlets ready for performing the selection for the desired agronomic traits were obtained one month after pollination, through in vitro maturation of embryos in B5 medium without growth regulators (Varotto et al., 2000)

6.3.2 Somatic Embryogenesis

Somatic embryogenesis is a process peculiar to the plant kingdom, by which somatic cells are able to behave like a zygote and differentiate a bipolar structure containing both root and shoot axes. Vasil et al. (1964, 1966) described somatic embryogenesis in cell suspensions of *C. endivia*, but they did not regenerate plants. Somatic embryos from *C. intybus* were regenerated by Heirwegh et al. (1985). Somatic embryoids and diploid plants have been obtained directly from anther culture of an hybrid *Cichorium intybus x Cichorium endivia* (hybrid 474) maintained by micropropagation (Guedira et al., 1989).

Cytological observations were able to demonstrate a single cell origin of embryoids derived from styles, (Dubois et al., 1988), root (Dubois et al., 1990) and leaves (Dubois et al., 1991) of hybrid 474. Original features characterizes *Cichorium* somatic embryogenesis obtained using inductive heat treatment: a callosic layer surrounds the organized cells of the developing embryos. The callosic wall disappears in growing proembryoids (Dubois et al., 1991). The hybrid 474 represents an alternative model to the well known *Daucus carota* model to study cytological and biochemical events during somatic embryogenesis (Rambaud et al., 1996).

Somatic embryoids differentiated from embryogenic callus produced by small fragments of stored root of Witloof chicory. Embryoids differentiation took place upon transfer on liquid medium, while germination of the embryos were obtained in liquid or semisolid medium (Schoofs and de Langhe, 1988). Radicchio types seem to be recalcitrant to somatic embryogenesis. In spite of the numerous efforts we have been undertaking using different plant tissues, inductive treatments and culturing media with various hormone combinations, we could not obtain somatic embryos. Sometimes we observed embryo-like structure in cell suspension cultures that were not able to develop further both in liquid or solid culture medium.

6.3.3 Androgenesis

Homozygosity may be achieved by recovering doubled haploids from culturing anther and/or microspores or ovary and/or ovules. Dihaploid plants are considered an ideal starting material for the production of F1 hybrids. Precise cytological investigations are prerequisite for further successful in vitro cultures. The chronological correspondence between male and female sporogenesis and gametogenesis in *C. intybus* was studied in several works (Louant and Logly, 1981; Pacini and Keizer, 1989; Varotto et al., 1996a).

To our knowledge, in spite of the interest that breeders would find in double haploid Chicory lines, a single work was published about androgenesis in *Cichorium*. Theiler-Hedtrich and Hunter (1994) reported the results obtained from microspore cultures from selected plants of Witloof, Red of Treviso and Robin (a commercial hybrid of Witloof x Red of Treviso). They were able to regenerate plantlets with different levels of ploidy, haploid to tetraploid and some mixoploid, with the tendency to spontaneously become polyploid *in vitro*. Microspore-derived plantlets revealed some differences in leaf phenotype, either in lamina size or form.

In order to induce the microspores of plants from a population of radicchio "Red of Chioggia" to produce haploid tissues, different media and incubating conditions were used. However, all the plants obtained via both organogenesis and differentiation from callus were diploid. The observation of some anther sections at the light microscope has shown that small buds and calli had derived from anther somatic tissues. It is interesting to underline that all the plants obtained in the experiments were not induced to flowering, even though *in vitro* culture of explants from induced plants usually give rise to plants that are themselves induced (Varotto et al., 1996b).

6.3.4 Genetic Transformation

The high capacity of *in vitro* regeneration via organogenesis from root or leaf explants has been repeatedly exploited in view of genetic transformation of Chicory. Genetic transformation of Witloof chicory conferring resistance to chlorosulfuron was achived by Vermulen et al. (1992), using the *Agrobacterium tumefaciens* system on leaf discs. *Agrobacterium rhizogenes* was used to convert Belgian endive from biennal to annual flowering (Sun et al., 1991). Genga et al. (1994) reported the transformation of Radicchio using *A. tumefaciens*.

Transformation experiments were carried out on shoot buds of *C. intybus* cv *sativum* using *Agrobacterium tumefaciens* carrying a reporter gene (uidA) coding for neomycin phosphotransferase conferring resistance to kanamycin (Frelleux et al., 1997). This system was proposed to introduce agronomic important traits in Chicory races of interest.

6.3.5 Protoplast Culture and Fusion

Plant protoplasts are the starting material for the production of somatic hybrids, through cell fusion procedures, and transgenic plants, through protoplast transformation.

The first successful experiments of plant regeneration from protoplast-derived calli in various *C. intybus* cultivars grown in greenhouse were reported by Crepy et al., (1982). The same method was then applied to Witloof to improve protoplast regeneration (Saksi et al., 1986). Routine preparation of protoplasts and plant regeneration were adapted to var. Magdebourg, allowing to obtain tetraploids (2n=36) by protoplast fusion (Rambaud et al., 1990a; Rambaud et al., 1992). Suitable protocols for plant regeneration from protoplast of different Radicchio types were also developed (Varotto et al., 1997).

Intergeneric protoplast fusion between var. Magdebourg and male-sterile *Helianthus annuus* gave some male-sterile chicories (Rambaud et al., 1993). The protoplasts of both species were fused using PEG method and the different colour of the cells (mesophyll green cells and uncoloured hypocotyl cells) was used to select the fusion products. A percentage between 3 and 20% of manually selected heterokaryocytes evolved in regenerating calli. All the plants obtained by protoplasts fusion showed a Chicory phenotype and some of them were sterile or male sterile. Analysis of mitochondrial genomes showed that a large part of the mtDNA of sunflower, but not all the genome, must have been incorporated into the chicory mtDNA (Rambaud et al., 1993). The genetic instability of the mitochondrial genome was also highlighted analyzing the fourth progeny derived from one CMS cybrid (Rambaud et al., 1997). Asymmetric somatic fusion procedures were applied in order to obtain cybrids between Radicchio and a male sterile sunflower line (Varotto et al., 2001). Chicory mesophyll proptoplasts were inactivated with iodoacetic acid and fused with hypocotyl sunflower protoplasts irradiated with γ rays: heterokaryons were manually selected thanks to the different cell colour. These experiments indicate that the efficiency of fusion is higher when protoplast are not inactivated. Southern analysis confirmed the hybrid nature of the regenerated plantlets that showed partial introgression of sunflower mitochondrial genome in chicory mitochondria. For a better description of the male sterile genotypes obtained with the different fusion procedures reported above, see section 5.6 on male-sterility.

7 Seed Production

Seed production represents the conclusive phase of the breeding activity and, as with any other cultivated species, is of paramount importance for the success of the crop. In both chicory and endive, the seed is the only plant material used for variety commercial propagation and planting and it often determines the quantity of yield and its qualitative commercial standard. Despite this, little research has been done on endive and chicory seed production. The fruit (seed) is an achene, obovoid to cylindrical, weakly ribbed, light brown to completely brown when ripe. Almost the whole of the endive and chicory commercial seed in Europe is produced in Netherlands, Northern France, and Italy, usually in areas where climatic conditions are sufficiently fair to permit an abundant differentiation of flowers, a long flowering period, a good presence of pollinating insects, available water during the filling period, and sufficiently dry conditions during the last part of the reproductive cycle, when the seeds need to mature and dry. In Italy, for instance, an important activity of seed production, handled by Italian and European seed companies as well, is present in the eastern portion of the Po Valley, south of the Po River (Romagna) and in the Marche region.

Although chicory and endive are completely different as far as their mating system and life cycle are concerned, and that endive is a little less cold tolerating than chicory, the seed production technique applicable to the two species is very similar and it does not deserve a separate description; possible specific features will emerge during the reading. Moreover, since this paper mostly concern the breeding aspects, in this paragraph emphasis has been put on the seed producing procedures which more directly may interest the breeder rather than to commercial seed producers.

In *C. endivia* selfing is the rule, but cross pollination due to visitor insects and cross fertilization are possible. Since most of the commercial seed is produced outdoors, spatial isolation is thus to be strictly respected both from different Endive varieties and, taking into account that chicory and endive are interfertile, from chicory fields (1500 to 2000 m) as well to avoid any possible contamination. The same is true for *C. intybus*.

The cultural practices applied to an endive and chicory seed crop during the vegetative phase are much the same as those used in growing them for the vegetable market, but a lower plant density (about 5 plants m^{-2}) is used. During the vegetative period fields are repeatedly inspected to remove off-type or diseased plants.

Although, being an annual, endive does not need vernalization to flower, the sowing can be made in autumn or, in areas with harsh winters, at the end of the cold season. In this case, in order to advance the cycle of the crop, getting more vigorous plants, and have a higher seed production, better results can be obtained if sowing is made in a protected environment, like a greenhouse, in January-February and transplanting in the open field takes place in March.

Chicory, being a biennial, needs vernalization to differentiate and produce the seed stalk. Thus, a seed crop has to be sown in autumn in order to give seed the next spring. If one wants the plants to give seed in the same year of sowing, the seeding operations have to take place in winter (end of January – beginning of February), better under protection, in order to let the small seed born or transplanted plants become naturally vernalized in the field by the cold temperatures of this period of the year. Since cultivars of different earliness have been selected within both Witloof chicory and Radicchio, it has to be pointed out that sowing has to be more anticipated for early than for late selections in as much as the cold requirement for flowering induction seems higher for the former.

The genetic control of flower induction and differentiation in endive, as well as in chicory, is however still unknown. Seed stalk emission may depend on the number of leaves which form the rosette, or day length, or both; surely enough, temperature does not seem to play a major role in endive flowering.

For both species flowering begins in May to June: the azure-blue flowers open early in the morning and under optimal light and temperature conditions anthesis is completed before 10 a.m. The flowers usually do not stay open longer than the whole morning unless conditions of moderate temperature and high humidity take place. Seeds mature 50 to 60 days after flowering and are collected in July to August. The seed heads on the plant do not mature uniformly and so, particularly under conditions of high temperature and dry air, some shattering can occur when the seeds are collected. However, seeds have to be harvested at a suitable developmental stage since, if they are not completely mature, germination may be poor. As known, besides of its genetic value, intrinsic and extrinsic properties of the seed depend upon the crop's growing conditions as well as on harvest techniques and, on the whole, determine its commercial quality and field performance. The harvesting is preferably carried out early in the morning by cutting the seed stalks at their base just before the seeds have dried out completely; as a matter of fact, the seeds may fall off the stalk and be lost if they are allowed to fully dry on the plant while still in the field. When the seed stalks are dry, plants are trashed using adequate equipment. Caution has to be taken during this process in order to offer the maximum guarantee in seed genetic purity and physical properties. Seed yield depends on plant density and architecture and, on average, is equal to 10 to 15 g per plant hence a production of 0.5-0.7 t ha^{-1} can be expected. The 1000 seed weight is 1.4 to 1.7 g (600 to 700 seeds per gram) depending on the cultivar.

Due to their angular shape, commercial endive and chicory seeds are often coated and pelletized with various materials to give them a more spherical shape. This brings about an easier use of planting machines, of pesticides employment against seed and soil borne pathogens, and/or of priming processing to improve the sowing performance.

Since in chicory F_1 hybrids have entered the seed market, special attention should be deserved to the multiplication of the parental lines and to the F_1 seed production. In chicory, *in vitro* techniques, or less sophisticated procedures based on the ability of sliced roots to produce numerous clones, may allow maintaining and adequately multiplying these parental lines without any risk of genetic contamination.

As far as F_1 hybrids are concerned, it might be worth to say that since they have been developed by private companies, no information are available on the adopted procedure. On the basis of what has been said on chicory's reproductive system, of our experience, and of private contacts (De Proft, p.c.) the European F_1 hybrid varieties are believed not to be true hybrids, but rather bulks of F_1 and inbred seeds as a consequence of the selection of self-compatible inbred lines as parents, and of the absence of a naturally induced male-sterility source. The situation is different in the U.S.A. where the use of male-sterile lines providing a reliable pollination control is authorized, thus allowing the SeedLink™ seed production system to be applied.

From the breeder's point of view, some observations may be added. As it has been previously said, *C. endivia* is an annual species which does not need vernalization to differentiate the floral bud and to elongate the seed stalk. Hence, when sown in spring, early enough to regularly complete its life cycle, it is able to produce mature and dry seed in only one growing season. In practical terms, this means that, if segregating progenies are evaluated in the field, selection and seed production of the selected plants can take place in the same season, thus respecting the annual cadence without any particular problem. The same situation occurs when the screening of the segregating material has to be made in greenhouse or in any other situation which implies artificial or protected growing conditions: at spring, the selected plants or progenies are transplanted in the field where they regularly flower and produce viable seed.

Things are quite different with *C. intybus*, which is a biennial species. This means that plants do not enter the reproductive stage without vernalization (Paulet, 1985; Van Stallen et al., 1999). This is the reason why both Witloof chicory and Radicchio, whose commercial product is a head of tighten leaves, are commonly grown as annual plants but have to be sown in late spring or in summer in order to give a head between October and February directly in the field or, as a response to the forcing procedure, sometime during the winter months according to the producer's choice. In any case, an anticipated or early flowering (*bolting*) as it may be caused by an early spring sowing or a late and sudden lowering of temperature, has to be accurately avoided in as much as, besides implying the shedding of unwanted seeds, resulting in weed chicory in the field of the following year, it causes a net loss of commercial production. This situation poses two problems: a) the need of non-bolting varieties able to render the crop at least partially independent of the temperature during the first stage of development and b) the need of technical procedures able to permit that selection and seed production of the selected plants may take place in the same growing season, early enough to proceed according to an annual cadence.

As for the first problem, breeding programs could help in developing new varieties resistant to early flowering. Unfortunately, the genetic control of flower induction and development in chicory are far from being understood. Information on the induction to flowering in chicory mainly derive from field (Gianquinto, 1997; Wiebe, 1997) and *in vitro* studies using Witloof explants (Demeulemeester and De Proft, 2000). Both in "Rosso di Chioggia" and Witloof chicory the cold requirement for flower induction can be substituted by other physical or chemical factors, such as long days or kinetin. A recent work on root chicory suggests an absolute vernalization requirement, since bolting never occurred in control plants which were not submitted to low temperatures but maintained in long days in the fields. However, the same study cannot definitively conclude that root chicory is an absolute cold-requiring plant since the effects of environmental parameters other than low temperature were not investigated (Dielen et al., 2005).

The exact requirements for floral initation in terms of duration and intensity of an effective vernalization treatment and the developmental stage in which seeds and/or plants are sensitive to cold treatments remain to be established. Some works suggest that an essential factor for chicory flowering is the post-vernalization requirement for long days (Gianquinto, 1997), but it is not known if temperature is as determinant as long days in this phase. It is quite evident that in chicory a better knowledge of the genetics governing flower induction and vernalization requirements is fundamental for developing varieties with known flowering behaviour.

In recent years, useful information on the genetic regulation of flowering are coming from the model plant *Arabidopsis thaliana* (Amasino, 2005). In our lab, we are currently investigating the molecular mechanisms that regulate the switch to flower in Radicchio, to verify whether such mechanism is the same that controls flowering in *Arabidopsis*, and finally, address the diversity among the classes of earliness to one of the cases known for this model plant. This aspect may be particularly interesting in view of the selection for earliness of commercial maturity. As a matter of fact, different classes of commercial earliness have been selected within each one of the various types of Radicchio currently grown and, at least to a certain extend, they show correspondence with as many classes of earliness in flowering time.

As for the second problem, it is noteworthy that the timing and cadence of the selection cycles might be quite different if one refers to the field head producing types of Radicchio or to the forced chicories like Witloof and the Radicchio Late Red of Treviso. For these two chicories, selection mainly occurs after forcing, on the basis of the observations made on the commercial product. The selected plants can be transplanted in pots and later on, after a period of acclimation, directly in the field or under an isolation cage.

The second problem mainly regards the other four types of Radicchio and what follows refers to the procedure most generally applied in their typical area of production. Three aspects have to be considered: a) the moment of selection, b) the conservation of the tap roots, and c) the timing of seed stalks development and seed production. The moment of selection strictly depends on the earliness of the material one is working with. In any case, the selection procedure, mainly based on field observations carried out during the autumn or winter months and which usually imply the destruction of the head, has to take into account the necessity of preserving the roots of the selected plants until the following spring, when they will be transplanted in the field to obtain seed.

Once selection has been completed, the selected plants are dig out and deprived of the leaves with a sharp cut at the base of the head, paying attention to preserve the upper portion of the root where axillary buds are present in great number. The tap roots, once they have been washed and treated with a fungicide, are transplanted in pots and maintained under plastic tunnels where temperature should not lower below 0 - 2°C. During conservation the roots start budding and it may be necessary to thin out the young sprouts.

Roots of early material, selected in September or October, are often cold stored, both before and/or after transplanting, to inhibit sprouting. At the beginning of January the preserved roots are transplanted under protected environment where, with the aim of accelerating the flowering process, the temperature should be raised to 18°C and artificial light (150-250 lux at the vegetation level) should be added from 1-2 hours before sunset until midnight. The adopted plant density is of 2 plants per m^2 which should bear upon a production of about 40 g of viable seed per plant. During this phase, a periodic defoliation of the basal portion of the seed stalk might be necessary to facilitate the air circulation and a careful disease and insect control should be carried out. As the seed stalk grows, plants should be clipped at 100-120 cm in order to favour earliness and contemporaneousness of flowering.

At the opening of the first flowers, pollinating agents (bees, flies or, more recently, bumblebees, according to the available volume) must be introduced under the isolation cage paying attention, at the same time, to drastically reduce and attentively control the use of any kind of chemical spraying. The flowering period may last two to three months with a peak between the third and the sixth week. This means that, independently of any strategy aimed to an anticipation of flowering, harvesting cannot occur later than the end of mid June, so to have the possibility of trashing the plants and prepare the seed in time for the mid July-mid August sowing.

8 Conclusive Remarks

It seems worth, at this point, to add a couple of observations related to the significance that Endive and Chicory have for horticulture within an European frame. The Witloof and Radicchio cultivar groups, on the whole, represent by far the largest quantity of chicory produced in Europe and outside. The first group is characterized by a strong uniformity mainly due to its very narrow genetic basis (de Proft et al., 2003). The Radicchio group, on the contrary, presents an almost astonishing phenotypic and genetic variability from which selection is still able to generate new and commercially valuable forms. Then, from a breeding point of view, the two groups may be considered at a different level. The first one, having been the object of public and private systematic research since many years, may be taken as a "refined genetic pool" where favourable genes are present at high frequency. The Radicchio group may represent a much more variable and possibly richer genetic pool for morphological and/or physiological variants are concerned, where, nevertheless, useful genes controlling some major quality traits or resistance to the most important pathogens are at very low frequency. As a matter of fact, farmers' selection has traditionally paid attention to morphological characters, on which the market value mainly relies, while little attention, if any, has been deserved to other important characters as biotic or abiotic stress resistance, bitterness, post harvest rotting, and others. In both cases, through an adequate use of the now available molecular tools, useful genes may be identified, isolated, and used through traditional or innovative breeding methods.

From this point of view, then, both the still available numberless farmers' populations of Radicchio and the improved genetic pool of Witloof chicory represent an invaluable reserve of genetic resources which must be adequately studied, analysed, classified, and compared in order to exploit them in breeding programs.

At this point, it may be recalled that *C. endivia* and *C. intybus* have the same number of chromosomes and are completely interfertile; the same degree of interfertility exists among different cultivars groups within each species and between the cultivar groups and the wild populations of both species. The complex *endivia-intybus* might thus be considered, from the breeder's point of view, as a unique genetic pool whose access, as far as we know, is in no way limited by sexual incompatibility. This observation, although trivial, rather than underlying the possibility of transferring useful genes across species, cultivars groups, or cultivated and wild types through traditional methods, i.e. selection of superior genotypes from segregating populations, may suggest the feasibility of breeding programs aimed to create horticultural novelties which might open new commercial perspectives or enlarge the already existing ones.

From a genetical point of view, the cultivar groups might be considered as different genetic pools usable as sources of selected genotypes or lines to be employed as parents for F_1 hybrid varieties. This possibility has to be entirely explored, possibly integrating traditional procedures, based on test crosses, with more sophisticated molecular approaches.

References

Amasino R.M. 2005. Vernalization and flowering time. Curr. Opin. Biotechnol. 16: 154-8.

Bachmann K. and Hombergen E.J. 1997. From phenotype via QTL to virtual phenotype in Microseris (Asteraceae): predictions from multilocus marker genotypes. New Phytol. 137: 9-18.

Badila P., Mikou K., and Paulet P. 1991. Two distinct modes of action of light on the in vitro development of root explants of *Cichorium intybus* L. J. Plant Physiol. 138: 370-375.

Barcaccia G., Varotto S., Soattin M., Lucchin M., and Parrini P. 2003. Genetic and molecular studies of sporophytic self-incompatibility in *Cichorium intybus* L. Proc. of the Eucarpia meeting on Leafy Vegetables Genetics and Breeding, March 19-21, 2003, Noordwijkerhout, The Netherland, p. 154.

Barcaccia G., Pallottini L., Soattin M., Lazzarin R., Parrini P., and Lucchin M. 2003. Genomic DNA fingerprints as a tool for identifying cultivated types of red chicory (*Cichorium intybus* L.) from Veneto, Italy. Plant Breed. 122: 178-183.

Bellamy A., Mathieu C., Vedel F., and Bannerot H. 1995. Cytoplasmic DNA and nuclear rDNA restriction fragment length polymorphisms in commercial witloof chicories. Theor. Appl. Genet. 91: 505-509.

Bellamy A., Vedel F., and Bannerot H. 1996. Varietal identification in *Cichorium intybus* L. and determination of genetic purity of F_1 hybrid seed samples, based on RAPD markers. Plant Breed. 115: 128-132.

Bianchedi A. 1961. I radicchi di Treviso. L'Italia Agricola. 1: 37-51.

Block G., Patterson B., and Subar A. 1992. Fruit, vegetables and cancer prevention. A review of the epidemiological evidence. Nutr. Cancer. 18: 1-29.

Caffaro L., Dameri R.M., Profumo P., and Bennici A. 1982. Callus induction and plantlets regeneration in *Cichorium intybus* L. I. A cytological study. Protoplasma 111: 107-112.

Castaño C.I., Demeulemeester M.A.C., and De Proft M.P. 1997. Incompatibility reactions and genotypic identity status of five commercial chicory (*Cichorium intybus* L.) hybrids. Sci. Hortic. 72: 1-9.

Conti F., Abbate G., Alessandrini A., and Blasi C. (EDS), 2005. An Annotated checklist of the Italian Vascular Flora. Palombi et Partner S.r.l. Roma.

Crepy L., Chupeau M.C., and Chupeau Y. 1982. The isolation and culture of leaf protoplast of *Cichorium intybus* L. Z. Pflanzenphysiol. 107: 123-131.

Demeulemeester M.A.C., and De Proft M.P. 2000. Use of chicory root explants in vitro as a model for the induction of different stem types. J. Plant Physiol. 156: 413-418.

De Simone M., Morgante M., Lucchin M., Parrini P., and Marocco A. 1997. A first linkage map of *Cichorium intybus* L. using a one-way pseudo-testcross and PCR-derived markers. Mol. Breed. 3: 415-425.

De Proft M.P., Van Stallen N., and Veerle N. 2003. Introduction: History of chicory breeding. Proc. of the Eucarpia meeting on Leafy Vegetables Genetics and Breeding, March 19-21, 2003, Noordwijkerhout, The Netherland, pp. 83-90.

d'Eeckenbrugge G.C., Van Hergk J.C., and Dutilleul P. 1989. A study of fructose yield components in chicory. Plant Breed. 102: 296-301.

d'Eeckenbrugge G.C. 1990. The progamic phase in *Cichorium intybus* L. Pollen tube growth in the style, incompatibility reaction and gametophytic competition. Euphytica 48: 17-23.

Dielen V., Notté C., Lutts S., Debavelaere V., Van Herck J.C., and Kinet J.M. 2005. Bolting control by low temperature in root chicory (*Cichorium intybus* var *sativum*). Field Crop Res. 94: 76-85.

DuPont M.S., Mondin Z., Williamson G., and Price K.R. 2000. Effect of variety, processing, and storage on the flavonoid glycoside content and composition of lettuce and endive. J. Agric. Food Chem. 48(9): 3957-3964.

Eenink A.H. 1981. Compatibility and incompatibility in witloof-chicory (*Cichorium intybus* L.). 2. The incompatibility system. Euphytica 30: 77-85.

Ernst-Schwarzenbach M. 1932. Zur genetik und fertilitat von *Lactuca sativa* L. und *Chicorium endivia* L. Archiv. der Julius Klaus-Stiftung, 7: 1-35.

Flamm G., Glinsmann W., Kritchevsky D., Prosky L., and Roberfroid M. 2001. Inulin and oligofructose as dietary fiber: a review of the evidence. Crit. Rev. Food Sci. Nutr. 41(5): 353-362.

Frese L., and Dambroth M. 1987. Research on the genetic resources of inulin-containing chicory (*Cychorium intybus* L.). Plant Breed. 99: 308-317.

Frese L., Dambroth M., and Bramm A. 1991. Breeding potential of root chicory (*Cichorium intybus* L. var. sativum). Plant Breed. 106: 107-113.

Frulleux F. Weyens G., and Michel J. 1997. *Agrobacterium tumefaciens*-mediated transformation of shoot-buds of chicory. Plant Cell Tissue Org. Cul. 50: 107-112.

Gautheret R.J. 1941. Recherches expérimentales sur la polarité des tissus de la racine d'Endive. C.R. Acad. Sci. Paris. 231: 37-39.

Gemeinholzer B., and Bachmann K. 2005. Examining morphological and molecular diagnostic character states of *Cichorium intybus* L. (Asteraceae) and *C. spinosum* L. P. System. Evol. 253: 105-123.

Genga A., Giansante L., Bernacchia G., and Allavena A. 1994. Plant regeneration from *Cichorium intybus* L. leaf explants transformed by *Agrobacterium tumefaciens*. J. Genet. & Breed. 48: 25-32.

Gianquinto G. 1997. Morphological and physiological aspects of phase transition in radicchio (*Cichorium intybus* L. var. *silvestre* Bisch.): influence of daylenght and its interaction with low temperature. Sci. Hortic. 71: 13-26.

Guedira M., Dubois-Tylski T., and Vasseur J. 1989. Embryogenèse somatique directe a partir de cultures d'anthères du *Cichorium* (Asteraceae). Can. J. Bot. 67: 970-976.

Hartman Th. 1956. After effects of low temperature on leaf morphology of *Cichorium intybus* L. Proc. K. Ned. Acad. Wet. 59: 677-684.

Hertog M.G.L., Hollman P.C.H., and Katan M.B. 1992. Content of potentially anticarcinogenic flavonoids of 28 vegetables and 9 fruits commonly consumed in The Netherlands. J. Agric. Food Chem. 40: 2379-2383.

Hishock S.J., and McInnis S.M. 2003. Pollen recognition and rejection during the sporophytic self-incompatibility response: *Brassica* and beyond. Trends Plant Sci 8: 606-613.

Horn R., Köhler R.H., and Zetsche K. 1991. A mitochondrial 16-kDA protein is associated with cytoplasmic male sterility in sunflower. Plant Mol. Biol. 17: 29-36.

Huyskes J.A. 1963. Breeding of witloof chicory for forcing without cover soil. Meded. Inst. Veredeling Tuinbouwgewassen, 202.

Jansen R.C., Van Ooijen J.W., Stam P., Lister C., and Dean C. 1995. Genotype-by-environment interaction in genetic mapping of multiple quantitative trait loci. Theor. Appl. Genet. 91: 33-37.

Johnson W.C., Jackson L.E., Ochoa O., Van Wijk R., Peleman J., St Clair D.A., and Michelmore R.W. 2000. Lettuce, a shallow-rooted crop, and Lactuca serriola, its wild progenitor, differ at QTL determining root architecture and deep soil water exploitation. Theor. Appl. Genet. 101: 1066-1073.

Kiers A.M., Mes T.H., van der Meijden R., and Bachmann K. 1999. Morphologically defined *Cichorium* (Asteraceae) species reflect lineages based on chloroplast and nuclear (ITS) DANN data. System. Bot. 24: 645-659.

Kiers A.M., Mes T.H., van der Meijden R., and Bachmann K. 2000. A search for diagnostic AFLP markers in *Cichorium* species with emphasis on endive and chicory cultivar groups. Genome 43(3): 470-476.

Koch G., and Jung C. 1997. Phylogenetic relationships of industrial chicory varieties revealed by RAPDs and AFLPs. Agronomie 17: 323-333.

Lapage E. 2002. Studiedag witloof op het Provinciaal Centrum in Rumbeke-Beitem. Proeftuinnieuws 5: 23-25.

Lavigne C., Millecamps J.L., Manach H., Cordonnier P., Matejicek A., Vasseur J., and Gasquez J. 1994. Monogenic semidominant sulfonylurea resistance in a line of white chicory. Plant Breed. 113: 305-311.

Lavigne C., Manac'h H., Guyard C., and Gasquez J. 1995. The cost of herbicide resistance in white chicory: ecological implications for its commercial use. Theor. Appl. Genet. 91: 1301-1308.

Louant B.P., and Longly B. 1981. Correspondence chronologiques entre les sporogenèses et gamétogeneses mâles et females chez *Cichorium intybus* L. (Chicorée de Bruxelles). Rev. Int. Biol. Veg. Bot. 4: 173-186.

Lucchin M., Barcaccia G., Lazzarin R., and Parrini P. 2003. E' in grado la selezione fenotipica di determinare differenziazione molecolare tra popolazioni in radicchio (*Cichorium intybus* L.). Atti del Congresso della Società Orticola Italiana, Gruppo di Lavoro Miglioramento Genetico: Lo stato dell'arte nel miglioramento genetico delle principali specie ortoflorofrutticole d'interesse mediterraneo, Valenzano (Bari), Italy. 25-26 Giugno 2002, a cura di L. Ricciardi, pp. 387-396.

Lucchin M., Tosini F., Barcaccia G., and Parrini P. 2007. Caratterizzazione e valorizzazione di popolazioni locali dei radicchi veneti. Italus Hortus (in press).

Mares D., Romagnoli C., Tosi B., Andreotti E., Chillemi G., and Poli F. 2005. Chicory extracts from *Cichorium intybus* L. as potential antifungals. Mycopath. 160(1): 85-91.

Mariani C., DeBeuckeleer M., Trueltner J., Leemans J., and Goldberg R.B. 1990. Induction of male sterility in plants by a chimaeric ribonuclease gene. Nature 347: 737-741.

Millecamps J.L. 1989. Sélection des Chicorées *Cichorium intybus* L. var Witloof résistantes aux sulfonylurées par cultures cellulaires. Thèse de Doctorat. Université de Lille.

Monegèr F., and Smart C.J. 1994. Nuclear restoration of cytoplasmic male sterility in sunflower is associated with the tissue-specific regulation of a novel mitochondrial gene. EMBO J. 13(1): 8-17

Monti A., Amaducci T., Pritoni G., and Venturi G. 2005. Growth, fructan yield, and quality of chicory (*Cichorium intybus* L.) as related to photosynthetic capacity, harvest time, and water regime. J. Exper. Bot. 56(415): 1389-1395.

Nasrallah J.B. 2000. Cell-cell signaling in the self-incompatibility response. Curr. Opin. Plant Biol. 3: 368-373.

Nenz E., Varotto S., Lucchin M., and Parrini P. 2000. An efficient e rapid procedure for plantlets regeneration from chicory mesophyll protoplasts. Plant Cell Tissue Org. Cul. 62(1): 85-88.

Nuding J. 1935. Einflufi von Standraum und ernahrung auf menge und glite der zichorienernte. Pflanzenbau. 11: 447-473.

Oltmann W., Burba M., and Bolz G. 1984. Die qualitat der zuckerrlibe. Bedeutung, Beurteilungskriterien und ziichterische Mafinahmen zuihrer Verbesserung. Fortschritte der Pflanzenziichtung, No. 12.

Pacini E., and Keijzer C.J. 1989. Ontogeny of intruding non-periplasmodial tapetum in the wild chicory *Cichorium intybus* (Compositae). Pl. Syst. Evol. 167: 149-164.

Paulet P. 1985. *Cichorium intybus* and *C. endiva*. In: A.H. Halevy (ed.), CRC Handbook of Flowering, Vol. II. CRC Press Inc., Boca Raton, FL. pp. 265-270.
Peters A.M., and Van Amerongen A. 1997. A pilot study on the effects of cultivation conditions of chicory (*Cichorium intybus* L.) roots on the levels of sesquiterpene lactones in chicons. Z. Lebensmittel Untersuchung und Forschung A205: 326-329.
Plocik A., Layden J., and Kesseli R. 2004. Comparative analysis of NBS domain sequences of NBS-LRR disease resistance genes from sunflower, lettuce, and chicory. Mol. Phylogenet. Evol., 31(1): 153-63.
Provvidenti R., Robinson R.W., and Shail J.W. 1979. Chicory: a valuable source of resistance to turnip mosaic for endive and escarole. J. Am. Soc. Horticult. Sci. 104: 726-728.
Rambaud C., Dubois J., and Vasseur J. 1993. Male-sterile chicory cybrids obtained by intergeneric protoplast fusion. Theor. Appl. Genet. 87: 347-352.
Rambaud C., Bellamy A. Dubreucq A., Bourquin J-C, and Vasseur J. 1997. Molecular analysis of the fourth progeny of plants derived fron cytoplasmic male sterile chicory cybrid. Plant Breed. 116: 481-486.
Reynaerts A., Van de Wiele H., de Sutter G., and Janssens J. 1993. Engineered genes for fertility control and their application in hybrid seed production. Sci. Hort. 55: 125-139.
Rice-Evans C.A., Miller N.J., Bolwell P.G., Bramley P.M., and Pridham J.B. 1995. The relative antioxidant activities of plant-derived polyphenolic flavonoids. Free Radical Res. 22: 375-388.
Rick C.M. 1953. Hybridization between chicory and endive. Proc. of the American Society for Hort. Sci. 62: 459-466.
Rossetto M., Lante A., Vanzani P., Spettoli P., Scarpa M., and Rigo A. 2005. Red chicories as potent scavengers of highly reactive radicals: a study on their phenolic composition and peroxyl radical trapping capacity and efficiency. J. Agric. Food Chem. 53(21): 8169-8175.
Ryder E.J. 1998. Lettuce, endive and chicory. Crop production science in horticulture series. CABI Publ., New York, USA.
Savonet G. 1961. The practical culture of chicory. [Witloofteelt in de praktijk]. Provinciale Landbouwdienst van Brabant, 64.
Schoofs J., and De Langhe E. 1988. Chicory (*Cichorium intybus* L.). In: Bajaj Y.P.S. (ed.). Biotechnology in forestry and agriculture. Vol. 6. Springer-Verlag, Berlin, Germany. pp. 294-321.
Sellin C. 1991. Recherche de Chicorées à café (*Cichorium intybus* L. var Magdebourg) résistantes aux herbicides (glyphosate et glufosinate) par cultures cellulaires. Thése de Doctorat. Universitè de Lille.
Sellin C., Forlani G., Dubois J., Nielsen E., and Vasseur J. 1992. Glyphosate tolerance in *Cichorium intybus* L. var Magdebourg. Plant Sci. 85: 223-231.
Sun L.Y., Tourand G., Charbonnier C., and Tepfer D. 1991. Modification of phenotype in belgian endive (*Cichorium intybus*) through genetic transformation by *Agrobacterium rhizogenes*: conversion from biennal to annual flowering. Transgenic Research, 1: 14-22.
Tesi R. 1965. Miglioramento genetico della scarola (*Cichorium endivia* L. var. *latifolium* Hegi). Primo contributo. Rivista di Ortoflorifrutticoltura italiana, 49: 257-268.
Tesi R. 1974. Miglioramento genetico della scarola (*Cichorium endivia* L. var. *latifolium* Hegi). IV Nuove varietà costituite. Riv. Ortoflorofruttic. Ital. 52: 330-341.
Theiler-Hedtrich R. and Hunter C.S. 1995. Regeneration of dihaploid chicory (*Cichorium intybus* L. var. *foliosum* Hegi) via microspore culture. Plant Breed. 114: 18-23.
Van Cutsem P., du Jardin P., Boutte C., Beauwens T., Jacqmin S., and Vekemans X. 2003. Distinction between cultivated and wild chicory gene pools using AFLP markers. Theor. Appl. Genet. 107(4): 713-718.
Van Melckebeke J. 1993. Teeltbescherming. In: M. de Baerdemaker (ed.), Witloofteelt, 3^{e} uitgave, Ministerie van Landbouw, Brussels. pp. 91-113.

Van Stallen N., Demeulemeester M.A.C., and De Proft M. 1999. An in vitro model for the study of hydroponic forcing in chicory (*Cichorium intybus* L.). Plant Cell Tissue Organ Cult. 55: 125-131.

Van Stallen N., Noten V., Demeulemeester M., and De Proft M. 2000. Identification of commercial chicory cultivars for hydroponic forcing and their phylogenetic relationships revealed by Random Amplified Polymorphic DNAs and Amplified Fragment Length Polymorphisms. Plant Breed. 119: 265-270.

Van Stallen N., Noten V., Neefs V., and De Proft M. 2001. The phylogenetic relationship between different *Cichorium intybus* cultivars and cultivar groups, as revealed by RAPDs. Plant Breed. 120: 425-428.

Van Stallen N., Vandenbussche B., Verdoodt V., and De Proft M.P. 2003. Construction of a genetic linkage map for witloof (*Cichorium intybus* L. var. *foliosum* Hegi). Plant Breed. 122: 521-525.

Van Stallen N., Vandenbussche B., Londers E., Noten V., and De Proft M. 2005. QTL analysis of production and taste characteristics of chicory (*Cichorium intybus* var. foliosum). Plant Breed. 124: 59-62.

Varotto S. 1995. Il sistema di Incompatibilità in *Cichorium intybus*. Osservazioni citologiche, istologiche, genetiche e possibili applicazioni della coltura *in vitro*. Tesi di Dottorato. Università di Padova.

Varotto S., Pizzoli L., Lucchin M., and Parrini P. 1995. The incompatibility system in italian red chicory (*Cichorium intybus* L.), Plant Breeding, 114: 535-538.

Varotto S., Parrini P., and Mariani P. 1996a. Pollen ontogeny in *Cichorium intybus* L. Grana, 35: 154-161.

Varotto S., Lucchin M., and Parrini P. 1996b. la coltura di antere in *Cichorium intybus* L. Italus Hortus, 3: 12-15.

Varotto S., Lucchin M., and Parrini P. 1997. Plant regeneration from protoplasts of Italian red chicory. J. Genet. & Breed. 51: 17-22.

Varotto S., Lucchin M., and Parrini P. 2000. Immature embryos culture in Italian red chicory. Plant Cell Tissue and Org. Cult. 62: 75-77.

Varotto S., Nenz E., Lucchin M., and Parrini P. 2001. Production of asymmetric somatic hybrid plants between *Cichorium intybus* and Helianthus annuus. Theor. Appl. Genet. 102: 950-956.

Vermeulen A., Desprez B., Lancelin D., and Bannerot H. 1994. Relationships among *Cichorium* species and related genera as determined by analysis of mitochondrial RFLPs. Theor. Appl. Genet. 88: 159-166.

Wricke G., and Weber W.E. 1986. Quantitative genetics and selection in plant breeding. Walter de Gruyter, Berlin-New York.

Wiebe H.J. 1997. Ursachen für die generative Entwicklung von Radicchio (*Cichorium intybus* var *foliosum*). Gartenbauwissenschaft, 62: 72-77.

Zitter T.A., and Guzman V.L. 1977. Evaluation of cost lettuce crosses, endive cultivars, and *Chicorium* introductions for resistance to bidens mottle virus. Plant Disease Rep. 61: 767-770.

Globe Artichoke and Cardoon

Sergio Lanteri[1] and Ezio Portis[1]

[1] University of Torino, Di.Va.P.R.A. Plant Genetics and Breeding, sergio.lanteri@unito.it

1 Introduction

Globe artichoke (*Cynara cardunculus* var. *scolymus* L.) contributes significantly to the Mediterranean agricultural economy, with an annual production of about 750 Mt (more than 60% of global production) from over 80kha of cultivated land. Italy is the leading world producer (about 470 Mt), followed by Spain (188 Mt), France (52.5 Mt) and Greece (35 Mt). In southern Europe, artichoke production is an important component of regional economic stability and social development and, thanks to its long growth cycle, its cultivation provides employment almost the whole year round. Globe artichoke is also cultivated, although to a lesser extent, in the Near East (Turkey and Iran), North Africa (Egypt, Morocco, Algeria, Tunisia), South America (Argentina, Chile and Peru), and the United States (mainly in California), and its cultivation is spreading in China (55 Mt in 2005) (FAO data 2005: http://faostat.fao.org/). The major importing countries of the fresh product are France, followed by Italy and Canada. Spain is the leading exporter of processed product, being responsible for about 85% of the traded market (Bianco, 2007). The edible parts of the globe artichoke are the immature composite inflorescences (heads or buds, more formally referred to as capitula), and these are consumed worldwide as both a fresh and a canned delicacy (Figure 1). At flowering, leathery green bracts encase a purple-blue flower. Each plant produces small, medium and large capitula, with the largest formed at the apex of the central stem (primary head). The smaller capitula (secondary heads) develop on the lateral branches. Both spiny and non-spiny types are cultivated in different areas.

The cultivation of cardoon (*C. cardunculus* var. *altilis* DC) is much less widespread than that of the globe artichoke; it remains of regional importance in Spain, Italy and the south of France, where it is used in traditional dishes. The edible parts of the plant are the fleshy stems (Figure 2) which are typically collected in late autumn-early winter. The leaves of the basal rosette are usually deeply divided, petiolate, very large (>50 x 35 cm), subcoriaceous, bright green and slightly tomentose on the upper surface, and white-tomentose on the lower one. Plant height can be over 2 m, and the flowers (blue, lilac or whitish) are grouped in globose capitula. Before cooking, the material is tied together, wrapped in straw, and/or buried for about three weeks in order to accentuate the flavour.

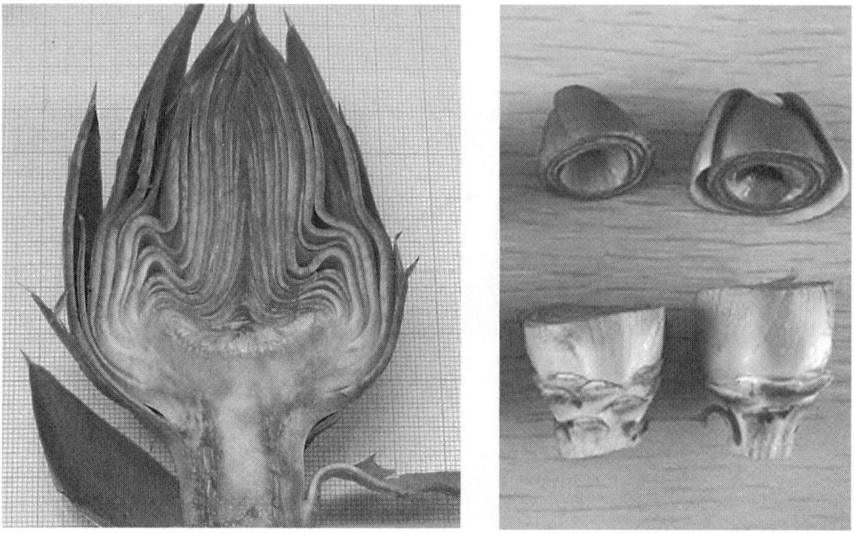

Fig. 1. Section of an immature capitulum of globe artichoke 'Spinoso sardo' (A) and the edible parte of the heads (B).

The major (but not the only) use of globe artichoke and cardoon is for human food. Seed yield in cardoon is about 2 t/ha and up to 0.8 t/ha in globe artichoke (at 5% w/v moisture), about 25% of which is oil of good alimentary quality (Foti et al., 1999). The oil has (i) a high and well balanced content of oleic and linoleic acids, (ii) a low content of free acids, peroxides, saturated and linoleic acids, and (iii) a favourable α-tocopherol content, which provides a good level of protection against oxidation (Maccarone et al., 1999). Furthermore, the seed material left after the extraction of oil can be used as a component of animal feed. Cardoon has been identified as a candidate for the production of lignocellulosic biomass and paper pulp (Quilho et al., 2004; Gonzàlez et al., 2004). Both species have long been known to represent a significant source of biopharmaceuticals (Adzet and Puigmacia, 1985; Debenedetti et al., 1993; Slanina et al., 1993; Wagenbreth, 1996; Sevcikova et al., 2002; Wang et al., 2003), and recent trials have demonstrated that some of the end products can be produced in worthwhile quantities even when the crop is grown in environments unsuitable for normal commercial production (Matthes and Honermeier, 2007). The roots and rhizomes, used also for brew or infusion, provide a source of inulin, a demonstrated enhancer of the human intestinal flora; and the leaves provide a source of antioxidants such as luteolin and di-caffeoylquinic acid, which (i) protect proteins, lipids and DNA from oxidative damage from free radicals (Gebhardt, 1997; Brown and Rice-Evans, 1998; Perez-Garcia et al., 2000), (ii) inhibit cholesterol biosynthesis and contribute to the prevention of arteriosclerosis and other vascular disorders (Kraft, 1997; Brown and Rice-Evans, 1998; Gebhardt, 1998; Pittlern and Ernst, 1998), (iii) inhibit HIV integrase, a key player in HIV replication and its insertion into host DNA (McDougall et al., 1998; Slanina et al., 2001), and (iv) have antibacterial activity (Martino et al., 1999). Extracts from the leaf and capitula can be used for the preparation of alcoholic drinks, beauty creams (Barbagallo et al., 2007), and milk coagulant used in the preparation of traditional cheeses such as the Algerian 'Djben' (Mouzali et al., 2004). It also is common as both an ornamental garden plant and source of cut flowers. The variously shaped capitula and the large blue or violet flowers are popular in floral compositions, and vase life ranges from 7 to 10 days depending on bud stage at harvest and the display conditions.

Fig. 2. Blanched stems of cardoon 'Gobbo di Nizza'.

2 Origin and Domestication

Cynara is a small genus belonging to the *Asteraceae* (daisy) family (Wiklund, 1992); it comprises seven species native to the Mediterranean basin - the crop complex *carduncculus*, which includes the globe artichoke, the cultivated cardoon and the wild cardoon (var. *sylvestris* (Lamk) Fiori), *C. syriaca* Boiss, *C. cornigera* Lindely (syn. *sibthorpiana* Boiss. et Heldr.), *C. algarbiensis* Cosson, *C. baetica* (Spreng.) Pau (syn. *alba* Boiss.), *C. humilis* L. and *C. cyrenaica* Maire & Weiller (Rottenberg and Zohary, 2005). The three *C. cardunculus* forms are fully cross-compatible with one another, and their F_1 hybrids are fertile. However, reproductive barriers separate the crop complex from the other wild *Cynara* species (Rottenberg and Zohary 1996). Crosses between *C. cardunculus* and *C. syriaca*, *C. algarbiensis*, *C. baetica* and *C. humilis* produce only very few seeds, and the F_1 hybrids are generally sterile. These four wild *Cynara* species are therefore regarded as members of the secondary wild gene pool (GP2) of the globe artichoke and cardoon (Rottenberg and Zohary, 2005). On both morphological (Wiklund, 1992) and cytogenetic (Zohary and Basnizki, 1975) grounds, the closest GP2 member to the cultivated complex is *C. syriaca.* Recently, the monophyly and evolution of *Cynara* has been investigated using ITS (Internal Transcribed Spacer region of nrDNA) sequence data (Robba et al., 2005). Evolutionary studies in five *Cynara* species, based on nucleotide sequence divergences of two internal spacers (ITS1 and ITS2) and an external transcribed spacer (ETS) of the 18S-5.8S-25S rDNA locus, suggest that the crop complex *C. cardunculus* is more differentiated and evolved with respect to the other wild species (Sonnante et al., 2007a).

The wild cardoon is by far the most widely distributed wild *Cynara* taxon, thriving in warm, dry and low altitude environments, but also colonizing man-made habitats such as the edges of cultivated fields and roadsides. It is native to the central and western parts of the Mediterranean basin, extending to the Canary Islands and Madeira. Isolated populations exist further to the east, such as in Cyprus and on the Turkish shore of the Black Sea (Rottenberg and Zohary, 2005). Wild *C. cardunculus* forms are also successful colonizers of extensive areas in the New World, in particular in the South American pampas and more recently in Mexico, California and Australia. The plant is allogamous and seed-propagated (achenes). The majority of seed is shed close to the parent plant, where it germinates rapidly following the first autumn rain, although germination can occur the whole year round under favourable conditions; flowers are usually produced by two year old plants. Like globe artichoke, the fleshy capitula, as well as the petioles and roots, if properly prepared, are edible. Portis et al. (2005a) have used microsatellite and amplified fragment length polymorphism (AFLP) assays to characterise the genetic variation in seven Italian populations of wild cardoon, collected from Sicily and Sardinia, and found that most of the genetic variation was present within, rather than between populations. Furthermore, as a result of geographical isolation, the the Sardinian and Sicilian populations proved highly differentiated, forming two distinct gene-pools.

DNA studies (Lanteri et al., 2004a; Acquadro et al., 2005a), cytogenetic- and isozyme-based (Rottenberg et al., 1996) have all confirmed wild cardoon to be the ancestor of both cultivated *C. cardunculus* forms, which evolved independently under the influence of distinct anthropogenic selection criteria - globe artichoke for its capitula, and cardoon for its fleshy leaves and stalks. The origin of cardoon and artichoke is presumed to date to the era of Theophrastus (371-287BCE), who described their cultivation in southern Italy and Sicily. The Ancient Greeks and Romans considered artichokes as both a delicacy and an aphrodisiac. In 77CE, the Roman naturalist Pliny the Elder mentioned their use for medicinal purposes, but it was most probably between 800 and 1500CE that the artichoke was domesticated and transformed, presumably in monastery gardens, into the plant which we know today. Although little is known of the process of domestication and subsequent diversification, the significant genetic differentiation between spiny and non-spiny globe artichoke types is suggestive that the two forms evolved in parallel (Lanteri et al., 2004a).

3 Varietal Groups

3.1 Globe artichoke

In Europe, globe artichoke commercial production is, at present, mainly based on the cultivation of perennial, vegetatively propagated clones, which guarantee high yields of marketable product. However a considerable number of new seed propagated cultivars has been developed in recent years, and these are gaining in popularity. The number of vegetatively propagated cultivars grown in the Mediterranean basin and elsewhere is difficult to determine with any accuracy, but is thought to be at least 120. A frequent complication is that a single cultivar can be known by different names in different locations. Around 10-20 clonally propagated cultivars are considered to be of major commercial importance (Basnizki and Zohary, 1994).

Varietal groups are generally identified on the basis of capitulum appearance (Dellacecca et al., 1976; Porceddu et al., 1976; Vanella et al., 1981): (I) the 'Spinosi' group has long sharp spines on its bracts and leaves; (II) the 'Violetti' group has medium-sized, violet-coloured and less spiny capitula; (III) the 'Romaneschi' group has spherical or sub-spherical non-spiny capitula; and (IV) the 'Catanesi' group has relatively small, elongated non-spiny capitula. A further classification is based on harvest time: early types can be forced to produce capitula between autumn and spring, if dormant underground shoots used for propagation are watered during summer, whereas late types produce capitula only during spring and early summer (Mauromicale and Ierna, 2000). Early flowering varieties are usually characterized by elongated and small capitula, compared to the globular and larger capitula of the late types (Lopez Anido et al., 1998).

In Italy, globe artichoke production is concentrated in Apulia, Sicily, Sardinia, Tuscany and Lazio. The most commonly grown traditional varieties are the spiny early types 'Spinoso sardo' and 'Spinoso di Palermo', and the non-spiny types 'Violetto di Provenza' and 'Violetto di Sicilia' (both early flowering), and 'Violetto di Toscana' and 'Romanesco' (late). More recently, other clonally propagated varieties have started to make an impact, notably 'Tema 2000', a very early, productive cultivar, which produces medium-sized, purple capitula, and 'Terom', a late cultivar producing large, violet capitula, which mature in a well synchronised fashion. In Spain, the most common vegetatively propagated cultivar is 'Blanca de Tudela' (synonym 'Blanca de España'), produced mainly in the south-east coastal region of Murcia and the Navarra; less widespread, but well-prized, is the French cultivar 'Violet de Provence'. In France, globe artichoke varieties are traditionally classified into three groups: (1) 'Brittany artichokes' with large, green, truncated capitula of spherical shape, among which the most commonly grown are 'Camus de Bretagne' and 'Gros Vert de Laon'; (2) 'Midi artichokes', originating from the south of France, with violet leaves and capitula (e.g., 'Violet de Provence', 'Violet de Hyères' and 'Violet du Gapeau'); and (3) secondary cultivars such as 'Blanc Hyerois', which are intermediate in form between 'Camus de Bretagne' and the purple cultivars. In Greece, the major areas of cultivation are the Peloponnese and Crete. The most important cultivars are 'Argos', globe shaped with tightly-closed green bracts, and 'Iodine of Attica', with purple bracts (Bianco, 2005). Outside of Europe, the most commonly grown cultivars are 'Bamafsigi', 'Baladi' and 'Violet de Provence' (the Nile delta region), the Turkish early 'Sakiz' and late 'Bayrampasa', grown, respectively on the Aegean coast and in Marmara (Ercan et al., 2004). In Algeria, Morocco and Tunisia, most cultivars have been introduced from other Mediterranean countries. This is also the case for Argentina, where the vegetatively propagated 'Violet de Provence' represents about 90% of production (Garcia et al., 2004). In the United States, the cultivar 'Green Globe', with large green capitula and thick fleshy scales, grown almost exclusively in coastal California, accounts for about 90% of production; another popular seed propagated cultivar, cultivated mainly in dry areas of California and Arizona, is 'Imperial Star'. In Chile, Peru and China, late seed-propagated cultivars predominate.

The clonally propagated varieties at present in cultivation are highly heterogeneous. DNA marker analysis of the three autochthonous Italian cultivars 'Spinoso sardo', 'Violetto di Sicilia' and 'Spinoso di Palermo' showed that all had a significant element of within-cultivar variation (Lanteri et al., 2001; Portis et al., 2005b). This reflects their multi-clonal composition, which is a direct consequence of the limited selection criteria applied by the farmers. In common practice plants are generally mown at the end of the season, and new propagative material is only collected some months later, without any selection for specific mother plants. An additional source of variation is spontaneous mutation, not necessarily detectable at the phenotypic level, and propagated over time in the absence of a meiotic sieve. Similarly, random amplified polymorphic DNA (RAPD) heterogeneity has been observed in two breeding populations of 'Green Globe' (Tivang et al., 1996) as well as in the cultivars 'Romanesco' and 'Locale di Mola' (Pagnotta et al., 2004; Lotti et al., 2003). Due to the high range of genetic variation found in cultivated populations, Lanteri et al. (2004a) suggested that the term 'varietal type' instead of 'variety' would be more appropriate in order to define the accessions of germplasm at present in cultivation.

Phenotypic chimerae have also been reported within cultivated materials (Pochard et al., 1969). In an AFLP-based genetic uniformity analysis of vegetatively propagated clones from 'Spinoso sardo', Lanteri et al. (2004b) identified three out of 120 individuals showing an electrophoretic profile distinct from that of the parent, with respect to one or two bands. As these variant patterns were generated from the individual DNAs with at least two AFLP primer combinations, while the profiles from other primer combinations were 'true to clone', the authors confirmed that some mother plants must have been chimeric.

3.2 Cardoon

Cardoon production is limited to Spain, Italy and to a limited extent, France. It is usually raised from seed and handled as an annual crop. Plants generated through root division tend to be tough and run rapidly up to flower, while seed-grown plants remain in rosette form and produce more tender leaf ribs. In Spain, the major production area is Navarra, and is mainly based on traditional cultivars, with most of the production (85%) used for canning and freezing. Fresh product is consumed locally from October to March (Macua et al., 2004). The most common varieties for industrial processing are 'Blanco de Peralta', a large sized type, with a wide, solid petiole, and 'Verde de Peralta', with irregular (wide and tight) but compact petioles, well liked on account of its low fibre content. The varieties destined for the fresh market are the highly productive 'Verde de Tafalla', 'Blanco de Valencia' (of smaller size and with white hollow petioles), 'Lleno de España'(medium to wide petioles), and 'Rojo de Corella' (reddish and solid petioles). In Italy, the most commonly grown varieties are 'Bianco avorio' (a vigorous type with thick, almost spineless stalks), and 'Gobbo di Nizza', an erect plant with thin ribs with some spines, both cultivated mainly in Piedmont (North-Western Italy) and often used uncooked in traditional dishes. In the Emilia region, 'Bianco pieno migliorato' has been selected for its solid stalks and leaves, along with 'Pieno inerme' and 'Gigante di Romagna' (synonym 'Gigante inerme a foglia intera'), which both have large and very tender stalks completely free of spines.

In France the most commonly grown varieties are 'Blanc ameliore', with non-spiny large stems; 'Plein blanc ameliore puvis', with broad and thick stalks, sporadic presence of spines and good storability; 'Vert vaulx velin' green, spineless and cold resistant and 'Rouge d'Alger'.

Microsatellite- and AFLP-based analyses have shown a considerable level of within-cultivar variation in both Italian as well as Spanish cultivars, presumably the result of the limited amount of selection that has been applied (Portis et al., 2005c). Seed is commonly mass selected on the basis of the phenotypic conformity of the maternal plant phenotype. Analogous results have been obtained in a RAPD-based study of 17 local Spanish cultivars (Itoiz et al., 2004).

4 Genetic Resources

4.1 Globe artichoke

Italy harbours the richest collection of globe artichoke primary cultivated germplasm and also houses the most abundant *in situ* diversity (Bianco, 1990). Most varietal types are cultivated in limited geographic areas and are identified by vernacular names. As a result, similar genotypes often carry different names in different localities. Furthermore, distinct varietal types differing from one another due to allelic variation at a small number of genes, can share a largely similar genetic background. Cluster analysis based on the discrimination of eight quantitative characters among 104 accessions was able to identify five major groups sharing similar characters and presumably a similar genome (Elia and Miccolis, 1996). DNA markers have proven to be informative for assessing the extent and distribution of genetic variation in living germplasm collections of globe artichoke, and for identifying suitable strategies for the establishment of core collections. Thus, for example, Sonnante et al. (2002) applied RAPD and AFLP (*ibid.* 2004) markers, but a more complete study was carried out by Lanteri et al. (2004a), who used AFLP markers to investigate patterns of genetic relatedness among 118 accessions, representative of worldwide cultivated germplasm (for three varieties clones of different provenances were included), using two accessions of cultivated cardoon and four of wild cardoon as out-groups. The resulting cluster analysis suggested that the traits selected by man have played an important role in determining variation and differentiation within cultivated artichoke germplasm. Specifically, two major clades could be defined (Figure 3, A and B), each of which was sub-divided into two further clusters. Within the first clade, one cluster (A1) was composed primarily of 'Catanesi' types (small elongated capitula), while the other (A2) included the 'Romaneschi' types (large spherical or sub-spherical non-spiny capitula and mainly cultivated in central Italy), together with clones of the dominant US variety 'Green Globe'. The second clade grouped the 'Spinosi' and 'Violetti' types with medium-small capitula, and all the Turkish accessions lay within one of the sub-clusters (B1). An important purpose of this analysis was to assemble a core collection of globe artichoke, able to take into account the hierarchical structure of the genepool. The amount of genetic variation between clones of the same variety was, in some cases, higher than that found between varietal types, and thus the lowest JSI (Jaccard Similarity Index) among clones can be taken as a reasonable threshold to identify material sharing the same genetic background.

A more recent study has assessed the level of genetic variation in autochthonous globe artichoke germplasm grown in family gardens in Sicily, at both the morphological and DNA level (Mauro et al., 2007). This collection consisted of 26 morphologically variable types, in particular differing for the number and size of the capitula. AFLP profling was able to demonstrate that 'Cimiciusa di Mazzarino' (many small capitula) is genetically the closest to the wild cardoon, and may therefore represent a transient form of domestication, which is believed to have been achieved in Sicily. The genetic variation in two landraces: 'Castellamare' and 'Campagnano' of the 'Romanesco' type has been assessed in a similar way to identify suitable strategies for preservation (Triofetti Nisini et al., 2007).

4.2 Cardoon

To date, we are aware of only a single attempt to evaluate the genetic variation of cardoon. This study applied microsatellite- and AFLP-profiling to a collection of the most widely grown cultivars in Italy and Spain, plus some local ecotypes, and a few globe artichoke and wild cardoon entries were used as an out-group (Portis et al., 2005c). The Spanish and Italian cultivars appeared to form two separate gene-pools, but a lack of information regarding the origin of the accessions prevented the establishment of a clear correlation between pedigree and genotype.

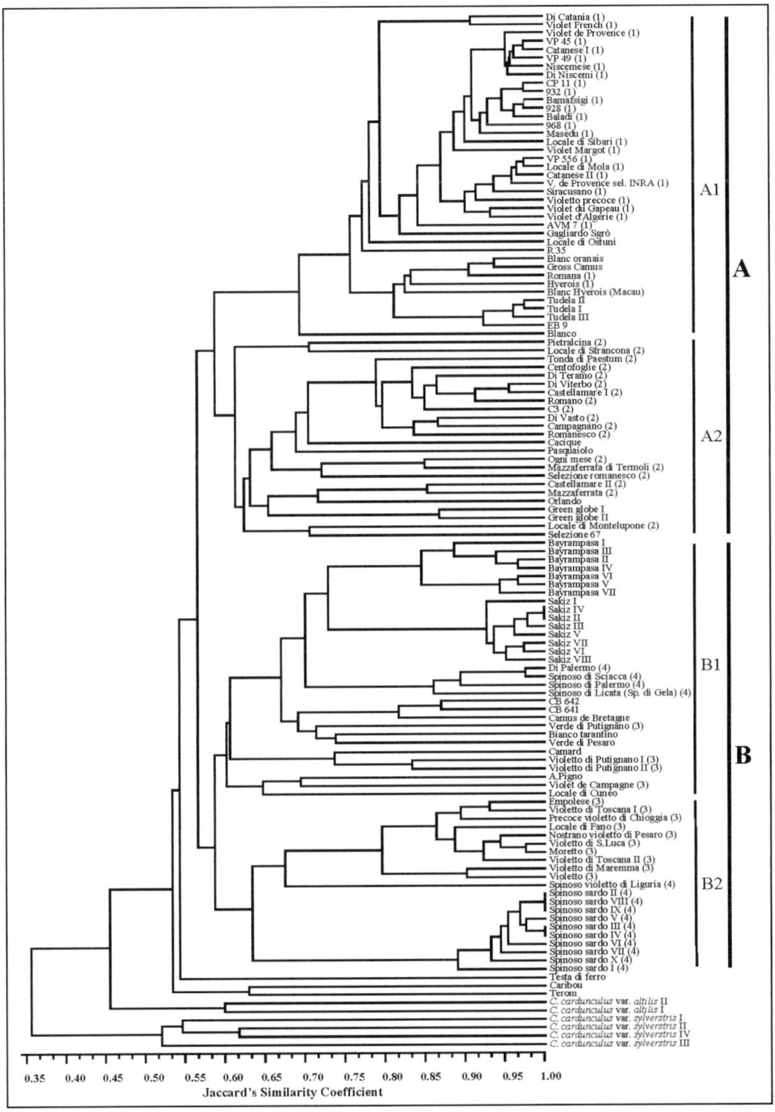

Fig. 3. Dendrogram obtained from UPGMA cluster analysis of AFLP data. (1): 'Catanesi' type;(2): 'Romaneschi' type; (3): 'Violetti' type; (4): 'Spinosi' type.

On the whole, a higher degree of genetic differentiation was present in globe artichoke than in the cultivated cardoon. This was not surprising, considering that globe artichoke is cultivated in many localities, but in a fragmentary fashion. As a result, it is likely that a range of selection criteria have been adopted to optimise genotypes both to match individual environments and to suit local tastes.

5 Breeding Methods and Techniques

5.1 Artichoke

5.1.1 Inheritance of Traits

Only few attempts have been made to uncover the inheritance of major capitulum traits (Pécaut, 1993; López Anido et al., 1998; Mauromicale et al., 2000). Most of the morphological and production traits (capitulum size, shape and weight, plant size and branchmg, peduncle length, earliness and harvest index) are governed by polygenic systems (Porceddu et al., 1976). The thornless bract character has been suggested to be controlled by two genes acting in epistasis, since the segregation between thorn absence and presence from the cross 'Spinoso sardo' x 'Romanesco' was observed to follow a 9:7 ratio (De Pace et al., 1976). However, in a different cross ('Romanesco' x 'Spinoso di Palermo'), its segregation was consistent with a 1:1 ratio, explicable with trait control being exerted by a single gene with two alternative alleles: dominant non-spiny (*Sp*) and recessive spiny (*sp*) (Lanteri et al., 2006). The 'Romanesco' parent was *Spsp* and 'Spinoso di Palermo' *spsp*. Other crosses between non-spiny and either spiny types or wild cardoon (Pochard et al., 1969; Basnizki and Zohary, 1994) have led to the same genetic model, and the *Sp* locus was successfully mapped to linkage group 16 in 'Romanesco', flanked by two microsatellite markers (Lanteri et al., 2006).

There is considerable variation in cultivated germplasm for the intensity of capitulum pigmentation, a character which is also very sensitive to temperature. Nonetheless, varieties are conventionally classified as either violet or green. Pochard et al. (1969) suggested that the genetic basis for anthocyanin pigmentation involves a series of modifiers in addition to one or two major genes. On the basis of segregation in a number of genetic backgrounds, Cravero et al. (2005) have proposed that capitulum colour is genetically determined as follows: *P-* allows anthocyanin production, resulting in purple bracts, while *pp* inhibits anthocyanin production resulting in green bracts; *U-* results in an uneven distribution of anthocyanin pigments encoded by *P*, while *uu* results in an even distribution of pigment in the presence of *P*. Other modifier genes or multiple alleles may also be involved, but the proposed model is sufficient for breeders seeking to enhance colour.

An important quality criterion for the capitulum is tightness, a property important for both the processing and the fresh markets (Macua, 1996). From crosses between soft and fairly compact capitulum types, Dellacecca et al. (1976) showed that the F_1 offspring all had fairly compact capitula; however, when the F_1 hybrids were further crossed with a compact capitulum type, the topcross progeny segregated

4:3:1 for compact: fairly compact: soft capitulum. There also appeared to be a pleiotropic relationship between capitulum form and tightness, with compact capitula being globular or sub-globular because the arrangement of bracts itself contributes to tightness. Cravero et al. (2005) were able to postulate the involvement of two independent loci, *C* and *H*, acting epistatically, such that *C*- genotypes give a compact capitulum, whatever the allelic state of *H*; while *H*- genotypes have tighter capitula in the presence of *cc*.

Lopez Anido et al. (1998) estimated levels of heritability for yield and phenological traits in 23 globe artichoke clones of diverse origin, and which showed extensive genetic variation for these traits, as reported previously (Marzi and Bianco, 1967; Pochard et al., 1969; Foury, 1969). Heritability ranged from 0.94 for capitulum height /diameter ratio, 0.83 for yield, 0.82 for the number of capitula harvested, to 0.48 for the capitulum base diameter. An analysis of variance of the components, broad sense heritabilities and genetic correlations indicated that it should be possible to increase yield and associated characters (i.e., capitula number, weight of secondary capitula, weight of primary capitulum) by 10-30% with a selection intensity of 5%. Lower estimates for broad sense heritabilities were identified in a clonal population of 'Violetto di Sicilia' (Abbate and Noto, 1981), presumably reflecting the lower genetic variation present in this material. The highly significant positive correlation established between capitulum mean diameter and mean weight suggested that selection should favour globular or truncated-conic forms, which have a height/diameter ratio of less than unity (Lopez Anido et al., 1998). The same study established that the weight of the primary capitulum was closely correlated with the base weight and diameter, both of which are important traits for the processing industry; and that many characters were affected by a significant clone x year interaction, as also indicated by Foury (1979) and Mauromicale and Copani (1989).

5.1.2 Breeding Achievements

Globe artichoke (2n=2x=34) is cross-pollinated, but (at least in Europe) is generally propagated via suckers, ovoli (underground dormant shoots with a limited root system), or by the division of rooted basal stem portions (De Vos, 1992; Pécaut, 1993). Cross-pollination is promoted by protandry, but self-pollination is not precluded. The stigmatic surface is receptive to pollen for 2-3 days following pollen shedding, and therefore the fertilisation of peripheral florets can be effected by the pollen of the more internal ones, since flowering progresses from the periphery to the centre of the capitulum. Some self-pollination is also possible via pollen transfer between capitula, as each plant produces, on average, four to six asynchronously flowering capitula. Simple strategies of pollen preservation and application allow for the straightforward generation of self-pollinated progeny (Mauromicale and Ierna, 2000), which typically segregate widely. In French and Italian cultivars, the proportion of the first inbred generation (I1) which conformed to the parental phenotype varied from just 2% in 'Violet de Provence' to 8% in 'Romanesco' (Pécaut, 1993); yet repeated selfing, combined with rigorous selection for conformation-to-type, allowed the extraction of true-breeding lines, with a phenotype comparable to that of the parental type (Foury, 1979).

Common breeding targets are the promotion of earliness, yield and capitulum quality, and selection to date has been based largely on intra-clonal variation (Deidda, 1967; Abbate and Noto, 1981; Pécaut, 1993; Mauromicale et al., 2000 Mallica et al., 2004). A list of clones and breeding lines selected in Italy, France and Spain has been presented by Bianco (2005). A few attempts to exploit segregation released by inter-varietal hybridization or selfing have been reported in the literature (Miller, 1975; Scarascia Mugnozza and Pacucci, 1976; Tesi, 1976; Basnizky and Zohary, 1987; 1994). Due to the high level of heterozygosity present in segregating populations, it is normally possible to identify genotypes with valuable agronomic characters, and these selections can then be maintained via vegetative propagation. Some examples of new varieties obtained in this way are 'Camerys', 'Caribou', 'Salanquet', 'Cacique', 'Carlite' and 'Terom' (Mauromicale, 1987). More recently, the green capitulum cultivars 'Galico', 'Castel', 'Capitan', 'BE15', 'Vertu', 'Polo' and the violet ones 'Velorus', 'Salambo', 'Satin', 'Vialin' have been released by French INRA stations Plougoulm de Saint-Pol de Léon and Avignon-Montfavet. In Italy, a significant area of the 'Romanesco' type cultivars 'Grato 1', Grato 2', 'Etrusco' and 'Moro di Corneto' has been grown (Graifenberg and Giustiniani, 1997; Papalini et al., 1997). Commonly, however, selection within segregating populations derived either by crossing or by selfing of early-flowering types results in late-flowering individuals, which appear to have lost the ability to be forced to produce capitula during the autumn. Thus, for example, Gil and Villa (2004) crossed the Spanish cultivar 'Tudela' with the American 'Imperial Star' in order to introduce earliness, but after three generations of recurrent selection were not able to obtain any improvement in earliness, which they concluded was under polygenic control. It is also possible that a cycle of sexual reproduction induces a kind of juvenility, with a consequent loss of earliness.

5.2 Cardoon

While globe artichoke is usually vegetatively propagated, cardoon is raised from seed and handled as an annual plant. Seeds are sown in late spring, and the plants over-summer in the vegetative state. Current cultivars have been obtained by mass selection in public or private nurseries. The criteria for cultivar selection are plant size, along with colour and consistency of the pulpy leaves. Commercial cultivars are genetically heterogeneous, as assessed by both RAPD (Itoiz et al., 2004) and AFLP (Portis et al., 2005c) fingerprinting. Further efforts should be carried out to restrict the genetic basis of the material in cultivation and obtain more uniform cultivars.

6 Current Goals of Breeding

The *C. cardunculus* genome is as yet poorly mapped. In order to move to a crossing strategy for breeding, a greater knowledge of artichoke and cardoon genetics will be essential. In particular, it will be advantageous to establish a framework of linkage relationships to allow the identification and localization of genes controlling important yield traits or resistance against pathogens.

Fig. 4. Head characteristics of the parents crossed for obtaining segregant F_1 population used for the construction of the genetic linkage maps: globe artichoke 'Romanesco clone C3' (A), globe artichoke 'Spinoso di Palermo' (B), cardoon (C) and wild cardoon (D). Examples of morphological variation of the heads observed in the F_1 segregant progenies.

The first genetic maps of globe artichoke, based on a two-way pseudo-testcross strategy, was recently generated by Lanteri et al. (2006). An F_1 population was created by crossing 'Romanesco clone C3' (a late-maturing, non-spiny type) with 'Spinoso di Palermo' (an early-maturing spiny type), and the progeny were genotyped using a number of marker types (Figure 4). The female map comprised 204 loci, spread over 18 linkage groups and spanned 1330.5cM with a mean marker density of 6.5cM. The equivalent figures for the male parent map were 180 loci, 17 linkage groups, 1239.4cM and 6.9cM. The presence of 78 loci in common to both maps allowed for the alignment of 16 of the linkage groups. The establishment of linkage relationships among such marker loci represents the initial step for the identification of chromosomal regions carrying genes of breeding interest, and their future targeting in breeding programmes via the incorporation of marker-assisted selection. Since globe artichoke is easily vegetatively propagated, the mapping populations are immortalised, and thus can be grown in contrasting environments to investigate genotype x environment interactions, which are known to be important for many commercial traits (see section 5.1.1., above).

Genetic maps are particularly powerful for the dissection of quantitative trait loci (QTL), which underlie the inheritance of many key agronomic characters. The present authors are currently planning the construction of genetic maps based on F_1 populations involving combinations between 'Romanesco clone C3' with either cultivated or wild cardoon accessions (Figure 4), as these will allow comparative QTL mapping studies. Wide cross populations of this type are suited for the investigation of the genetic control of quantitative characters in exotic genetic backgrounds. The wild cardoon represents the most straightforward wild resource to exploit for globe artichoke improvement, since it is not genetically isolated from the cultivated gene pool. The other six members of the genus *Cynara* are more problematical, although Rottenberg and Zohary (2005) have succeeded in producing a small number of viable hybrids between four of these and *C. cardunculus*. Possible traits for introduction from exotics include flavour (*C. syriaca*, *C. humilis*), earliness, medicinal properties, dwarfness (*C. cornigera*), white flowers (*C. baetica*, *C. cornigera*) and resistance to pests and diseases, the most damaging being the soil-borne pathogens *Sclerotinia sclerotiorum*, *S. rolfsii*, *Rhizoctonia solani* and *Verticillium daliae.*

7 Integration of New Biotechnologies in Breeding Programmes

7.1 Development and Application of Molecular Markers

To date, molecular marker studies in *C. cardunculus* have been carried out using RAPD (Tivang et al., 1996; Lanteri et al., 2001; Sonnante et al., 2002), AFLP (Lanteri et al., 2004a; 2004b; Sonnante et al., 2004; Portis et al., 2005a; 2005b; 2005c) and Inter-simple sequence repeats (ISSR; Lanteri et al., 2004b). A set of specific microsatellites were developed by Acquadro et al. (2003), using both published and *de* novo acquired (from an enriched genomic library) DNA sequences. Acquadro et al. (2005a) later used the novel "microsatellite amplified library"

(MAL) approach to derive a further set of microsatellite assays. This technique, which represents a combination of AFLP and a primer extension-based enrichment, provides a rapid means to increase the efficiency of microsatellite identification, avoiding the requirement for a hybridization enrichment step (Figure 5). Further *C. cardunculus* microsatellites were then developed using a two-step 'primer extension' procedure, based on the microsatellite-AFLP technique (Acquadro et al., 2005b). For this approach, amplicons highly enriched for microsatellite sequences are produced and forward primers directed towards the microsatellite motif are designed from the sequences of bands isolated from the gel profile. Thereafter the opposite microsatellite flanking sequence is isolated via a nested strategy based on a template of restricted-ligated genomic DNA. The extent of the polymorphism uncovered by the total set of 32 microsatellites was explored in a survey of both cultivated and wild accessions. A sequence-specific amplified polymorphism (S-SAP) assay, based on the CYRE-5 Ty1-copia type retrotransposon sequence, was generated by Acquadro et al. (2006) and its effectiveness in assessing genetic variation across 22 *C. cardunculus* accessions, including both cultivated and wild types, was compared to that achievable with AFLP fingerprinting. Finally, 29 of these S-SAP loci were incorporated into the core genetic map, confirming their dispersed distribution across the globe artichoke genome.

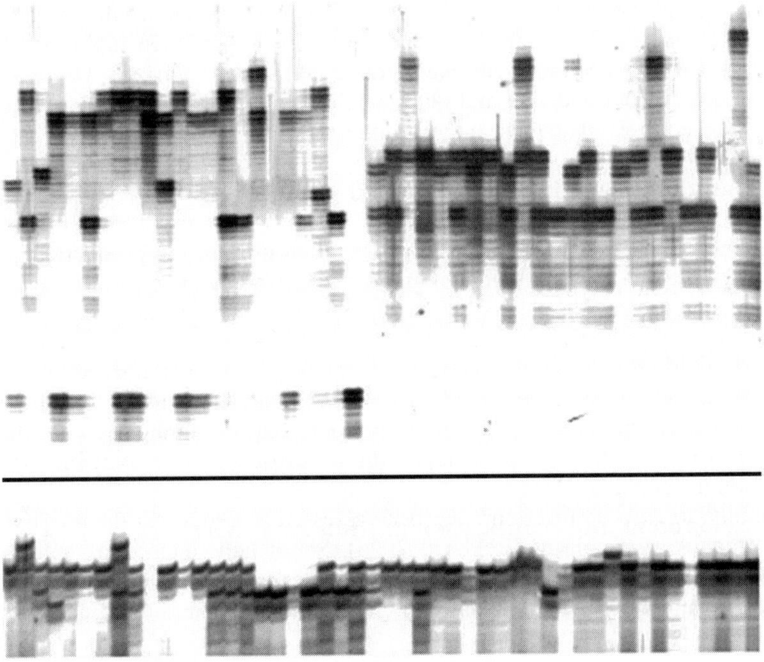

Fig. 5. Polymorphisms observed in a wild cardoon population by applying two microsatellites developed thorough the novel approach MAL (microsatellite amplified library) at the Di.Va.P.R.A. Plant Genetics and Breeding University of Turin.

7.2 *In vitro* Tissue Culture

Propagation through meristem culture has been widely applied in globe artichoke late types (producing capitula from spring to early summer); the obtained mother virus free plants may represent a source for the production of sanitary controlled propagative material (Barba et al., 2004; Papanice et al., 2004). Plants obtained in this way have shown improved field performance with respect to both qualitative and quantitative traits, and this can compensate for the higher cost of the planting material (Saccardo et al., 2007). On the other hand, the micro-propagation of early-flowering types has frequently produced regenerants which are not true-to-type, specifically in that they often revert to the juvenile stage with a consequent loss of earliness (Cadinu et al., 2004; Frau et al., 2004; Tavazza et al., 2004; Elia et al., 2007). This phenomenon has been ascribed to epigenetic modification of genes involved in the flowering process. A similar epigenetic mechanism may also explain variants arising from vegetative propagation (Pochard et al., 1969; Pécaut and Martin, 1993), such as the 'bull' type described in 'Violet de Provence' and Blanca de Tudela' (Esteva and Martìncz, 2004). Recently, Gallitelli et al. (2006) have described how two cycles of meristem culture of the early cultivar 'Brindisino', separated by an *in vitro termotheraphy* treatment, allowed for the selection of virus-free plants which retained their earliness under field conditions.

In vitro mutagenesis has been induced by the gamma irradiation of micro-propagated shoots of 'Romanesco clone C3', which were then multiplied for two cycles, rooted and acclimatised (Stamigna et al., 2005). Variants with respect to earliness, bud colour and size and plant height were identified, but further analyses are still needed to validate these putative mutants.

The availability of an efficient protocol for the *in vitro* production of haploid plants and subsequent diploidisation would greatly speed the development of the homozygous material needed for F_1 hybrid breeding. The first reported attempts to culture anthers from five Italian cultivars resulted only in the production of callus (Motzo and Deidda, 1993). Although microspores can now be reproducibly cultured, development beyond the second division has not yet been attained, presumably because of non-optimal culture conditions (Stamigna et al., 2004a; Chatelet et al. 2005). Haploid production via gynogenesis has been also been unsuccessful (Motzo and Deidda, 1993; Babes, 1997), although *in situ* gynogenesis using fertilisation with irradiated pollen has been reported by the INRA station Maraîchères de Montfavet (France). However, this method is at present not sufficiently reproducible for general use (Stamigna et al., 2004a).

Finally, *in vitro* tissue culture has been assessed as a route for the preservation of globe artichoke germplasm (Bekheet, 2006). Shoot buds and callus cultures were successfully stored for 15 months at 5°C in the dark, and plantlets regenerated from cultures were successfully adapted to field conditions after simple acclimatization procedures.

7.3 Improvement of Bio-Components of Interest

Modern biotechnology allows the role of genes involved in specific biosynthetic pathways to be defined, and this understanding can potentially increase or modulate the content of key molecules in plant tissue. The chemical composition of globe artichoke leaves and capitula has been extensively studied (Nichiforescu, 1970; Adzet and Puigmacia, 1985; Debenedetti et al., 1993; Slanina et al., 1993; Dranik et al., 1996; Wagenbreth, 1996; Sevcikova et al., 2002; Wang et al., 2003; Di Venere et al., 2007, Mabeau et al., 2007; Melilli et al., 2007). The major phenolic molecules present are the di-caffeoylquinic acids (such as cynarin), which are largely restricted to *Cynara* spp., along with their precursor chlorogenic acid (CGA). The high polyphenol content is the reason why capitula brown after harvest, in a process catalysed by the polyphenol oxidases (oxidoreductases) (Todaro et al., 2006). The anti-oxidative properties of *C. cardunculus* extracts have been tested in rat hepatocytes and shown to posses anti-proliferative and apoptotic effects on cancer cells (Miccadei et al., 2006).

Only pioneer studies have yet been carried out to elucidate biosynthetic pathways of secondary metabolites. A gene sequence(s) encoding a hydroxycinnamoyltransferase (HCT) involved in the synthesis of CGA has recently been isolated in globe artichoke (Comino et. al., 2007). *In silico* analysis revealed that this sequence shares homology with one of the five main acyltransferase groups (i.e. anthranilate N-hydroxycinnamoyl/ benzoyltransferase). Heterologous expression of the full-length HCT cDNA in *E. coli* demonstrated that the recombinant enzyme efficiently synthesizes p-coumaroyl quinate from p-coumaroyl-CoA and quinic acid, confirming its identity as an hydroxycinnamoyl-CoA: quinate HCT. Variable levels of HCT expression were observed among wild and cultivated forms of *C. cardunculus*. The level of expression was correlated with CGA content, consistent with the predicted involvement of HCT in the biosynthesis of CGA. In an analogous approach, 7et al. (2006b) isolated sequences with high similarity to the PAL (phenylalanine ammonia-lyase) gene family, which encodes enzymes involved in the first step of phenylpropanoid biosynthesis pathway by catalyzing the de-amination of phenylalanine. These sequences differed from one another with respect to intron length, and were differentially expressed in various globe artichoke organs. Root inulin content has been evaluated in wild and cultivated forms of *C. cardunculus*, varying in relation to genotype (higher in wild and cultivated cardoon than in globe artichohe) and stage of plant development (maximum content just before beginning of flower formation) (Melilli and Raccuia, 2007).

8 Globe Artichoke Seed Production

Artichoke seeds (Figure 6) express little dormancy and germinate readily in the range 10-25°C, although higher temperatures do impair germination (Basnizki and Mayer, 1985; Foury 1987). Light inhibits germination, in a cultivar-dependent manner (Basnizki and Mayer, 1985). When stored at room temperature in sealed containers, seeds maintain viability for up to five years (Basnizki and Zohary, 1994).

The effective production of seed of high yielding cultivars is an important objective for artichoke breeding programmes, and could bring about a substantial improvement in both quality and quantity of artichoke production. Seed propagation would allow (i) a reduction in the cost of planting, (ii) a reduction in the spread of pathogens (mainly viruses), (iii) a reduction in fertilizer use and irrigation requirement, since seed-propagated plants develop deeper root systems than do the adventitious roots produced by suckers, and do so in a shorter time, (iv) the selection of varieties whose product can be designed for the industry of transformation (at present predominantly the secondary smaller heads are used in industry), (v) the transformation of artichoke from a perennial to annual crop, which would support crop rotation and thereby limit environmental impact, and (vi) an expansion in the choice of cultivar to the producer. Pioneering studies to obtain seed-propagated cultivars were initiated in France in the 1960's, and began to bear fruit after about twenty years. The first European commercialised seed-propagated cultivars were 'Talpiot', '044', 'Giorgio' and 'Agribas', which produce green capitula between April and May (Basnizky and Zohary, 1987; Mauromicale et al., 1989) and are only slightly sensitive to treatment with gibberellic acid (GA_3), which is commonly widely applied to promote earliness. In the same years, cultivars obtained through mass selection from the cultivar 'Green Globe' were released in USA.

Only limited effort to date has been dedicated to the intra-varietal selection of clones to be used as parents of seed-propagated synthetic varieties. The general and specific combining abilities of several clones have been derived to determine modes of gene action for yield-related and morphological traits (Cravero et al., 2004). In two sets of diallel crosses, with four clones in each, significant differences were established between the mean values of several traits. Most of the variables evaluated were controlled by additive genetic effects, so that recurrent selection should be effective in increasing the mean value of these variables.

F_1 hybrid seed varieties are a highly desirable commodity. First, however, it is necessary to extract uniform, homozygous lines by repeated self-pollination and selection, and to perform a qualitative and quantitative test of F_1 combinations between inbred lines. Globe artichoke suffers, however, from considerable inbreeding depression. Overall plant vigour decreases with increased homozygosity, with deleterious effects observed for leaf area, stem height, the number and size of marketable capitula, the quality and quantity of pollen, and the number of viable seed (Foury, 1979; Pécaut, 1993; Basnizki and Zohary, 1994). Inbreeding depression can appear as early as I2, and in some cases, the effects of inbreeding were so marked that it was impossible to continue selfing beyond I3 or I4 (Chatelet et al., 2005). Pécaut (1993) found out that , on average, the I4 generation represents the best compromise between vigour, seed production and homogeneity. By inter-crossing I3 or I4 selections, an increase in total average yield of up to 81% was achievable (Basnizki and Zohary, 1994).

The production of F_1 seed propagated varieties would be eased by the identification and incorporation of male sterility (MS) into lines to be used a female parents. A monogenic recessive MS allele, which results in brownish and dry anthers at flowering was identified by Principe (1984), but this source of MS was not fully expressed in other genetic backgrounds (Foury et al., 2005). MS has been identified in F_2 populations obtained from a cross between a French male sterile clone and both 'Romanesco' and 'Violet de Provence'. This MS generated fully sterile pollen but showed no visible anomalies during tetrad formation, so a post-meiotic mechanism was presumed to be responsible for pollen sterility. It is probably under the control of two recessive genes, since a segregation ratio of 15 male fertile: 1 male sterile was detected in the F_2 of the cross 'MF' x 'MS' (Stamigna et al., 2004b). The first highly vigorous F_1 hybrids on the market were 'H137' and 'H223', producing green capitula with shades of violet from April to May, and lacking sensitivity to GA_3 treatment. More highly GA_3 sensitive F_1 hybrids such as 'Orlando' are now on the market. Following GA_3 treatment, production can begin in November, just a few weeks later than is achieved from clonally propagated early types (Mauromicale and Ierna, 2000).

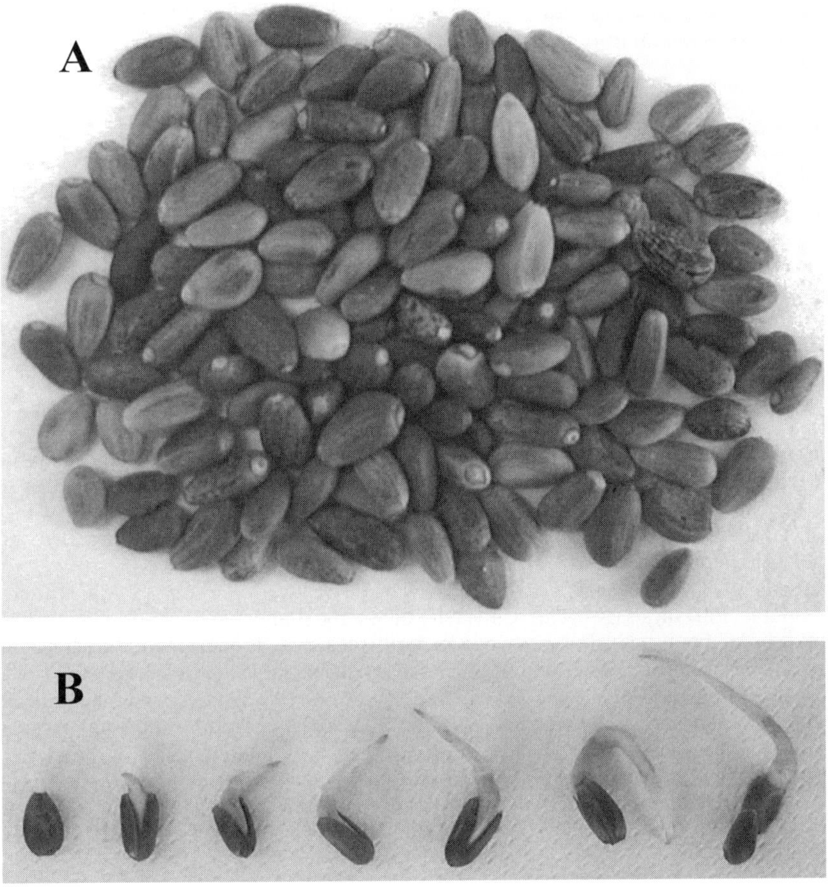

Fig. 6. Dry (A) and germinating (B) seeds of globe artichoke.

Seed-propagated cultivars and F_1 hybrids are now becoming increasingly popular, and account for almost the whole of production in Chile, Peru and China. Numerous trials have been carried out to test their environmental stability. Leskovar et al. (2007) described an evaluation of the seed-propagated cultivars 'Emerald', 'Experimental Red', 'Imperial Star', 'Green Globe' and 'Purple Romagna' in southern Texas, while cultivars 'Concero' and 'Opal', together with some selections commercialized by Nunhems, were tested by Calabrese et al.(2007) in southern Italy and by Miguel et al. (2004a) and Garcia et al. (2004) in southern Spain. No concrete data have yet been obtained regarding the seed propagation of most of the commonly grown early varietal groups. This is mainly because appropriate treatments with GA_3 can only partially advance capitulum production (Miguel et al., 2004b), and in addition, the treatment often induces plant abnormalities such as a reversion to the vegetative state (Condés et al., 2007). The next years are likely to see a large-scale substitution of local landrace types with cultivars selected abroad, with the consequent loss of an important source of local germplasm.

References

Abbate, V., and Noto, G. 1981. Variabilità ambientale e genotipica in popolazioni siciliane di *Cynara scolymus* ed isolamento di nuovi cloni di Violetto di Sicilia, in: *Atti III Congresso Inernazionale Studi sul Carciofo*, ed. Laterza, Bari, pp. 797-807.

Acquadro, A., Portis, E., and Lanteri, S. 2003. Isolation of microsatellite loci in artichoke (*Cynara cardunculus* L. var. *scolymus* L.), Mol. Ecol. Notes 3:37-39.

Acquadro, A., Portis, E., Alberini, E., and Lanteri, S. 2005a. M-AFLP based protocol for microsatellite loci isolation in *Cynara cardunculus* L. (*Asteraceae*). Mol. Ecol. Notes 5:272-274.

Acquadro, A., Portis, E., Lee, D., Donini, P., and Lanteri, S. 2005b. Development and characterisation of microsatellite markers in *Cynara cardunculus* L. Genome 48:217-225.

Acquadro, A., Portis, E., Moglia, A., Magurno, F., and Lanteri, S. 2006. Retrotransposon based S-SAP as a platform for the analysis of genetic variation and linkage in globe artichoke. Genome 49:1149-1156.

Adzet, T., and Puigmacia, M. 1985. High-performance liquid chromatography of caffeoylquinic acid derivatives of *Cynara scolymus* L. leaves. J. Chromatogr. 348:447-452.

Babes, G. 1997. *Impiego di Tecnologie Non Convenzionali Per La Moltiplicazione Del Carciofo*, PhD thesis (X cycle), University of Viterbo, Italy.

Barba, M., Di Lernia, G., Babes, G., and Citrulli, F. 2004. Produzione e conservazione di germoplasma di carciofo di tipo 'Romanesco' esente da virus. Italus Hortus 11:5-10.

Barbagallo, R., Chisari, M., Spagna, G., Ierna, A., Patanè, A., Occhipint, A., and Mauromicale, G. 2007. Casein activity expression in flowers of *Cynara cardunculus* spp. Acta Hort. 730:195-199.

Basnizki, J., and Mayer, M. 1985. Germination in *Cynara* seeds: Effects of light and temperature on the function of the endosperm. Agronomie 5:529-532.

Basnizki, J., and Zohary, D. 1987. A seed planted cultivar of globe artichoke. HortScience 22:678-679.

Basnizki, J., and Zohary, D. 1994. Breeding of seed planted artichoke. Plant Breed. Rev. 12:253-269.

Bekheet, S. 2006. In vitro preservation of globe artichoke germplasm, in: *Proceeding VI International Symposium on Artichoke, Cardoon and Their Wild Relatives,* Lorca-Murcia, Spain, pp. 44.
Bianco, V. V. 1990. Carciofo (*Cynara scolymus*). In *Orticoltura,* ed. V. V. Bianco and F. Rimpini, Patron Editore, Bologna.
Bianco, V. V. 2005. Present situation and Future Potential of Artichoke in the Mediterranean Basin. Acta Hort. 681:39-55.
Bianco, V. V. 2007. Present and prospects of utilization of fresh and processed artichoke. Acta Hort. 730:23-37.
Brown, J. E., and Rice-Evans, C. A. 1998. Luteolin rich artichoke extract protects low density lipoprotein from oxidation in vitro. Free. Rad. Res. 29:247-255.
Cadinu, M., Repetto, A., Frau, A., Beneventi, S., and Meloni, S. 2004. Influence of The Explant Type on The Phenoptypic Changes in Micropropagated Plants of Artichoke. Acta Hort. 660:373-380.
Calabrese, N., De Palma, E., and Damato, G. 2007. Harvest time and yield of artichoke cultivars propagated vegetatively or by seed. Acta Hort. 730:345-350.
Chatelet, P., Stamigna, C., and Thomas, G. 2005. Early development from isolated microspores of *Cynara cardunculus* var. *scolymus* (L.) Fiori, Acta Hort. 681:375-380.
Comino, C., Lanteri, S., Portis, E., Acquadro, A., Romani, A., Hehn, A., Larbat, R., and Bourgaud, F. 2007. Isolation and functional characterization of a cDNA coding a hydroxycinnamoyltransferase involved in phenylpropanoid biosynthesis in *Cynara cardunculus* L. BMC Plant Biology 7:14.
Condés, L. F., Pato, A., and Jiménez, J. 2007. Evaluation of the floral induction and early .production of Madrigal F1 artichoke, grown from seed, subjected to different GA3 treatments. Acta Hort. 730:171-175.
Cravero, V. P., Picardi, L. A., and Cointry, E. L. 2005. An approach for understanding the heredity of two quality traits (head color and tightness) in globe artichoke (*Cynara scolymus* L.). Genet. Mol. Biol. 28:431-434.
De Pace, C., Porceddu, E., and Pacucci, G. 1976. Ulteriori risultati di una serie di incroci diallelici nel carciofo, in: *Proceedings II International Congress on Artichoke*, ed. Minerva Medica, Torino, pp. 657-665.
De Vos, N. E. 1992. Artichoke production in California. HortTech. 2:438-444.
Debenedetti, S. L., Palacios, P. S., Wilson, E. G., and Coussio, J. D. 1993. HPLC analysis of caffeoylquinic acids contents in Argentine medicinal plants. Acta Hort. 333:191-199.
Deidda, M. 1967. Contributo al miglioramento genetico del carciofo, in: *Proceedings I International Congress on Artichoke*, ed. Minerva Medica, Torino, pp. 157-174.
Dellacecca, V. V., Magnifico, V., Marzi, V., Porceddu, E., and Scarascia Mugnozza, G. T. 1976. Contributo alla conoscenza delle varietà di carciofo coltivate nel mondo, in: *Proceedings II International Congress on Artichoke*, ed. Minerva Medica, Torino, pp. 119-316.
Di Venere, D., Linsalata, V., Pieralice, M., Cardinali, A., Sergio, L., and Crinò, P. 2007. Biochemical characterization of clones from two 'Romanesco' landraces. Acta Hort. 730:443-448.
Dranik, L.I., Dolganenko, L. G., Slapke, J., and Thoma, H. 1996. Chemical composition and medical usage of *Cynara scolymus* L. Rastit. Resur. 32:98-104.
Elia, A., and Miccolis, V. 1996. Relationship among 104 artichoke (*Cynara scolymus* L.) accessions using cluster analysis. Adv. Hort. Sci. 10:158-162.
Elia, A., Conversa, G., Montervino, C., Di Brita, S., and Lotti, C. 2007. Micropropagation of the early artichoke cultivar 'Violet de Provance'. Acta Hort. 730:127-134.
Ercan, N., Onus, A. N., Polat, E., Ayar, F., Temirkaynak, M., and Şensoy, A. S. 2004. Determination of optimum GA3 concentrations and awakening irrigation time for globe

artichoke (*Cynara scolymus* L. cv. Sakýz) grown in mediterranean region of Turkey. Acta Hort. 660:197-200.

Esteva, J., and Martìnez, J. 2004. Evaluation of yield, earliness and head characteristics of bull variant plants in globe artichoke varieties 'Blanca de Tudela' and 'Violet de Provence' at Murcia. Acta Hort. 660:117-121.

Foti, S., Mauromicale, G., Raccuia, S. A., Fallico, B., Fanella, F., and Maccarone, E. 1999. Possible alternative utilization of *Cynara* spp. I. Biomass, grain yield and chemical composition of grain. Ind. Crop Prod. 10:219-228.

Foury, C. 1969. Étude de la biologie florale de l'artichaut *Cynara scolymus* L.: Application à la sélection 2. Étude des descendances obtenues en fécondation contrôllèe. Ann. Amélior. Plantes 19:23-52.

Foury, C. 1979. Quelques aspects pratiques de la sélection généalogique de l'artichaut I: presentation, creation de lignées. Ann. Amélior. Plantes 29:383-418.

Foury, C. 1987. *Quelques Aspects du Dèvelopement de l'Artichaut (Cynara scolymus L.) Issu de Semences; Analyse Plus Particuliere de la Floraison en Conditions Naturelles*, These Doctoral, University Orsay, Paris, pp. 189.

Foury, C., Martin, F., Vaissière, B., Morison, N., and Corre, J. 2005. Advantages et Difficultes de la Creation d'Hybrides F1 d'Artichaut à Semer, Acta Hort. 681:315-322.

Frau, A., Mallica, G., Baghino, L., Cadinu, M., and Repetto, A. 2004. La micropropagazione del carciofo 'Spinoso sardo': un valido strumento per aumentare la produttività degli impianti. Italus Hortus 11:38-41.

Gallitelli, D., Papanice, M., Campanale, A., Bottalico, G., and Sumerano, P. 2006. Risanamento nel carciofo 'Brindisino', in: *Atti del Convegno Conclusivo Progetto MIPAF 'Carciofo'*, Roma, pp. 22-24.

García, S. M., Cointry, E. L., Firpo, I. T., López, F. S., Cravero, V. P., and Asprelli, P. 2004. Influence of sowing dates over seed-grown artichoke production. Acta Hort. 660:387-390.

Gebhardt, R. 1997. Antioxidative and protective properties of extracts from leaves of artichoke (*Cynara scolymus* L.) against hydroperoxide induced oxidative stress in cultured rat hepatocytes. Toxicol. Appl. Pharmacol. 144:279-286.

Gebhardt, R. 1998. Inhibition of cholesterol biosynthesis in primary cultured rat hepatocytes by artichoke (*Cynara scolymus* L.) extracts, J. Pharmacol. Exp. Ther. 286:1122-1128.

Gil, R., and Villa, F. 2004. Breeding for earliness on seed propagated globe artichoke. Acta Hort. 660:35-37.

González, J., Pérez, F., Fernández, J., Lezaun, J.A., Rodríguez, D., Perea, F., Romero, C., Ochoa, M.J., and García, M. 2004. Study of *Cynara cardunculus* L. lignocelullosic biomass production in dry conditions. Acta Hort. 660:221-228.

Graifenberg, A., and Giustiniani, L. 1997. Problematiche colturali e valorizzazione del carciofo. Informatore Agrario 27:53-57.

Itoiz, R., Chocarro, A., and Royo, J. B. 2004. Genetic Variability of Cardoon Populations Evaluated Using RAPD. Acta Hort. 660:249-251.

Kraft, K. 1997. Artichoke leaf extract. Recent findings reflecting effects on lipid metabolism, liver and gastrointestinal tracts. Phytomedicine 4:369-378.

Lanteri, S., Di Leo, I., Ledda, L., Mameli, M. G., and Portis, E. 2001. RAPD, variation within and among populations of globe artichoke (*Cynara scolymus* L.), cv. 'Spinoso sardo'. Plant Breed. 120:243-247.

Lanteri, S., Saba, E., Cadinu, M., Mallica, G. M., Baghino, L., and Portis, E. 2004a. Amplified fragment length polymorphism for genetic diversity assessment in globe artichoke. Theor. Appl. Genet. 108:1534-1544.

Lanteri, S., Acquadro, A., Saba, E., and Portis, E. 2004b. Molecular fingerprinting and evaluation of genetic distances among selected clones of globe artichoke (*Cynara cardunculus* L. var. *scolymus* L.) 'Spinoso sardo'. J. Hort. Sci. Biotech. 79:863-870.

Lanteri, S., Acquadro, A., Comino, C., Mauro, R., Mauromicale, G., and Portis, E. 2006. A first linkage map of globe artichoke (*Cynara cardunculus* var. *scolymus* L.) based on AFLP, S-SAP, M-AFLP and microsatellite markers. Theor. Appl. Genet. 112:1532-1542.

Leskovar, D., Goreta, S., Piccinini, G., and Yoo, K. 2007. Strategies for globe artichoke introduction in South Texas. Acta Hort. 730:157-163.

Lopez Anido, F. S., Firpo, I. T., García, S. M., and Cointry, E. L. 1998. Estimation of genetic parameters for yield traits in globe artichoke (*Cynara scolymus* L.). Euphytica 103:61-66.

Lotti, C., De Giovanni, C., Fanizza, G., and Ricciardi, L. 2003. L'analisi AFLP nell'identificazione varietale del carciofo (*Cynara cardunculus* L. var. *scolymus*). Italus Hortus 4:246-248.

Mabeau S., Baty Julienne, C., Anne-Blandine, H., Chodosas, O., Metra, P., and Chesne, C. 2007. Antioxidant activity of artichoke extracts and by-products. Acta Hort. 730:491-496.

Maccarone, E., Fallico, B., Fanella, F., Mauromicale, G., Raccuia, S. A., and Foti, S. 1999. Possible alternative utilization of *Cynara* spp. II. Chemical characterization of their grain oil. Ind. Crop Prod. 10: 229-237.

Macua, J. I., Lahoz, I., Malumbre, A., Garnica, J., Urmeneta, I., and Arrondo, M. A. 2004. Commercial varieties of cardoon in Navarra. Acta Hort. 660:215-221.

Macua, J. I. 1996. Collecciòn de variedades de alcachoha, in: *Actas I Jornadas Técnicas de Alcachofa,* ed. ITGA, Tudela – Navarra, pp. 151-161.

Mallica, G., Baghino, L., Cadinu, M., and Repetto, A. 2004. Risultati della selezione clonale sulla cultivar di carciofo 'Spinoso sardo'. Italus Hortus 11:25-28.

Martino, V., Caffini, N., Phillipson, J. D., Lappa, A., Tchernitchin, A., Ferraro, G., Debenedelli, S., Schilcher, H., and Acevedo, C. 1999. Identification and characterization of antimicrobial components in leaf extracts of globe artichoke (*Cynara scolymus* L.). Acta Hort. 501:111-114.

Marzi, V., and Bianco, V. V. 1967. Risultati di prove di confronto tra varietà di carciofo di provenienze diverse in Puglia e Lucania, in: *Proceedings I International Congress on Artichoke*, ed. Minerva Medica, Torino, pp. 97-116.

Matthes, C., and Honermeier, B. 2007. Cultivation of the artichoke as medicinal plant under temperate climate conditions in Germany. Acta Hort. 730:483-489.

Mauro, R., Ierna, A., Portis, E., Lanteri, S., and Mauromicale, G. 2007. Morphological and molecular characterization of autochthonous Sicilian globe artichokes grown in family gardens. Acta Hort. 730:113-121.

Mauromicale, G. 1987. Panorama varietale del carciofo e sua prevedibile evoluzione. Informatore Agrario 5:69-75.

Mauromicale, G., Basnizki, Y., and Cavallaro, V. 1989. Primi risultati sperimentali sulla propagazione del carciofo (*Cynara scolymus* L.) per seme. Rivista di Agronomia 23: 417-423.

Mauromicale, G., and Copani, V. 1989. Caratteristiche biologiche e produzione di cloni diversi di carciofo isolati in popolazioni siciliane di Violetto di Sicilia. Tecnica Agricola 41:3-17.

Mauromicale, G., Morello, N., Santoiemma, G., and Ierna, A. 2000. Nuove varietà per migliorare la cinaricoltura siciliana. Informatore Agrario 26:47-51.

Mauromicale, G., and Ierna, A. 2000. Panorama varietale e miglioramento genetico del carciofo. Informatore Agrario 26:39-45.

Mcdougall, B., King, P. J., Wu, B. W., Hostomsky, Z., Manfred, G., and Robinson, W. E. 1998. Dicaffeoylquinic acid and dicaffeoyltartaric acid are selective inhibitors of human immunodeficiency virus type 1 integrase. Antimicrob. Agents *Ch.* 42:140-146.

Melilli, M. G., and Raccuia, S. A. 2007. Inulin and water-soluble –sugars variations in *Cynara* roots during the biological cycle. Acta Hort. 730:475-481.

Melilli, M. G., Trigali, S., Riggi, E., and Raccuia, S. A. 2007. Screening of genetic variability for some phenolic constituens of globe artichoke head. Acta Hort. 730:85-91.

Miccadei, S., Di Venere, D., Cardinali, A., Linsalata, V., Bugianesi, R., Foddai, M. S., Valentini, S., Fraioli, R., and Maiani, G. 2006. Azione antiossidante e apoptotica di frazioni polifenoliche da Cnara scolymus in epatociti di ratto e in cellule di epatoma umano, in: *Atti del Convegno Conclusivo Progetto MIPAF 'Carciofo'*, Roma, pp. 104-106.

Miguel, A., Baixauli, C., Aguilar, J. M., Giner, A., Maroto, J. V., Lòpez, S., and Pascual, B. 2004a. Cultivar Trials of Seed Propagated Artichoke. Acta Hort. 660:111-116.

Miguel, A., Baixauli, C., Aguilar, J. M., Giner, A., Maroto, J. V., Lòpez S., and Pascual, B. 2004b. Gibberellic Acid Concentrations in Seed propagated Artichoke. Acta Hort. 660:167-172.

Miller, T. 1975. New artichoke clones. New Zealand J. Agr. 131:33-35.

Motzo, R., and Deidda, M. 1993. Anther and ovule culture in globe artichoke. J. Genet. Breed. 47:263-266.

Mouzali, L., Aziza, M., Bensiameur-Touati, K., and Hellal-Benateya, A. 2004. Cardoon (*Cynara cardunculus* L.) used as vegetable rennet in an algerina traditional cheese making 'Djben'. Acta Hort. 660:207-213.

Nichiforescu, E. A. 1970. Composition of caffeoylquinic acid derivatives of artichoke (*Cynara scolymus* L.). Plant Med. Phytother. 4:56-62.

Pagnotta, M. A., Cardarelli, M. T., Rey Muñoz, R., Tucci, M., and Saccardo, F. 2004. Assessment of genetic variation in artichoke of 'Romanesco' type by molecular markers. Acta Hort. 660:99-104.

Papalini, P., Cinquanta, A., and Ercolani, R. 1997. Etrusco e Moro di Corneto due nuove cultivar di carciofo. Informatore Agrario 27:59-61.

Papanice, M. A., Campanale, A., Bottalico, G., Sumerano, P., and Gallitelli, G. 2004. Produzione di germoplama risanato di carciofo brindisino. Italus Hortus 11:11-15.

Pécaut, P. 1993. *Globe Artichoke Cynara scolymus L. Genetic Improvements of Vegetable Crops*, ed. Kallo G. and Bergh B.D., Pergamon, Oxford, pp. 737-746.

Pécaut, P., and Martin, P. 1993. Variation occurring after natural and in vitro multiplication of early Mediterranean cultivars of globe artichoke (*Cynara scolymus* L.). Agronomie 13:909-919.

Perez-Garcia, F., Adzet, T., and Canigueral, S. 2000. Activity of artichoke leaf extract on reactive oxygen species in human leukocytes. Free Radical Res. 33:661-665.

Pittlern, M. H., and Ernst, E. 1998. Artichoke leaf extract for serum cholesterol reduction. Perfusion 11:338-340.

Pochard, E., Foury, C., and Chambonet, D. 1969. Il miglioramento genetico del carciofo, in: *Proceedings I International Congress on Artichoke*, ed. Minerva Medica, Torino, pp. 117-143.

Porceddu, E., Dellacecca, V., and Bianco, V. V. 1976. Classificazione numerica di cultivar di carciofo, in: *Proceedings II International Congress on Artichoke*, ed. Minerva Medica, Torino, pp. 1105-1119.

Portis, E., Acquadro, A., Comino, C., Mauromicale, G., Saba, E., and Lanteri, S. 2005a. Genetic structure of island populations of wild cardoon [*Cynara cardunculus* L. var. *sylvestris* (Lamk) Fiori] detected by AFLPs and SSRs. Plant Science 169:199-210.

Portis, E., Mauromicale, G., Barchi, L., Mauro, R., and Lanteri, S. 2005b. Population structure and genetic variation in autochthonous globe artichoke germplasm from Sicily Island. Plant Science 168:1591-1598.

Portis, E., Barchi, L., Acquadro, A., Macua, J. I., and Lanteri, S. 2005c. Genetic diversity assessment in cultivated cardoon by AFLP (amplified fragment length polymorphism) and microsatellite markers. Plant Breed. 124:299-304.

Principe, J. A. 1984. Male sterility in artichoke. HortScience 19:864.

Quilho, T., Gominho, J., and Pereira, H. 2004. Anatomical characterisation and variability of the thistle *Cynara cardunculus* in view of pulping potential. Iawa J. 25:217-230.

Robba, L., Carine, M. A., Russell, S. J., and Raimondo, F. M. 2005. The monophyly and evolution of *Cynara* L. (*Asteraceae*) sensu lato: evidence from the Internal Transcribed Spacer region of nrDNA. Plant Syst. Evol. 253:53-64.

Rottenberg, A., and Zohary, D. 1996. The wild relatives and the wild ancestry of the cultivated artichoke. Gen. Res. Crop Evol. 43:53-58.

Rottenberg, A., Zohary, D., and Nevo, E. 1996. Isozyme relationships between cultivated artichoke and the wild relatives. Gen. Res. Crop Evol. 43:59-62.

Rottenberg, A., and Zohary, D. 2005. Wild genetic resources of cultivated artichoke. Acta Hort. 681:307-311.

Saccardo, F., Micozzi, F., Di Lernia, G., Piccioni, C., Barba, M., and Pagnotta, M. A. 2007. Virus free artichoke germplasm: quali-quantitative response of globe artichoke. Acta Hort. 730:375-379.

Scarascia Mugnozza, G. T., and Pacucci, G. 1976. Tipi di potenziale valore pratico isolati nell'ambito di un programma per il miglioramento genetico del carciofo, in: *Proceedings II International Congress on Artichoke*, ed. Minerva Medica, Torino, pp. 117-143.

Sevcikova, P., Glatz, Z., and Slanina, J. 2002. Analysis of artichoke (*Cynara cardunculus* L.) extract by means of micellar electrokinetic capillary chromatography. Electrophoresis 23:249-252.

Slanina, J., Taborska, E. and Musil, P. 1993. Determination of cynarine in the decoctions of the artichoke (*Cynara cardunculus* L.) by the HPLC method. Cesko-SloV. Farm. 42: 265-268.

Slanina, J., Taborska, E., Bochorakova, H., Slaninova, I., Humpa, O., Robinson, W. E., and Schram, H. 2001. New and facile method of preparation of the anti-HIV-1 agent, 1,3-dicaffeoylquinic acid. Tetrahedron Lett. 42:3383-3385.

Sonnante, G., De Paolis, A., Lattanzio, V., and Perrino, P. 2002. Genetic variation in wild and cultivated artichoke revealed by RAPD markers. Gen. Res. Crop Evol. 49:247-252.

Sonnante, G., De Paolis, A., and Pignone, D. 2004. Relationships among artichoke cultivars and some related wild taxa based on AFLP markers. Plant Genet. Res. 1:125-133.

Sonnante, G., Carluccio, A. V., Sonnante, G., Vilatersana, L. R., and Pignone, D. 2007a. Variable RDNA regions provide suggestions on artichoke domestication and evolution history. Acta Hort. 730:123-125.

Sonnante, G., De Paolis, A., and Pignone, D. 2007b. Isolation, characterization and expression of PAL gene family in artichoke. Acta Hort. 730:81-84.

Stamigna, C., Crinò, P., Chatelet, P., and Saccardo, F. 2004a. Induction of embryogenesis in isolated microspores of artichoke (*Cynara scolymus* L.). Acta Hort. 660:139-14.

Stamigna, C., Micozzi, F., Pandozy, G., Crinò, P., and Saccardo, F. 2004b. Produzione di ibridi F1 di carciofo mediante impiego di cloni maschiosterili. Italus Hortus 11:29-33.

Stamigna, C., Saccardo, F., Pandozy, G., Ancora, G., and Crinò, P. 2005. *In vitro* mutagenesis of globe artichoke (cv. Romanesco). Acta Hort. 681:403-410.

Tavazza, R., Papacchioli, V., and Ancora, G. 2004. An improved medium for in vitro propagation of globe artichoke (*Cynara scolymus* L.) cv. Acta Hort. 660:91-97.

Tesi, R. 1976. Primi risultati del miglioramento genetico delle varietà toscane di *Cynara cardunculus*, var. *scolymus*, in: *Proceedings II International Congress on Artichoke*, ed. Minerva Medica, Torino, pp. 747-763.

Tivang, J., Skroch, P. W., Nienhuis, J., and De Vos, N. 1996. Randomly Amplified Polymorphic DNA (RAPD) variation among and within artichoke (*Cynara scolymus* L.) cultivars and breeding populations. J. Am. Soc. Hort. Sci. 121:783-788.

Todaro, A., Iernia, A., Peluso, O., Mauromicale, G., and Spagna, G. 2006. Determination of polyphenol oxidase and studies on browing inactivation in three artichoke cultivars, in: *Proceeding VI International Symposium on Artichoke, Cardoon and their Wild Relatives,* Lorca-Murcia, Spain, pp. 42

Trionfetti Nisini, P., Crinò, P., Pagnotta, M. A., Gavazza, R., and Ancora, G. 2007. Recovery and characterization of Italian artichoke traditional landraces 'Romanesco' type. Acta Hort. 730:101-106.

Vanella, B., Porceddu, E., and De Pace, C. 1981. Applicazioni di metodi di analisi numerica per il miglioramento genetico del carciofo, in: *Atti III Congresso Int. Di Studi sul Carciofo*, ed. Laterza, Bari, pp. 797-807.

Wagenbreth, D. 1996. Evaluation of artichoke cultivars for growing and pharmaceutical use. Beitr. Zuchtungsforsch 2:400-403.

Wang, M. F., Simon, J. E., Aviles, I. F., He, K., Zheng, Q. Y., and Tadmor, Y. 2003. Analysis of antioxidative phenolic compounds in artichoke (*Cynara scolymus* L.). J. Agr. Food Chem. 51:601-608.

Wiklund, A. 1992. The genus *Cynara* L. (*Asteraceae-Cardueae*). *Bot. J. Linn. Soc.* 109: 75-123.

Zohary, D. and Basnizki, J. 1975. The cultivated artichoke *Cynara scolymus*. Its probable wild ancestors. Econ. Bot. 29:233-235.

Lettuce

Beiquan Mou[1]

[1] United States Department of Agriculture, Agricultural Research Service, bmou@pw.ars.usda.gov

1 Introduction

Lettuce (*Lactuca sativa* L.) is a major fresh vegetable and its leaves are commonly found in salad mixtures and sandwiches. In some eastern countries like China and Egypt, stems instead of leaves of lettuce are consumed, either cooked, raw, pickled, dried, or as a sauce. Some less common uses for lettuce include a cigarette without nicotine made from lettuce leaves, edible oil extracted from seeds of a primitive lettuce, and a sedative made of dried latex contained in stems and other tissues. Lactucarium, the dried latex produced from a wild lettuce relative, *Lactuca virosa* L., is used to make a sleep-inducing medicine (Ryder, 1986).

As a cool-season crop, lettuce is extensively grown on all the continents, particularly in temperate and subtropical regions. World lettuce production was at more than 22 million tons on about 1 million ha in 2005 (Table 1). About two-thirds of the total production area in the world is in Asia. China is the largest lettuce producer and accounts for about half of the world's total production and area, primarily for lettuce stems. The United States has the largest production of lettuce as a salad crop, and produced 22% of the world's lettuce supply on only 13% of the production area due to its higher yield. Western Europe claims about 13% of the total lettuce production and area in the world. Lettuce is also grown in large areas in India, Japan, Mexico, and Turkey.

Lettuce is a self-pollinated annual plant. It forms a deep taproot with largely horizontal lateral roots, most densely near the soil surface for water and nutrient absorption. Nearly sessile leaves are spirally arranged in a dense rosette on the often shortened stem. There is considerable diversity in colour, shape, surface, margin, and texture of leaves among different types and forms of lettuce. Leaf margins may be entire, lobed, incised, indented, or undulating. Leaf surface can be smooth, savoy, or

Table 1. Harvested area (hectare), production (thousand metric tons), and yield in major lettuce-producing countries and regions in 2005.

Country or region	Harvested area (ha)	Production (1000 m tons)	Yield (kg/ha)
Australia	6,121	127.2	20,786
Austria	1,600	61.9	38,688
Bangladesh	7,689	32.0	4,162
Belgium	2,204	80.0	36,298
Canada	3,891	92.4	23,739
Chile	6,600	90.0	13,636
China	500,250	11,005.0	21,999
Egypt	6,000	140.0	23,333
France	16,500	526.0	31,879
Germany	8,200	200.0	24,390
Greece	4,200	80.0	19,048
Guatemala	1,900	38.0	20,000
India	120,000	790.0	6,583
Iran	3,600	90.0	25,000
Israel	950	40.0	42,105
Italy	43,604	846.8	19,420
Japan	22,000	530.0	24,091
Jordan	1,278	35.0	27,383
Mexico	11,290	243.4	21,559
Netherlands	2,000	73.0	36,500
New Zealand	1,300	31.0	23,846
Niger	4,400	40.0	9,091
Peru	2,900	33.0	11,379
Portugal	4,400	95.0	21,591
South Africa	2,194	33.0	15,003
South Korea	7,000	210.0	30,000
Spain	39,000	920.0	23,590
Switzerland	1,160	35.0	30,172
Syria	2,500	53.0	21,200
Turkey	19,700	375.0	19,036
United Kingdom	5,514	135.0	24,483
United States	131,280	4,976.9	37,910
Venezuela	1,500	31.0	20,667
Africa	15,484	270.6	17,477
Asia	689,699	13,213.3	19,158
Eastern Europe	5,558	57.6	10,354
Western Europe	130,425	3,098.8	23,759
N & C America	150,838	5,374.4	35,630
South America	15,174	198.4	13,078
World	1,015,159	22,382.3	22,048

Source: Food and Agricultural Organization of the United Nations (FAO, 2005).

crinkled. The colour of lettuce leaves ranges from yellow to dark green with different degrees of shininess; anthocyanin may mix in to cover all or part of the leaves, or in a spotting pattern or just along the margin. Stem elongation signals the end of the vegetative stage and the reproductive phase begins. A single stem is usually formed to bear the inflorescence, which is a dense corymbose panicle composed of many capitula, each consisting of many florets. The number of florets usually ranges from 12 to 20, but can be as few as seven and as many as 35 (Feráková, 1977). Each floret produces a single-seeded achene, which is ribbed and topped with a pappus hair. Seed colour varies, including white, yellow, brown, gray, and black. Newly harvested seeds usually have a short period of dormancy, and most cultivars exhibit varying levels of thermodormancy.

2 Origin and Domestication

Prior to domestication by humans, lettuce grew wild. It is still not exactly clear which species were involved in the evolution that led to modern-day lettuce. However, it is certain that *L. serriola* is one of or the only direct ancestor(s) (Lindqvist, 1960a; de Vries, 1990, 1997; Kesseli et al., 1991). The chromosomes of *L. sativa* and *L. serriola* are very similar morphologically (Feráková, 1977), and they cross freely with each other. The two taxa are considered by some as subspecies of the same species. It is most likely that changes in *L. serriola* caused by mutations led to the appearance of forms that were favoured by humans, particularly forms without spines on stems and leaves and with large seeds. They were then selected for use and further modified to fit human needs. These early forms would have been suitable for animal consumption and for oil from the seeds for domestic use. Several primitive forms still exist and are used for these purposes in Egypt today (Harlan, 1986: Ryder, 1999). Most of these grow and develop rapidly and have non-reflexed involucres to prevent seed loss, large seeds, and high oil content in the seeds (35%, Boukema et al., 1990). One of these landraces, known as USDA (the U.S. Dept. of Agriculture) Plant Introduction (PI) 251245 from Egypt, is used for seed oil (Fig. 1).

The existence of these primitive forms in the Middle East provides strong support for the idea that lettuce probably originated in the eastern Mediterranean basin. Another piece of evidence is found in Egyptian tomb paintings dated from the Middle Kingdom, about 4,500 years ago. The stylized illustrations appear to be bundles of stem lettuce with elongated heads and lanceolate, pointed leaves, similar to the type still grown in Egypt today. From Egypt, cultivated lettuce spread to Greece and Rome and throughout the Mediterranean region (Lindqvist, 1960a). Around the Mediterranean basin, the romaine type of lettuce, also known as cos type (which may suggest its early use on Kos island near Turkey), predominated and still is the most common lettuce type today. They most closely resemble the stem lettuces and thus probably have evolved from the stem types. The first written records of lettuce cultivation are credited to Herodotus who mentioned that about 550 B.C. a cos-like lettuce was eaten at the Persian Court (de Vries, 1997). The cos lettuces are distributed close to the likely centre of origin and have considerable variation in leaf shape and length, flat and erect stature, open and closed heads, texture, and colour. It is likely that leaf or cutting lettuces, butterhead, Latin types, and Batavia-type crisphead were all selected from this rich source of variability (Ryder, 1999).

Fig. 1. A primitive form of lettuce, PI 251245, from Egypt.

Later, lettuce-growing areas further expanded from the Mediterranean region to the rest of Europe. The first indication of lettuce cultivation in Northwestern Europe is found in the herbal of Schöffer (1485) who described four lettuce types (de Vries, 1997). Pieter made a painting in 1553 (Jesus in the house of Martha and Maria, Museum Booymans van Beuningen, Rotterdam, the Netherlands) that shows a butterhead lettuce. The oldest cultivars in the largest lettuce germplasm collection in Europe (CGN, the Netherlands) are two French cultivars: 'Passion Blonde a Graine Blanche' from 1755 and 'Palatine' from 1771 (Boukema et al., 1990).

Lettuce was brought to the New World by Christopher Columbus. Peter Martyr reported its presence on Isabela Island in 1494 (Ryder, 1997), just two years after Columbus's first voyage. In the next 400 years following its introduction, an assortment of leaf types was grown in America, including the loosely packed, soft-headed crisp lettuce, known as Batavia types. At the beginning of the 20th century, most popular lettuces were butterhead types. In the early part of the century, however, the Batavia type of crisphead lettuce began to predominate, because it could be grown on irrigated large farms in the western USA and could maintain good quality for 10-12 days for the shipment to the rest of the country. In the 1940s, T.W. Whitaker crossed a Batavia cultivar 'Imperial' with an heirloom variety 'Brittle Ice' and developed the first true iceberg lettuce, 'Great Lakes', which is a compact firm-headed crisphead lettuce. Soon the iceberg type gained popularity and became dominant in the United States. Iceberg lettuce was introduced back to Europe and has become the most important lettuce type in the U.K. and the Scandinavian countries. The iceberg type has also become popular in Spain, Germany, Australia, and Japan, among others, and is now a major item in these countries.

Cultivated lettuce was introduced into China between 600 and 900 A.D. (de Vries, 1997). The Chinese selected lettuce for the bulky succulent non-bitter stem and long narrow leaves for consumption of the stem mainly as a cooked vegetable.

Domestication of the wild types of lettuce has resulted in the loss of prickles from leaves and stems, less latex and tissue bitterness, reduced suckering, slow bolting except for stem lettuce, an increase in seed size, and non-reflexed involucres to prevent seed shattering. Human selection and later breeding efforts have also led to changes in size, shape, colour, texture, and taste of leaves and plants, heading habits, resistance to diseases and insects, and adaptation to different geographic areas and environments.

3 Horticultural Types

There is a great diversity of shape, size, and colour among lettuce cultivars, which are classified into types mainly based on leaf shape, size, and texture, head formation, and stem type. Six generally recognized types are crisphead, butterhead, romaine, leaf (Fig. 2), stem (Fig. 3), and Latin (Fig 4).

Fig. 2. Four principal types of lettuce in the world except China: crisphead, butterhead, romaine, and leaf.

Fig. 3. Stem lettuce, the major lettuce type produced in China.

Fig. 4. Latin lettuce, cultivar 'Little Gem'.

CRISPHEAD Commonly called "iceberg" or head lettuce, the crisphead type produces a spherical firm head that weighs between 700 and 1000 g when grown in the field. The early leaves are elongated at the rosette stage, and gradually increase in width with each successive leaf until they are broader than long at maturity. After 10-12 leaves, leaves change to cup-shaped and begin overlapping each other and enclosing later leaves to form a head structure. New leaves continue to appear and expand from inside to fill the head, which becomes large and firm. Overmature heads become hard, and leaves may rupture and taste bitter. The head may then burst with the elongation of the seed stalk. Outer leaves are bright green or dull green, and the interior colour changes progressively from lighter green to whitish or creamy yellow towards the centre. The tightly folded inner leaves are rugose, brittle, and crispy with a mild taste. Included in this type is a crisp subtype called Batavia that forms a less dense and softer head and weighs about 500 g when mature. A cultivar named 'Iceberg' is actually a Batavia type of lettuce.

In the modern history of lettuce breeding, four crisphead lettuce groups have been developed in the United States with different colour, leaf texture and shape, head size and shape, and butt appearance (Ryder, 1986). These include: (1) Imperial (light- or medium-green leaves that are serrated or wavy, relatively soft texture, and varied in butt colour and ribbiness), (2) Great Lakes (bright-green leaf colour in various shades, serrated very crisp leaves, and whitish butt with prominent ribs), (3) Empire (deeply serrated light-green leaves, very crisp, heads often conical in shape, and whitish butt with flat ribs), and (4) Vanguard (dull-green leaves, softer texture than Great Lakes, wavy leaf margins, and green butt with flat ribs). Many cultivars with pest resistance and other traits were developed within each group until a new improved type appears. Since the release of the cultivar 'Vanguard' by R.C. Thompson in 1958, the Vanguard group has become the predominant one in the United States.

BUTTERHEAD Sometimes called cabbage lettuce, butterhead lettuce produces a smaller and less compact head than the crisphead type. Leaves are broad, crumpled, relatively thin, and tender with a soft oily texture. The colour of outer leaves is lighter than most crisphead lettuces, and the inner colour is yellowish. Some cultivars have red pigmentation on the outer leaves. Taste varies from bland to relatively sweet. In Europe, there are two subtypes based on day-length sensitivity and growing season. Day-neutral types are grown outdoors in summer and weigh about 350 g. They have firmer heads at maturity and are slower bolting than the short-day types that are commonly grown in protective shelters in winter and may weigh 150-200 g. There are also two subtypes in the U.S. based on head appearance and size. The Boston type is larger, lighter in colour, and has a closed head. The Bibb type is smaller and darker green, and the head is relatively open at the top.

ROMAINE The romaine type is also known as roman or cos lettuce. It has elongated, coarse, and relatively crispy textured leaves with prominent broad midveins. Plants tend to have an upright stature and form a loaf-shaped head after the rosette stage. The heads are either closed or relatively open at the top. The closed type is used for "romaine heart" production, in which only the closed inner leaves are harvested. Outer leaves are usually light to dark green, and interior leaves are yellowish. Romaine heads may weigh as much as 750 g. They have a sweeter and stronger taste than that of crisphead lettuces. Some cultivars have red leaves, which are usually harvested while young and small to produce a value-added product called mesclun (spring mix), a mixture of baby leaves including lettuce, endive, chicory, spinach, and other vegetables.

LEAF Leaf, or cutting, lettuces have considerable variation in leaf size, shape, margin, colour, and texture. They form a bunch or rosette of leaves that may have broad, elongated, or lobed shape like oak leaves, smooth or frilled margins, and yellow, green, or red colours in varying shades. Texture ranges from crispy to soft. Its relatively open growth habit gives it fewer bleached leaves and a stronger taste than the crisphead types. Leaf lettuce may weigh up to 0.5 kg when all leaves are harvested. Leaf lettuces may form a loose head when over mature.

STEM Stem lettuce is also called celtuce, stalk, or asparagus lettuce. Plants are grown mainly for the erect thickened stem (4-10 cm in diameter) that may be as long as 50-60 cm. The leaves are usually long and narrow, but may be as broad as romaine leaves. Except for the young foliage, leaves are not palatable due to their high latex content and bitterness, and are removed before consumption. Stems are peeled and the soft translucent green core is cooked or eaten raw.

LATIN Latin lettuce is also known as grassé lettuce. It has an upright stature and somewhat resembles romaine lettuce, but the leaves are shorter and less crispy. Heads are sometimes closed and leaf texture resembles Bibb-type butterhead.

Currently, 62% of the lettuce production in the United States is of the crisphead type, 23% is romaine, and 15% is leaf and butterhead types (USDA-NASS, 2006). More than 90% of the American production is concentrated in two western states, of which California accounts for 3/4 and Arizona contributes 1/4.

In Great Britain, about 75% of the production is crisphead type, 15% is butterhead, and 10% is romaine (Ryder, 1999). This is a striking contrast to the time period in the 1970s and earlier when 80-90% of the lettuce grown there was the butterhead type. Lettuce production in Germany consists of about 2/3 butterhead and 1/3 crisphead. A large portion of lettuce produced in the Netherlands, mostly butterhead type, is in greenhouses, while the summer outdoor crops include also crisphead and leaf lettuces. In Belgium, about 2/3 of lettuce, primarily butterhead, is grown under cover and 1/3 outdoors. France produces mostly butterhead and Batavia lettuces for domestic consumption and iceberg-type crisphead for export. Iceberg, romaine, and Latin lettuces are grown in Spain, with romaine type mainly for home consumption and iceberg lettuce for export. More than half of the lettuce grown in Italy is butterhead, about 1/3 is romaine, and the rest is crisphead and leaf lettuces.

In Australia, most production is of the iceberg type, and to a lesser extent, of romaine and leaf lettuces. Japan produces mostly iceberg, leaf, and butterhead lettuces. About 2/3 of lettuce grown in Israel is romaine and the remainder is iceberg. Stem lettuce is grown mainly in China and Egypt. Latin lettuce has some popularity in Mediterranean regions and to some extent also in Argentina and Chile, with small areas grown in the U.S.

4 Genetic Resources

Lettuce belongs to the family Asteraceae (Compositae), tribe Cichoreae, and genus *Lactuca*. Of about 100 species of *Lactuca*, only four can be crossed to each other by conventional hybridization methods and thus form the most important breeding group. They include *L. sativa* L., *L. serriola* L., *L. saligna* L., and *L. virosa* L.. They are all self-fertilized diploids with 2n = 2x = 18 chromosomes.

L. serriola, common wild lettuce or prickly lettuce, is distributed on every continent where lettuce is grown. It tends to form a rosette and has relatively wide stem leaves, either entire or toothed, which are held vertically (Fig. 5). There are usually spines on leaves and stems. The inflorescence is composed of pedicelled flowers on a pyramidal panicle producing small seeds. Crosses between *L. sativa* and *L. serriola* can be made easily, and therefore the two forms are considered by some as subsections of the same species. *L. serriola* serves as a source of valuable traits for lettuce such as disease resistance: anthracnose (*Microdochium panattonianum*; Brandes, 1918), aster yellows (phytoplasma; Thompson, 1944), corky root (*Sphingomonas suberifaciens*; Brown and Michelmore, 1988; Mou and Bull, 2004), downy mildew (*Bremia lactucae*; Farrara et al., 1987; Bonnier et al., 1992), lettuce drop (*Sclerotinia minor*; Abawi et al., 1980; *Sclerotinia sclerotiorum*; Whipps et al., 2002), and lettuce dieback (lettuce necrotic stunt and tomato bushy stunt viruses; Grube et al., 2005b); and insect resistance: cabbage looper (*Trichoplusia ni*; Kishaba et al., 1973; Kishaba et al., 1980), leafminer (*Liriomyza langei*; Mou and Liu, 2003, 2004), and root aphid (*Pemphigus bursarius*; Ellis et al., 2002), as well as carotenoid content (β-carotene and lutein; Mou, 2005), and flowering time (Ryder and Milligan, 2005).

Fig. 5. *Lactuca serriola*, PI 491093.

L. saligna is mainly distributed in Europe and western and central Asia, although it can also be found in North America and Australia. It differs from *L. serriola* in having very narrow toothed stem leaves and bearing sessile flowers on a spikelike panicle (Fig. 6). *L. saligna* can be crossed with *L. sativa or L. serriola* with little difficulty, but the progenies of the crosses are often completely or partially sterile. *L. saligna* may provide resistance to diseases: aster yellows (Thompson, 1944), corky root (Brown and Michelmore, 1988), cucumber mosaic virus (Provvidenti et al., 1980), downy mildew (Bonnier et al., 1992; Lebeda and Reinink, 1994; Jeuken and Lindhout, 2002, 2004), leaf spot (*Stemphylium botryosum*; Netzer et al., 1985), lettuce dieback (Grube et al., 2005b), lettuce infectious yellows virus (McCreight, 1987), and tomato spotted wilt virus (Hartmann, 1991); to insects: cabbage looper (Whitaker et al., 1974), leafminer (Mou and Liu, 2003, 2004), and root aphid (Ellis et al., 2002); and the trait carotenoid content (β-carotene and lutein; Mou, 2005).

Fig. 6. *Lactuca saligna*, PI 509525.

L. virosa is found in Western Europe and North Africa. It has blue-green leaves and exhibits a strong tendency to form a rosette (Fig. 7). Plants show annual as well as biennial growth habit and produce panicles resembling *L. serriola. L. virosa* can be crossed with *L. sativa* or *L. serriola* in both directions, but the F_1 hybrids are highly sterile. However, the hybrids may be made fertile by the treatment of the flowers with colchicine to double the chromosome number, generating an amphidiploid (Thompson and Ryder, 1961). Fertile plants may also be produced by *in vitro* embryo rescue of the hybrids between *L. sativa* and *L. virosa* (Maisonneuve et al., 1995). An attempt to make crosses between *L. saligna* and *L. virosa* in either direction was unsuccessful (Lindqvist, 1960b). *L. virosa* has been used as sources of resistance to diseases: beet western yellows virus (Maisonneuve et al., 1991), broad bean wilt virus (Provvidenti et al., 1984), big vein (Mirafiori lettuce big-vein virus and Lettuce big-vein associated virus; Bos and Huijberts, 1990; Hayes et al., 2006), lettuce mosaic virus (Maisonneuve et al., 1999), corky root (Brown and Michelmore, 1988), downy mildew (Bonnier et al., 1992), lettuce dieback (Grube et al., 2005b), and lettuce drop (*S. minor*; Abawi et al., 1980); and to insects: green peach aphid (*Myzus persicae*; Reinink and Dieleman, 1989), leafminer (Mou and Liu, 2003, 2004), lettuce aphid (*Nasonovia ribisnigri*; Eenink et al., 1982a), and root aphid (Ellis et al., 2002); and carotenoid content (β-carotene and lutein; Mou, 2005).

Fig. 7. *Lactuca virosa*, IVT-280.

There are many germplasm collections of lettuce, both cultivars and wild species, around the world. The major ones (with more than 500 accessions) include Centre for Genetic Resources, Wageningen, the Netherlands (CGN, www.genebank.nl, March 8, 2006); Institute of Plant Genetics and Crop Plant Research, Gatersleben, Germany (IPK, www.ipk-gatersleben.de, March 8, 2006); N.I. Vavilov Institute of Plant Industry, St. Petersburg, Russia (VIR, www.vir.nw.ru, March 8, 2006); GEVES, Brion, France; INRA, Versailles, France; Research Institute of Crop Production, Olomouc, Czech Republic; Horticulture Research International, Wellesbourne, United Kingdom (HRI); Institute of Agrobotany, Tápiószele, Hungary; Institute of Crop Germplasm Resources, Beijing, China; University of California, Davis, USA; USDA Western Regional Plant Introduction Station, Pullman, USA (www.ars-grin.gov/cgi-bin/npgs/html/site.pl?W6, March 8, 2006);

and USDA Crop Improvement and Protection Unit, Salinas, USA. Passport and evaluation information about the accessions can be searched on-line at the websites provided and seeds may be requested from the institutions.

The CGN maintains a searchable database for lettuce, the International *Lactuca* Database (www.genebank.nl/collections/ildb, March 8, 2006), which currently contains information of more than 12,000 accessions from 21 institutes and genebanks in 17 countries. This represents all major *Lactuca* collections in the world except China. The Department of Plant Sciences at University of California, Davis keeps an electronic database, Lettcv (http://compositdb.ucdavis.edu/database/lettcv2/display, March 8, 2006), which archives genetic, passport, and performance data available for over 4,500 lettuce cultivars. Another on-line database is curated by the Department of Horticultural Science, North Carolina State University, Raleigh, USA (http://cuke.hort.ncsu.edu/cucurbit/wehner/vegcult/vgclintro.html, March 8, 2006), which provides cultivar descriptions for about 500 lettuce cultivars grown in North America.

5 Major Breeding Achievements

There is no doubt that breeding and selection of lettuce were practiced by growers and seedsmen for many years, as evidenced by the existence of many cultivars of several types and subtypes in the latter part of the 19th century. However, these efforts had largely been undocumented until the 1920s. Notable accomplishments of modern lettuce breeding have been made in the areas of disease/insect resistance, improved quality, and increased yield.

5.1 Resistance to Diseases and Insects

The first documented success of breeding for disease resistance was achieved by Jagger of USDA in 1926 when he released the first brown-blight-resistant Imperial series (Jagger et al., 1941). Brown blight disease (etiology unknown) caused yellowing, necrosis, and death of lettuce plants, and devastated the lettuce industry in California that had become the principal area of lettuce production and shipment to all parts of the U.S. by that time. Jagger selected healthy survivors from many cultivars he screened, and developed the first three Imperial cultivars (Imperial 2, 3, and 6). These cultivars were then crossed with a French variety 'Blonde Lente á Monter' to add resistance to downy mildew, resulting in a second series of Imperial cultivars. As the cultivars became susceptible to new downy mildew races, new resistant cultivars were bred, such as 'Imperial 410', released in 1945 and 'Valverde', in 1959. The Imperial group dominated lettuce production in the U.S. from the 1930s to the early 1950s.

Downy mildew is also a serious problem in Europe and much of the breeding effort there has concentrated on the disease. New resistant cultivars have been continuously bred as new virulent strains of the pathogen emerge in an endless stream. Major public and private breeding programs have been in the Netherlands, France, and England.

Lettuce mosaic is a virus disease found in most lettuce production areas of the world. Resistant cultivars have been released both in the U.S. and in Europe. 'Vanguard 75' was the first mosaic-resistant cultivar released in the U.S., mainly for the desert areas in California and Arizona (Ryder, 1979b).

Big vein, another virus disease, was first reported in California in 1934 (Jagger and Chandler, 1934) and has subsequently been found around the world. Ryder used the low level of resistance identified in a cultivar 'Merit' in breeding to develop three cultivars with greater resistance: 'Thompson', 'Sea Green', and 'Pacific' (Ryder and Robinson, 1995).

Corky root, a bacterial disease also found in lettuce growing areas worldwide, has the potential to cause serious yield losses. Dickson (1963) screened 44 accessions of *L. sativa* and identified three resistant ones including a line from Turkey, PI 171669. Sequiera (1978) released the first group of resistant cultivars: 'Marquette', 'Montello', and 'Green Lake' from crosses to PI 171669.

Lettuce aphid is a major insect pest in Europe and has become a serious problem in the U.S. in recent years. It is protected from insecticides because the insect likes to feed on younger leaves inside the lettuce head. Resistance was found in several *L. virosa* accessions in 1978 and the first resistant cultivar 'Dynamite' was released in 1996 (van der Arend et al., 1999).

5.2 Quality Improvement

Improvement in lettuce quality is often associated with the development of new types or subtypes. The creation of romaine, leaf, butterhead, stem, and Latin types was all accomplished by unknown breeders in the centuries before crop breeding became a science. Only the iceberg subtype of crisphead lettuce was developed by modern plant breeding. Whitaker of USDA crossed an 'Imperial' cultivar of the soft-headed Batavia subtype with an heirloom variety 'Brittle Ice' and released in 1941 the first true iceberg lettuce, 'Great Lakes', which has a crispy firm head (Ryder, 1999). 'Great Lakes' group of cultivars rapidly replaced the existing group of 'Imperial' strains in the U.S., resulting in a more dependable and better quality product for consumers. The most remarkable cultivar of this group was the downy mildew resistant 'Calmar', which remained the industry standard for 15 years since its release in 1960.

Thompson made a cross between PI 125130 (*L. virosa*) and a line derived from a complex *L. sativa-L. serriola* cross and treated the F_1 with colchicine. The resulting fertile amphidiploid was backcrossed to cultivated lettuce to bring it back to the diploid state, leading to the release of 'Vanguard' in 1958 (Thompson and Ryder, 1961). 'Vanguard' was the first cultivar derived from a cross with *L. virosa*, which gives it several high-quality attributes unmatched by earlier crisphead lettuces: dull green colour, relatively soft texture, flat ribs, less bitterness and deterioration, resistance to tipburn, and superior flavour. Along with 'Empire', 'Merit', and 'Climax', these cultivars and their derivatives have been dominant in the desert lettuce-producing areas in California and Arizona from the early 1960s to the present. Ryder crossed a Vanguard-like breeding line with 'Calmar' and released the cultivar 'Salinas' (Fig. 8; Ryder, 1979a). It combines the excellent quality traits of 'Vanguard' with the vigour, size, and wide adaptability of 'Calmar', and matures uniformly. Since its release in 1975, the Vanguard-Salinas group has dominated crisphead lettuce production worldwide ('Salinas' is known as 'Saladin' in Europe).

Fig. 8. 'Salinas', the most widely grown cultivar in the history of lettuce production.

5.3 Yield Increase

Yield of lettuce is the product of head weight and the number of heads harvested per unit area. Besides cultivar differences, yield is influenced by plant density, cultural practices, disease/insect control, and the percentage of plants harvested by the shipper. Shippers differ substantially in their harvesting standards: some cut heads with considerable variation in shape, size, and maturity, while others only take heads with optimum appearance and maturity. However, if the market is strong with high demand, shippers may be less selective at harvest time than in a poor market. Nevertheless, plant breeding contributes greatly to the yield advance: improved root system for more efficient use of water and nutrients, increased head size and weight, resistance to diseases, insects, tipburn, and bolting, and more uniform maturity that results in higher harvest percentage.

From 1953 to 1977, lettuce yield increased from 17,051 kg/ha to 24,609 kg/ha in California (Ryder and Whitaker, 1980). This was mainly the result of the cultivar changes from the Imperial and Great Lakes groups to larger and heavier 'Calmar' in the coastal districts and Vanguard and Empire groups in the desert areas. In 2005, lettuce yield in the U.S. was 37,910 kg/ha (Table 1), despite a large increase in the production of romaine, leaf, and butterhead lettuces that tend to have lighter weight. This represents a 122% yield improvement from 1953.

6 Current Goals of Breeding

There are often too many present and potential problems for a breeding program to solve. Breeders have to identify the major problems affecting the lettuce production and quality in their target area that can be solved by breeding, prioritize them, and work on as many projects as their budgets, labour, and facilities will allow. Goals of today's breeding programs can be divided into three general areas: resistance to diseases and insects, enhancement of production, and improvement in quality. These goals may overlap, e.g., better disease or insect resistance may enhance production and quality, but they are discussed separately for clarity.

6.1 Resistance to Diseases, Insects, and Other Disorders

Lettuce suffers from many diseases and insects. The symptoms (with colour photographs), causal organism, disease cycle and epidemiology, and control of lettuce diseases are fully described in the *Compendium of Lettuce Diseases* (Davis et al., 1997) and *A Colour Atlas of Diseases of Lettuce and Related Salad Crops, Observation, Biology and Control* (Blancard et al., 2006). The importance of a specific disease or insect varies between production regions. Some common diseases and pests of lettuce are discussed here.

6.1.1 Fungus Diseases

Downy mildew (Fig. 9) is caused by *Bremia lactucae* Regel and is favoured by relatively low temperatures and high humidity. Dominant genes in lettuce known as *Dm* genes match the avirulence genes in the pathogen in a gene-for-gene fashion and confer race-specific immunity to seedlings as well as adult plants. So far, more than 20 *Dm* or *R* (resistance factor) genes have been identified. Virulent isolates of the pathogen have historically appeared soon after deployment of the corresponding resistance genes in widely grown lettuce cultivars due to mutation or sexual recombination of the fungus, leading to rapid loss of resistance. Breeders are trying two strategies to make resistance more durable. One is to deploy different resistant genes, especially those from exotic sources, in different or the same cultivars. *L. saligna* has been identified as a non-host for lettuce downy mildew (Bonnier et al., 1992). The other strategy is to utilize, alone or in combination with *Dm* genes, field (or partial) resistance that is race-nonspecific and displays a longer latent period, fewer and smaller lesions on fewer affected leaves, and a slower rate of disease progress. Cultivars 'Iceberg' (syn. 'Batavia Blonde à Bord Rouge') and 'Grand Rapids' are known to possess this type of resistance (Grube and Ochoa, 2005).

Fig. 9. Sporulating downy mildew lesions on leaf of crisphead lettuce (Courtesy R. C. Grube).

Lettuce drop can be caused by two related species of *Sclerotinia*, *S. minor* Jagger (Fig. 10) or *S. sclerotiorum* (Lib.) de Bary. Despite extensive screening efforts, only partial resistance has been described among wild *Lactuca* species, PI accessions, breeding lines, and cultivars (Newton and Sequiera, 1972; Abawi et al., 1980; Whipps et al., 2002; Grube and Ryder, 2004). An association between growth habit and resistance to *S. sclerotiorum* has been observed for lettuce (Newton and Sequiera, 1972; Whipps et al., 2002), and Grube and Ryder (2004) reported that those with the highest levels of *S. minor* resistance were primitive, stem, or romaine-type lettuces with upright growth habits, many of which were early-bolting. This suggests that the resistance may be due to a morphological escape mechanism rather than physiological resistance. However, no association between horticultural type and resistance was detected in two greenhouse studies with *S. minor* (Abawi et al., 1980; Subbarao, 1998).

Fig. 10. Drop of crisphead lettuce caused by *Sclerotinia minor* (Courtesy R. C. Grube).

Verticillium wilt is an emerging disease first identified in a field near Watsonville, California in 1995 and has subsequently spread into the most important lettuce production area in the U.S., the Salinas Valley, resulting in crop loss as high as 80%. Two races have been identified in the causal organism, *Verticillium dahliae* Kleb., which can be transmitted through seeds (Vallad et al, 2005). Sources of resistance to race 1 have been found among different lettuce types except the iceberg type (Ryder et al., 2003), but no resistance to race 2 has been identified yet.

Fusarium wilt, another emerging disease of lettuce, was first observed in Japan in 1955, and has since been discovered in the United States (California in 1990, Arizona in 2001), Iran (1995), Taiwan (1998), and Italy (2001). Three races of the pathogen *Fusarium oxysporum* f. sp. *lactucae* have been identified in Japan and three race differentials have been established: 'Patriot' is susceptible to races 1, 2, and 3; 'Costa Rica No.4' is resistant to race 1 and susceptible to races 2 and 3; and 'Banchu Red Fire' is resistant to race 2 and susceptible to races 1 and 3 (Fujinaga et al., 2003). The isolates from the U.S. have been confirmed to be race 1 (McCreight et al., 2005). No lettuce germplasm has been found resistant to race 3, although different types of lettuce with resistance to races 1 and 2 have been identified (Fujinaga et al., 2001; Matheron et al., 2005; McCreight et al., 2005). Seed transmission of the pathogen has been reported by Garibaldi and coworkers (2004).

6.1.2 Virus Diseases

Lettuce mosaic, caused by lettuce mosaic potyvirus (LMV), is the most important viral disease of lettuce in the world. The virus is transmitted within fields primarily by the green peach aphid (*Myzus persicae*) in a nonpersistent manner. The disease is controlled mainly by use of virus-indexed seed in the U.S. (0 infected seed in 30,000 tested), and by a combination of seed indexing and resistant cultivars in Europe. The resistant cultivars may not be virus-free, but the rate of virus multiplication in the plant and symptom expression are drastically reduced. The virus may be seed-borne; usually 1-3% of seeds on an infected susceptible plant will contain the virus and the percentage may range up to 15% (Couch, 1955), while the seed transmission in a resistant plant is reduced by 90% (Ryder, 1973). Several resistance genes have been identified, including two recessive alleles at a single locus, $mo\text{-}1^g$ ($mo1^1$, from 'Gallega de Invierno', which has been introduced into resistant cultivars in Europe) and $mo\text{-}1^e$ ($mo1^2$, from PI 251245 from Egypt, which has been incorporated in resistant cultivars in the U.S.), a dominant gene *Mo2* that is very effective against certain isolates from Greece and the Middle East (Blancard et al., 2006), and two partially dominant alleles, *Mo3* (from a *L. virosa* accession PIVT1398 and effective against all the pathotypes identified; Maisonneuve et al., 1999; Maisonneuve, 2003) and *Mi'* (from 'Balady Aswan Green'; Ryder, 2002) that gives a mild reaction to LMV and may be combined with other resistance genes to increase the level of resistance. The $mo\text{-}1^g$ and $mo\text{-}1^e$ alleles were found to code for forms of the eukaryotic translation initiation factor eIF4E and have been cloned and sequenced (Nicaise et al., 2003). Several resistance-breaking isolates able to overcome $mo\text{-}1^g$ and $mo\text{-}1^e$ have been found around the world in recent years.

Big vein, another virus disease of lettuce, is caused by *Mirafiori lettuce big-vein virus* (MLBVV) and vectored by a soilborne fungus *Olpidium brassicae. Lettuce big-vein associated virus* (LBVaV) is also frequently isolated from symptomatic plants, but no causal relationship has been confirmed yet. Worldwide in its distribution, the disease is favoured by cool and wet soil conditions and thus may occur at high levels in production fields during spring in California's coastal regions and during winter in Arizona. Partial resistance has been found and used to breed big-vein resistant cultivars (Ryder and Robinson, 1995), although MLBVV infection was found in asymptomatic resistant plants (Hayes et al., 2006). Complete resistance has been identified in accessions of *L. virosa* (Bos and Huijberts, 1990; Ryder et al., 2003). One of those accessions (IVT280) remained symptomless and virus free in inoculated greenhouse tests (Hayes et al., 2006).

6.1.3 Bacterial Disease

Corky root of lettuce has been observed in major lettuce-producing areas of the world, including North America, Western Europe, Australia, and New Zealand. Infected plants develop yellow to brown lesions on the roots that later become longitudinal corky ridges. In severely infested fields in California and Florida, yield losses from reduced head size can reach 30-70% (Fig. 11). The pathogen most commonly isolated from diseased roots is the bacterium *Sphingomonas suberifaciens* (Yabuuchi et al., 1999), formerly *Rhizomonas suberifaciens*, although several other bacterial species have been isolated (van Bruggen, 1997). The use of cultivars resistant to *S. suberifaciens* has been and should continue to be an important management strategy for corky root. The resistance to corky root is conferred by a recessive allele (*cor*) at a single locus (Brown and Michelmore, 1988), which has been deployed in resistant lettuce cultivars.

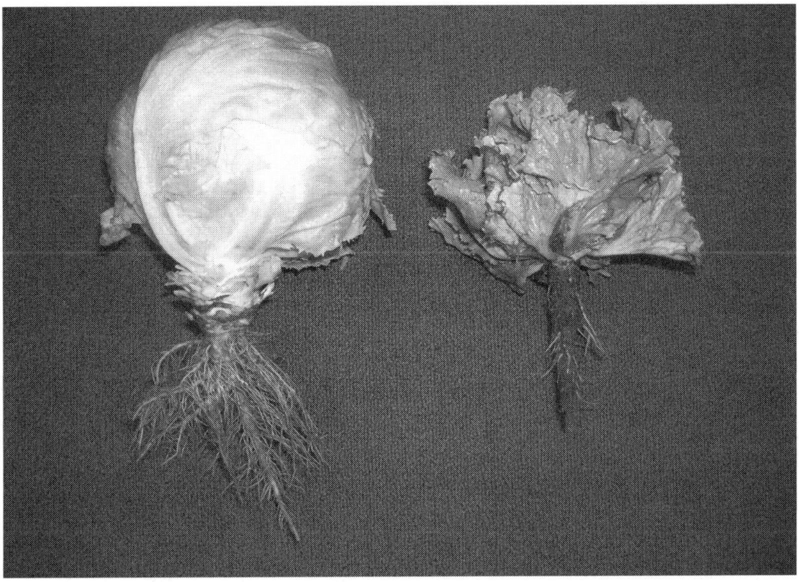

Fig. 11. Crisphead lettuce affected by corky root disease – susceptible cultivar (right) and resistant cultivar (left).

6.1.4 Insect Pests

The lettuce aphid, *Nasonovia ribisnigri*, is the most important aphid pest in Great Britain and Holland, and has become a serious problem in the U.S. in recent years. The insect tends to colonize inside the lettuce head, and therefore is protected from insecticides. Resistance was found in *L. virosa* and has been transferred to *L. sativa* by using *L. serriola* as a bridge species (Eenink et al., 1982a). The resistance is controlled by a single, incompletely dominant gene, *Nr* (Eenink et al., 1982b), which also confers partial resistance to the green peach aphid. The resistance is based on antibiosis, although aphids are not killed but forced to migrate to susceptible plants (van der Arend, 2003). Resistance due to antibiosis may put pressure on the insect to develop resistance-breaking biotypes. In anticipating this to happen, efforts to search for new sources of resistance have led to the discovery of a resistant *L. serriola* line, PI 491093. Data from preliminary experiments suggest that resistance in PI 491093 is genetically different from the *Nr*-mediated resistance (Grube et al., 2005a).

Infestation of leafminer (mainly *Liriomyza. trifolii, L. huidobrensis*, and *L. langei*) insects causes stand reductions at the seedling stage of lettuce and results in reduced crop quality and crop contamination at harvest. Mou and Liu (2003, 2004) screened more than 200 lettuce germplasm lines and identified sources of resistance to leafminers. Different mechanisms of resistance (antixenosis and antibiosis) exist in lettuce, and resistant genotypes from choice tests remained resistant to leafminers under no-choice conditions (Mou and Liu, 2004). The heritability estimates for leafminer sting density in lettuce were relatively high (Mou and Liu, 2004). Lower sting density in a *L. saligna* line is controlled by a single dominant gene (Mou, unpublished).

6.1.5 Other Disorders

Tipburn is the collapse and necrosis at leaf margins of rapidly expanding inner leaves caused by calcium deficiency. This physiological disorder often happens in warm weather and is especially a problem for heading lettuce types, the closed heads of which reduce transpiration and therefore the calcium supply from the water flow. Relatively resistant crisphead cultivars such as 'Tiber' and 'Salinas' are available, although the genetic base is unknown.

Rib blight is also called brown rib; it is a brown or black discoloration of the lower midrib and/or sections of vascular branches off the midrib. The cause is not known, but it seems related to the genetic background of the cultivars.

6.2 Production Traits

A major production related trait is yield. Yield of lettuce may be measured by the weight of the product per unit area of land, but more often by the number of harvested heads (or cartons) with adequate size and weight per acre or hectare. Plant spacing is usually standardized within each production area, so the maximum number of heads per unit area is fixed. The key to yield improvement is to increase the percentage of harvestable heads. Harvestable heads require certain size, weight, shape, and minimum disease, insect, and physical damage. Crisphead lettuce in the U.S. is expected to be large and heavy, but consumers in Europe desire smaller heads. Heads that are too large or out of shape are more difficult to pack in a carton. The most desirable shape for crisphead and butterhead lettuces is spherical. Romaine lettuce with excessive or brittle ribs or vase-shaped plants may be easily damaged by packing and should be avoided. Some leaf lettuce cultivars tend to form rudimentary heads, especially under cool weather or past optimum maturity, and this habit is considered undesirable. Premature seed stalk elongation (early bolting) results in the formation of a large core inside the head, increased bitterness, and poor head shape, and may prevent head formation in heading types.

Uniformity is a component of yield, as a larger portion of uniformly mature plants leads to higher harvest percentage. A more uniform field reduces the number of cuttings required and increases harvest efficiency and cost-effectiveness. This is more critical for crisphead type than for other types of lettuce. Breeders can select crisphead lines with a greater percentage of heads that are well filled at the same time and feel firm and not loose and puffy when squeezed by hand.

Lettuce is grown in a wide range of geographic areas of the world with vastly different daylengths, temperatures, humidities, soil types, and other climatic and environmental conditions. Varieties with wide adaptations, such as 'Great Lakes

659' and 'Salinas', are exceptions rather than the rule. Often cultivars are developed in specific regions to fit into different production areas and "slots" of the growing seasons. Breeding lines at later generations may be tested in as many locations and growing seasons as possible to determine their range of adaptability. Non-crisphead lettuce types tend to be more widely adapted, due to their shorter growing period and less stringent heading requirements.

6.3 Quality Traits

Quality traits for lettuce may be divided into horticultural quality and nutritional quality. Horticultural quality may include leaf colour, texture, and taste. A dark green (dull or bright) exterior is desirable for the crisphead type. The colour of butterhead lettuce ranges from yellow green to dark green, but most cultivars are light green. Romaine lettuces usually have dark green colour, although they tend to be light or yellow green in southern Europe and the eastern Mediterranean basin. The colour of leaf lettuces varies greatly from yellow to dark green. Red colour appears in varying intensity and distribution, although it is limited to the exterior leaves of the heading types as anthocyanin synthesis is light regulated.

Crisphead lettuce should have a crisp and crunchy texture, while the butterhead type is usually soft and oily. Romaine type is crisper than leaf lettuces, and Latin lettuces are intermediate between butterhead and romaine types. Although all lettuce types have a rather mild taste, they should be selected toward sweetness and against bitterness.

Although crisphead lettuce, a staple food, is only twenty-sixth in nutritional value (vitamins and minerals) among 39 common fruits and vegetables, it was ranked fourth in nutritional contribution in the U.S. behind tomato, orange, and potato, when consumption is considered (Stevens, 1974). The nutritional value of lettuce varies greatly with type. Leaf and romaine lettuces have much higher vitamin and mineral contents than the crisphead type (Table 2). As the synthesis and absorption of many nutrients are light dependent, the lower nutrient content of crisphead lettuce is largely due to the enclosure of its leaves in the head structure (Mou and Ryder, 2004). Mou (2005) found that the wild species (*L. serriola, L. saligna*, and *L. virosa*) and primitive forms of lettuce (90% moisture content) had higher β-carotene (provitamin A) and lutein contents than cultivated lettuce (95% moisture content) on fresh weight bases, mainly as a result of the higher moisture content of the cultivated ones which dilutes the concentration of these carotenoids. There was also significant variation in carotenoid content among and within different types of lettuce. β-Carotene and lutein concentrations were highly correlated, and the concentrations of both nutrients were highly correlated with chlorophyll concentration, suggesting that the carotenoid content may be improved indirectly by selecting for chlorophyll content or green colour.

In northern Europe, lettuce is grown in greenhouses under low light and temperature conditions in winter. These conditions may result in excessive accumulation of nitrate in the plant, which may be converted to carcinogenic nitrosamines after ingestion or to nitrites, responsible for a condition in the very young known as "blue baby". Therefore, low nitrate content may be a breeding objective in those areas for greenhouse lettuce.

Table 2. Nutrient content of different lettuce types, per 100 g of edible product.

Nutrient	Crisphead	Butterhead	Red leaf	Green leaf	Romaine	Stem
Water, g	95.6	95.6	95.6	95.1	94.6	
Energy, kcal	14	13	16	15	17	
Protein, g	0.90	1.35	1.33	1.36	1.23	0.60
Total lipid (fat), g	0.14	0.22	0.22	0.15	0.30	0.10
Carbohydrate, g	2.97	2.23	2.26	2.79	3.28	1.90
Dietary fiber, g	1.2	1.1	0.9	1.3	2.1	
Total sugar, g	1.76	0.94	0.48	0.78	1.19	
Calcium, mg	18	35	33	36	33	7
Iron, mg	0.41	1.24	1.20	0.86	0.97	2.00
Magnesium, mg	7	13	12	13	14	
Phosphorus, mg	20	33	28	29	30	31
Potassium, mg	141	238	187	194	247	
Sodium, mg	10	5	25	28	8	
Zinc, mg	0.15	0.20	0.20	0.18	0.23	
Copper, mg	0.025	0.016	0.028	0.029	0.048	
Manganese, mg	0.125	0.179	0.203	0.250	0.155	
Selenium, μg	0.1	0.6	1.5	0.6	0.4	
Vitamin A, IU*	502	3312	7492	7405	5807	33
Vitamin B-6, mg	0.042	0.082	0.100	0.090	0.074	
Vitamin C, mg	2.8	3.7	3.7	18.0	24.0	1.0
Vitamin E, mg	0.18	0.18	0.15	0.29	0.13	
γ-Tocopherol, mg	0.09	0.27	0.24	0.37	0.36	
Vitamin K, μg	24.1	102.3	140.3	173.6	102.5	
Folate, μg	29	73	36	38	136	
Lutein+zeaxanthin, μg	277	1223	1724	1730	2312	
Niacin, mg	0.123	0.357	0.321	0.375	0.313	0.500
Pantothenic acid, mg	0.091	0.150	0.144	0.134	0.142	
Riboflavin, mg	0.025	0.062	0.077	0.080	0.067	0.020
Thiamin, mg	0.041	0.057	0.064	0.070	0.072	0.030

*IU, International Unit. Source: USDA, Agricultural Research Service (USDA-ARS, 2005) except for stem lettuce (Lu, 2000).

7 Breeding Methods and Techniques

Hybridization is a technique most often used in a crop breeding program. The inbred nature of lettuce dictates the three principal breeding strategies used to improve the crop: pedigree method, backcross, and single-plant or mass selection. Attempts to produce F_1 hybrids commercially have not been successful. Lettuce pollen is heavy and sticky, and therefore is not easily transferred by wind. There is a dearth of pollinating insects for lettuce plants. Hand pollination cannot produce large numbers of seeds efficiently.

7.1 Crossing Technique

To ensure the synchronization of flowering time, planting of parents that differ in flowering time should be staggered. Lettuce is a long-day plant, so lighting may be added in greenhouses to increase photoperiod and promote bolting, especially in wintertime when daylength is short. Each lettuce flower consists of 10 to 20 florets that make up the capitulum (composite flower, Fig. 12). On sunny, warm days, flowers may open in the early morning, while on cloudy, cool days, they may not open until late morning. Some wild species may flower earlier than cultivated lettuce during the day; the early flowering parents can be kept in a cool dark room or growth chamber to "wait" for the late flowering ones. The floret opens its five (usually linked) petals and exposes the five anthers that fuse to form a tube through which the pistil emerges. The anthers shed pollen inside the stamen tube, and the elongating style with stigma sweeps the pollen along as it emerges from the top of the tube. The stigmatic lobes then begin to split, gradually form a "V" shape, and finally curl backwards, signalling the end of pollen reception.

Fig. 12. Lettuce flowers.

Several emasculation methods have been proposed over time, but most lettuce breeders use a method devised by Oliver (1910), or a modification of it. The open flower can be washed with a stream of water using a wash bottle several times during the emergence of pistils to remove adhering pollen. Excess water may be removed by a stream of air or blown by mouth. The flower is left to air dry for a short time, and pollination is made by taking an open flower from the male parent and pressing it onto the emasculated flower before its stigmatic lobes begin to curl back. If it is desirable to remove pollen from a large number of flowers, a fine spray nozzle may be placed over the plants and the entire panicle can be washed intermittently or continuously during the pistil emergence period (Ryder and Johnson, 1974).

To ensure the seeds from crossed flowers are hybrids and not selfed ones, the male plants should carry dominant marker genes, preferably expressed in the seedling stage. The hybrid plants will show the dominant trait while the selfed ones will show the recessive. In lettuce, anthocyanin, shade of green, leaf shape, and resistance to downy mildew may be used as markers. Both parents should be grown for comparison. Later stage traits such as seed colour (black is dominant over white colour) are of less use, as plants have to be kept to maturity before hybrid identification can be made.

7.2 Pedigree Method

Pedigree breeding is usually used to combine favourable characters from two or more parents and to eliminate unfavourable traits from the parents. Records are kept for each generation of the cross and subsequent selfing, so any selected plant can be traced back to the original cross. The F_2 population should be as large as possible, as it offers the best chance for trait recombination and selection. The F_2 plants may be selected in the field or in the greenhouse (e.g., large number of seedlings may be inoculated with a pathogen and screened for disease resistance). Field planting should use spacing similar to commercial fields. Field evaluated traits are selected according to market demands and the breeder's personal preference. Colour, size, shape, disease resistance, and other traits are all gauged by eye as breeders walk through the field. For crisphead lettuce, for example, we select plants with optimum size, spherical and symmetric head shape, dark green exterior leaves that overlap nicely, flat ribs, and disease resistance. The head may be squeezed by hand near the top and around all sides to confirm shape and firmness. For crisphead lettuce to be grown in the desert districts of California and Arizona, plants with upright wrapper leaves (leaves surrounding the head) are selected to protect the heads from sunburn. Tentatively accepted plants may be marked by placing coloured stakes next to them. Then the head is cut twice at right angles across the top to a depth of about 75 mm (care should be taken to avoid damage to the growing point), and the quarters are pulled apart and inspected for proper interior colour, regular overlapping leaf arrangement, and absence of tipburn and other blemishes. Most leaves are then stripped off, and the plants are dug up, transported into the greenhouse, and transplanted into pots for seed production or further crosses.

With each generation of selection and inbreeding, differences between lines and similarities within lines will become more pronounced. In F_3 and later generations, uniformity traits become more important and lines are selected for the highest proportions of firm, well shaped heads with acceptable appearance and resistances to tipburn and diseases. Individual plants are selected within the uniform rows until F_6 or F_7 generation, when seeds may be massed or plants may be mass selected for field trials.

The purposes of the field trials are to measure yield, evaluate horticultural traits, assess disease resistance, and determine range of adaptation. Field trials may be conducted on the breeding station and, preferably, on farm, with replications and commercial cultivars as control. The yield may be assessed by sampling plants for head weight measurement and/or by calculating harvest percentage. To derive harvest percentage, plants are divided into harvestable and nonharvestable categories (Ryder, 1986). The nonharvestable category includes (1) diseased plants, (2) off-type plants, (3) late germinating plants that will not form heads, and (4) physically damaged plants. Harvestable plants include (1) overmature plants that will be discarded, (2) mature, well shaped plants, (3) undermature plants needing several more days of growth, and (4) poorly shaped plants that are relatively undesirable. The percentage of plants in the harvestable category that are actually harvested by a commercial harvesting crew is the harvest percentage.

7.3 Backcross

The backcross is a useful method when the objective is to introduce a specific character controlled by one gene, such as disease resistance, from one (donor) parent to the other (recurrent) parent. The process starts with a cross and is followed by backcrossing to the recurrent parent for several generations. With each backcross the proportion of genes from the donor parent is reduced by half. After four generations of backcrossing, 97% of the genome is from the recurrent parent, theoretically. Plants selected for backcross at each generation should possess the gene being transferred and some resemblance to the recurrent parent, which may reduce the number of generations required to reach a given state of similarity to the recurrent parent. Alternatively, molecular markers linked or unlinked to the desired gene may be used to select plants with a minimum proportion of genomic material from the donor parent (Michelmore, 1995). Crisphead cultivars 'Vanguard 75' and 'Salinas 88' were both developed using backcross programs to introduce the gene for LMV resistance, $mo\text{-}1^e$, into 'Vanguard' and 'Salinas', respectively (Ryder, 1979b, 1991).

7.4 Selection

When a new cultivar is released at F_6 or F_7 generation, it may still contain some residual variability. Mutational changes may also accumulate in an existing cultivar. Selection may be practiced within the cultivar to produce variants that are considered new even though they are similar to the original cultivar. This practice is often used by seed companies to develop new cultivars from cultivars released by public breeding programs. Some growers also carry out "on-farm breeding" to select from a cultivar plants that are best adapted to their local environment or growing conditions.

8 Integration of New Biotechnologies in Breeding Programs

Biotechnology is developing rapidly and provides great opportunities and potential for lettuce improvement. It creates additional variability and gives breeders more tools to manipulate the genetics and genome of lettuce to fit our needs. Biotechnology has already made significant impact on today's lettuce breeding programs and will certainly play an even more important role in the future. These technologies may be divided into four general categories: tissue culture techniques, molecular markers, the introduction of transgenes, and genomic approach.

8.1 Tissue Culture

Doerschug and Miller (1967) first reported the successful *in vitro* regeneration of lettuce shoots from various tissues. Since then, tissue culture techniques have become routine in lettuce. Embryo rescue is an important application of this technique to increase the access to wild lettuce germplasm. It has been used to transfer resistance to downy mildew, LMV, and beet western yellows virus from *L. virosa* and *L. saligna* into cultivated lettuce (Maisonneuve et al., 1995, 1999; Maisonneuve, 2003). Protoplast fusion has also been used to gain access to exotic, sexually incompatible germplasm. Mazier et al. (1999) obtained hybrids between *Lactuca sativa* and two wild *Lactuca* species, *L. tatarica* and *L. perennis* using protoplast technique. However, the regenerated plants had very low fertility even after several backcross generations, which limited the production of useful plant material.

Propagation from axillary and apical buds of lettuce plants can also be routinely carried out. This procedure has been used to regenerate plants from stored heads selected for resistance to a postharvest disorder – russet spot (Koevary et al., 1978; Bloksberg and Saltveit, 1986) and to rescue plants that were selected in the field but subsequently became infected with *Botrytis* spp. after being transferred into the greenhouse for seed production (Pink and Carter, 1987). If propagation through tissue culture can be done efficiently on a large scale, it may be used to produce F_1 hybrid plants and fix the hybrid vigour. Somaclonal variation provides another source of genetic variability, although it has not been fully exploited in lettuce.

8.2 Molecular Markers

Molecular markers are highly reliable selection tools and may greatly enhance the efficiency of lettuce breeding. They are not influenced by environmental conditions and may be scored at early stages of development (e.g. seedling stage) in the greenhouse or the laboratory. Plants with appropriate markers do not need to be inoculated with a pathogen for the selection of disease resistance. Molecular markers will be a routine component of lettuce breeding programs in the near future as more useful markers are identified and the technology becomes easier to use and less expensive.

Different types of molecular markers have been developed over the years. Isozymes are differently charged protein molecules that can be separated using electrophoresis (usually starch gel) and can provide genetic information as codominant markers (Markert and Moller, 1959). However, the paucity of isozyme loci and the fact that they are subject to post-translational modifications often restrict their utility (Staub et al., 1982). Another codominant marker, restriction fragment length polymorphism (RFLP) is produced by digesting genomic DNA with restriction enzymes and thus yielding variable-size DNA fragments that are separated by electrophoresis, transferred to nitrocellulose or nylon membranes, and hybridized to radioactively (usually P^{32}) or nonradioactively labelled probe DNA for visualization on photographic film (Landry et al., 1987). There are disadvantages in the use of RFLPs, including the radioactive materials often required for detection, the large amount of plant tissue needed, and the laborious and time-consuming procedures. Subsequently, several DNA marker systems based on polymerase chain reaction (PCR) became available. Random amplified polymorphic DNA (RAPD), a dominant marker, is generated by PCR amplification of random genomic DNA segments that are flanked by the annealing sites of the same but opposite-oriented primers (usually 10 nucleotides long) of arbitrary sequence (Williams et al., 1990). Utility of a desired RAPD marker can be increased by sequencing its termini and, based on the sequence information, designing longer primers (e.g. 24 nucleotides) for specific amplification of markers (Paran and Michelmore, 1993). Although such sequence characterized amplified regions (SCAR) are usually also dominant markers, SCARs are more reproducible than RAPDs. Production of amplified fragment length polymorphisms (AFLPs) is based on selective amplification of restriction enzyme-digested DNA fragments (Vos et al., 1995). The ability of AFLP technology to generate many markers with minimum primer testing, and the high resolution of the bands on denaturing polyacrylamide gels make this dominant marker attractive. The increasingly available sequence information makes the use of sequence-based marker systems possible. These include two types of codominant markers, microsatellites and single nucleotide polymorphisms (SNPs). Microsatellites, also called simple sequence repeats (SSRs), are DNA sequences with repeat lengths of a few base pairs (Witsenboer et al., 1997; van de Wiel et al., 1999). SNPs are individual sites of nucleotide base variation in a genome (Moreno-Vázquez et al., 2003).

There has been considerable effort to develop molecular markers for marker-assisted selection (MAS) in lettuce, especially markers closely linked to disease resistance genes. Most markers produced so far are for disease resistance controlled by a single gene. Moreno-Vázquez et al. (2003) developed SCAR markers that are located as close as 2.3 centimorgans (cM) from the *cor* gene that confers resistance to corky root disease (Fig.13). SCAR and RAPD markers flanking the LMV resistance gene *mo-1* at down to 3.2 cM were identified by Irwin et al. (1999). Subsequently, the cloning and sequencing of the *mo-1* gene (Nicaise et al., 2003) made possible the development of SNP markers, the "perfect markers" based directly on the sequence variation of *mo-1* alleles (Michelmore and Truco, 2005).

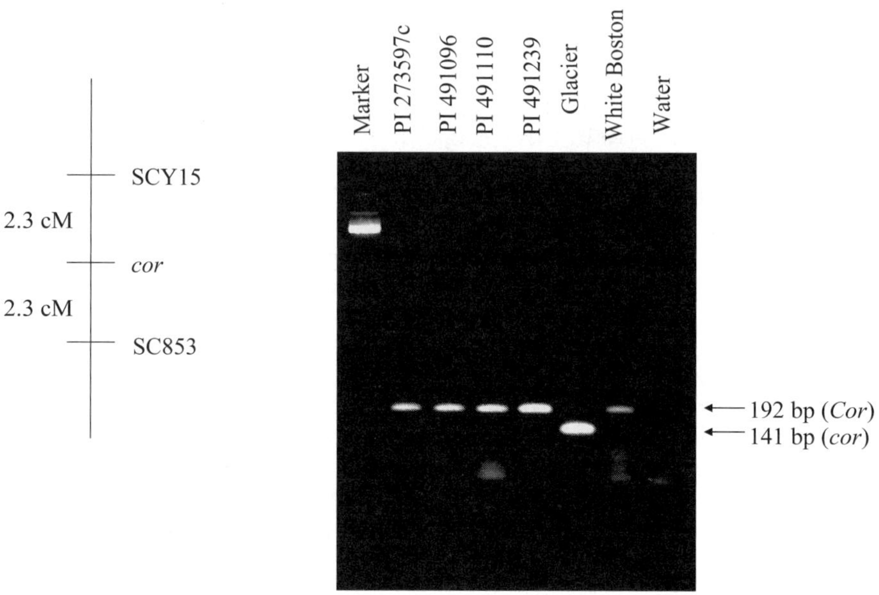

Fig. 13. Analysis of lettuce germplasm for resistance to corky root using SC853 SCAR marker. A 192-basepair (bp) DNA fragment was associated with the susceptible allele *Cor*, and a 141-bp fragment was associated with the resistant allele *cor* (Mou and Bull, 2004).

Much interest has been generated in marker-assisted identification of quantitative trait loci (QTL) that appear to follow complex, polygenic inheritance patterns with each gene having a small effect on the trait, such as field resistance to downy mildew and resistance to big-vein in lettuce. Jackson (1998) reported the association of taproot characters and other morphological QTL with molecular markers in a cross between 'Salinas' and a *L. serriola* line, accounting for 37-83% of the total variation. Using AFLP analysis, Jeuken and Lindhout (2002) found that three QTLs explained 51% of the quantitative resistance against a downy mildew race NL14 in a resistant *L. saligna* x susceptible *L. sativa* cross. Molecular markers also helped to identify a single QTL that described 23-25% of the total phenotypic variation for high temperature germination traits in lettuce (Argyris et al., 2005). These studies may aid in unravelling the complexity of QTL and eventually breeding of these important traits by MAS.

Molecular markers will allow pyramiding different disease resistance genes (e.g. different *Dm* genes against downy mildew) in a cultivar that would not otherwise be possible. The use of DNA markers will also permit combinations of resistances against multiple diseases and insects.

Different marker systems have been used to characterize lettuce germplasm collections. Variation of banding patterns at RFLP loci revealed relationships among 67 accessions of *L. sativa* and five related wild species (Kesseli et al., 1991). Witsenboer et al. (1997) analyzed microsatellite loci of 58 accessions, including both cultivars and related species. AFLPs and microsatellites were employed to characterize the entire lettuce collection of the Centre for Genetic Resources, The Netherlands (CGN), 2,323 accessions of lettuce and related wild species (van Hintum, 2003).

The identity and purity of seeds are already tested commercially using molecular marker technology. It allows DNA fingerprinting and the differentiation among lettuce cultivars and wild germplasm accessions, thus may be used for Plant Variety Protection (PVP). Kesseli and Michelmore (1986) separated 17 cultivars and 15 PI accessions of lettuce and wild species, using 42 isozyme systems. Nine nearly identical butterhead cultivars were differentiated by using 13 RAPD primers (Waycott and Fort, 1994). Using a new marker system, target region amplification polymorphism (TRAP), Hu et al. (2005) were able to discriminate 53 lettuce cultivars and six accessions of *L. serriola* and *L. saligna* with 10 fixed primers and four arbitrary primers.

Molecular markers may accelerate the introgression of desired genes from wild species. Fingerprinting of wild germplasm allows selection against and therefore rapid removal of unwanted chromosomal regions originating from the wild donor parent, which should speed up a backcross program. Identification of recombination between markers flanking a desirable gene can help break the association with undesirable traits and remove so called "linkage drag". The resistance to lettuce aphid from *L. virosa* was closely linked to recessive traits of yellow leaves and a greatly reduced head. Despite many rounds of backcrossing, the product was still of extremely poor quality. It was decided to use DNA markers flanking the introgression to select for recombination in the vicinity of the gene. This approach eventually led to the selection of an individual bearing recombinant events very close to each side of the gene, thereby removing the linkage drag (Peleman and van der Voort, 2003).

8.3 Plant Transformation

Plant transformation is potentially a powerful tool for lettuce breeding as it can break the species barrier and bring in favourable traits from other gene pools that are not easily accessible through traditional breeding techniques. Transformation of lettuce has been accomplished in two ways: by *Agrobacterium* and by electroporation.

Transformation using *Agrobacterium tumefaciens* has become routine in lettuce. Lettuce transformation was first reported by Michelmore et al. (1987) who transferred a NPT II gene for kanamycin resistance into plant cells by using Ti plasmids of *A. tumefaciens* and found that a butterhead cultivar 'Cobham Green' had much higher transformation rate than several crisphead cultivars. Later, an iceberg lettuce cultivar 'South Bay' was successfully transformed with a β-glucuronidase (GUS) reporter gene (Torres et al., 1993). Curtis et al. (1994) were able to transform 13 cultivars of crisphead, butterhead, leaf, and romaine lettuces, although the number of shoots produced differed among the genotypes.

Electroporation is a method of direct transfer of foreign DNA into protoplasts with the help of electrical pulses. Chupeau et al. (1989) transformed lettuce protoplasts with the kanamycin resistance gene using electroporation and successfully regenerated transgenic plants.

Since lettuce suffers from many diseases in its growth and development, disease resistance has been a major focus of transgenic research in lettuce. Several genes encoding enzymes that hydrolyze fungal cell walls, such as chitinases and endoglucanases, enzymes for phytoalexin synthesis, and other anti-fungal proteins have been cloned and introduced into different crops. A β-1,3-glucanase gene from *Arthrobacter* spp. was transformed into two lettuce cultivars and different levels of resistance to downy mildew were observed (Dede, 1998). Dias et al. (2006) introduced an oxalate decarboxylase gene from edible mushroom into lettuce and obtained two lines highly resistant to *Sclerotinia sclerotiorum* in a detached-leaf assay, although they have not been evaluated in the field.

Cloned components of pathogens can also be used to induce resistance response. The method most often used is the expression of a virus coat protein gene, which has provided high levels of resistance against many different viruses. The coat protein gene of the lettuce mosaic virus was transferred into lettuce and different levels of protection against the virus accumulation in plants were observed (Dinant et al., 1997). In another study, the transformation of 'Cobham Green' lettuce with LMV coat protein gene sequences generated some resistance in the R_2 generation but failed to express the resistance in later generations (Gilbertson, 1998; Michelmore, 1998). Transgenic lettuce plants expressing the nucleocapsid protein gene of the lettuce isolate of tomato spotted wilt virus (TSWV) were protected against the isolate to varying degrees in different lines (Pang et al., 1996).

Another transgenic research area is the development of herbicide resistance in lettuce. Torres et al. (1999) reported the successful transformation of 'South Bay' lettuce with the gene for glyphosate (Round-Up) herbicide resistance. Stable expression of the bialaphos herbicide resistance (*bar*) gene in a lettuce cultivar 'Evola' was observed up to R_2 generation (Mohapatra et al., 1999).

Research has been conducted to enhance the tolerance of lettuce to environmental stresses, such as high/low temperature, drought, and salinity conditions. Transformation of 'Grand Rapids' lettuce with the gene coding for an enzyme in the proline biosynthesis pathway resulted in plants with resistance to high (37°C) and low (0°C) temperatures (Pileggi et al., 2001). Over-expression of the ABF3 gene from *Arabidopsis* improved tolerance to cold (-4°C) and drought stresses in transgenic lettuce (Vanjildorj et al., 2005). Park et al. (2005) introduced a late embryogenesis abundant protein gene from rapeseed into lettuce and significantly improved the growth characteristics of transgenic plants in drought and salt stress tests.

There is considerable interest in the improvement of horticultural and nutritional quality of lettuce through genetic engineering. 'Evola' lettuce transformed with an ipt gene under control of the senescence-specific SAG12 promoter from *Arabidopsis* significantly delayed development and leaf senescence in mature heads, raising the hope of increased shelf life (McCabe et al., 2001). Lettuce plants expressing a nitrate reductase gene from tobacco had lower nitrate accumulation in the leaves (Curtis et al., 1999). Transformation of lettuce with the gene for the iron-binding protein ferritin from soybean resulted in plants with iron levels ranging from 1.2 to 1.7 times that of the control plants (Goto et al., 2000). 'Salinas 88' lettuce transformed with a mutant gene for metal-binding protein metallothionein from mouse had significantly higher zinc content than control plants (Zuo et al., 2002).

Transgenic lettuce has not been commercialized despite demonstrated benefits and considerable research being conducted on the crop. Some of the limitations are technical. Transgenes tend to be switched off in later generations following their introduction into lettuce (Michelmore, 1997). Transgene silencing might be due to DNA methylation that is mostly restricted to transposons and high copy repeats (Martienssen, 2004), chromatin restructuring, or other unknown mechanism. McCabe et al. (1999) suggested using promoters from plant genes and screening for plants with single, high expressing inserts for stable transgene expression in lettuce. Other limiting factors are commercial. Consumer concern over genetically modified (GM) food has resulted in the reluctance of processors and marketers to accept the biotech products already developed. The limited acreage of lettuce makes it more difficult to recover the costs of research, development, and segregation of GM and non-GM lettuce commodities. Current practices in patenting and intellectual property protection have added barriers to the use of biotechnology for the creation and commercialization of new lettuce varieties. Meanwhile, science-based research is needed to assess the impact of genetic engineering on biodiversity, environmental safety, and human health. The future of transgenic lettuce may largely depend on the progress in these areas.

8.4 Genomic Approach

With the completion of the sequencing projects for human, *Arabidopsis thaliana*, rice, and other model organisms, we have entered the genomic era. Genomic research may impact lettuce breeding by providing a better understanding of the structural organization and functional properties of genes and genomes, allowing gene-based selection through identification of molecular markers and high-throughput genotyping technologies, and increasing the genepool available by supplying detailed characterization of germplasm, which includes new sources of desired traits, or transgenes.

As part of a large comparative genomics project involving lettuce, sunflower, and *Arabidopsis*, over 68,000 lettuce expressed sequence tags (ESTs) and about 19,000 distinct DNA sequences have been obtained from 'Salinas' and a *L. serriola* line UC96US23, representing approximately a third to a half of all the genes in lettuce (Michelmore, 2005). These sequences have been searched for SNPs and motifs characteristic of candidate genes for disease resistance and horticultural traits, and 140 potential resistance gene candidates have been identified. Work is under way to generate 160,000 additional EST sequences from five species of *Lactuca* (*L. sativa*, *L. serriola*, *L. saligna*, *L. virosa*, and *L. perennis*). Based on the sequence information obtained, an oligonucleotide chip is being developed to allow massive parallel SNP discovery, genotyping, and expression analysis. The EST sequence data can be searched in an online database http://cgpdb.ucdavis.edu/ (April 19, 2006).

Based on sequences of cloned disease resistance genes from a variety of plant species, degenerate oligonucleotide primers were designed to identify resistance gene candidates (RGCs) in lettuce by PCR (Shen et al., 1998; Linden et al., 2004). Over 70 RGCs have been sequenced, of which 40 have been mapped. Comparison of the EST sequences to the RGC sequences suggests that there are at least 21 RGC families in lettuce. The functions of the RGCs are being studied by using gene silencing through RNA interference (Michelmore, 2005).

Information from six intra- and inter-specific mapping populations of lettuce have been integrated to develop a consensus genetic map that comprises over 2,700 molecular and phenotypic markers (isozyme, RFLP, RAPD, AFLP, SSR, disease resistance, and morphological markers) in nine linkage groups. Recombinant inbred lines derived from the reference mapping population 'Salinas' x UC96US23 (*L. serriola*) are being analyzed to identify QTLs controlling morphological and developmental traits in lettuce (Michelmore, 2005). A gene for downy mildew resistance, *Dm3*, was cloned by a combination of map-based cloning, mutation analysis and RGC approaches (Shen et al., 2002). Efforts to clone other *Dm* genes are in progress. A searchable database, Compositdb (http://compositdb.ucdavis.edu/, April 19, 2006), supplies genetic and DNA sequence information as well as gel images for markers and descriptions of RFLP probes and primers for PCR-based markers. A map viewer (http://cgpdb.ucdavis.edu/database/genome_viewer/viewer/, April 19, 2006) has been developed to provide easier access to marker and phenotypic data and to allow comparison among different genetic maps.

9 Seed Production

For a public breeding program, lettuce seeds are increased in the later stage of the cultivar development to generate sufficient amount of seeds for large-scale trials and for release to seed companies. This is usually done in a greenhouse to protect plants from wind, rain, insects, and diseases. For seed companies, seed production also starts in a greenhouse to produce breeder's seed. Breeder's seed is used to produce foundation seed that may be grown outdoors in progeny rows and checked for uniformity. Foundation seed is then used to produce stock seed in the field for sale to growers for market production.

Lettuce seeds are produced in areas that provide isolation and a dry climate to minimize the impact of pathogens, minimal wind movement to prevent loss of seed, and sufficient warmth for seed maturation. Two major locations of commercial lettuce seed production are the San Joaquin Valley (between Fresno and Bakersfield, Fig. 14) of California and the Griffith-Hay District in New South Wales, Australia. Seeds are also produced in Chile, China, southern France, Italy, and several other locations in the USA. California seed is planted in early May and harvested in September, while Australian seed is sown in late November and harvested in May, thus allowing two seed crops in one year if needed (Ryder, 1999).

Fig. 14. A lettuce seed production field near Fresno, California, U.S.A. (Courtesy Y. Peng).

In California, a lettuce seed crop is planted either as a single row on a 76 cm (30 inch) wide raised bed or as double rows on a 152 cm (60 inch) wide bed. The sowing and growing of lettuce for seed is similar to lettuce for market, and adequate supplies of nitrogen and moisture are essential. Hawthorn and Pollard (1956) found an increase of seed yield associated with increasing N up to about 90 kg ha^{-1} but not beyond that. Izzeldin et al. (1980) showed that best moisture levels for high seed yield led to smaller seed with lower vigour. An intermediate water deficit produced the best combination of yield and quality. Different cultivars also differ in seed yield. The fields are inspected and rogued regularly to remove off-types and plants with diseases, particularly lettuce mosaic.

Non-heading cultivars are allowed to complete their vegetative phase of growth and produce a seed stalk. Crisphead lettuce, however, forms a dense head, and the seed stalk may not be able to penetrate the top of the head and may be forced to grow in a circular manner inside the head or break. To prevent this circumstance, the top of the head may be cut open or sliced off to allow the stalk to elongate through. Alternatively, sharp downward pressure may be exerted on the head to break the leaf ribs and allow the leaves surrounding the growing point to be pulled off. This is done by hand in a continuous motion, before the head becomes densely packed. These procedures may need to be done more than once to ensure most seed stalks will come through easily, particularly for those bolting-resistant cultivars. Some lettuce fields are treated with gibberellic acid to promote seed-stalk formation before heading occurs. Application of gibberellin at the rosette stage increased the percentage of bolting and seed yield, as compared with manual deheading or no treatment (Harrington, 1960; Globerson and Ventura, 1973). Gray et al. (1986) found that gibberellin also increased uniformity of both bolting time and seed production over cutting the tops of mature heads. Butterhead lettuce has a shorter growing period and forms a less compact head than the crisphead type, thus its seed stalk usually can go through the head without much difficulty.

Lettuce plants flower over a period of 50 - 70 days and seeds mature 12 - 17 days after flowering. The flowering and seed maturation usually occur in three peak periods. Soffer and Smith (1974a) found that seeds harvested from the first two peaks were heavier than those from the third peak, and over 90% of the seed yield came from the first 35 days of a 70-day flowering period. Seed weight was the best of several predictors of seed vigour that was defined by radicle length 3 days after germination (Smith et al., 1973; Soffer and Smith, 1974b).

Lettuce seeds may be mechanically harvested in two ways. One is the "shake method", in which seeds are usually harvested at two to three intervals by bending the seed head over and shaking the seeds into a container. The second, "cut method", is to combine heads at a given time, curing seeds on canvas sheets and then harvesting the seeds from seed heads. The "shake" method has the advantage of maximizing seed yield and allowing the separation of seed lots based on time of maturity, while the "cut" method is an once-over and therefore more rapid harvest operation.

Harvested seeds are cleaned, sized on an air-screen device to remove light seeds, and stored in bins in a relatively cool, dry area. Seeds stored in ambient conditions will remain viable for 1 - 3 years, depending upon the temperature and relative humidity. The temperature for medium-term to long-term storage should not exceed 10°C. The optimum storage for the long term (up to 25 years or longer) is at -20°C and a low relative humidity or at -196°C (in liquid nitrogen).

Lettuce seeds are narrow and angular, and most raw seeds are coated with various materials to make nearly spherical seed pellets, which allows planting at uniform spacing by precision planters and subsequently an easier thinning operation. The coating material may also contain nutrients and pesticides to promote emergence and early growth. Seed priming is used to enhance the germination of coated seeds in areas under high temperature conditions, such as western deserts of the United States. The procedure involves soaking seeds in water, salt solutions, or polyethylene glycol to initiate the germination process, which is halted before the radicle penetrates through the pericarp. The seeds are then dried and stored for later use. Seeds may be tested for germination percentage and seed-borne pathogens like lettuce mosaic virus, and then packaged in containers for sale on the market.

References

Abawi, G.S., Robinson, R.W., Cobb, A.C., and Shail, J.W. 1980. Reaction of lettuce germplasm to artificial inoculation with *Sclerotinia minor* under greenhouse conditions. Plant Dis. 64: 668-671.

Argyris, J., Truco, M.J., Ochoa, O., Knapp, S.J., Still, D.W., Lenssen, G.M., Schut, J.W., Michelmore, R.W., and Bradford, K.J. 2005. Quantitative trait loci associated with seed and seedling traits in *Lactuca*. Theor. Appl. Genet. 111: 1365-1376.

Blancard, D., Lot, H., and Maisonneuve, B. 2006. A Colour Atlas of Diseases of Lettuce and Related Salad Crops, Observation, Biology and Control, Academic Press, Burlington, MA, USA.

Bloksberg, L.N., and Saltveit, M.E. 1986. Regeneration of plants from axillary buds of harvested and stored heads of field-grown iceberg lettuce. HortScience 21: 1201-1203.

Bonnier, F.J.M., Reinink, K., and Groenwold, R. 1992. New sources of major gene resistance in *Lactuca* to *Bremia lactucae*. Euphytica 61: 203-211.

Bos, L., and Huijberts, N. 1990. Screening for resistance to big vein disease of lettuce. Crop Prot. 9: 446-452.

Boukema, I.W., Hazekamp, Th., and van Hintum, Th. J.L. 1990. CGN collection reviews: the CGN lettuce collection, Centre for Genetic Resources, the Netherlands (CGN), Wageningen, pp. 27.

Brandes, E.W. 1918. Anthracnose of lettuce caused by *Marssonina panattoniana*. J. Agr. Res. 13: 261-280.

Brown, P.R., and Michelmore, R.W. 1988. The genetics of corky root resistance in lettuce. Phytopathol. 78: 1145-1150.

Chupeau, M.C., Bellini, C., Guerche, P., Maisonneuve, B., Vastra, G., and Chupeau, Y. 1989. Transgenic plants of lettuce (*Lactuca sativa*) obtained through electroporation of protoplasts. Bio/Technol. 7: 503-508.

Couch, H.B. 1955. Studies on seed transmission of lettuce mosaic virus. Phytopathol. 45: 63-70.

Curtis, I.S., Power, J.B., Blackhall, N.W., de Laat, A.M.M., and Davey, M.R. 1994. Genotype-independent transformation of lettuce using *Agrobacterium tumefaciens*. J. Exp. Bot. 45: 1441-1449.

Curtis, I.S., Power, J.B., de Laat, A.M.M., Caboche, M., and Davey, M.R. 1999. Expression of a chimeric nitrate reductase gene in transgenic lettuce reduces nitrate in leaves. Plant Cell Rept. 18: 889-896.

Davis, R.M., Subbarao, K.V., Raid, R.N., and Kurtz, E.A., ed. 1997. *Compendium of Lettuce Diseases*, APS Press, St. Paul, MN, USA.

de Vries, I.M. 1990. Crossing experiments of lettuce cultivars and species (*Lactuca* sect. *Lactuca*, Compositae). Pl. Syst. Evol. 171: 233-248.

de Vries, I.M. 1997. Origin and domestication of *Lactuca sativa* L. Genet. Resour. Crop Evol. 44: 165-174.

Dede, Y. 1998. Development of the downy mildew pathogen *Bremia lactucae* on transgenic lettuce expressing a bacterial β-1,3-glucanase. Turkish J. of Agric. and Forestry 22: 313-321.

Dias, B.B.A., Cunha, W.G., Morais, L.S., Vianna, G.R., Rech, E.L., de Capdeville, G., and Aragâo, F.J.L. 2006. Expression of an oxalate decarboxylase gene from *Flammulina* sp. in transgenic lettuce (*Lactuca sativa*) plants and resistance to *Sclerotinia sclerotiorum*. Plant Pathol. 55: 187-193.

Dickson, M.H. 1963. Resistance to corky root rot in head lettuce. Proc. Amer. Soc. Hort. Sci. 82: 388-390.

Dinant, S., Maisonneuve, B., Albouy, J., Chupeau, Y., Chupeau, M.C., Bellec, Y., Gaudefroy, F., Kusiak, C., Souche, S., Robaglia, C., and Lot, H. 1997. Coat protein gene-mediated protection in *Lactuca sativa* against lettuce mosaic potyvirus strains. Molec. Breed. 3: 75-86.

Doerschug, M.R., and Miller, C.O. 1967. Chemical control of organ formation in *Lactuca sativa* explants. Amer. J. Bot. 54: 410-413.

Eenink, A.H., Groenwold, R., and Dieleman, F.L. 1982a. Resistance of lettuce (*Lactuca*) to the leaf aphid *Nasonovia ribisnigri*. 1. Transfer of resistance from *L. virosa* to *L. sativa* by interspecific crosses and selection of resistant breeding lines. Euphytica 31: 291-299.

Eenink, A.H., Groenwold, R., and Dieleman, F.L. 1982b. Resistance of lettuce (*Lactuca*) to the leaf aphid *Nasonovia ribisnigri*. 2. Inheritance of the resistance. Euphytica 31: 301-304.

Ellis, P.R., McClement, S.J., Saw, P.L., Phelps, K., Vice, W.E., Kift, N.B., Astley, D., and Pink, D.A.C. 2002. Identification of sources of resistance in lettuce to the lettuce root aphid, *Pemphigus bursarius*. Euphytica 125: 305-315.

Farrara, B.F., Ilott, T.W., and Michelmore, R.W. 1987. Genetic analysis of factors for resistance to downy mildew (*Bremia lactucae*) in species of lettuce (*Lactuca sativa* and *L. serriola*). Plant Pathol. 36: 499-514.

Feráková, V. 1977. *The Genus Lactuca L. in Europe*, Universita Komenskeho, Bratislava.

Food and Agricultural Organization of the United Nations (FAO), 2005, (February 12, 2006); http://faostat.fao.org

Fujinaga, M., Ogiso, H., Tsuchiya, N., and Saito, S. 2001. Physiological specialization of *Fusarium oxysporum* f. sp. *lactucae,* a causal organism of fusarium root rot of crisp head lettuce in Japan. J. Gen. Plant Pathol. 67: 205-206.

Fujinaga, M., Ogiso, H., Tsuchiya, N., Saito, S., Yamanaka, S., Nozue, M., and Kojima, M. 2003. Race 3, a new race of *Fusarium oxysporum* f. sp. *lactucae* determined by a differential system with commercial cultivars. J. Gen. Plant Pathol. 69: 23-28.

Garibaldi, A., Gilardi, G., and Gullino, M.L. 2004. Seed transmission of *Fusarium oxysporum* f. sp. *lactucae*. Phytoparasitica 32: 61-65.

Gilbertson, R.L. 1998. Evaluation of coat protein-mediated resistance for lettuce mosaic virus, Annual Report, California Lettuce Research Board, 1997-98, pp. 71-74.

Globerson, D., and Ventura, J. 1973. Influence of gibberellins on promoting flowering and seed yield in bolting-resistant lettuce cultivars. Israel J. Agric. Res. 23: 75-77.

Goto, F., Yoshihara, T., and Saiki, H. 2000. Iron accumulation and enhanced growth in transgenic lettuce plants expressing the iron-binding protein ferritin. Theor. Appl. Genet. 100: 658-664.

Gray, D., Steckel, J.R.A., Wurr, D.C.E., and Fellows, J.R. 1986. The effects of applications of gibberellins to the parent plant, harvest date and harvest method on seed yield and mean seed weight of crisphead lettuce. Ann. Appl. Biol. 108: 125-134.

Grube, R.C., Hayes, R., Mou, B., and McCreight, J.D. 2005a. Lettuce breeding, USDA-ARS, Annual Report, California Lettuce Research Board, pp. 29-63.

Grube, R.C., and Ochoa, O.E. 2005. Comparative genetic analysis of field resistance to downy mildew in the lettuce cultivars 'Grand Rapids' and 'Iceberg'. Euphytica 142: 205-215.

Grube, R., and Ryder, E. 2004. Identification of lettuce (*Lactuca sativa* L.) germplasm with genetic resistance to drop caused by *Sclerotinia minor*. J. Amer. Soc. Hort. Sci. 129: 70-76.

Grube, R.C., Wintermantel, W.M., Hand, P., Aburomia, R., Pink, D.A.C., and Ryder, E.J. 2005b. Genetic analysis and mapping of resistance to lettuce dieback: a soilborne disease caused by tombusviruses. Theor. Appl. Genet. 110: 259-268.

Harlan, J.R. 1986. Lettuce and the sycomore: sex and romance in ancient Egypt. Econo. Bot. 40: 4-15.

Harrington, J.F. 1960. The use of gibberellic acid to induce bolting and increase seed yield of tight-heading lettuce. Proc. Amer. Soc. Hort. Sci. 75: 476-479.

Hartmann, R.W. 1991. Breeding lettuce for resistance to tomato spotted wilt virus in Hawaii, Hawaii Institute of Tropical Agriculture and Human Resources, Honolulu, H.I., pp. 7.

Hawthorn, L.R., and Pollard, L.H. 1956. Production of lettuce seed as affected by soil moisture and fertility, Bulletin 386, Utah State Agricultural College, Logan, Utah, USA, pp. 2-23.

Hayes, R.J., Wintermantel, W.M., Nicely, P.A., and Ryder, E.J. 2006. Host resistance to *Mirafiori lettuce big-vein virus* and *Lettuce big-vein associated virus* and virus sequence diversity and frequency in California. Plant Dis. 90: 233-239.

Hu, J., Ochoa, O.E., Truco, M.J., and Vick, B.A. 2005. Application of the TRAP technique to lettuce (*Lactuca sativa* L.) genotyping, *Euphytica* 144: 225-235.

Irwin, S.V., Kesseli, R.V., Waycott, W., Ryder, E.J., Cho, J.J., and Michelmore, R.W. 1999. Identification of PCR-based markers flanking the recessive LMV resistance gene *mo1* in an intraspecific cross in lettuce. Genome 42: 982-986.

Izzeldin, H., Lippert, L.F., and Takatori, F.H. 1980. An influence of water stress at different growth stages on yield and quality of lettuce seed. J. Amer. Soc. Hort. Sci. 105: 68-71.

Jackson, L.E. 1998. Plant-soil relationships in lettuce, Annual Report, California Lettuce Research Board, 1997-98, pp. 213-227.

Jagger, I.C., and Chandler, N. 1934. Big vein of lettuce, *Phytopathology* 24: 1253-1256.

Jagger, I.C., Whitaker, T.W., Uselman, J.J., and Owen, W.M. 1941. The Imperial strains of lettuce, No. 596, U.S. Department of Agriculture, Washington, DC, USA, pp. 15.

Jeuken, M., and Lindhout, P. 2002. *Lactuca saligna*, a non-host for lettuce downy mildew (*Bremia lactucae*), harbors a new race-specific *Dm* gene and three QTLs for resistance. Theor. Appl. Genet. 105: 384-391.

Jeuken, M.J.W., and Lindhout, P. 2004. The development of lettuce backcross inbred lines (BILs) for exploitation of the *Lactuca saligna* (wild lettuce) germplasm. Theor. Appl. Genet. 109: 394-401.

Kesseli, R.V., and Michelmore, R.W. 1986. Genetic variation and phylogenies detected from isozyme markers in species of *Lactuca*. J. Hered. 77: 324-331.

Kesseli, R., Ochoa, O., and Michelmore, R. 1991. Variation at RFLP loci in *Lactuca spp.* and origin of cultivated lettuce (*L. sativa*). Genome 34: 430-436.

Kishaba, A.N., McCreight, J.D., Coudriet, D.L., Whitaker, T.W., and Pesho, G.R. 1980. Studies of ovipositional preference of cabbage looper on progenies from a cross between cultivated lettuce and prickly lettuce. J. Amer. Soc. Hort. Sci. 105: 890-892.

Kishaba, A.N., Whitaker, T.W., Vail, P.V., and Toba, H.H. 1973. Differential oviposition of cabbage loopers on lettuce. J. Amer. Soc. Hort. Sci. 98: 367-370.

Koevary, K., Rappaport, L., and Morris, L.L. 1978. Tissue culture propagation of head lettuce. HortScience 13: 39-41.

Landry, B.S., Kesseli, R.V., Farrara, B., and Michelmore, R.W. 1987. A genetic map of lettuce (*Lactuca sativa* L.) with restriction fragment length polymorphism, isozyme, disease resistance, and morphological markers. Genetics 116: 331-337.

Lebeda, A., and Reinink, K. 1994. Histological characterization of resistance in *Lactuca saligna* to lettuce downy mildew (*Bremia lactucae*). Physiol. Mol. Plant Pathol. 44: 125-139.

Linden, C.G. van der, Wouters, D.C.A.E., Mihalka, V., Kochieva, E.Z., Smulders, M.J.M., and Vosman, B. 2004. Efficient targeting of plant disease resistance loci using NBS profiling. Theor. Appl. Genet. 109: 384-393.

Lindqvist, K. 1960a. On the origin of cultivated lettuce. Hereditas 46: 319-350.

Lindqvist, K. 1960b. Cytogenetic studies in the serriola group of *Lactuca*. Hereditas 46: 75-151.

Lu, G. 2000. *Lettuce Cultivation Techniques*, Jindun Publishing, Beijing (in Chinese).

Maisonneuve, B. 2003. *Lactuca virosa,* a source of disease resistance genes for lettuce breeding: results and difficulties for gene introgression, in: *Eucarpia Leafy Vegetables 2003*, T.J.L. van Hintum, A. Lebeda, D.A. Pink, and J.W. Schut, ed., Proceedings of the Eucarpia Meeting on Leafy Vegetables Genetics and Breeding, Noordwijkerhout, the Netherlands, 19-21 March, 2003, Centre for Genetic Resources, Wageningen, the Netherlands, pp. 61-67.

Maisonneuve, B., Bellec, Y., Souche, S., and Lot, H. 1999. New resistance against downy mildew and lettuce mosaic potyvirus in wild *Lactuca* spp., in: *Eucarpia Leafy Vegetables '99*, A. Lebeda, and E. Krístková, ed., Proceedings of the Eucarpia Meeting on Leafy Vegetables Genetics and Breeding, Olomouc, the Czech Republic, 8-11 June, 1999, Palacky University Olomouc, Olomouc, pp. 191-197.

Maisonneuve, B., Chovelon, V., and Lot, H. 1991. Inheritance of resistance to beet western yellows virus in *Lactuca virosa* L. HortScience 26: 1543-1545.

Maisonneuve, B., Chupeau,, M.C., Bellec, Y., and Chupeau, Y. 1995. Sexual and somatic hybridization in the genus *Lactuca*. Euphytica 85: 281-285.

Markert, C.L., and Moller, F. 1959. Multiple forms of enzymes: Tissues, ontogenetic, and species specific patterns. Proc. Natl. Acad. Sci. USA 45: 753-763.

Martienssen, R.A. 2004. Crop plant genome sequence: What is it good for? Crop Sci. 44: 1898-1899.

Matheron, M. E., McCreight, J.D., Tickes, B.R., and Porchas, M. 2005. Effect of planting date, cultivar, and stage of plant development on incidence of Fusarium wilt of lettuce in desert production fields. Plant Dis. 89: 565-570.

Mazier, M., Maisonneuve, B., Bellec, Y., Chupeau, M.C., Souche, S., and Chupeau, Y. 1999. Interest of protoplasts for lettuce breeding, in: *Eucarpia Leafy Vegetables '99*, A. Lebeda, and E. Krístková, ed., Proceedings of the Eucarpia Meeting on Leafy Vegetables Genetics and Breeding, Olomouc, the Czech Republic, 8-11 June, 1999, Palacky University Olomouc, Olomouc, pp. 239-244.

McCabe, M.S., Garratt, L.C., Schepers, F., Jordi, W.J.R.M., Stoopen, G.M., Davelaar, E., van Rhijn, J.H.A., Power, J.B., and Davey, M.R. 2001. Effects of P_{SAG12}-*IPT* gene expression on development and senescence in transgenic lettuce. Plant Physiol. 127: 505-516.

McCabe, M.S., Mohapatra, U., Power, J.B., and Davey, M.R. 1999. Instability of transgene expression in lettuce (*Lactuca sativa*): possible causes and prevention, in: *Eucarpia Leafy Vegetables '99*, A. Lebeda, and E. Krístková, ed., Proceedings of the Eucarpia Meeting on Leafy Vegetables Genetics and Breeding, Olomouc, the Czech Republic, 8-11 June, 1999, Palacky University Olomouc, Olomouc, pp. 251-256.

McCreight, J.D. 1987. Resistance in wild lettuce to lettuce infectious yellows virus. HortScience 22: 640-642.

McCreight, J.D., Matheron, M.E., Tickes, B.R., and Platts, B. 2005. Fusarium wilt race 1 on lettuce. HortScience 40: 529-531.

Michelmore, R.W. 1995. Isolation of disease resistance genes from crop plants. Current Opinion in Biotech. 6: 145-152.

Michelmore, R.W. 1997. Applications of biotechnology to disease resistance in lettuce, in: *Compendium of Lettuce Diseases*, R.M. Davis, K.V. Subbarao, R.N. Raid, and E.A. Kurtz, ed., APS Press, St. Paul, pp. 11-13.

Michelmore, R.W. 1998. Genetic variation in lettuce, Annual Report, California Lettuce Research Board, 1997-98, pp. 54-59.

Michelmore, R.W. 2005. Genetic variation in lettuce, Annual Report, California Lettuce Research Board, 2004-05, pp. 90-100.

Michelmore, R., Marsh, E., Seely, S., and Landry, B. 1987. Transformation of lettuce (*Lactuca sativa*) mediated by *Agrobacterium tumefaciens*. Plant Cell Report 6: 439-442.

Michelmore, R.W., and Truco, M.J. 2005. Breeding leaf lettuce, Annual Report, California Lettuce Research Board, 2004-05, pp. 80-87.

Mohapatra, U., McCabe, M.S., Power, J.B., Schepers, F., van der Arend, A.A.D., and Davey, M.R. 1999. Expression of the *bar* gene confers herbicide resistance in transgenic lettuce. Transgenic Res. 8: 33-34.

Moreno-Vázquez, S., Ochoa, O., Faber, N., Chao, S., Jacobs, J.M.E., Maisonneuve, B., Kesseli, R.V., and Michelmore, R.W. 2003. SNP-based codominant markers for a recessive gene conferring resistance to corky root rot (*Rhizomonas suberifaciens*) in lettuce (*Lactuca sativa*). Genome 46: 1059-1069.

Mou, B. 2005. Genetic variation of beta-carotene and lutein contents in lettuce. J. Amer. Soc. Hort. Sci. 130: 870-876.

Mou, B., and Bull, C. 2004. Screening lettuce germplasm for new sources of resistance to corky root. J. Amer. Soc. Hort. Sci. 129: 712-716.

Mou, B., and Liu, Y. 2003. Leafminer resistance in lettuce. HortScience 38: 570-572.

Mou, B., and Liu, Y. 2004. Host plant resistance to leafminers in lettuce. J. Amer. Soc. Hort. Sci. 129: 383-388.

Mou, B., and Ryder, E.J. 2004. Relationship between the nutritional value and the head structure of lettuce. Acta Hort. 637: 361-367.

Netzer, D., Globerson, D., Weintal, C.H., and Elyassi, R. 1985. Sources and inheritance of resistance to Stemphylium leaf spot of lettuce. Euphytica 34: 393-396.

Newton, H.C., and Sequiera, L. 1972. Possible sources of resistance in lettuce to *Sclerotinia sclerotiorum*. Plant Dis. Rpt. 56: 875-878.

Nicaise, V., German-Retana, S., Sanjuan, R., Dubrana, M.-P., Mazier, M., Maisonneuve, B., Candresse, T., Caranta, C., and LeGall, O. 2003. The eukaryotic translation initiation factor 4E controls lettuce susceptibility to the potyvirus lettuce mosaic virus. Plant Physiol. 132: 1272-1282.

Oliver, G.W. 1910. New methods of plant breeding, Bulletin 167, U.S. Bureau of Plant Industry, Washington, DC, USA, pp. 12-13.

Pang, S.Z., Jan, F.J., Carney, K., Stout, J., Tricoli, D.M., Quemada, H.D., and Gonsalves, D. 1996. Post-transcriptional transgene silencing and consequent tospovirus resistance in transgenic lettuce as affected by transgene dosage and plant development. The Plant J. 9: 899-909.

Paran, I., and Michelmore, R.W. 1993. Development of reliable PCR-based markers linked to downy mildew resistance genes in lettuce. Theor. Appl. Genet. 85: 985-993.

Park, B.J., Liu, Z., Kanno, A., and Kameya, T. 2005. Increased tolerance to salt- and water-deficit stress in transgenic lettuce (*Lactuca sativa* L.) by constitutive expression of LEA. Plant Growth Regul. 45: 165-171.

Peleman, J.D., and van der Voort, J.R. 2003. The challenges in marker assisted breeding, in: *Eucarpia Leafy Vegetables 2003*, T.J.L. van Hintum, A. Lebeda, D.A. Pink, and J.W. Schut, ed., Proceedings of the Eucarpia Meeting on Leafy Vegetables Genetics and Breeding, Noordwijkerhout, the Netherlands, 19-21 March, 2003, Centre for Genetic Resources, Wageningen, the Netherlands, pp. 125-130.

Pileggi, M., Pereira, A.A.M., Silva, J.D.S., Pileggi, S.A.V., and Verma, D.P.S. 2001. An improved method for transformation of lettuce by *Agrobacterium tumefaciens* with a gene that confers freezing resistance. Brazilian Archives Biol. Techno. 44: 191-196.

Pink, D.A.C., and Carter, P.J. 1987. Propagation of lettuce (*Lactuca sativa*) breeding material by tissue culture. Ann. Appl. Biol. 110: 611-616.

Provvidenti, R., Robinson, R.W., and Shail, J.W. 1980. A source of resistance to a strain of cucumber mosaic virus in *Lactuca saligna* L. HortScience 15: 528-529.

Provvidenti, R., Robinson, R.W., and Shail, J.W. 1984. Incidence of broad bean wilt virus in lettuce in New York State and sources of resistance. HortScience 19: 569-570.

Reinink, K., and Dieleman, F.L. 1989. Comparison of sources of resistance to leaf aphids in lettuce. Euphytica 40: 21-29.

Ryder, E.J. 1973. Seed transmission of lettuce mosaic virus in mosaic resistant lettuce. J. Amer. Soc. Hort. Sci. 98: 610-614.

Ryder, E.J. 1979a. 'Salinas' lettuce. HortScience 14: 283-284.

Ryder, E.J. 1979b. 'Vanguard 75' lettuce. HortScience 14: 284-286.

Ryder, E.J. 1986. Lettuce breeding, in: *Breeding Vegetable Crops*, M.J. Bassett, ed., AVI Publishing, Westport, pp. 433-474.

Ryder, E.J. 1991. 'Salinas 88' lettuce. HortScience 26: 439-440.

Ryder, E.J. 1997. Introduction, in: *Compendium of Lettuce Diseases*, R. M. Davis, K.V. Subbarao, R.N. Raid, and E.A. Kurtz, ed., APS Press, St. Paul, pp. 1-8.

Ryder, E.J. 1999. *Lettuce, Endive and Chicory*, CABI Publishing, New York.

Ryder, E.J. 2002. A mild systemic reaction to lettuce mosaic virus in lettuce: Inheritance and interaction with an allele for resistance. J. Amer. Hort. Sci. 127: 814-818.

Ryder, E.J., Grube, R.C., Subbarao, K.V., and Koike, S.T. 2003. Breeding for resistance to diseases in lettuce: successes and challenges, in: *Eucarpia Leafy Vegetables 2003*, T.J.L. van Hintum, A. Lebeda, D.A. Pink, and J.W. Schut, ed., Proceedings of the Eucarpia Meeting on Leafy Vegetables Genetics and Breeding, Noordwijkerhout, the Netherlands, 19-21 March, 2003, Centre for Genetic Resources, Wageningen, the Netherlands, pp. 25-30.

Ryder, E.J., and Johnson, A.S. 1974. Mist depollination of lettuce flowers HortScience 9: 584.

Ryder, E.J., and Milligan, D.C. 2005. Additional genes controlling flowering time in *Lactuca sativa* and *L. serriola*. J. Amer. Soc. Hort. Sci. 130: 448-453.

Ryder, E.J., and Robinson, B.J. 1995. Big-vein resistance in lettuce: Identifying, selecting, and testing resistance cultivars and breeding lines. J. Amer. Soc. Hort. Sci. 120: 741-746.

Ryder, E.J., and Whitaker, T.W. 1980. The lettuce industry in California: A quarter century of change, 1954-1979. Hort. Rev. 2: 164-207.

Sequiera, L. 1978. Two root rot resistant varieties of head lettuce, Wis. Agr. Expt. Sta. Rpt., pp. 2.

Shen, K.A., Chin, D.B., Arroyo-Garcia, R., Ochoa, O.E., Lavelle, D.O., Wroblewski, T., Meyers, B.C., and Michelmore, R.W. 2002. *Dm3* is one member of a large constitutively expressed family of nucleotide binding site-leucine-rich repeat encoding genes. Mol. Plant-Microbe Interact. 15: 251-261.

Shen, K.A., Meyers, B.C., Islam-Faridi, M.N., Chin, D.B., Stelly, D.M., and Michelmore, R.W. 1998. Resistance gene candidates identified by PCR with degenerate oligonucleotide primers map to clusters of resistance genes in lettuce. Mol. Plant-Microbe Interact. 11: 815-823.

Smith, O.E., Welch, N.C., and Little, T.M. 1973. Studies on lettuce seed quality: I. Effect of seed size and weight on vigor. J. Amer. Soc. Hort. Sci. 98: 529-533.

Soffer, H., and Smith, O.E. 1974a. Studies on lettuce seed quality: III. Relationships between flowering pattern, seed yield, and seed quality. J. Amer. Soc. Hort. Sci. 99: 114-117.

Soffer, H., and Smith, O.E. 1974b. Studies on lettuce seed quality: IV. Individually measured embryo and seed characteristics in relation to continuous plant growth (vigor) under controlled conditions. J. Amer. Soc. Hort. Sci. 99: 270-275.

Staub, J.E., Kuhns, L.J., May, B., and Grun, P. 1982. Stability of potato tuber isozymes under different storage regimes. J. Amer. Soc. Hort. Sci. 107: 405-408.

Stevens, M.A. 1974. Varietal influence on nutritional value, in: *Nutritional Qualities of Fresh Fruits and Vegetables*, P.L. White, and N. Selvey, ed., Futura Publications, Mt Kisco, New York, pp. 87-109.

Subbarao, K.V. 1998. Progress towards integrated management of lettuce drop. Plant Dis. 82: 1068-1078.

Thompson, R.C. 1944. Reaction of *Lactuca* species to the aster yellows virus under field conditions. J. Agr. Res. 69: 119-125.

Thompson, R.C. and Ryder, E.J. 1961. Descriptions and pedigrees of nine varieties of lettuce, Technical Bulletin No. 1244, Agricultural Research Service, U.S. Dept. of Agriculture, Washington, D.C., pp. 19.

Torres, A.C., Cantliffe, D.J., Laughner, B., Bieniek, M., Nagata, R., Ashraf, M., and Ferl, R.J. 1993. Stable transformation of lettuce cultivar South Bay from cotyledon explants. Plant Cell Tissue Organ Cult. 34: 279-285.

Torres, A.C., Nagata, R.T., Ferl, R.J., Bewick, T.A., and Cantliffe, D.J. 1999. In vitro assay selection of glyphosate resistance in lettuce. J. Amer. Soc. Hort. Sci. 124: 86-89.

U.S. Dept. of Agriculture, Agricultural Research Service (USDA-ARS), 2005, USDA National Nutrient Database for Standard Reference, Release 18, Nutrient Data Laboratory Home Page, (April 11, 2006), http://www.nal.usda.gov/fnic/foodcomp/Data/.

U.S. Dept. of Agriculture, National Agricultural Statistics Service (USDA-NASS), 2006, Vegetables 2005 Summary, (March 3, 2006), http://usda.mannlib.cornell.edu/reports/nassr/fruit/pvg-bban/vgan0106.txt.

Vallad, G.E., Bhat, R.G., Koike, S.T., Ryder, E.J., and Subbarao, K.V. 2005. Weedborne reservoirs and seed transmission of *Verticillium dahliae* in lettuce. *Plant Dis*. 89: 317-324.

van Bruggen, A.H.C. 1997. Corky root, in: *Compendium of Lettuce Diseases*, R.M. Davis, K.V. Subbarao, R.N. Raid, and E.A. Kurtz, ed., APS Press, St. Paul, pp. 28-29.

van de Wiel, C., Arens, P., and Vosman, B. 1999. Microsatellite retrieval in lettuce (*Lactuca sativa* L.). Genome 42: 139-149.

van der Arend, A.J.M. 2003. The possibility of *Nasonovia ribisnigri* resistance breaking biotype development due to plant host resistance: a literature study, in: *Eucarpia Leafy Vegetables 2003*, T.J.L. van Hintum, A. Lebeda, D.A. Pink, and J.W. Schut, ed., Proceedings of the Eucarpia Meeting on Leafy Vegetables Genetics and Breeding, Noordwijkerhout, the Netherlands, 19-21 March, 2003, Centre for Genetic Resources, Wageningen, the Netherlands, pp. 75-81.

van der Arend, A.J.M., Ester, A., and van Schijndel, J.T. 1999. Developing an aphid resistant butterhead lettuce 'Dynamite', in: *Eucarpia Leafy Vegetables '99*, A. Lebeda, and E. Krístková, ed., Proceedings of the Eucarpia Meeting on Leafy Vegetables Genetics and Breeding, Olomouc, the Czech Republic, 8-11 June, 1999, Palacky University Olomouc, Olomouc, pp. 149-157.

Van Hintum, T. 2003. Molecular characterization of a lettuce germplasm collection, in: *Eucarpia Leafy Vegetables 2003*, T.J.L. van Hintum, A. Lebeda, D.A. Pink, and J.W. Schut, ed., Proceedings of the Eucarpia Meeting on Leafy Vegetables Genetics and Breeding, Noordwijkerhout, the Netherlands, 19-21 March, 2003, Centre for Genetic Resources, Wageningen, the Netherlands, pp. 99-104.

Vanjildorj, E., Bae, T.W., Riu, K.Z., Kim, S.Y., and Lee, H.Y. 2005. Overexpression of *Arabidopsis ABF3* gene enhances tolerance to drought and cold in transgenic lettuce (*Lactuca sativa*). Plant Cell Tissue Organ Cult. 83: 41-50.

Vos, P., Hogers, R., Bleeker, M., Reijans, M., van de Lee, T., Hornes, M., Frijters, A., Pot, J., Peleman, J., Kuiper, M., and Zabeau, M. 1995. AFLP: A new technique for DNA finger-printing. Nucleic Acids Res. 23: 4407-4414.

Waycott, W., and Fort, S.B. 1994. Differentiation of nearly identical germplasm accessions by a combination of molecular and morphologic analyses. Genome 37: 577-583.

Whipps, J.M., Budge, S.P., McClement, S., and Pink, D.A.C. 2002. A glasshouse cropping method for screening lettuce lines for resistance to *Sclerotinia sclerotiorum*. Eur. J. Plant Pathol. 108: 373-378.

Whitaker, T.W., Kishaba, A.N., and Toba, H.H. 1974. Host-parasite interrelations of *Lactuca saligna* L. and the cabbage looper, *Trichoplusia ni* (Hubner). J. Amer. Soc. Hort. Sci. 99: 74-78.

Williams, J.G.K., Kubelik, A.R., Livak, K.J., Rafalski, J.A., and Tingey, S.V. 1990. DNA polymorphisms amplified by arbitrary primers are useful as genetic markers. Nucleic Acids Res. 18: 6531-6535.

Witsenboer, H., Vogel, J., and Michelmore, R.W. 1997. Identification, gene localization, and satellite diversity of selectively amplified microsatellite polymorphic loci in lettuce and wild relatives (*Lactuca* spp.). Genome 40: 923-936.

Yabuuchi, E., Kosako, Y., Naka, T., Suzuki, S., and Yano, I. 1999. Proposal of *Sphingomonas suberifaciens* (van Bruggen, Jochimsen, and Brown 1990) comb. Nov., *Sphingomonas natatoria* (Sly 1985) cob. Nov., *Sphingomonas u*rsincola (Yurkov et al. 1997) comb. Nov., and emendation of the genus *Sphingomonas*. Microbiol. Immunol. 43: 339-349.

Zuo, X., Zhang, Y., Wu, B., Chang, X., and Ru, B. 2002. Expression of the mouse metallothionein mutant ββ-cDNA in the lettuces (*Lactuca sativa* L.). Chinese Sci. Bulletin 47: 558-562.

Family Brassicaceae (=Cruciferae)

Cabbage and Kale

Amando Ordás[1] and M. Elena Cartea[1]

[1] Misión Biológica de Galicia, Spanish National Research Council (CSIC), aordas@mbg.cesga.es

1 Introduction

Cabbage and kale are two closely related vegetables that are grown all over the world, although cabbage is cultivated to a much greater extent. Both belong to the species *Brassica oleracea* L., a species of the *Brassicaceae* (= *Cruciferae*) family comprising several crops known under the generic name of cole crops. There are some types of kale that are grown for fodder.

Modern classifications of the family *Brassicaceae* are based on a review by Schultz (1936) who reduced the number of tribes from 19 to 15; nowadays the number of tribes commonly accepted is 13. A problem arises when trying to determine the number of genera included within the family. Hickey and King (1997) indicate 390 genera, Paterson et al. (2000) 360, and the USDA (2006) 107. The position of the family in the general classification of the kingdom of plants is as follows (USDA, 2006):

Kingdom *Plantae*
- Subkingdom *Tracheobionta* (vascular plants)
  - Superdivision *Spermatophyta* (seed plants)
    - Division *Magnoliophyta* (flowering plants)
      - Class *Magnoliopsida* (dicotyledons)
        - Subclass *Capparales*
          - Family *Brassicaceae*

The taxonomy of the genus *Brassica* is complicated. Gómez-Campo (1999) points out that a chapter on the taxonomy of *Brassica* and allies would have benefited from being postponed perhaps a decade or so to receive –and digest– inputs from other fast developing fields to become more useful and complete. This author presents a complete classification of the genus *Brassica* and its allied genera, indicating subgenera, sections, species, and subspecies. More than 70 years ago, U (1935) studied the cytology of the genus and established the relationships among the genomes of the six species with greatest agricultural importance all over the world (Figure 1). *Brassica rapa* comprises turnip, turnip rape, and Chinese cabbage; *B. oleracea* comprises the so called cole crops; the amphidiploid between them, i.e. *B. napus*, includes rape and swede. The other species in the triangle are several types of mustards.

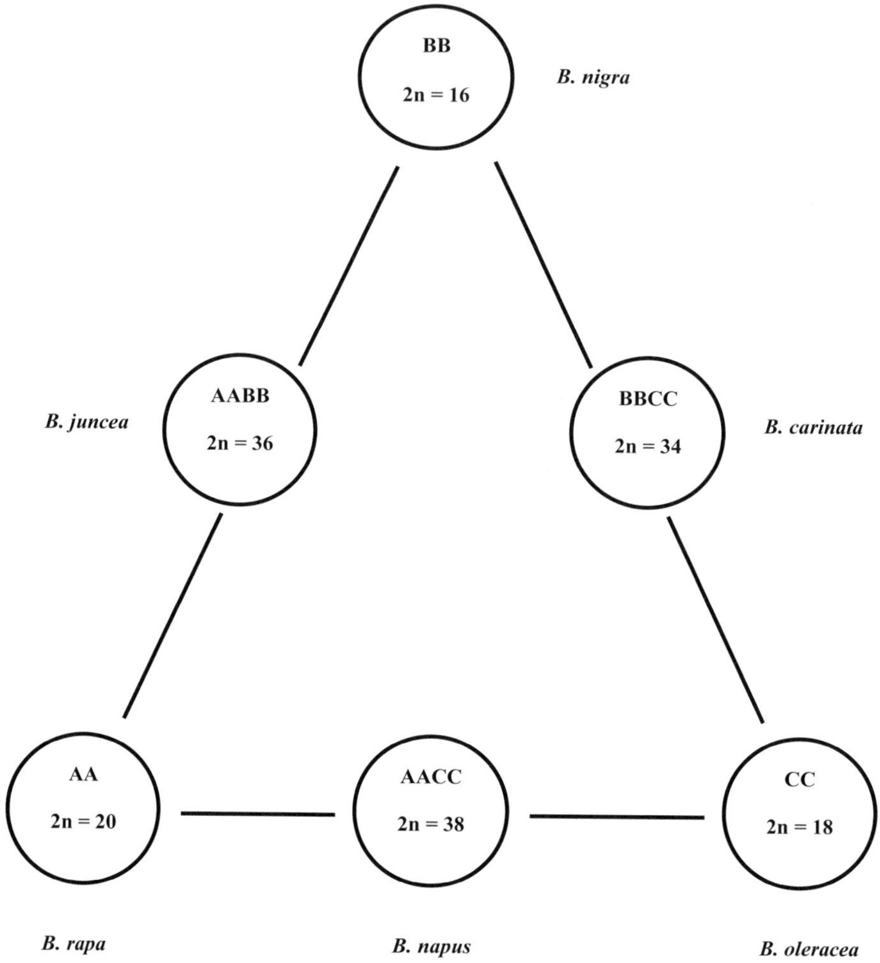

Fig. 1. U triangle showing the relationships between the cultivated species of the genus *Brassica* (U, 1935).

Brassica oleracea is a versatile species that under human selection has generated several crops (Table 1), each targeting to a different organ of the plant (leaves along the stem: kales; leaves surrounding the terminal bud: cabbage; enlarged axillary buds: Brussels sprouts; inflorescences: cauliflower and broccoli; swollen stem: kohlrabi and marrow stem kale). There are several types of cabbages: white cabbage, red cabbage, and savoy. Portuguese cabbage forms a loose head and can be considered an intermediate form between typical white cabbages and kale. These different crops have been classified as varieties or convarieties, but under modern cultonomic terms they would be cultivar-groups (Spooner et al., 2003).

Table 1. Crops from the species *Brassica oleracea.*

Cultivar group	Common name
Acephala	Kale and collards
Alboglabra	Chinese kale
Botrytis	Cauliflower
Capitata	Cabbage
Costata	Portuguese cabbage (tronchuda)
Gemmifera	Brussels sprouts
Gongylodes	Kohlrabi
Italica	Broccoli
Medullosa	Marrow stem kale
Ramosa	Thousand head kale
Sabauda	Savoy

The species within the *Brasicaceae* have a floral structure consisting of four sepals, four petals, six stamens (four long and two short) and a gynoecium formed by two carpels and a superior ovary. The fruit is a capsule (siliqua) divided into two locula by a septum. Under normal growing conditions cabbages and kales are biennial plants, i.e. flowers develop in the second year. Alteration of the growing conditions, for instance by vernalizing seedlings, can induce flowering.

Cabbage was grown in 129 countries in 2005 on a total surface area of 3,218,971 ha according to FAO (2006). China is the country where cabbage occupies the biggest area in the world: 1,719,450 ha in 2005, followed, but by far, by India (280,000 ha) and Russia (168,000 ha). It must be pointed out that under the name "cabbage" FAO includes red, white, and savoy cabbages; Chinese cabbages; Brussels sprouts; green kale; and sprouting broccoli. All these crops belong to the species *B. oleracea*, with the exception of Chinese cabbage which, as mentioned above, is *B. rapa*. The biggest yields, however, do not correspond to any of these countries. The differences in yield are not as big as the differences in acreage. The best yielder, according to the same data (FAO, 2006), is South Africa (64,012 kg/ha), followed closely by South Korea (63,462 kg/ha).

Cabbage –and also kale when it is used for human consumption– is usually consumed after boiling, a process that tenderizes the leaves, releases sugars, and produces a characteristic aroma. Cabbage is also consumed as sauerkraut, which is cabbage cut into fine slices that undergo lactic acid fermentation in a brine made of its own juice with salt. Sauerkraut was traditionally made at home, but nowadays it is mainly an industrial product.

B. oleracea is an allogamous species due to self-incompatibility. There are two general schemes of self-incompatibility in higher plants: (1) gametophytic or haplo-diplo schemes in which incompatibility depends on the genotype of the gametophyte and (2) sporophytic or diplo-diplo schemes in which incompatibility is impressed on the gametophyte by its sporophytic parent (Allard, 1999). The latter system is usually controlled by a series of alleles at a single locus and this is the system forcing allogamy in *B. oleracea*. One S locus with 50 to 70 alleles controls incompatibility in *B. oleracea*. The example in Table 2, involving the cross $S_1S_3 \times S_1S_2$, demonstrates the complex reaction of this system (Chiang et al., 1993).

Table 2. Sporophytic self-incompatibility system: $S_1S_3 \times S_1S_2$ (Chiang et al., 1993).

Pollen reaction	Pistil reaction	Compatibility
Independent	Independent	Incompatible
S_1 dominant to S_2	Independent	Incompatible
S_2 dominant to S_1	Independent	Compatible
Independent	S_1 dominant to S_3	Incompatible
Independent	S_3 dominant to S_1	Compatible
S_1 dominant to S_2	S_1 dominant to S_3	Incompatible
S_1 dominant to S_2	S_3 dominant to S_1	Compatible
S_2 dominant to S_1	S_1 dominant to S_3	Compatible
S_2 dominant to S_1	S_3 dominant to S_1	Compatible

At least 15 molecular genetic maps have been constructed for the genus *Brassica* including all the major cultivated species. These maps include 935 different publicly available probes, many shared by multiple maps (Paterson et al., 2000). Slocum et al. (1990) presented a detailed genetic linkage map for *B. oleracea* based on the segregation of 258 restriction fragment length polymorphisms. They defined nine linkage groups, covering 820 recombination units. Colinearity –i.e. the conservation of gene content, order and orientation between chromosomes of different species or between non-homologous chromosomes within a single species– has been shown to occur almost over the entire genomes of *B. nigra, B. oleracea* and *B. rapa*, and virtually all available data support the notion that these diploid genomes have been through one or more cycles of extensive chromatin duplication fairly recently in their evolutionary history (Newbury and Paterson, 2003).

2 Origin and Domestication

As Gómez-Campo and Prakash (1999) point out, there are two aspects in the origin of any cultivated plant that should be considered: (1) the origin of the taxon itself before domestication as a result of the evolutionary process in the wild and (2) its origin in cultivation, i.e. the history of its domestication and subsequent use and diversification.

In 1960, Röbbelen, as a result of studies on chromosome morphology, revealed that the genome of *B. oleracea* can be represented by ABBCCDEEF; in other words, it is tetrasomic for chromosomes B, C, and E, meaning that the basic chromosome number is six. The duplicated chromosomes lost homology during the evolution of the species. Mizushima (1980) and Prakash and Hinata (1980) proposed that *B. oleracea*, as well as *B. rapa* and *B. nigra,* were aneuploids, evolving in ascending order from a common ancestor with a basic chromosome number of n=6 (7). After a study carried out on 33 diploid taxa of the subtribe *Brassicinae* using chloroplast DNA, these proposals were refuted by Warwick and Black (1991) who had the objective, among others, of testing taxonomic classifications, and species and cytodeme relationships within the genus *Brassica.*

Snogerup (1980) identified ten wild *Brassica* species which commonly occur as more or less isolated populations in maritime habits. Wild *B. oleracea* has been found in the coasts of northern Spain, western France and southern and southwestern Britain (Lázaro and Aguinagalde, 1998; Snogerup, 1980). It is biennial or perennial.

It has been generally accepted that the early evolution of the different cultivated brassicas occurred in the Mediterranean area. Writings from the classical period of Western civilization show evidence that certain kinds of *B. oleracea* had been cultivated by the ancient Greeks, as cited by Hodgkin (1995). For instance, Theophrastus (372-287 BC) lists three varieties in common use; Cato (c. 200 BC) also identifies three kinds of brassicas –a large type with smooth leaves and thick stem, a curly variety called ‘parsley cabbage’, and a mild type which was tender and had a small stalk; Pliny (1st century AD) lists six types, crops that from their description seem to be cabbage, kohlrabi, and sprouting broccoli.

The present state of knowledge suggests that modern crops derived from the wild *B. oleracea* and not from the wild Mediterranean species, even though much of the early selection of the different crop types may have occurred in the Mediterranean area (Hodgkin, 1995). This author also points out that it seems unlikely that wild *B. oleracea* was ever distributed in the Mediterranean region; he considers that the evidence currently indicates that early cultivated forms of *B. oleracea* were brought from the Atlantic coast to the Mediterranean where selection for many of the early crop types occurred (Figure 2). However, the issue is not closed as other opinions exist. For instance, Swarup and Brahmi (2005) state that all cole crops originated from the Mediterranean region, mainly by mutation and introgression from wild species during evolution or by human selection.

3 Varietal Groups

There are many different crops known as cabbage or kale, with a lot of confusion on their names and the differences between them. For instance, a bulletin from the Wisconsin Fast Plants program (Anonymous, 1987) states that collards belong to the variety *sabellica* and kales to the variety *acephala*. However, kale and collards are more commonly assigned to the same *B. oleracea acephala* group. Much of the confusion is because many cole crops have been around since ancient times and they are spread worldwide, so they have given different names throughout history, many of the names being the same for different crops.

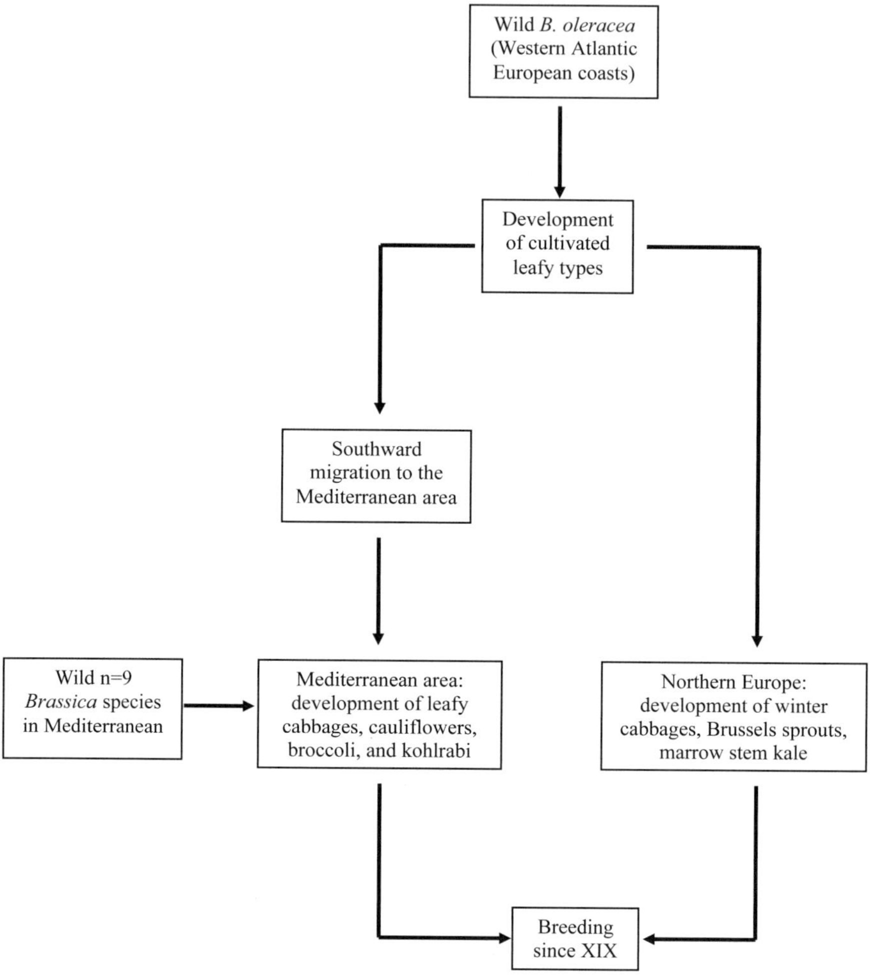

Fig. 2. Evolution of the cultivated forms of *Brassica oleracea* (adapted from Hodgkin, 1995).

What can be firmly stated is that there are two main groups among the cole crops used by their leaves: those that form a head (cabbages) and those that do not (kales). Among cabbages we find white cabbage (the most common type), red cabbage (a small, round headed type with dark red leaves), and savoy (a type of cabbage with a round head with crinkled leaves). In the group not forming a head we have collards (distinguished from the other kales by the shape of the leaves –smooth in texture and broad in collards, smaller and with frilly edges in kales) and several types of kales like curly kale (also known as borecole), thousand-head kale (used for forage), marrow stem kale (with engrossed stems and used in some countries as forage for cattle and other farm animals), and kales without any epithet.

A difficult crop to classify is Portuguese cabbage or 'tronchuda', that forms open heads or even no heads at all and that might be considered a variety of collard. A special type of *B. oleracea* is the crop called Brussels sprouts, characterised by forming many small heads along the stem.

4 Genetic Resources

The importance of the genetic resources for the continuous improvement of crops is something universally admitted nowadays. Germplasm banks are the reservoirs from where plant breeders get variability for their improvement programs. In all the species not only the cultivated varieties, but also the wild ancestors when known and available, must be maintained.

Since 1982 the International Board for Plant Genetic Resources (now International Plant Genetic Resources Institute, IPGRI) has sponsored several missions to collect wild *Brassica* species. According to IPGRI policy, each sample is split into three parts, which are stored in the Polytechnic University of Madrid (Spain), the University of Tohoku (Sendai, Japan), and in a seed bank in the country where each sample was collected (Chiang et al., 1993).

In the United States, the National Plant Germplasm System (NPGS) is a cooperative effort by public (State and Federal) and private organizations to preserve the genetic diversity of plants. The NPGS comprises several centres scattered all over the country, each one specialized in a specific crop or crops, or in a task. As of 16 July 2006 the NPGS holds a total of 471,318 accessions that represent 216 families, 1,914 genera, and 11,756 species. Most *Brassica* species are maintained at the Plant Genetic Resources Unit (PGRU) located at the Geneva campus (New York) of Cornell University. The PGRU was formed in 1986 by merging the Northeast Regional Plant Introduction Station and the National Clonal Germplasm Repository for Apple, Tart Cherry and Grape, and maintains 1,480 accessions of *B. oleracea*. Among them, there are 857 cabbages, 12 Portuguese cabbages, 68 accessions of Brussels sprouts, and 9 of medular kale. There are also some accessions from wild brassicas. The seeds were stored in the past at 20% RH and 0° C, but presently the main seed storage area has undergone a renovation which has dropped the temperature to -18° C, which eliminates the need for a dehumidifier.

In 1982 Prof. Williams, at the University of Wisconsin, established the Crucifer Genetics Cooperative (CrGC) to acquire, maintain, and distribute seed stocks and pollen of *Brassicaceae*. Presently, the program provides seed stocks and information of rapid cycling brassicas, concretely the six species of the U triangle.

In Western Europe a first program, financed by the European Community, run from 1981 through 1984, had the main objective of collecting local landraces of cruciferous crops in the countries that formed the Community at that time. The total number of accessions was approximately 3,500. Under the aegis of the European Cooperative Program for Crop Genetic Resources Networks (ECP/GR) a working group on brassicas was established in 1991. One of the main efforts of this group was to set up a European brassica database (Bras-EDB), which was developed by the Centre for Genetic Resources The Netherlands (CGN) (Boukema et al., 1995).

The objective of the Bras-EDB is to support rationalization of genetic resources activities in brassica; its purposes are (Boukema et al. 1995):

- -make an inventory of the European brassica holding
- -trace duplicate accessions
- -trace gaps in the European brassica germplasm holding
- -coordinate activities such as collecting missions and seed regeneration/seed increase programs

and it can be accessed at http://documents.plant.wur.nl/cgn/pgr/brasedb/.

Major updates of Bras-EDB were done in 2001 and 2005, supported financially by the European Commission by the project RESGEN CT99 109-112: "*Brassica* collections for broadening agricultural use, including characterising and utilising genetic variation in *Brassica carinata* for its exploitation as an oilseed crop". CGN only manages the database; requests for material listed in the database should be directed to the institute holding the material. The database contains passport data of about 19,600 accessions of the genus *Brassica* and includes 36 collections from 22 countries.

The project RESGEN CT99 109-112 ran from 1 January 2000 to 31 December 2003 and included 16 participants from 9 out of the 15 countries forming the European Union (EU) then. This project was an important attempt to unify efforts on brassica germplasm within the EU. The participant countries included France, Germany, Greece, Italy, the Netherlands, Portugal, Spain, Sweden, and the United Kingdom. Besides documenting brassica collections, other objectives of the project were to create a core collection of the four species included in the project (*B. oleracea, B. rapa, B. napus*, and *B. carinata*); characterization of the *Brassica* species using minimum descriptors; regeneration of collections with priority on accessions included in the core collection; evaluation for different properties with priority also in material included in the cores; rationalization and safety duplication of collections, and recommendations for further collecting (van Soest et al., 2004). The core collection of *B. oleracea* was established with a total of 396 accessions.

Other European countries that maintain collections of cabbages and/or kales are Bulgaria, Croatia, Czech Republic, Hungary, Poland, Russia, Switzerland, and Turkey. From a historical point of view, the N. I. Vavilov Research Institute for Plant Industry (VIR) at St. Petersburg (Russia) stands out. In 2006 its collection of plant genetic resources encompassed 320,000 accessions of 155 botanical families, 425 genera, and 2,532 species. Its brassica collection in 1996 comprised 6,879 accessions, 3,519 of them being *B. oleracea*. The number of cabbages in the collection was 1,760, with 927 having passport data. The basic working collections are stored in St. Petersburg in laminated aluminium bags at room temperature; in the National Seed Storage Facility located on the Kuban Experiment Station, the seeds are preserved at 4° C. These conditions were obviously not very good. In 1995

IPGRI provided assistance to acquire equipment for long-term storage, where the seeds are kept at -15 to -20° C (Shashilova, 1997).

There are also collections of cole crops in Israel, Ethiopia, South Africa, India, Philippines, and Taiwan (Swarup and Brahmi, 2005).

One of the main problems that germplasm curators must face is to maintain collections in good conditions of viability, especially in active banks, to minimize the need for regeneration. For instance, in the Misión Biológica de Galicia (Spain) a control made on brassica accessions after three years of storage at 0 to 2° C and at about 50% RH showed that viability had not changed significantly (Baladrón and Ordás, 1988); however, ten years later it was found that the viability of several samples had gone down to very low levels (unpublished data).

Gómez-Campo (2002) tested 40 different types of containers for their ability to exclude water vapour, using silica gel with a cobalt indicator. Only sealed brass cans, 'Kilner' jars with rubber seals, laboratory bottles normally used for liquid chemicals, or flame-sealed glass ampoules were considered safe for use in long-term preservation. The 36 remaining containers allowed moisture to enter within 2 to 3 years or less. It must be pointed out that samples at the Misión Biológica de Galicia were kept in paper bags; presently, a sample of each accession has been placed in flame-sealed ampoules as described by Gómez-Campo (2002) to ensure long-term viability. This method is quite convenient for small-sized seeds as those of brassicas.

5 Major Breeding Achievements

Breeding strategy and goals are dependent on market trends. Breeding and objectives reached in the last decades were addressed to satisfy both the grower and the consumer and therefore, they could be considered in terms of crop improvement and product improvement. The main criteria used for crop improvement are yield, resistance to biotic (diseases and pests) and abiotic stresses, and uniformity of the crop. Appearance, shelf life, taste, commercial quality, and nutritional value are part of product improvement.

5.1 Agronomic Value and Appearance: Introduction of F_1 Hybrids

In cabbage breeding, the major advance in the last decades has been the introduction of F_1 hybrids using the natural system of self-incompatibility. Their use improves not only uniformity and production, but also early maturity because of the notable vigorous development consequence of the heterosis. The optimum market qualities required today can only be obtained from F_1 hybrid cultivars, which, therefore, are the only ones produced by all major seed companies. The two systems currently used for seed production have been the incorporation of male sterility and the selection for incompatibility. Both have been a major concern of cabbage breeding programs.

In the past, brassica hybrid breeding has been done using the sporophytic self-incompatibility mechanism since there is no cytoplasmic male sterility in *B. oleracea*. The instability and complex inheritance of the self-incompatibility mechanism makes it difficult to use, contributing to a low quality of F_1 hybrids.

However, the production of brassica F_1 hybrids is now developing faster, using cytoplasmic male sterility introduced from *Raphanus sativus* into *B. oleracea* and double haploids parent lines obtained through microspore culture. In the last decades, an increase of male sterility in the production of new hybrids was reported since the system is simpler to develop and less influenced by the environment than the incompatibility system.

The most important quantitative traits desirable in cabbage breeding have been: head weight (around 1.0-1.5 kg), short growing period, maturity, frost hardiness, storage ability, and morphological traits related to head development (de Moel and Everaats, 1990). Both genotype and growing season influence the yield and head development (Sundstrom and Story, 1984; de Moel and Everaats, 1990; Kleinhenz and Wszelaki, 2003; Wszelaki and Kleinhenz, 2003).

Growers are looking for high quality commercial products in terms of characters that contribute favourably towards the vegetable's appearance. Traits relative to appearance include head shape and colour, the number of outer leaves, firmness, attractive green colour of the wrapper leaves, and core dimensions (de Moel and Everaats, 1990). These are essential traits in cabbage breeding, constituting the major objectives in some cases. Firm heads with short internal stems and late forms not subject to frost damage or rotting outer leaves are traits to be considered. In the last decades, commercial cabbage seed market demand that the new cultivars must be essentially round with small compact frames (Dickson and Wallace, 1986). Heads with dark green and well-developed wrapper leaves have been considered necessary for fresh market. Positive genetic correlations between these characters will help the breeder during selection.

In kales, the most desirable agronomic traits are related to early vigour, lodging resistance, winter hardiness, uniformity, forage yield, and digestible organic matter for use as forage crop (Bradshaw and Mackay, 1985). In the last decades, the main objective in kale breeding has been crop uniformity since a uniform brassica field reduces harvest time and makes grading much easier. Most kale varieties grown today are open-pollinated varieties, making high uniformity almost impossible to attain owing to the cross-pollination habit of the crop. Phenotypic uniformity is commonly achieved by making F_1 hybrids, relying on the homozygosity of the inbred parents. However, this improved uniformity is gained at the expense of a reduced genetic base, with the accompanying risks of susceptibility to diseases and pests. Although many of the kale cultivars would not be commercially viable today, many have specific traits which could still be useful. Kale cultivars may yet provide new sources of resistance to different diseases and pests.

As kale is also grown to feed cattle and sheep, the improvement of digestible organic matter yield in the dry matter has been an important selection criterion of any breeding program for this crop. In 1971, a kale breeding program started at the Scottish Crop Research Institute. A population improvement by half-sib family selection was used to increase the digestible organic matter yield in fodder kale and to reduce the content of indole glucosinolates (Bradshaw and Mackay, 1985; Bradshaw, 1987).

5.2 Disease and Pest Resistance

The increase in the economic importance of vegetable brassica crops has led to an increase in research on insects and pathogens that attack crops and on the plant host responses. A number of pests and diseases affect cabbage and kales in the field and in storage. A significant success in cabbage and kale breeding was the incorporation of insect resistance (especially lepidopterous larvae) and multiple disease resistance (especially clubroot and black rot). A great effort in breeding programs was to combine disease and pest resistance in cultivars suitable for fresh market production and for processing and storage.

Pest resistance
Brassica crops are attacked by a wide range of insects that feed on their roots, stems, leaves, and reproductive parts (Table 3). Most of the pests are crucifer specialists, along with a few generalist herbivores. The most common ones are lepidopterous pests which can cause significant damage in cabbages and kales because their optimum consumption period concurs with the maximum number of larvae present in the fields. Larvae feed on their leaves, affecting the value of the crop for human consumption. Several works have reported the differences in cultivar response to the attack of various insects by cruciferous crops, including cabbages (Dickson and Eckenrode, 1980; Hoy and Shelton, 1987) and kales (Pimentel, 1961; Radcliffe and Chapman 1966; Picoaga et al., 2003).

Larvae of imported cabbage worm (*Pieris rapae*), cabbage looper (*Trichoplusia ni*), and diamondback moth (*Plutella xylostella*) are probably the most common lepidopterous pests of brassica crops worldwide. Other insects such as cabbage root fly (*Delia radicum*), cabbage aphid (*Brevicoryne brassicae*), and flea beetles (*Phyllotreta* spp.) can also cause severe damages to cabbage and kales.

Insect pests are controlled by using insecticides, although there is risk of residues remaining in the crops. The use of resistant cultivars could benefit growers by reducing insecticide use and decreasing the rate at which insects develop resistance to insecticides. Research focused on developing resistant *B. oleracea* cultivars as a method for providing sustainable long-term control of pests and particularly to three lepidopterous species: *P. rapae*, *P. xylostella*, and *T. ni* has been extensively reported (Radcliffe and Chapman, 1966, Dickson and Eckenrode, 1980; Lamb, 1989; Stoner, 1990; Godin and Boivin, 1998). However, until now, breeding for resistance to lepidopterous insects has yielded very little success.

Resistance in some USA cabbage lines to *P. rapae*, *P. xylostella, M. brassicae,* and *E. forficalis* has been found (Dickson and Eckenrode, 1975, 1980; Dickson et al., 1984). Recently, Picoaga et al. (2003) found resistance to *M. brassicae* in Galician kale landraces. Studies of genetic resistance to *P. rapae* have shown that resistance is quantitative, with additive and dominance effects. Several resistance mecha nisms have been studied. For instance, glossy leaves and high concentrations of glucosinolates are preferred leaf traits for oviposition of some insect pests (Stoner, 1990, 1992). For example, the diamondback moth preferred to oviposit on glossy varieties of collards and cabbages; red cabbage cultivars were less preferred for oviposition by *P. rapae* than green cultivars (Dickson and Eckenrode, 1975; Stoner, 1992).

Table 3. Common insect pests of cabbage and kale (based on Chiang et al., 1993; Lamb, 1989).

Common name	Scientific name	Family	Distribution
Order Lepidoptera			
Imported cabbage worm	*Pieris rapae*	*Pieridae*	North America and Europe
Large cabbage white butterfly	*Pieris brassicae*	*Pieridae*	Europe
Cabbage looper	*Trichoplusia ni*	*Noctuidae*	North America and Europe
Diamond-back moth	*Plutella xylostella*	*Plutellidae*	Worldwide
Cabbage moth	*Mamestra brassicae*	*Noctuidae*	Europe
Bertha armyworm	*Mamestra configurata*	*Noctuidae*	North America
Garden pebble moth	*Evergestis forficalis*	*Pyralidae*	Europe
Order Diptera			
Cabbage root fly	*Delia radicum*	*Anthomyidae*	North America and Europe
Order Homoptera			
Cabbage aphids	*Brevicoryne brassicae*	*Aphididae*	Worldwide
Whiteflies	*Aleyrodes* spp.	*Aleyrodidae*	Southern US and Europe
Order Coleoptera			
Flea beetles	*Phyllotreta* spp.	*Chrysomelidae*	Worldwide
Cabbage stem flea beetles	*Psylliodes* spp.	*Chrysomelidae*	Europe and Canada

Moderate levels of resistance to *D. radicum* have been identified in cabbage and kale cultivars, and other forms of *B. oleracea,* but the level of resistance has not been considered high enough for a practical exploitation in breeding programs (Dosdall et al., 2000). Resistance may be linked to the differences in root morphology between horticultural forms. Strong correlation exists between infestation by cabbage root fly and root rot fungi (primarily *Fusarium* ssp.) with both agents contributing to decreased crop yields.

There are no commercial brassica crops, available in Europe, resistant to *B. brassicae*. The resistance of those few *B. oleracea* accessions that have been reported as resistant to *B. brassicae* results from the agronomically unacceptable glossy leaf trait. This trait also makes plants extremely susceptible to flea beetle attack (Stoner, 1992). In response to this, a research program was carried out which identified a number of sources of resistance to *B. brassicae* in both wild and cultivated species of *Brassica* (Singh et al., 1994; Ellis et al., 1996, 1998).

Disease resistance

Several diseases attack cabbage and kale crops worldwide (Table 4). Black rot (*Xanthomonas campestris* pv. *campestris*) and clubroot (*Plasmodiophora brassicae* Woron.) could be considered the most important diseases. Both diseases are hard to prevent by cultural practices and chemical treatments are not generally applicable. The identification and utilization of resistance genes to pests and diseases would be of benefit in reducing the need for chemical substances, thus reducing adverse environmental effects and decreasing costs. A great effort has been made in the last decades to improve plant resistance against black rot and clubroot. To develop reliable host resistance, brassica breeders have been searching for resistant sources with high yield potential.

Table 4. Common diseases of cabbage and kale (based on Dickson and Wallace, 1986; Tewari and Mithen, 1999).

Common name	Scientific name	Distribution
Clubroot	*Plasmodiophora brassicae*	Worldwide
Black rot	*Xanthomonas campestris* pv. *campestris*	Worldwide
Alternaria leaf spot	*Alternaria* ssp.	Worldwide
Blackleg or stem canker	*Leptosphaeria maculans*	Worldwide
Downy mildew	*Peronospora parasitica*	Worldwide
White rust	*Albugo candida*	North America and Europe
Fusarium yellows	*Fusarium oxysporum* f. *conglutinens*	North America and Europe
Bottom rot	*Rhizoctonia solani*	Worldwide
Blackleg or stem canker	*Leptosphaeria maculans*	Worldwide

Resistance against black rot has been identified in different genotypes of *B. oleracea*, including commercially available cultivars of cabbage and kales (Hansen and Earle, 1995; Taylor et al., 2002; Tonguç and Griffiths, 2004). However, none of these sources have provided complete resistance to the disease and its quantitative genetic control complicates its use to produce resistant hybrid varieties. Although several sources of resistance against clubroot are available (Crute et al., 1980; Crute, 1986; Crisp et al., 1989; Dias et al., 1993), the results of resistance breeding have been largely disappointing and few breeding programs for resistance have been successful.

The cultivars carrying clubroot resistance have found limited applications because of their insufficient level of resistance or because of their insufficient quality. Resistance has come primarily from kale and rutabaga. High levels of resistance to clubroot within the kale group have been reported (Crisp et al., 1989; Monteiro and Williams, 1989) but it has been difficult to incorporate this resistance into desired morphotypes of *B. oleracea*. The biennial character of 'Galega kale' accessions evaluated by Monteiro and Williams (1989) could have been the main factor in the accumulation of clubroot resistance. Resistance in *B. oleracea* usually involves several genes for control of each race. A group of major genes showing dominance effects in the resistance to race 6 of clubroot in cabbage was reported by Chiang and Crête (1976). Resistance to clubroot is quantitative under polygenic control and involves recessive (Crute et al., 1980, 1983; Voorrips, 1995) or dominant (Grandclément et al., 1996) alleles. A detailed review of inheritance of clubroot resistance in *B. oleracea* has been done by Voorrips et al. (2003). Considerable efforts have been made in the last decade to introduce major resistance genes to clubroot into *B. oleracea* from other *Brassica* species (Chiang and Crête, 1983).

Alternaria brassicae and *Leptosphaeria maculans* generally attack oilseed brassica crops causing notable economic loss. Most breeding programs have been conducted on *B. napus* and *B. rapa* cultivars, although some research has been performed on certain vegetable brassica crops. For instance, *A. brassicae* can cause black spot disease in cabbages and kales by developing spotting symptoms on any part of the plant and is capable of significantly reducing the market value of the product. Resistance to *Alternaria* ssp. was not found in *B. oleracea* whereas high resistance was observed in the wild relatives of *Brassica* outside the tribe *Brassicaceae*. *Leptosphaeria maculans* is also a critical pathogen of horticultural *Brassica* species, although the search for sources of resistance on kale and cabbage cultivars has not been as extensive as that on rapeseed. Other fungal diseases, such as fusarium yellows, downy mildew, powdery mildew, and bottom rot, have been described in cabbage by Dickson and Wallace (1986).

5.3 Nutritional Quality: Glucosinolates and Their Role in Pest and Disease Resistance

The rapid transformation of the vegetable market has forced producers of brassicas to increase the quality of their products. Nowadays, consumers are aware of the need for a constant supply of phytochemicals contained in plants to get optimal health benefits and there is a growing tendency to demand quality products with a higher added value. In this aspect, cabbage and kale crops are becoming more popular because of their nutritional value and anti-cancer properties. Marketable yield, traditionally considered as the main priority for breeders, is not of prevailing importance in cabbage and kale breeding. Because vegetables form an essential part of a well-balanced diet, breeders have paid in the last years a careful attention to the chemical composition of the marketed vegetables. Brassica vegetables including cabbage and kale are very popular and highly nutritious, being consumed in enormous quantities all over the world. For instance, they contain high amounts of calcium, carotene, vitamin C, and vitamin E. They are also rich in total antioxidants (Nilsson et al., 2006), with kale rated as the second vegetable showing the highest level of antioxidants among 22 tested (Cao et al., 1996).

In the past decade, much interest has been devoted to the positive effects of glucosinolates, a class of phytochemicals whose breakdown products are reported to possess cancer preventive activity (Rosa, 1999; Mithen et al., 2000; Fahey et al., 2001; Finley, 2005; Smith et al., 2005). Research impetus was initially generated when breakdown products were found to be partly responsible for the characteristic flavour of brassica vegetables (Fenwick et al., 1983b). More recently, numerous studies have related the decreased risk of cancer to a diet rich in brassica vegetables. The toxic and biological effects of glucosinolates have been widely reviewed (Fenwick et al., 1983a; Rosa et al., 1997).

Glucosinolates are the major class of secondary metabolites found in brassica crops. The molecule comprises a β-thioglucoside N-hydroxysulfate containing a side chain and a β-D-glucopyranose moiety. The biosynthesis of glucosinolates has been the subject of several comprehensive reviews (Giamoustaris and Mithen, 1996; Rosa et al., 1997; Fahey et al., 2001; Halkier and Gershenzon, 2006). Glucosinolates can be grouped into three chemical classes: aliphatic, indolyl, and aromatic, according to whether their amino acid precursor is methionine, tryptophan or an aromatic amino acid (tyrosine or phenylalanine), respectively (Giamoustaris and Mithen, 1996). The most important glucosinolates found in brassica vegetables are methionine-derived glucosinolates (Mithen et al., 2003). In the edible parts of cabbage and kales several glucosinolates belonging to the three chemical classes have been identified. Sinigrin, glucobrassicin, and glucoiberin have been identified as the major glucosinolates in these crops, with sinigrin making the major contribution of glucosinolates in kales while glucobrassicin or glucoiberin do so in cabbage leaves.

When tissue is damaged, the glucosinolates are hydrolyzed to various bioactive breakdown products by the endogenous plant enzyme myrosinase (thioglucoside glucohydrolase, EC 3.2.1.147). These breakdown products include isothio cyanates, thiocyanates, epithonitriles, oxazolidine, and nitriles (Fenwick et al., 1983a), depending on the substrate, pH conditions, availability of ferrous ions, and level and activity of specific protein factors such as the epithiospecifier protein (ESP) (Halkier and Gershenzon, 2006). At physiological pH, isothiocyanates are the major products whereas at more acid pH, nitriles are formed.

Certain glucosinolates have been identified as potent cancer-prevention agents due to the ability of some hydrolysis products to induce phase II detoxification enzymes, such as quinone reductase, glutathione-S-transferase, and glucuronosyl transferases. Isothiocyanates have different potential in cancer chemoprevention. Sulforaphane, the isothiocyanate derived from glucoraphanin found in broccoli, has been the object of numerous studies (Zhang et al., 1992; Farnham et al., 2000, 2004; Brown et al., 2002). Sulforaphane and other isothiocyanates may prevent tumour growth by blocking the cell cycle and promoting apoptosis (Keum et al., 2004). There may be additive and/or synergistic effects of isothiocyanates absorbed in combination from the diet. The most important glucosinolates found in the leaves of cabbages and kales are summarized in Table 5.

Table 5. Principal glucosinolates identified in leaves of cabbage and kale (based on Rosa, 1999).

Trivial name	Glucosinolate side chain	Common name	Reference
Sinigrin	2-Propenyl-	White cabbage	VanEtten et al. (1976, 1980)
		Savoy cabbage	VanEtten et al. (1976, 1980)
		Red cabbage	VanEtten et al. (1976, 1980)
		Kale	Carlson et al. (1987)
			Kushad et al. (1999)
		Collards	Carlson et al. (1987)
Glucoiberin	3-Methylsul-finylpropyl-	White cabbage	VanEtten et al. (1976, 1980)
		Savoy cabbage	VanEtten et al. (1976, 1980)
		Red cabbage	VanEtten et al. (1976, 1980)
		Kale	Carlson et al. (1987)
		Collards	Carlson et al. (1987)
Glucobrassicin	Indol-3-ylmethyl-	White cabbage	Sones et al. (1984)
		Savoy cabbage	Sones et al. (1984)
		Kale	Carlson et al. (1987)
		Collards	Carlson et al. (1987)
Progoitrin	2-Hydroxybut-3-enyl-	Savoy cabbage	VanEtten et al. (1976, 1980)
		Red cabbage	VanEtten et al. (1976, 1980)
		Kale	Carlson et al. (1987)
Gluconapin	But-3-enyl-	Red cabbage	VanEtten et al. (1976, 1980)
		Kale	Carlson et al. (1987)
Glucoraphanin	4-methylsul-finylbutyl-	Red cabbage	VanEtten et al. (1976, 1980)
		Kale	Kushad et al. (1999)

Although the primary function of glucosinolates in plants is presently unknown, several studies have suggested that they play an allelopathic role in plant resistance against fungi, nematodes, herbivores, and weeds (Rosa et al., 1997). In the past decades, the importance of these plant secondary metabolites has increased after the discovery of their potential in cancer prevention, as biofumigants in agriculture, and in crop protection. Moreover, the presence of glucosinolates in *Arabidopsis* has promoted a great research effort into the study of these metabolites (Halkier and Gershenzon, 2006).

6 Current Goals of Breeding

The current goals of breeding programs are focused on the incorporation of multiple diseases and insect resistance and on increasing the nutritional quality of the product by modifying the profile and content of glucosinolates.

6.1 Incorporation of Multiple Disease and Insect Resistance

The development of insect and pest resistant varieties is a current priority in brassica breeding programs. The evaluation and screening of new sources of resistance to the most common pests and pathogens, the study of the genetic basis of resistance, and the identification of genes for resistance are present goals of breeding programs in cabbage and kale. Nevertheless, the genetic basis of multiple disease resistance is complicated by environmental interactions with pathogens and correlated gene expression of different resistance traits. Notable advances have been made in the use of molecular methods for locating the position of resistance genes on genetic maps of brassicas.

Sources of resistance to most common diseases have been identified in cabbages and kales but have not been transferred into commercial varieties yet. The most important diseases for kales and cabbages, clubroot and black rot, are presently maintained below the threshold of economic damage with agronomic practices such as crop rotation and application of fungicides. The interest in releasing environmentally friendly varieties involving a minimum use of chemicals is related to the increasing present demand of crops resistant to several pests and diseases. Breeding for clubroot resistance, especially in *B. oleracea*, has been troubled by the complex of the plant-pathogen interaction and by inadequate characterization of the resistance nature.

The introgression of the resistance genes to black rot and to clubroot into *B. oleracea* from other *Brassica* species has been widely studied by several authors (Hansen and Earle, 1995; Taylor et al., 2002; Tonguç and Griffiths, 2004). Resistance was achieved through the development of interspecific and intergeneric hybrids (by using the embryo rescue technique) and backcross plants between accessions and cultivars of *B. oleracea* crops. For example, Hansen and Earle (1995) introduced a dominant resistance to black rot from *B. carinata* into *B. oleracea* by means of protoplast fusion although inheritance of this resistance was unstable. Tonguç and Griffiths (2004) introduced resistance to races 1 and 4 found in *B. juncea* into *B. oleracea* and all hybrids and backcrosses obtained were reported as resistant to both races.

Achievements on the inheritance of resistance to black rot (Tonguç and Griffiths, 2004) and to clubroot (Voorrips et al., 2003) have been recently reviewed. For the first time, resistance genes were identified based on gene-for-gene interaction with different races of the pathogen.

Despite the intense effort made to identify sources of resistance or tolerance to lepidopterous pests, only a few resistant cultivars have been reported. In recent years, several cultural and chemical control strategies have been identified for reducing losses due to insects such as cabbage root maggot and lepidopterous infestations. But in spite of implementing these strategies to limit the impact of some insect pests, these can still cause substantial losses. Breeding programs aimed at the introduction of resistance to these insects are presently scarce, and as a result, moderate levels of resistance have been reported for most insect pests. Consequently, research on the identification of new sources of resistance useful for cabbage and kale improvement as well as for transfer to other brassica crops is reviewed. In recent years, researchers have started to explore the possibility of transferring resistance traits to the cabbage aphid from wild to cultivated *Brassica* species (Pink et al., 2003) and to cabbage root maggot from *Sinapis alba* to *B. napus* (Dosdall et al., 2000). F_2 populations obtained from hybrids are currently used to develop molecular markers to map the locations of the genes for resistance. Once markers are identified, this information can be used to transfer genes for insect resistance to commercial cultivars.

The future prospects to control brassica pathogens and pests using all the available molecular tools are quite promising. A significant advance in understanding the genetic basis of resistance has been obtained with the mapping of disease resistance genes. Several *Brassica* linkage maps based largely on restriction fragment length polymorphism (RFLPs) markers have been developed to identify major quantitative trait loci (QTLs) for clubroot resistance in *B. oleracea* (Voorrips et al., 1997; Rocherieux et al., 2004). Landry et al. (1992) were the first to report genetic markers for resistance to clubroot race 2 and they found two QTLs associated to resistance to this disease. RAPD markers associated to this trait were detected by Grandclément et al. (1996). In another study, Moriguchi et al. (1999) identified three major QTLs for clubroot resistance, each in a separate linkage group, in a population derived from a cross between a clubroot-susceptible inbred cabbage line and a resistant inbred kale line.

Despite the existence of resistant cultivars, little is known about the genetic control of resistance of *B. oleracea* to *X. campestris* pv. *campestris*. Quantitative trait loci controlling juvenile and adult plant reactions of *B. oleracea* to this disease were mapped using RFLPs from a resistant cabbage × susceptible broccoli cross (Camargo et al., 1995). Two genomic regions on linkage groups 1 and 9 were associated with both young and adult plant resistance. The authors also found two additional QTLs on linkage group 2 associated only with young plant resistance.

The role of glucosinolates on pest and disease resistance is another major aim of current breeding programs. Many secondary metabolites found in plants have a role in defence against herbivores, pests and pathogens (Bennett and Wallsgrove, 1994). Extensive research work has been done on glucosinolates as plant defence compounds and their potential for the control of pests (Giamoustaris and Mithen, 1995; Hopkins et al., 1998; Agrawal and Kurashige, 2003) and diseases (Rosa and Rodrigues, 1999; Tierens et al., 2001). These compounds may act as feeding deterrents or oviposition stimuli for the major pests of cabbage and kales. Isothiocyanates, along with indolyl glucosinolates, have also been shown to be major determinants in the feeding response of the brassica specialists. Specialist feeders have clearly turned the glucosinolate defence mechanism against the plant, using the presence of the defence compounds to identify and locate the host.

Crucifer-specialized herbivores such as *P. xylostella*, *P. rapae*, and *B. brassicae* often use glucosinolates or their breakdown products as an attractive signal for the identification of suitable host plants (Giamoustaris and Mithen, 1995; Reddy and Guerrero, 2000). In addition, they are able to detoxify glucosinolates present in these crops. For instance, *P. xylostella* larvae feed exclusively on crucifers and produce a gut sulphatase that cleaves glucosinolates to form inactive desulfoglucosinolates (Ratzka et al., 2002). A different detoxification mechanism is found in *P. rapae* in the form of a gut enzyme that directs glucosinolate breakdown toward nitriles, which appear to be less toxic than isothiocyanates (Wittstock et al., 2004). Larvae of *T. ni*, a generalist lepidopteran herbivore, tend to feed on the older, less well-defended parts of the plant. *Brevicoryne brassicae* a crucifer-feeding specialist sequesters plant-derived glucosinolates and produces its own myrosinase as a defence against predators.

6.2 Development of Brassica Crops with Modified Glucosinolate Content

There is currently an increasing interest in the alteration of levels of specific glucosinolates in crop plants as certain glucosinolates have desirable properties in cancer prevention and crop protection. The study of inheritance of genes controlling glucosinolate biosynthesis will allow for manipulation of these genes and the development of lines of cabbage and kale with specific glucosinolate profiles. The biology and biochemistry of glucosinolates has been widely reviewed (Halkier and Gershenzon, 2006) and significant progress has been made in understanding the genetics of glucosinolate biosynthesis. Recent studies have identified key genes responsible for structure regulation and glucosinolate content in different *Brassica* species.

Research on *Arabidopsis* has lead to a notable increase in the knowledge on the biology and biochemistry of glucosinolates (Halkier and Gershenzon, 2006). The identification of glucosinolate biosynthetic genes in *Arabidopsis* has allowed the modification of the glucosinolate profile in this species (Li and Quiros, 2003). Thus, the cloning of some key genes provides the opportunity to engineer brassica crops with specific glucosinolate content by modifying their glucosinolate side-chains. This manipulation can have several possible applications. For example, down-regulation of *BoGSL-ALK*, which controls side-chain desaturation and cosegregates with *BoGSL-OH*, responsible for side-chain hydroxylation, could produce brassica varieties lacking the antinutrient progoitrin and would simultaneously produce plants accumulating glucoraphanin as a source of anticarcinogens. Although considerable progress is being made in that field, presently, it is not possible to identify the genes responsible for the variations in glucosinolate content. These genes will be essential in attempts to initiate molecular strategies for modifying the level of glucosinolates with the aims of improving flavour and nutritional aspects and to study interactions with pests. To date, most effort has been devoted to the modification of the expression of CYP79 enzymes, which catalyze the conversion of amino acids to aldoximes (Halkier and Gershenzon, 2006).

Most of the genes controlling glucosinolate breakdown are still unknown. In the near future, the isolation of genes of glucosinolate metabolism could facilitate the development of plants with modified glucosinolate profiles. Indeed, manipulation of glucosinolate metabolism will allow the study of the potential of glucosinolates to improve human nutrition and to improve the pest resistance.

7 Breeding Methods and Techniques

7.1 Breeding Methods

The breeding methods used with kales and cabbages have been determined by the mode of reproduction of these crops, the type of variety, and the inheritance of the particular trait to be improved. Kales and cabbages are outbreeding species, and consequently, various methods of population improvement have been used. Strategies destined to obtain improved populations have been applied mainly on kales and strategies destined to obtain hybrids have been applied mainly on cabbages. In many cases, mass selection has been successful, but this simple method was replaced by advanced types of recurrent selection. New population cultivars which were previously obtained by mass selection were raised by means of single plant selection and consecutive progeny testing. The breeding methods including intrapopulation phenotypic selection (mass selection and individual selection) have been described in these crops (Kristofferson, 1927; Bradshaw, 1984; Bradshaw and Mackay, 1985; Chiang et al., 1993).

The easiest type of family selection scheme for the kale breeder is half-sib family selection on a biennial cycle, in which the half-sib families are produced by the insect pollinators in an isolation cage. The production of full-sib families of kale involves more work than the production of half-sib families, but the rate of population improvement that could be achieved is faster.

The production of cabbage has profited considerable from hybrid breeding. Cabbage breeders have used the advantage that hybrid heterosis provide to increase yield and uniformity maturity. For breeding of hybrid cultivars, various requirements are necessary: an available system to large scale of seed production and the heterosis must be sufficient to allow the identification of hybrids with high combining ability. The diallel cross has been frequently used to evaluate the magnitude of heterosis in F_1 hybrids (Chiang et al., 1993). In 1960, all cabbage cultivars were basically open-pollinated. By 1980 most cultivars were hybrid except some old cultivars that still perform well in special locations. Today, most cabbage varieties grown are F_1 hybrids (Zhiyuan et al., 1999).

7.2 Tissue Culture and Transgenic Technology

Brassica species have pioneered the development of ‘in vitro’ culture techniques to produce interspecific hybrids by protoplast fusion, for mass propagation, and to produce double haploid lines by microspore and anther cultures. Anther culture and microspore culture have been reliable techniques of rapid homozygous plant production to derive double haploid lines. Double haploid lines are particularly useful in breeding and in genetic studies since they allow the rapid establishment of homozygous lines from wide crosses which may cover a considerable range of phenotypic variation. Successful anther and microspore cultures have been reported for several crops in *B. oleracea* including cabbages (Kuginiki et al., 1999) and kales (Arnison and Keller, 1990). The value of microspore culture technology in brassicas is evident in outcrossing self-incompatible crops, such as cabbages and kales. Double haploids have been valuable in the mapping of genes and in determining their linkage relationship to other genes important in plant breeding. Double haploids have been also used to detect QTLs and to analyze traits within populations.

Transgenic cabbage and kale cultivars with enhanced resistance or horticultural breeding are currently available (Christey and Braun, 2004). Most of the work on *Brassica* transformation has dealt with *B. napus* because of its economic importance. However, some transgenic plants have been also recovered from major brassica vegetables including cabbages and kales. While transgenic canola is widely grown, there are few reports of vegetable brassica field trials and there are no commercial genetically-modified vegetable brassica crops.

Various gene transfer methods have been highly effective. The most general approach for genetic transformation is to use *Agrobacterium tumefaciens* and *Agrobacterium rizhogenes*. Some examples of cabbage and kale transformation using *Agrobacterium* are shown in Table 6. *Agrobacterium rhizogenes* will also transfer the T-DNA of binary vectors in trans, thereby enabling the production of transgenic plants containing foreign genes after regeneration from hairy roots. Production of fertile transgenic brassicas via Ri-mediated transformation has been also reported in kales (Hosoki et al., 1989; Christey and Sinclair, 1992; Cogan et al., 2001) and cabbages (Berthomie and Jouanin, 1992; Christey et al., 1997; Cogan et al., 2001).

Table 6. Cabbage and kale transformation using *Agrobacterium tumefaciens* and *A. rizhogenes* (reviewed by Christey and Braun, 2004).

Species	Crop	Explant	Reference
A. tumefaciens			
	Cabbage	leaf petiole	Metz et al. (1995)
	Cabbage	hypocotyl	Jin et al. (2000)
	Cabbage	hypocotyl	Lee et al. (2000, 2002)
A. rizhogenes			
	Cabbage	hypocotyl	Cogan et al. (2001)
	Cabbage	leaf	Christey et al. (1997)
	Red cabbage	leaf petiole	Berthomie and Jouanin (1992)
	Kale	hypocotyl	Hosoki et al. (1989), Cogan et al. (2001)
	Forage kale	leaves	Christey and Sinclair (1992)

The location and exploitation of genes for pest and disease resistance is a current goal in cabbage and kale breeding. For the future, it is important to identify genes playing key roles in defence systems that are functional in transgenic crops under field conditions. A combination of various breeding strategies and transgenes might be a way to provide more durable disease and pest resistance. Strategies for the generation of transgenic plants with increased resistance to insect herbivory and diseases have been performed. Two main types of transgenic insect resistant brassicas have been developed: those expressing *Bacillus thuringiensis* (Bt) genes and those expressing proteinase inhibitor (PI) genes, capable of controlling a wide spectrum of insect pests.

Most of the insect resistance transgenic *Brassica* produced until now expresses genes encoding insecticidal crystal proteins from *B. thuringiensis*. Bt genes have been expressed in all the major groups of brassica crops including kales and cabbages. In addition to the material published, private companies have created and evaluated various Bt brassicas for possible commercial use. However, to date no Bt brassicas have been released commercially because of the general concerns about public acceptance of all genetically-modified crops, especially in Europe.

Recently, transgenic cabbage resistant to *P. xylostella* has been developed through *A. tumefaciens*-mediated transformation with *B. thuringiensis* (Bt) *cry* genes. Insect resistant crops expressing *cry* genes were first grown commercially in 1996 and since then various *cry* genes have been introduced into several *B. oleracea* crops. Several *cry* genes introduced into cabbage and kale have been shown to control successfully insect pests such as *P. xylostella*. Some examples of transgenic cabbage and kale plants expressing *B. thuringiensis* (Bt) genes and proteinase inhibitor (PI) genes are shown in Table 7. Transgenic cabbage plants transformed with a synthetic Bt gene *cry1Ab3* were resistant to larvae of diamondback moth whereas plants transgenic for *cry1Ia3*, a wild type Bt gene, were susceptible (Jin et al., 2000). Larvae of several lepidopterous (*P. xylostella* and *P. rapae*) developed more slowly and caused less leaf damage on transgenic plants than on control. However, resistance was only partial with delay of insect development rather than insect mortality.

Table 7. Transgenic cabbage and kale plants expressing *Bacillus thuringiensis* (Bt) genes and with proteinase inhibitor (PI) genes.

Common name	Insect	Genes	Reference
Cabbage	*P. xylostella*	Bt (*cry1Ac)*	Metz et al. (1995)
		Bt (*cry1Ab3*)	Jin et al. (2000)
		Bt (*cry1Ab)*	Bhattacharya et al. (2002)
	P. rapae	PI (CpTI)	Fang et al. (1997)
Kale/collard	*P. xylostella*	Bt (*cry1C)*	Cao et al. (1999)

8 Integration of New Biotechnologies in Breeding Programmes

The integration of new biotechnologies in breeding programs has focused on the use of genetic marker technologies. During the last decades, big progress has been made in brassica breeding due to a better understanding of the plant genome and the application of techniques for its manipulation at molecular level. Notable advances in the improvement of cabbage and kale crops should be achieved by using genetic marker technologies. However, most of the new cultivars produced until now still result from traditional procedures.

Molecular markers have been widely used to assist breeding and selection procedures in brassica crops. Marker development in brassicas has been an active subject since the early 1980s with the development of the first RFLP linkage maps for *B. oleracea*, *B. rapa*, and *B. napus* (reviewed by Quiros, 2001). Several maps have been developed independently for *B. oleracea* involving crosses between different crops. The development of DNA-based genetic markers and genetic mapping in recent years has made possible the study of structure, origin, and evolution of *Brassica* genomes (Quiros et al., 2001). *Brassica* species have complex genomes, highly duplicated with intra-and inter-genomic conservation of linkage blocks, which permits homoeologous recombination.

Most of the work using molecular markers in brassica breeding has been based until now on genetic mapping in single segregation populations generated to locate specific traits using various DNA marker systems. Maps are being extensively used to tag genes of interest, including QTLs of economic importance. In *B. oleracea*, the linkage maps have been applied to identify markers linked to morphological traits, disease resistance, and vernalization requirement for flower induction. Molecular markers linked to several agronomic traits have been reported and some of them have been successfully integrated into breeding programs. However, for quantitative traits, the success of mapping information to carry out marker-assisted selection has not achieved the first expectations. Mapping quantitative trait loci is often not sufficient to develop efficient markers for identification of genes or for trait introgression.

A detailed list of several studies involving application of DNA markers as effective tools for brassica breeding has been reported by Quiros (2001). A number of useful genes which may have economic impact have been mapped in brassica crops. Cytoplasmic male sterility restorers and disease resistance are the major genes mapped on *B. oleracea*. Linked markers to these genes could be used for molecular manipulation and characterisation, marker-assisted selection, and gene isolation. Intense efforts have been made to use a large range of biotechnological methods for the development of efficient systems of male sterility for the production of hybrid varieties. The area of glucosinolate composition has been especially studied due to the biological activity of the isothiocyanates, derived from these compounds, some of which have cancer-protecting properties, as previously described.

New technologies, including allele-trait association studies, candidate gene approaches, and single nucleotide polymorphism markers (SNPs), have been reviewed by Quiros and Paterson (2004) and developed mainly in *B. napus*. These approaches will be applied in the next years to cabbage and kale breeding.

The adoption of *Arabidopsis* as a plant model system for molecular genetics has cleared the way to study with detail the complex structure of *Brassica* genomes, since *Arabidopsis* is a close relative of *Brassica* species. Recently, ESTS (expressed sequenced tags) from *Arabidopsis* have been used as RFLP markers on *B. oleracea* for comparative alignment of the genomes of both species (Babula et al., 2003). Significant progress will be achieved in the coming years through integration of candidate gene approaches in brassica crops using the detailed information now available for the *Arabidopsis* genome.

9 Seed Production

The primary methods for production of commercial hybrid seed based on incompatibility are as follows (Chiang et al., 1993).

To produce a single-cross F_1 hybrid, two self-incompatible inbred lines with different S alleles are planted in alternate rows. The ratio of male to female is 1:1, 1:2, or 1:3, depending on the pollen-producing capacity of the male inbred. For a double-cross, four incompatible inbreds are needed. The crossing scheme is: $[(S_1S_1 \times S_2S_2) \times (S_3S_3 \times S_4S_4)]$. Finally, for a top-cross a good open-pollinated cultivar is used as the pollen parent and a single incompatible inbred as the female parent.

To cross plants in the glasshouse by hand, medium or fine tweezers are the one essential tool needed; emasculating scissors may also be used (Downey et al., 1980). Buds which are about to open or will open the following day are selected for emasculation. Such flowers give usually the best fertilization rate.

The production of inbreds must overcome self-incompatibility. A younger bud three to four days prior to natural flower opening will have the least self-incompatibility and still be large enough for bud pollination. From six to eight buds can be opened at one time with a pointed object such as a toothpick or a forceps, and pollen from an older open flower from the same plant can be transferred to the stigma and seed obtained. Pollen can be transferred with a small brush (Dickson and Wallace, 1986). Another way to obtain a large quantity of self-pollinated seed is CO_2 gas treatment. This technique is as effective and reliable as bud pollination, but less laborious (Chiang et al., 1993).

Another task the breeder must frequently face is to multiply populations. This can be accomplished by crossing plants by hand as described above. A more efficient system, used presently in the Misión Biológica de Galicia, consists in planting the population to multiply in an insect-isolated crib and putting inside a hive of *Bombus terrestris*. These hives are inexpensive and work efficiently for about a month.

References

Agrawal, A. A., and Kurashige, N. S. 2003. A role for isothiocyanates in plant resistance against the specialist herbivore *Pieris rapae*. J. Chem. Ecol. 29:1403-1415.

Allard, R. W. 1999. *Principles of Plant Breeding*, 2nd ed., Wiley, New York.

Anonymous. 1987. Around the world with *Brassicas*, Wisconsin Fast Plants, University of Wisconsin, Madison.

Arnison, P. G., and Keller, W. A. 1990. A survey of anther culture response of *Brassica oleracea* L. cultivars grown under field condition. Plant Breed. 104:125-133.

Babula, D., Kaczmarek, A., Barakat, A., Delseny, M., Quiros, C. F., and Sadowski, J. 2003. Chromosomal mapping of *Brassica oleracea* based on ESTs from *Arabidopsis thaliana*: complexity of the comparative map. Mol. Genet. Genom. 268:656-665.

Baladrón, J. J., and Ordás, A. 1988. Monitoring viability in the genus *Brassica*. Cruciferae Newsl. 13:124-125.

Bennett, R. N., and Wallsgrove, R. M. 1994. Secondary metabolites in plant defence mechanisms. New Phytol. 127:617-633.

Berthomieu, P., and Jouanin, L. 1992. Transformation of rapid cycling cabbage (*Brassica oleracea* var. *capitata*) with *Agrobacterium rhizogenes*. Plant Cell Rep. 11:334-338.

Bhattacharya, R. C., Viswakarma, N., Bhat, S. R., Kirti, P. B., and Chopra, V. L. 2002. Development of insect-resistant cabbage plants expressing a synthetic *cryIA(b)* gene from *Bacillus thuringiensis*. Curr. Sci. 83:146-150.

Boukema I. W., Jongen, M. W. M., and van Hintum, T. J. L. 1995. in: *Report of a Working Group on Brassica (Second Meeting, 13-15 November 1994, Lisbon, Portugal)*, T. Gass, M. Gustafsson, D. Astley, and E. A. Frison, compilers, International Plant Genetic Resources Institute, Rome, pp. 4-7.

Bradshaw, J. E. 1984. Computer simulation of family selection schemes suitable for kale (*Brassica oleracea* L.), involving half-sib and selfed families. Theor. Appl. Genet. 68: 503-508.

Bradshaw, J. E. 1987. The choice of selection index in kale (*Brassica oleracea* L.) population improvement. Theor. Appl. Genet. 75:165-169.

Bradshaw J. E., and Mackay, G. R. 1985. Half-sib family selection for yield of digestible organic matter in kale (*Brassica oleracea* L.). Euphytica 34:201-206.

Brown A. F., Yousef, G. G., Jeffery, E. H., Klein, B. P., Walling, M. A., Kushad, M. M., and Juvik, J. A. 2002. Glucosinolate profile in broccoli: variation in levels and implications in breeding for cancer chemoprotection. J. Amer. Hort. Sci. 127:807-813.

Camargo, L. E. A., Williams, P. H., and Osborn, T. C. 1995. Mapping of quantitative trait loci controlling resistance to *Brassica oleracea* to *Xanthomonas campestris* pv *campestris* in the field and the greenhouse. Phytopathology 85:1296-1300.

Cao, G., Sofic, E., and Prior, R. L. 1996. Antioxidant capacity of tea and common vegetables. J. Agr. Food Chem. 44:3426-3431.

Cao, J., Tang, J. D., Strizhov, N., Shelton, A. M., and Earle, E. D. 1999. Transgenic broccoli with high levels of *Bacillus thuringienses* Cry1C protein control diamondback moth larvae resistant to Cry1A or Cry1C. Mol. Breed. 5:131-141.

Carlson, D. G., Daxenbichler, M. E., VanEtten, C. H., Kwolek, W. F., and Williams, P. H. 1987. Glucosinolates in crucifer vegetables: broccoli, Brussels sprouts, cauliflower, collards, kale, mustard greens and kohlrabi. J. Amer. Hort. Sci. 112:173-178.

Chiang, M. S., and Crête, R. 1976. Diallel analysis of the inheritance of resistance to race 6 of *Plasmodiophora brassicae* in cabbage. Can. J. Plant Sci. 56:865-868.

Chiang, M. S., and Crête, R. 1983. Transfer of resistance to race 2 of *Plasmodiophora brassicae* from *Brassica napus* to cabbage (*Brassica oleracea* ssp. *capitata*), V. The inheritance of resistance. Euphytica 32:479-483.

Chiang, M. S., Chong, C., Landry, L. S., and Crête, R. 1993. Cabbage, in: *Genetic improvement of vegetable crops*, K. Kallooo, and Bergh, B. O., eds, Pergamon Press, Oxford, UK, pp. 113-155.

Christey, M. C., and Sinclair, B. K. 1992. Regeneration of transgenic kale (*Brassica oleracea* var. *acephala*), rape (*B. napus*) and turnip (*B. campestris* var. *rapifera*) plants *via Agrobacterium rhizogenes* mediated transformation. Plant Sci. 87:161-169.

Christey, M. C., and Braun, R. H. 2004. Production of transgenic vegetable Brassicas, in: *Biotechnology in Agriculture and Forestry*, vol. 54, *Brassica*, E. C. Pua, and Douglas, C. J., eds, Springer-Verlag, Berlin Heidelberg New York, pp. 169-194.

Christey, M. C., Sinclair, B. K., Braun, R. H., Wyke, L. 1997. Regeneration of transgenic vegetable brassicas (*Brassica oleracea* and *Brassica campestris*) via Ri-mediated transformation. Plant Cell Rep. 16:587-593.

Cogan, N., Harvey, E., Robinson, H., Lynn, J., Pink, D., Newbury, H. J., and Puddephat, I. 2001. The effects of anther culture and plant genetic background on *Agrobacterium rhizogenes*-mediated transformation of commercial cultivars and derived doubled-haploid *Brassica oleracea*. Plant Cell Rep. 20:755-762.

Crisp, P., Crute, I. R., Sutherland, R. A., Angell, S. M., Bloor, K., Burgess, H., and Gordon, P. L. 1989. The exploitation of genetic resources of *Brassica oleracea* in breeding for resistance to clubroot (*Plasmodiophora brassicae*). Euphytica 42:215-226.

Crute, I. R. 1986. The relationship between *Plasmodiophora brassicae* and its hosts. The application of concepts relating to variation in interorganismal associations. Adv. Plant Pathol. 5:1-52.

Crute, I. R., Gray, A. R., Crisp, P., and Buczacki, S. T. 1980. Variation in *Plasmodiophora brassicae* and resistance to clubroot disease in *Brassicas* and allied crops. A critical review. Plant Breed. Abstr. 50:91-104.

Crute, I. R., Phelps K., Barnes A., Buczacki S. T., and Crisp, P. 1983. The relationship between genotypes of three *Brassica* species and collections of *Plasmodiophora brassicae*. Plant Pathol. 32:405-420.

de Moel, C. P., and Everaats, A. P. 1990. Growth, development, and yield of white cabbage in relation to time of planting. Acta Hort. 267:279-288.

Dias, J. S., Ferreira, M. E., and Williams, P. H. 1993. Screening of Portuguese cole landraces (*Brassica oleracea* L.) with *Peronospora parasitica* and *Plasmodiophora brassicae*. Euphytica 67:135-141.

Dickson, M. H., and Eckenrode, C. J. 1975. Variation in *Brassica oleracea* resistance to cabbage looper and imported cabbage worm in the greenhouse and field. J. Econ. Entomol. 68:757-760.

Dickson, M. H., and Eckenrode, C. J. 1980. Breeding for resistance in cabbage and cauliflower to cabbage looper, imported cabbage worm and diamond back moth. J. Amer. Soc. Hort. Sci. 105:782-785.

Dickson, M. H., and Wallace, D. H. 1986. Cabbage breeding, in: *Breeding Vegetable Crops*. Bassett, M. J. ed., AVI Publishing Company, Inc. Westport, Connecticut, pp. 395-432.

Dickson, M. H., Eckenrode, C. J., and Blamble, A. E. 1984. NYIR 9602, NYIR 9605, and NYIR 8329 lepidopterous pest-resistant cabbage breeding lines. HortScience 19:311-312.

Dosdall, L. M., Good, A., Keddie, B. A., Ekuere, U., and Stringam, G. 2000. Identification and evaluation of root maggot (*Delia* spp.) (Diptera: Anthomyiidae) resistance within Brassicaceae. Crop Protection 19:247-253.

Downey, R. K., Klassen, A. J., and Stringam, G. R. 1980. Rapeseed and mustard, in: *Hybridization of Crop Plants*. Fehr, W. R., and Hadley, H. H. eds., American Society of Agronomy-Crop Science Society of America, Madison, Wisconsin, pp. 495-509.

Ellis, P. R., Singh, R., Pink, D. A. C., Lynn, R., and Saw, P. L. 1996. Resistance to *Brevicoryne brassicae* in horticultural brassicas, *Euphytica* 88:85-96.

Ellis, P. R., Pink, D. A. C., Phelps, K., Jukes, P. L., Breeds, S. E., and Pinnegar, A. E. 1998. Evaluation of a core collection of *Brassica oleracea* accessions for resistance to *Brevicoryne brassicae*, the cabbage aphid. Euphytica 103:149-160.

Fahey, J. W., Zalcmann, A. M., and Talalay, P. 2001. The chemical diversity and distribution of glucosinolates and isothiocyanates among plants. Phytochemistry 56:5-61.

Fang, H. J., Li, D. L., Wang, G. L., Li, Y. H., Zhu, Z., and Li, X. H. 1997. An insect resistance transgenic cabbage plant with cowpea trypsin inhibitor (CpTI) gene. Acta Bot. Sin. 39:940-945.

FAO, 2006, FAOSTAT data, Rome (June 2, 2006); http://faostat.fao.org.

Farnham, M. W., Stephenson, K. K., and Fahey, J. W. 2000. Capacity of broccoli to induce a mammalian chemoprotective enzyme varies among inbreed lines. J. Amer. Soc. Hort. Sci. 125:482-488.

Farnham, M. W., Wilson, P. E., Stephenson, K. K., and Fahey, J. W. 2004. Genetic and environmental effects on glucosinolate content and chemoprotective potency of broccoli. Plant Breed. 123:60-65.

Fenwick, G. R., Heaney, R. K., and Mullin, W. J. 1983a. Glucosinolates and their breakdown products in food plants, CRC. Crit. Rev. Food Sci. Nutr. 18:123-201.

Fenwick, G. R., Griffiths, N. M., and Heaney, R. K. 1983b. Bitterness in Brussels sprouts (*Brassica oleracea* L. var. *gemmifera*): the role of glucosinolates and their breakdown products. J. Sci. Food Agr. 34:73-80.

Finley, J. W. 2005. Proposed criteria for assessing the efficacy of cancer reduction by plant foods enriched in carotenoids, glucosinolates, polyphenols and seleno compounds. Ann. Bot. 95:1075-1096.

Giamoustaris, A., and Mithen, R. 1995. The effect of modifying the glucosinolate content of leaves of oilseed rape (*B. napus* ssp. *oleifera*) on its interaction with specialist and generalist pests. Ann. Appl. Biol. 126:347-363.

Giamoustaris, A., and Mithen, R. 1996. Genetic of aliphatic glucosinolates. Side-chain modification in *Brassica oleracea*. Theor. Appl. Genet. 93:1006-1010.

Godin, C., and Boivin, G. 1998. Seasonal occurrence of lepidopterous pests of cruciferous crops in southwestern Quebec in relation to degree-day accumulations. Can. Entomol. 130:173-185.

Gómez-Campo, C. 1999. Taxonomy, in: *Biology of Brassica Coenospecies,* C. Gómez-Campo, ed., Elsevier, Amsterdam, pp. 3-32.

Gómez-Campo, C. 2002. Long term seed preservation: the risk of selecting inadequate containers is very high, *Monographs ETSIA, Univ. Politécnica de Madrid* 163:1-10.

Gómez-Campo, C., and Prakash, S. 1999. Origin and domestication, in: *Biology of Brassica Coenospecies,* C. Gómez-Campo, ed., Elsevier, Amsterdam, pp. 33-58.

Grandclément, C., Laurent, F., and Thomas, G. 1996. Detection and analysis of QTLs based on RAPD markers for polygenic resistance to *Plasmodiophora brassicae* Woron. in *Brassica oleracea* L. Theor. Appl. Genet. 93:86-90.

Halkier, B. A., and Gershenzon, J. 2006. Biology and biochemistry of glucosinolates. Ann. Rev. Plant Biol. 57:303-333.

Hansen, L. N., and Earle, E. D. 1995. Transfer of resistance to *Xanthomonas campestris* pv. *campestris* into *Brassica oleracea* L. by protoplast fusion. Theor. Appl. Genet. 91: 1293-1300.

Hickey, M., and King, C. 1997. *Common Families of Flowering Plants*, Cambridge University Press, Cambridge, UK.

Hodgkin, T. 1995. Cabbages, kales, etc., in: *Evolution of Crop Plants,* 2nd ed., J. Smartt and N. W. Simmonds, eds., Longman, Burnt Mill, Harlow, UK, pp. 76-82.

Hopkins, R. J., Griffits, D. W., Birch, A. N. E., and McKinlay, R. G. 1998. Influence of increasing herbivore pressure on modification of glucosinolate content of swedes (*Brassica napus* spp. *rapifera)*. J. Chem. Ecol. 24:2004-2019.

Hosoki, T., Shiraishi, K., Kigo, T., and Ando, M. 1989. Transformation and regeneration of ornamental kale (*Brassica oleracea* var. *acephala* DC) mediated by *Agrobacterium rhizogenes*. Sci. Hort. 40:259-266.

Hoy, C. W., and Shelton, A. M. 1987. Feeding response of *Artogeia rapae* (Lepidoptera: Pieridae) and *Trichoplusia ni* (Lepidoptera: Noctuidae) to cabbage leaf age. Environ. Entomol. 16:680-682.

Jin, R. G., Lui, Y. B., Tabashnik, B. E., and Borthakur, D. 2000. Development of transgenic cabbage (*Brassica oleracea* var. *capitata*) for insect resistance by *Agrobacterium tumefaciens*-mediated transformation. In vitro Cell. Develop. Biol. Plant 36:231-237.

Keum, Y. S., Jeong, W. S., and Kong, A. N. T. 2004. Chemoprevention by isothiocyanates and their underlying molecular signaling mechanisms. Mutat. Res. Fundam. Mol. Mech. Mutagen. 555:191-202.

Kleinhenz, M. D., and Wszelaki, A. 2003. Yield and relationships among head traits in cabbage as influenced by planting date and cultivar. I. Fresh market. HortScience 38: 1349-1354.

Kristofferson, K. B. 1927. Contributions to the genetics of *Brassica oleracea*, II. Hereditas 9:343-348.

Kuginiki, Y., Miyajima, T., Masuda, H., Hida, K., and Hirai, M. 1999. Highly regenerative cultivars in microspore culture in *B. oleracea* L. var. *capitata*. Breed. Sci. 49:251-256.

Kushad, M. M., Brown, A. F., Kurilich, A. C., Juvik, J. A., Klein, B. K., Wallig, M. A., and Jeffery, E. H. 1999. Variation of glucosinolates in vegetable subspecies of *Brassica oleracea*. J. Agr. Food Chem. 47:1541-1548.

Lamb, R. J. 1989. Entomology of oilseed *Brassica* crops, *Ann. Rev. Entomol.* 34:211-229.

Landry, B. S., Hubert, N., Crete, R., Chiang, M. S., Lincoln, S. E., and Etoh, T. 1992. A genetic map of *Brassica oleracea* based on RFLP markers detected with expressed DNA sequences and mapping of resistance genes to race 2 of *Plasmodiophora brassicae* (Woronin). Genome 35:409-420.

Lázaro, A., and Aguinaglade, I. 1998. Genetic diversity in *Brassica oleracea* L. (Cruciferae) and wild relatives ($2n = 18$) using isozymes. Ann. Bot. 82:821-828.

Lee, Y. H., Lee, S. B., Suh, S. C., Byun, M. O., and Kim, H. I. 2000. Herbicide resistant cabbage (*Brassica oleracea* ssp. *capitata*) plants by *Agrobacterium*-mediated transformation. J. Plant Biotechnol. 2:35-41.

Lee, Y. H., Yoon, I. S., Suh, S. C., and Kim, H. I. 2002. Enhanced disease resistance in transgenic cabbage and tobacco expressing a glucose oxidase gene from *Aspergillus niger*. Plant Cell Rep. 20:857-863.

Li, G., and Quiros, C. F. 2003. In planta side-chain glucosinolate modification in *Arabidopsis* by introduction of dioxygenase *Brassica* homolog *BoGSL-ALK*. Theor. Appl. Genet. 106:1116-1121.

Metz, T. D., Dixit, R., and Earle, E. D. 1995. *Agrobacterium tumefaciens*-mediated transformation of broccoli (*Brassica oleracea* var. *italica*) and cabbage (*Brassica oleracea* var. *capitata*). Plant Cell Rep. 15:287-292.

Mithen, R. F., Dekker, M., Verkerk, R., Rabot, S., and Johnson, I. T. 2000. The nutritional significance, biosynthesis and bioavailability of glucosinolates in human foods. J. Sci. Food Agric. 80:967-984.

Mithen, R. F., Faulkner, K., Magrath, R., Rose, P., Williamson, G., and Marquez, J. 2003. Development of isothiocyanates-enriched broccoli, and its enhanced ability to induce phase 2 detoxification enzymes in mammalian cells. Theor. Appl. Genet. 106:727-734.

Mizushima, V. 1980. Genome analysis in *Brassica* and allied genera, in: *Brassica Crops and Wild Allies*, C. Gómez-Campo (ed.), Japan Scient. Soc. Press, Tokyo, pp. 89-105.

Monteiro, A. A., and Williams, P. H. 1989. The exploration of genetic resources of Portuguese cabbage and kale for resistance to several *Brassica* diseases. Euphytica 41:215-225.

Moriguchi, K., Kimizuka, C., Takagi, C., Ishii, K., and Nomura, K. 1999. A genetic map based on RAPD, RFLP, isozyme, morphological markers and QTL analysis for clubroot resistance in *Brassica oleracea*. Breed. Sci. 49:257-265.

Newbury, H. J., and Paterson, A. H. 2003. Genomic colinearity and its application in crop plant improvement, in: *Plant Molecular Breeding*, H. J. Newbury, ed., Blackwell, Oxford, UK, pp. 60-81.

Nilsson, J., Olsson, K., Engqvist, G., Ekvall, J., Olsson, M., Nyman, M., and Akesson, B. 2006. Variation in the content of glucosinolates, hydroxycinnamic acids, carotenoids, total antioxidant capacity and low-molecular-weight carbohydrates in *Brassica* vegetables. J. Sci. Food Agric. 86:528-538.

Paterson, A. H., Bowers, J. E., Burow, M. D., Draye, X., Elsik, C. G., Jiang, C., Katsar, C. S., Lan, T., Lin, Y., Ming, R., and Wright, R. J. 2000. Comparative genomics of plant chromosomes. Plant Cell 12:1523-1539.

Picoaga, A., Cartea, M. E., Soengas, P., Monetti, L., and Ordás, A. 2003. Resistance of kale populations to lepidopterous pests in northwestern Spain. J. Econ. Entomol. 96:143-147.

Pimentel, D. 1961. An evaluation of insect resistance in broccoli, Brussels sprouts, cabbage, collards, and kale. J. Econ. Entomol. 54:56-158.

Pink, D. A. C., Kift, N. B., Ellis, P. R., McClement, S. J., Lynn, J., and Tatchell, G. M. 2003. Genetic control of resistance to the aphid *Brevicoryne brassicae* in the wild species *Brassica fruticulosa*. Plant Breed. 122:24-29.

Prakash, S., and Hinata, K. 1980. Taxonomy, cytogenetics, and origin of crop *Brassica*, a review. Opera Bot. 55:1-57.

Quiros, C. F. 2001. DNA-based marker *Brassica* maps, in: *Advances in Cellular and Molecular Biology of Plants,* vol. I, DNA based marker in plants, R. L. Phillips, and Vasil, I. K., eds, Kluwer, Dordrecht, pp. 201-238.

Quiros, C. F., and Paterson, A. H. 2004. Genome mapping and analysis, in: *Biotechnology in Agriculture and Forestry*, vol. 54, *Brassica*, E. C. Pua, and Douglas, C. J., eds, Springer-Verlag, Berlin Heidelberg New York, pp. 31-42.

Quiros, C. F., Grellet, F., Sadowski, J. Suzuki, T., Li, G., and Wroblewski, T. 2001. *Arabidopsis* and *Brassica* comparative genomics: sequence, structure and gene content in the *ABI1-Rps2-Ck1* chromosomal segment and related regions. Genetics 157:1321-1330.

Radcliffe, E. B., and Chapman, R. K. 1966. Varietal resistance to insect attack in various cruciferous crops. J. Econ. Entomol. 59:120-125.

Ratzka, A., Vogel, H., Kliebenstein, D. J., Mitchell-Olds, T., and Kroymann, J. 2002. Disarming the mustard oil bomb. Proc. Natl. Acad. Sci. USA 99:11223-11228.

Reddy, G. V. P., and Guerrero, A. 2000. Behavioral responses of the diamondback moth, *Plutella xylostella*, to green leaf volatiles of *Brassica oleracea* subsp. *Capitata.* J. Agric. Food Chem. 48:6025-6029.

Röbbelen, G. 1960. Beitrage zur Analyse des *Brassica*-Genomes. Chromosoma 11:205-228.

Rocherieux, J., Glory, P., Giboulot, A., Boury, S., Barbeyron, G., Thomas, G., and Manzanares-Dauleux, M. J. 2004. Isolate-specific and broad-spectrum QTLs are involved in the control of clubroot in *Brassica oleracea*. Theor. Appl. Genet. 108:1555-1563.

Rosa, E. A. S. 1999. Chemical composition, in: *Biology of Brassica Coenospecies*, C. Gómez-Campo, ed, Elsevier Science B. V., Amsterdam, pp. 315-357.

Rosa, E. A. S., and Rodrigues, P. F. 1999. Towards a more sustainable agriculture system. The effect of glucosinolates on the control of soilborne diseases. J. Hort. Sci. Biotech. 74:667-674.

Rosa, E. A. S., Heaney, R. K., Fenwick, G. R., and Portas, C. A. M. 1997. Glucosinolates in crop plants. Hort. Rev. 19:99-215.

Schulz, O. E. 1936. Cruciferae, in: *Die natürlichen Pflanzenfamilien*, Vol. 17b, A. Engler and P. Prantl, eds., Duncker and Humblot, Berlin, pp. 227-658.

Shashilova, L. I. 1997. The Russian *Brassica* collection, in: *Report of a Working Group on Brassica, Third Meeting, 27-29 November 1996, Rome, Italy*, International Plant Genetic Resources Institute, Rome, pp. 56-57.

Singh, R., Ellis, P. R., Pink, D. A. C., and Phelps, K. 1994. An investigation of the resistance to cabbage aphid in *Brassica* species. Ann. Appl. Biol. 125:457-465.

Slocum M. K., Figdore, S. S., Kennard, W. C., Suzuki, J. Y., and Osborn, T. C. 1990. Linkage arrangement of restriction fragment length polymorphism loci in *Brassica oleracea*. Theor. Appl. Genet. 80:57-64.

Smith, T. K., Lund, E., Clarke, R., Bennet, R. N., and Johnson, I. T. 2005. Effects of Brussels sprout juice on the cell cycle and adhesion of human colorectal carcinoma cells (HT29) in vitro. J. Agric. Food Chem. 53:3895-3901.

Snogerup, S. 1980. The wild forms of the *Brassica oleracea* group (2n=18) and their possible relations to the cultivated ones, in: *Brassica Crops and Wild Allies*, S. Tsunoda, K. Hinata and C. Gómez-Campo, eds., Japan Scient. Soc. Press, Tokyo, pp. 120-132.

Snogerup, S., Gustafsson, M., and von Bothmer, R. 1990. *Brassica* sect. *Brassica (Brassicaeae)* 1. Taxonomy and variation. Willdenowia 19:271-365.

Sones, K., Heaney, R. K., and Fenwick, G. R. 1984. The glucosinolate content of UK vegetables-cabbages (*Brassica oleracea*), swede (*B. napus*) and turnip (*B. rapa*). Food Additives Contaminants 1:289-296.

Spooner, D. M., Hetterscheid, W. L. A., van den Berg, R. G., and Brandenburg, W. A. 2003. Plant nomenclature and taxonomy. Hort. Rev. 28:1-60.

Stoner, K. A. 1990. Glossy leaf wax and plant-resistance to insects in *Brassica oleracea* under natural infestation. Environ. Entomol. 19:730-739.

Stoner, K. A. 1992. Density of imported cabbageworms (Lepidoptera: Pieridae), cabbage aphids (Homoptera: Aphididae), and flee beetles (Coleoptera: Chrysomelidae) on glossy and trichome bearing lines of *Brassica oleracea*. J. Econ. Entomol. 85:1023-1030.

Sundstrom, F. J., and Story, R. N. 1984. Cultivar and growing season effects on cabbage head development and weight loss during storage. HortScience 19:589-590.

Swarup, V., and Brahmi, P. 2005. Cole crops, in: *Plant Genetic Resources: Horticultural Crops*, B. S. Dhillon, R. K. Tyagi, S. Saxena, and G. J. Randhawa (eds.), Narosa Publishing House Pvt. Ltd., New Delhi, pp. 75-88.

Taylor, J. D., Conway, J., Roberts, S. J., Astley, D., and Vicente, G. J. 2002. Sources and origin of resistance to *Xanthomonas campestris* pv. *campestris* in *Brassica* genomes. Phytopathology 92:105-111.

Tewari, J. P., and Mithen, R. F. 1999. Diseases, in: *Biology of Brassica Coenospecies*, C. Gómez-Campo, ed, Elsevier Science B. V., Amsterdam, pp. 375-411.

Tierens, K. F., Thomma, B. P., Brower, M., Schmidt, J., Kistner, K., Porzel, A., Mauch-Mani, B., Cammue, B. P. A., and Broekaert, W. F. 2001. Study of the role of antimicrobial glucosinolate-derived isothiocyanates in resistance of *Arabidopsis* to microbial pathogens. Plant Physiol. 125:1688-1699.

Tonguç, M., and Griffiths, P. D. 2004. Development of black rot resistant interspecific hybrids between *Brassica oleracea* L. cultivars and *Brassica* accession A19182, using embryo rescue. Euphytica 136:313-318.

U, N. 1935. Genomic analysis in *Brassica* with special reference to the experimental formation of *B. napus* and peculiar mode of fertilization, *Japan J. Bot.* 7:389-452.

USDA, 2006. The PLANTS Database, National Plant Data Center, Baton Rouge (July 12, 2006); http://plants.usda.gov/classification.html.

VanEtten, C. H., Daxenbichler, M. E., Williams, P. H., and Kwolek, F. 1976. Glucosinolates and derived products in cruciferous vegetables. Analysis of the edible part from twenty-two varieties of cabbage. J. Agric. Food Chem. 24:452-455.

VanEtten, C. H., Daxenbichler, M. E., Tookey, H. L., Kwolek, F., Williams, P. H., and Yoder, O. C. 1980. Glucosinolates: potential toxicants in cabbage cultivars. J. Amer. Soc. Hort. Sci. 105:710-714.

van Soest, L. J. M., Boukema, I. E., and Bas, N. 2004. The achievements of the EU GENRES CT99 109-112 project *Brassica*, including *B. carinata*. Cruciferae Newsl. 25:117-119.

Voorrips, R. E. 1995. *Plasmodiophora brassicae*. Aspects of pathogenesis and resistance in *Brassica oleracea*. Euphytica 83:139-146.

Voorrips, R. E., Jongerius, M. C., and Kanne, H. J. 1997. Mapping of two genes for resistance to clubroot (*Plasmodiophora brassicae*) in a population of doubled haploid lines *Brassica oleracea* by means of RFLP and AFLP markers. Theor. Appl. Genet. 94:75-82.

Voorrips, R. E., Jongerius, M. C., and Kanne, H. J. 2003. Quantitative trait loci for clubroot resistance in *Brassica oleracea*, in: *Biotechnology in Agriculture and Forestry, vol. 52, Brassicas and legumes*, T. Nagata, and Tabata, S., eds, Springer, Berlin Heidelberg, New York, pp. 87-104.

Warwick, S. I., and Black, L. D. 1991. Molecular systematics of *Brassica* and allied genera (Subtribe Brassiciane, Brassiceae) - chloroplast genome and cytodeme congruence. Theor. Appl. Genet. 82:81-92.

Wittstock, U., Agerbirk, N., Stauber, E. J., Olsen, C. E., and Hippler, M. 2004. Successful herbivore attack due to metabolic diversion of a plant chemical defence. Proc. Natl. Acad. Sci. USA 101:4859-4864.

Wszelaki, A., and Kleinhenz, M. D. 2003. Yield and relationships among head traits in cabbage as influenced by planting date and cultivar. II. Fresh market. HortScience 38:1355-1359.

Zhang, Y., Talalay, P., Cho, C. G., and Posner, G. H. 1992. A major inducer of anticarcinogenic protective enzymes from broccoli: isolation and elucidation of structure. Proc. Nat. Acad. Sci. USA 89:2399-2403.

Zhiyuan, F., Wang, X., Dongyu, Q., and Guangshu, L. 1999. Hybrid seed production in cabbage. J. New Seeds 1:109-129.

Cauliflower and Broccoli

Ferdinando Branca[1]

[1] Università di Catania, Dipartimento di OrtoFloroArboricultura e Tecnologie Agroalimentari (DOFATA), fbranca@unict.it

1 Introduction

Cauliflower (*Brassica oleracea* var. *botrytis*) and broccoli (*Brassica oleracea* var. *italica*) are traditional European crops that have become widespread in Asia in recent decades whereas their presence in Europe has been quite stable. Statistical data on cauliflower are available, whereas for broccoli they are merged with those of cauliflower and with cabbage, and so its trends are not easy to determine.

During the 1999-2005 period the harvesting area of cauliflower increased by about 28% mainly as a consequence of the significant increase in China, from 219.000 ha to 363.000 ha, and in India, from 250.000 ha to 280.000 ha, (FAOSTAT, 1999; FAOSTAT, 2005). These two countries account for about 70% of the total Asian harvesting area with a production of about 7.4 and 4.8 million tons for China and India respectively, covering about 93% of Asian cauliflower production, which amounts to 13 million tons. On the other hand, during the same period the harvesting area in Europe decreased by about 9%, with a production of about 2.4 million tons. In Italy, a slight increase in harvesting area was observed, with a total production of about 513.000 tons (FAOSTAT, 1999; FAOSTAT, 2005). World cauliflower yield is about 18.3 tons ha^{-1} and varies greatly among the main producer countries; from 22.1 to 15.4 tons ha^{-1}, respectively in Italy and in France (FAOSTAT, 2005).

The interest in cauliflower and broccoli cultivation has grown in recent years due to the genetic improvement programs carried out in several countries, mainly in Asia, Europe and the USA, and due to the new opportunities offered by the food industry in exploiting traditional and new phenotypes in new transformation processes (IV and V gamme). Also, the healthy compounds contained in the produce of several brassicas which allows them to be defined as functional foods, are important for increasing the consumption of cauliflower and broccoli. The great diversity still present for cauliflower and broccoli in several germplasm banks, which could be exploited to provide new horticultural items, is important for breeding programmes aimed at satisfying new consumer requirements.

In this context, recent scientific results, in terms of knowledge and understanding of the genetic resources available for Brassicaceae, the traditional and new breeding techniques, and the current breeding tasks are summarized here.

2 Origin and Domestication

The elucidation of the origin of cauliflower and broccoli, as for all other crops belonging to *Brassica oleracea,* is still an important task for several researchers. Initial studies on the DNA polymorphism suggested a monophyletic origin (Song at al., 1990), whereas for other authors the domestication of several of the cultigroups of *Brassica oleracea* L. is polyphyletic and is strictly related to several wild *Brassica* species which represent a common complex gene pool (Snogerup et al., 1990). The latter authors hypothesize that several crops originate from *B. olearacea* subsp. *oleracea,* widespread from the European Atlantic coast to the Nordic sea. This subspecies is biennial or perennial with woody stem, up to 3 m tall, large leaves and high glucosinolates content. The plant is diploid (2n=18), cross-pollinated, self-incompatible, with high tendency to mutations, and crosses freely with several wild *Brassica* species widespread in the Mediterranean basin.

The origin of cauliflower and broccoli crops from the *Brassica oleracea* group seems to be located in the Mediterranean basin and linked to others relatives which are likely to be *B. cretica, B. incana, B. insularis, B. macrocarpa, B. montana, B. rupestris and B. villosa* (Gómez-Campo and Gustafsson, 1991).

The evolution of cauliflower and broccoli would seem to have taken place in the Mediterranean basin, in particular in its east coast (Gray, 1982). The intense trading relationships between numerous countries of the Mediterranean area in Roman times supported the spread and exchange of genetic materials in several regions. The evolution processes probably led to adaptation to different soil-climatic conditions by the several, the cultivation and selection of genotypes with interesting agronomical and qualitative traits, and permitted the identification of several types and forms of cauliflower and broccoli (Nuez et al., 1999).

Several studies on the relationship between cauliflower and broccoli have been performed over recent years on the basis of bio-morphological, anatomical, biochemical, and molecular traits. On the basis of the reproductive processes a wider gene pool was suggested to exist for cauliflower, as compared to broccoli (Gray, 1982; Crisp and Gray, 1984; Gray, 1989). The DNA polymorphism for cauliflower showed a direct origin from broccoli and an indirect origin from wild *Brassica* types with introgression of the broccoli gene pool (Song et al., 1988, 1990; Smith and King, 2000). This latter hypothesis is supported by the morphological characteristics of broccoli which are much more similar to several wild *Brassica* species rather than to cauliflower (Nuez et al., 1999).

Cauliflower crops may have been introduced in western Europe a much longer time after their origin on the east side of the Mediterranean coast, due to the difficulties in producing seed in environmental conditions that were different to those of their places of origin. In any case the origin of these crops is uncertain and approximate due the confusion arising from the identical terminology used to describe the plants and crops. Some forms of *B. oleracea* were cultivated in Greece as early as the IV century B.C. whereas crop diversification with other types and forms started later.

Teofrasto (III-IV B.C.) described three types of brassicas, one of which was wild, characterized by smooth leaves, acrid taste, and utilized for medicinal purposes; the two others were cultivated, and differentiated by the crisp leaves in one of the types and by smooth leaves and reproductive difficulties in the other; the latter type could be related to one of the ancient initial forms of cauliflower which showed less flower sterility. Among these types, forms with sprouts development were also identified. It would seem, at that time, that several types of *B. oleracea* were grown in the west Mediterranean basin.

Plinio (I B.C.) mentioned types which seem intermediate forms between cauliflower and broccoli, and which showed variation in terms of curd/head compactness, symmetry and uniformity. These types, very similar to cauliflower and broccoli, were grown in the Roman age, especially above all on the Italian peninsula, where the differentiation process took place (Nuez et al., 1999).

Cauliflower, as said before, seems to originate along the east coast of the Mediterranean basin from wild forms of *B. oleracea* which were widespread in Italy, where the introgression of the broccoli gene pool probably took place. On other hand some authors put forward the hypothesis that cauliflower originated in the central Mediterranean basin directly from broccoli and, only after that, spread to the other areas (Crisp, 1982; Gray, 1982).

Abu-Zacarìa (Ibn-el-Awan) described two forms of "berza siriacas" (Syriac kale) during the XII century, one more similar to cauliflower and the other to broccoli (Hervé, 2003). The absence of cold requirements for summer cauliflower, which was the first to be grown in Europe, supported the hypothesis that the east Mediterranean basin was the area of origin of cauliflower. In any case it is thought that the cauliflower was introduced into Italy by Genoa citizens from the west or from Cyprus around 1490 and after this its seed production started in Campania.

In 1578 Dodoneus described a type of cauliflower, agamously propagated, from material from Cyprus, called *B. cypria,* which in other countries did not produce seed because it was highly sensitive to the cold.

The cauliflower crop, widespread along the Italian peninsula since the XV century, appeared in France and in Great Britain during the XVI century with the name of "Cyprus kale"; in the same period European writers described the crop in Egypt and in Turkey. Cauliflower spread during the XVII and XVIII centuries throughout all Europe (Boutelou and Boutelou, 1801; Hyams, 1971). Miller (1724) suggested the cauliflower was introduced from Cyprus and the winter cauliflower (white cauliflower) from Sicily. In Great Britain the cauliflower appeared in London markets in 1619 but it did not adapt well to cultivation till 1660 when the Ertfurt type (snow ball) was selected in Germany. This type permitted the establishment of Northern European cauliflower seed production, which till that date concentrated in the Mediterranean basin (Cyprus). This was possible with the Dutch method to propagate vegetatively some plants parts in greenhouse to overcome the winter conditions, which induced flowering, during the second growing year. With that propagation method the selection of types suited to environmental conditions that were different to Mediterranean ones took place.

During the XIX and XX centuries, trade development and human migration among several continents spread the crop to all sides of the world, as a consequence of the new cultivars and hybrids mainly developed in European and Asian countries; in India, the crop was introduced from Great Britain in 1822.

The cultivation of the first forms of broccoli along the Italian peninsula determined the selection of several types of broccoli characterized mainly by the aptitude for sprouting and by a wide range of head sizes (Giles, 1994). The more compact type, similar to cauliflower, was named Calabrese (from Calabria, a southern Italian region). During the XVII century, broccoli was grown in Great Britain and in the northern European countries and in the XVIII century in Spain (Boutelou and Boutelou, 1801). From this time cultivars were established and named in relation to the colour or to the harvest period (Crisp, 1982). Broccoli crops were widespread during the XIX century in North America, introduced by the Italian community emigrating from Calabria and Sicily, and in the XX century throughout the world, above all during the colonial period (Buck, 1956).

3 Varietal Groups

Cauliflower and broccoli are differentiated from other crops of *B. oleracea* by the morphological modification of the reproductive organs which represent the produce. Both crops are characterized by hypertrophy of flower branches in different phases of their development.

For cauliflower this hypertrophy starts at the early development stage, before the elongation of flower stems and for this reason the primordial reproductive buds present at this stage began sterile. The flower branches represent the sink organ where the reserve substances accumulate, resulting in their abnormal development which determines a continuing branching and the amplitude of curd angle. The flower meristems are reversed in inflorescence ones due to the presence of an homeotic gene in homozygous conditions (Carr and Irish, 1997). Broccoli heads are harvested in correspondence of a more developed flowering stage characterized by the presence of fertile flower buds.

The produce represents about 20-30% of the whole cauliflower plant and 30-40% of the broccoli one. The development of reproductive organs is arrested by some genes which determine the hypertrophy of the flower branching stems, arising as a consequence of the accumulation of reserves in parenchyma cells (Ruffio-Chable and Hervé, 1994).

Cauliflower cvs. are distinguished mainly by harvesting time in summer and overwinter cultivars, and by their cold requirement for flower induction. Differences also exist among cultivars for curd morphology (Figure 1).

Fig. 1. Different types of cauliflower curds.

In Italy at the beginning of the XX century, Viani (1929) described cauliflower crops saying they were derived by improving process of the heading broccoli ("cavolo a testa di broccoli") for more compact curd shape as a consequence of the reduced length of curd branches (floret stems), for the fine grain, strictly related to flower buds size, and for the absence of pigment in the curd. This author recalls the use of this crop in the Roman age when two varieties were identified , one to obtain early production and the other one for late production (Plinio, I B.C.); the former probably is related to *Brassica cumana* and the latter to *Brassica cyferia*. The varietal groups distinguished by Viani were the white or yellow curded ones and the violet ones. The former were catalogued in relation to the early ('Primaticcio di Toscana', 'cavolfiore di Jesi', 'precocissimo d'Ertfurt', 'gigante di Napoli', 'd'Algeri' and 'di Malta') and to the late harvesting time ('natalini', 'carnevalini', 'marzuoli', 'Pasqualini' or 'Apriloti', 'Pisano', 'Gigante tardivo di Napoli' or 'Gennarese', 'd'Olanda', 'Duro d'Inghilterra', 'Stadthold', 'Walcheren') cultivars, whereas the latter were widespread in Sicily ('Agostina', 'Sammartinara', 'Natalisca', 'Gennarolo', 'Febbrarolo', 'Marzuddu' and 'Apriloto').

Among the early cvs. of cauliflower, Tamaro (1916) described 'Primaticcio di Toscana', with yellowish curds, 'di Jesi', 'Gigante di Napoli', 'd'Algeri' and 'Primaticcio di Malta' (all these cvs. were sown in May for harvesting in October), whereas for the late ones cited the 'cavolfiore nano d'Ertfurt', 'd'Olanda', 'duro d'Inghilterra', 'Stadholdt', 'Valcheren', 'Pisano' and 'Carnevalesco', all of them characterized by compact white curds. Forti (1929) confirmed the presence in Italy of the cvs. described by Tamaro and cited also the 'cavolfiore nero of Sicily' which is very similar to the 'cavolfiore d'Algeri'.

Allegra farm, a Sicilian horticultural nursery, in the first few decades of the last century listed among the white curded cvs. 'Candia', 'd'Algeri precoce', 'Primus', 'Secundus', 'Gigante di Napoli precoce', 'Gigante di Napoli tardivo', 'Castità', 'Tardivo di Malta', and 'Tardivo Metropole'. Whereas among the violet curded Sicilian cvs. listed 'Natalino', 'Gennaiolo' and 'Marzotico' (Allegra, 1934). The most utilized white curded cvs. were the 'Gigante di Napoli' with its selections, which are harvested from December to March ('Natalino', 'Gennarese', 'Marzanico', 'Aprilatico'), 'Tardivo di Fano' and 'Precoce di Jesi'.

Most of the cauliflower cultivars are white curded, but in some Italian regions traditional green or violet curded cultivars are still grown. In particular, the green curded cultivars are grown mainly in Lazio, in Marche and in west Sicily, whereas the violet curded ones are grown in east Sicily. These cvs. are utilized mainly in home gardens and in peri-urban farms but sometimes also in open fields.

Violet cauliflower is growing mainly in Catania province, probably in relation to the particular environmental conditions which allow the full expression of the violet colour. This is determined mainly by the presence of the anthocyanin pigments, the synthesis of which is favoured by wide temperature oscillation and high solar radiation levels. Like the other types of cauliflower, the plant is herbaceous, both annual (summer type) and biennale (overwinter type), and compared to the most widespread commercial white curded cultivars is much more vigorous, taller, and with a high number of leaves. The anthocyanin pigment can also be present in the stem, leaves and above all in the midrib. The leaf is generally elliptic, entire; the curd from pink to dark violet, convex in shape, less compact, great grain (flower bud size), and it is not covered by leaves.

In the group of violet cauliflower, a culton (taxonomic unit describing distinct plants originating in or maintained in cultivation) named "ciurietto" has been identified, characterized by a dark violet colour and big grain, which seems to originate from a cross between cauliflower and broccoli (Branca and Iapichino, 1997).

Regarding broccoli, the evolution of flower organs is more regular, curd branching and its angle are reduced and the flower buds are fertile in comparison to cauliflower. The curd is usually green due to the presence of chlorophyll, although in the Southernmost Italian regions there is also a widespread presence of landraces with curds light red as a consequence of the presence of anthocyanin pigments.

In Italy and in Sicily several types of broccoli are widespread whereas the crop, as said before, was introduced into Great Britain during 1700 and a century later in the USA where it was grown by Italian emigrants near Boston and New York. At this time, the USA is the top broccoli producer in the world also due to the high demand from the food industry for processing.

In North America, both open pollinated cvs. and F1 hybrids have been selected from 'Snowball' cvs. that derive from the Ertfurt type, the parent of the most widespread cultivars in Northern Europe, and which is characterized by smaller and compact plants and white compact curd. The mid-season cultivars utilized in the USA derived from French ones, whereas the late autumn cultivars derived from Italian types; the former are characterized by larger plants, and the latter by larger curds in comparison with the early cultivars. During recent decades, the F1 hybrid cvs. are replacing the open-pollinated ones because of their uniformity and high curd quality.

Broccoli cultivars show great variability in the Southernmost Italian regions where they are distinguished mainly by the harvesting time and cold requirements for flower induction, sprouting habit, leaf shape and colour, head size, grain, colour and angle of curvature. Some cultivars do not produce big heads, but are appreciated for their aroma and are utilized for their young and tender leaves. Broccoli is less sensitive to temperature extremes than cauliflower and forms heads more readily. As with cauliflower, the wide diversity of forms and types is still underutilized, and in Italy in recent decades F_1 hybrid cultivars are replacing traditional landraces. In addition, seed samples of the commercialized F_1 hybrids have contributed to genetic pollution of the traditional landraces resulting in a great part of germplasm being gradually lost. In the last decade, this germplasm has been included in the monitoring of broccoli landraces with the aim to safeguard, characterize and exploit them by some public Italian Institutions (Branca and Iapichino, 1997; Branca et al., 2007; Branca and Candido, 2007).

At the moment broccoli world production is mainly oriented to green head F1 hybrids, referred to as calabrese type, differentiated for their seasonal growing period. These cultivars are represented by very uniform and small sized plants, with a principal first harvest of the main head and only one or two secondary heads which often are not harvested, and characterized by short vegetative time and big head size. The most utilized cultivars in recent decades have been 'Marathon' and derived cultivars with big curd size that are highly appreciated by consumers worldwide.

These cultivars are differentiated by growing cycle, plant size, leaf colour, secondary flower branching (sprouting), resistance to biotic and abiotic stress, and curd size, colour, grain, firmness, and harvesting uniformity.

In relation to the growing cycle, the cultivars are distinguished in early ones, with a growing cycle of 70 days, and in late ones with a cycle of about 160 days. Of course the cycle length is related to the sowing or transplanting date and above all to environmental conditions. The new cvs. are F1 hybrids characterized by very early production, big size of the principal head which is very compact, and absence of secondary shoots and heads.

Viani (1929) described broccoli as a form originating from a wild cabbage which with successive improvement gave rise to branched forms with green-violet or white heads. At the beginning of the last century the broccoli sprouting types (*cavolo broccolo asparagio*) and the headed types (*cavolo broccolo a testa*) were identified.

The sprouting broccoli cultivated in Sicily at the beginning of the XX century were named *broccoli a foglie verde-scuro* (with entire leaves, violet-green headed and stem with few buds), *broccolo a foglie rotondeggianti* (glaucous leaves, with small green heads and stem with many buds), *broccoli Marte di Bordeaux* (with branched round heads, violet in colour) and Sprouting broccoli (English cultivar with several secondary shoots with small heads, yellow-green coloured). The *broccoli a testa* was an improved type of the sprouting ones which was characterized by the differentiation of only one big head, violet coloured.

Tamaro (1916) described two cvs., the *Broccoli bianco ordinario* with hard white heads and the *Broccolo romano* characterized by branching plant with small violet-grey heads. Forti (1929) described broccoli as differing from cauliflower only because they were over-winter crops and for the high number of their small leaves; anyway, they seem to be cauliflower like the cvs. 'Broccolo bianco commune', 'Broccolo di Pasqua', 'Broccolo bianco Roscoff', 'Broccolo bianco Mammoth', 'Broccolo viola' and 'Broccolo verde dei castelli Romani'; these two latter two were much appreciated for their flavour. The catalogue of a Sicilian nursery in the first decades of the last century lists the broccoli cvs. 'broccolo precoce', 'Mammut bianco' and 'Broccoli rapa', whereas the cvs. 'Broccolo di Sicilia Natalino', 'Gennaiolo and Marzotico' were listed among branching broccoli cvs. (Allegra, 1934).

In the European common catalogue of varieties of vegetable crops of 2002, about 600 cultivars were listed for cauliflower, of which 40% registered for the Netherlands and only 8% for Italy, and about 200 for broccoli of which 45% registered for the Netherlands and 21% for Italy (2002). In particular, about 10% of the listed cultivars were synonymous and the F1 hybrids registered were 184 for cauliflower, and 113 for broccoli. The F1 hybrids grown are characterized by wide adaptation to environmental conditions, disease resistance, size and harvest uniformity, leaf covering, and curd suitable for industrial requirements.

4 Taxonomy, Cytogenetics and Genetic Resources

Cauliflower and broccoli belong to the species *Brassica oleracea* L. and they are considered two distinct botanical varieties, respectively var. *botrytis* and var. *italica.*

B. oleracea is a member of the complex Brassicaceae family and specifically to the *Brassica* genus. In fact, this species is strictly related to other species of the genus and indirectly shares a gene pool with them. The taxonomic and cytogenetic studies identified six species with different chromosome numbers belonging to *Brassica* crops (Prakash and Hinata, 1980). The cytogenetic relationship of the species is given by the U triangle (1935) in which the three vertices are represented by *B. nigra* (n=8), *B. oleracea* (n=9) and *B. campestris* (n=10), which by intercrossing gave rise to *B. carinata* (n=17), *B. juncea* (n=18) and *B. napus* (n=19). As a consequence of these occasional intercrosses new amphidiploid species originated. The genome A was attributed to *Brassica campestris* L., the genome B to *B. nigra* L. and genome C to *B. oleracea* L.

The genome A is represented by turnip, turnip rape, Chinese cabbage and sarson crops, which on morphological basis are assigned to the three main types named oleiferous, leafy and rapiferous. The first type is economically important in North America for oilseed production, the second one is widespread in Asia, and recently also in Europe, for salad and for pickling purposes, and the third one represents minor crops in Europe and in New Zealand and its produce is generally utilized for food purposes (McNaughton, 1995a). According to the International Code of Botanical Nomenclature, Oost et al. (1987) utilized the name of *B. rapa* rather than *B. campestris,* and since then this variation has often been adopted. The genome B is represented by black mustard which was widespread in the Middle Ages in Europe as condiment but now is widespread in the same areas as weed. Finally, the genome C is represented by cabbage which as a consequence of several domestication processes gave rise to several botanical varieties and related crops, such as var. *acephala* (kale), var. *botrytis* (cauliflower), var. *capitata* (cabbage), var. *gemmifera* (Brussels sprout), var. *gongylodes* (kohl-rabi), var. *italica* (broccoli) and var. *sabauda* (Savoy cabbage). Different authors instead of giving the taxonomical rank of botanical varieties, gave those of subspecies or subvarieties; the taxonomic history is complicated and several taxonomists have tried to offer an appropriate classification of them (Lamarck, 1784; Linnaeus, 1753; De Candolle, 1821).

Among the amphidiploid species, *B. carinata* (Ethiopian mustard), derived from the union of the BB and CC genomes, is widespread in Abyssinian Plateau, *B. juncea* (mustard), derived from the union of AA and BB genomes, is grown in Asia and has several forms, and *B. napus* (oilseed rape), derived from the union of the AA and CC genomes, is widely grown in Asia, Europe and North America (McNaughton, 1995b).

The genetic resources available for breeding of *Brassica* crops are strictly related to the boundaries of their primary, secondary and tertiary gene pools and to their evolutionary history (Harlan, 1975) (Figure 2). The primary gene pool is represented by *B. oleracea* itself but several studies have been carried out to investigate the other gene pools and their potential utilization.

The studies of the pachytene chromosome morphology proved the basic genomes of *Brassica* crops - AA (2n=20) for *B. campestris*, BB (2n=16) for *B. nigra* and CC (2n=18) for *B. oleracea.* Their origin is explained by the process of polyploidy from a common ancestor with chromosome number x=6 (Röbbelen, 1960). As a consequence of these studies the chromosomes AABCDDEFFF for *B. campestris,* ABCDDEFF for *B. nigra,* and ABBCCDEEF for *B. oleracea* were proposed. Crossing the three species with each other and observing the meiotic pairings of the related amphidiploids, it was deduced that the genome B is more distant than both A and C genomes, but probably all these genomes are in part homologous and evolved from a common one (Attia and Röbbelen, 1986). The common ancestor with x=6 chromosome probably originated different poly-aneuploids species by rearranging the original genome and duplicating chromosome segments (Quiros et al., 1987; Quiros et al., 1988). Recent studies on genomic libraries of *B. napus* and *B. oleracea* have shown shared fragments among A, B and C genomes, proving the partial homology of them and confirming the origin of the amphidiploid species *B. napus, B. carinata* and *B. juncea* from the parental diploid ones (Hosaka et al., 1990;

Slocum et al., 1990). The phylogenetic studies confirm the hypothesis that the evolution of *Brassica* and allied genera started from a n=6 prototype and evolved into increasing the chromosome number (Prakash and Hinata, 1980; Song et al., 1990). These studies proved a close relationship among A, B, and C genomes of *Brassica* species, partially homologous, which represent the secondary gene pool.

Fig. 2. Some Mediterranean wild *Brassica* species (n=9): *B. rupestris* (left) and *B. macrocarpa* (right).

The tertiary gene pool was defined by Harbed (1976), who grouped species and genera related to *Brassica* crops in 36 cytodemes potentially capable of exchanging genetic materials. The cytodemes are characterized by a common specific chromosome number (between 7 and 13) and by high interfertility within them; crossing between the cytodemes is possible, by utilizing special techiniques. The principal cytodemes are represented among *Diplotaxis, Enarthrocarpus, Eruca, Erucastrum, Hirschfeldia, Rhynchosinapis, Sinapis, Sinapodendron, Trachystoma* genera.

During the last decade several genomic studies have started to clarify the role played by two floral homeotic genes which determine the developing of the curd instead of the simple inflorescence. The genes involved in changing reproductive structures, which arrest the development at the inflorescence meristem stage, were found in *Arabidopsis thaliana* and named APETALA 1 - AP1 - and CAULIFLOWER – CAL - (Mandel et al., 1992; Gustafson-Brown et al., 1994; Kempin et al., 1995). AP1 mutation determines loss of sepals and petals with leaf-like structures, whereas CAL participates in specification of flower meristem identity. The genotypes which are mutant for both genes stop their development at the inflorescence meristem stage (Kempin et al., 1995). The orthologues genes, named *BoAP1* (Anthony et al., 1993; Anthony et al., 1996; Carr and Irish, 1997) and *BoCAL* (Kempin et al., 1995), were found in *B. oleracea* to segregate in both wild and cultivated varieties carrying nonsense mutations and are fixed in the var. *botrytis* and *italica* (Purugganan et al., 2000). These two genes are strictly associated with floral organ identity, switching from inflorescence to floral meristem fate, and are closely related to members of the MADS-box gene family for flowering transcription factors (King, 2003). In particular, *BoCAL* probably represents a selection target for the evolutionary domestication and it shows polymorphism in some cultivated varieties of *B. oleracea* such as var. *acephala* and *oleracea* (Purugganan et al., 2000).

On the basis of segregation of recessive alleles of BoCAL and BoAP1, which arrest inflorescence and/or flower development at different stages, a simple genetic model was proposed (Smith and King, 2000). Although the BoCAL mutant allele seems to be implicated in arresting inflorescence development, and it is present in a relatively high frequency in cauliflower, it is not determinant for curding phenotype (Labate et al., 2003). However, *BoCAL-a* allele on linkage group O3 seem to have been (unconsciously) selected by farmers during cauliflower domestication because of the modified inflorescence structure in *B. oleracea* (Purugganan et al., 2000). The association of *BoAP1-a* and *BoAP1-c* with the self-incompatible locus S seems to have reduced the number of S-alleles within the gene-pool of modern cauliflower cultivars (King, 2003).

The genetic improvement of *Brassica* crops is an important task for many researchers as a consequence of the economic importance of the resulting produce. These crops are located in different areas and economic systems, and often play an important social role in relation to new consumer requirements and food technologies.

Cauliflower and broccoli genetic resources are strictly associated to other *Brassica* crops, both belonging to other varieties of *B. oleracea* utilized mainly as vegetable (primary gene pool) or to other species such as *B. carinata, B. juncea, B. napus, B. nigra* and *B. rapa* utilized mainly as condiment, medicinal, oilseed and fodder crops (secondary gene pool).

The wide diversity present in *Brassica* and its availability is very important for breeders for setting up new cultivars to satisfy new requirements. Of course, this diversity is widespread mainly at their region of origin, which for cauliflower and broccoli is located in the Mediterranean basin. Besides, the high level of polymorphism of *B. oleracea,* and in general of *Brassica* genera, supports the wide range of utilization as a consequence of human selection of several cultivars often strictly adapted to different environmental conditions and uses.

Within this framework it is important to mention the role played by germplasm collections held in several regional and national genebanks, universities, and public and private companies. For this reason since 1989, a European working group, coordinated by the International Plant Genetic Resources Institute (IPGRI), was set up. This working group has the aim of developing an integrated system of collection, characterization, evaluation, documentation and duplication of genetic resources, and to trace new strategies for long term conservation (Gass et al., 1995; Maggioni et al.,

1997; Maggioni, 1998) (Figure 3). The *Brassica* working group (BWG) is one of the older groups of the European Cooperative Program for Crop Genetic Resources Network (ECP/GR) managed by IPGRI. Within the ECP/GR, the European Database for *Brassica* was established (Bras-EDB), managed by the Centre for Genetic Resources of Wageningen (the Netherlands), which includes 36 collections from 22 countries, with about 19,600 accessions (Hintum and Boukema, 1993; Boukema and Hintum, 1998). This database also holds the characterization and evaluation results of the last European joint project (EU project GEN RES 109-112), which are very useful for future research. Besides, BWG traced the guidelines for safety-duplication of the accessions held in the European collections and for their regeneration, and the strategy for *in situ* conservation of wild *Brassica* species widespread in Europe.

Fig. 3. Controlled pollination in the field for the maintenance of the genetic integrity of *Brassica* accessions.

In the USA the several accessions held by Regional Plant Introduction Stations are included in the database for Germplasm Resources Information Network (GRIN) at the Beltsville Agricultural Research Centre, and the collections are duplicated and stored for long term conservation at the National Seed Storage Laboratory at Fort Collins (ARS-GRIN, 1997). In China, a national network of regional genebanks and the Institute of Crop Germplasm Resources has the responsibility for the long-term conservation of genetic resources, including *Brassica* crops, for which the data are stored in the Chinese Genetic Resources Information System – CGRIS - (Chaoyo and Xu, 1992).

5 Major Breeding Achievements

The genetic improvement of cauliflower and broccoli carried out during recent decades dealt with modifying the harvest index by reducing the vegetative portions to increase the reproductive ones. The commercial portion is represented by 20-25% (cauliflower) and 30-35% (broccoli) of the plant which allows about 10-15 t ha^{-1} of produce to be obtained depending on the cvs. used, the growth period, the agricultural techniques, etc. (Hervé, 1992).

Curd and head firmness have been studied as a way to increase the post-harvest quality of the produce and this was achieved by reducing the length of floret stems and reducing the grain size. Moreover, the inflorescence density is another parameter that has been utilized for plant selection to identify types resistant to compression stress.

The cauliflower curd colour selected up to some years ago was only white or creamy white, as this was the request of consumers and traders. Instead of waiting for completion of long growing cycles to detect the unfavourable curd pigmentation, as a consequence of anthocyanins, carotenoids and chlorophyll in curd tissues, a technique was set up to select against the purple defect in aseptic culture (Crisp and Walkey, 1974; Crisp et al., 1975 a).

As already said, cauliflower curd and broccoli head consist of many compressed and branched peduncles with several pre-floral and floral meristems. Under certain environmental conditions some genotypes show bracts which grow through the curd/head surface, reducing the produce value and, in consequence, the plant breeder discarded them according to the level of expression of this character (Crisp et al., 1975b).

The development of new F1 hybrids of cauliflower and broccoli allowed harvest uniformity to be achieved which is an important factor not only for the produce utilized as fresh vegetable but also for that used for industrial purposes. For example, size uniformity of curds/heads, and of their florets, is requested by the freezing industry (Acciarri et al., 1997).

Screening for disease and pest resistance of *Brassica* germplasm has permitted identification and exploitation of tolerant genotypes utilized for genetic improvement programs. Several studies have been carried out, such as for clubroot (*Plasmodiophora brassicae*) for which sources of resistance were found to be under the control of an incomplete dominant and a recessive gene, *pb1 pb1, pb2 pb2* (Vriesenga and Honma, 1971; Chiang and Crete, 1970).

Regarding black root (*Xanthomonas campestris*) resistance this was found to be controlled mainly by a major gene which in heterozygous conditions is influenced by one recessive and one dominant modifier gene in cabbage (Sharma et al., 1972; Williams et al, 1972). In India several cultivars resistant to *Xanthomonas campestris* were selected, and this resistance is dominant and polygenic in several Indian landraces (Bianco and Pimpini, 1990).

Concerning insect resistance, several studies focus on cabbage aphid (*Brevicorne brassicae*), cabbage looper (*Trichoplusia ni*), cabbage moth (*Mamestra brassicae*) and white butterfly (*Pieris brassicae*). The 'Rubin' cv. of Brussels sprout, characterized by red leaves, was much more resistant than other cvs. with green foliage to white butterfly and cabbage moth (Dun and Kempton, 1976). Tolerance sources to cabbage aphid were found in the cultivars 'Darkmar 21', 'Vremo Inter' and 'Seven Hills', although differences in the reproductive capabilities of different biotypes of *B. aphid* were observed and no definite sources have been identified (Dun and Kempton, 1971). Sources of resistance to cabbage looper were found for the cabbage lines PI296133 and PI302985 and in the cauliflower lines PI234599 and PI343483 (Dickson and Eckenrode, 1975). No source of resistance to cabbage root fly (*Delia brassicae*) has been found, although some resistances were observed in the recombinants from European biennale and European annual cauliflower (Crisp et al., 1977).

Some glucosinolates, such as sinigrin which is a feeding deterrent of *Myzus persicae* in cole crops, are able to stop insect damage (David and Gardiner, 1966) or the aglycone 2-phenylethyl isothiocyanate which in turnip crops deterred *Drosophila melanogaster* feeding (Lichtenstein et al., 1962). The genetic improvement in *B. oleracea* often deals with a reduction of glucosinolates content to obtain produce with a smooth taste, but in several cases the cultivar obtained was more sensitive to insect attack (Finch et al., 1975; Cole, 1978).

Genes for resistance to high temperature and rainfalls permitted the selection of cultivars suitable to tropical and subtropical areas of India, Brazil, Hawaii, Israel, and Taiwan, where cauliflower can be grown at 10-15 C both in the northern and southern latitudes (Bianco, 1990). In the USA, a male sterile line (NY7642A), which maintains the white colour of the curd also in full sunlight due to the low peroxidase enzyme content, was selected from an Egyptian population (Bianco and Pimpini, 1990)

Regarding broccoli, several breeding programs carried out since the 1960s in the USA aimed to develop disease resistant broccoli for mechanical harvesting.

In recent years several cultivars have been developed by crossing the var. *italica* and the var. *botrytis* of *B. oleracea* followed by several back crosses aimed to enlarge the curd size and to increase the insertion angle of florets, to reduce curd stem, to improve curd compactness and colour, to regulate flower bud opening, to increase bruise resistance during harvesting and post harvest stages, to improve harvesting uniformity and to increase the green colour after cooking, to evaluate the nutritive value and resistance to biotic stress. In addition current goals for developing improved F1 hybrids are a shortening of the growing cycle, a reduced plant size for increasing crop density, an easy floret separation for freezing, resistance for downy mildew and black rot, absence or reduction in the number of marketable secondary shoots, heat tolerance, long shelf life quality, high uniformity and productivity, and mechanical harvesting capability (Nonnecke, 1989).

6 Current Goals of Breeding

Over the last few years new breeding objectives have dealing with an increase in the quality of the produce been identified. This crop, in fact, has shown up to now that several traits such as curd colour and shape, and presence and amount of antioxidant compounds, for which several authors have proved that there is great variability which has not been fully exploited (Crisp, 1982; Crisp and Gray, 1984; Massie et al., 1996; Branca and Iapichino, 1997; Branca et al., 2002; Genna et al., 2004). In particular, great interest is devoted to the traditional Italian germplasm and landraces which show great variability for the above-cited characters. Specific projects have been carried out over recent decades in some European countries for using Italian pigmented cauliflower germplasm to develop new cultivars (Crisp, 1982; Crisp and Tapsell, 1993; Massie, 1998).

Pigmented curd cauliflowers are distinguished from the ordinary white curded ones by the relative content of some pigments (chlorophyll, anthocyanins and carotenoids) which determine the curd colour (Figure 4). In this way, in the green curded cultivars chlorophyll is present but not the other pigments, in the violet curded ones anthocyanins are always present, associated in some cases with chlorophyll and/or carotenoids; the presence of carotenoids and absence of other pigments determines the expression of a yellowish-orange curd in some genotypes of landraces in Sicily.

Fig. 4. Colour diversity in cauliflower.

Anthocyanin synthesis is stimulated by high levels of solar radiation and wide temperature fluctuation; these conditions occur also along the east Sicilian coast where several brassica landraces rich in this compound are widespread. This area is traditionally focused on violet cauliflower selection and cultivation probably because in these conditions, strictly affected by the presence of the sea and of Etna volcano (more than 3.000 m over sea level), the colour expression is affected by the great variations in temperature between day and night. This climatic characteristic in past probably permitted the maximum expression of this character which was selected by man (Acciarri et al., 2005). Both anthocyanins and carotenoids are antioxidant compounds which show a positive antiradical action. Besides, in addition to the presence of these compounds in the green and violet curd tissues there are significant amounts of other antioxidants, such as ascorbic acid and glucosinolates. Several authors confirmed the great interest of the coloured cauliflower, which could be considered as a functional food because it not only has a nutritional function, but prevents diseases.

Glucosinolates (GLs), atypical β-thioesters, are represented by more than one hundred compounds presents in *Capparales* plant tissues and in particular in Brassicaceae species. Their amount and composition show significant variation depending on several factors (Rosa et al., 1997), like the species, organs, environmental conditions, and cultivar. GLS are hydrolysed by myrosinase enzyme (MYS) in isothiocyanates (ITCs) upon ingestion by humans, compounds which have been shown to prevent and cure several human diseases (Fahey e Talalay, 1999; Vang et al., 2001; Zhu e Loft, 2003; Jeffery et al., 2003; Johnson, 2002; Keum et al., 2004; Zhang et al., 2005). Among ITCs the sulfurafane, produced by hydrolysis of glucoraphanin (4-methylsulfinylbutyl), shows great antioxidant activity and protective effects against several types of cancer, such as prostrate cancer (Fahey et al., 1997; Verhoeven et al., 1997; Poppel van et al., 1999; Gamet-Payrastre et al., 2000; Kristal and Lampe, 2002; Finley, 2005).

Glucoraphanin is mainly present in broccoli tissues, whereas in cauliflower it is mainly present in the pigmented curd landraces, while in the commercial white curded cultivars this compound is often not present (Branca et al., 2002). On the contrary, in the latter the presence has been reported, in low amounts, of sinigrin, and its isotiocyanate, the allil isotyaciante, which shows toxic activity for several animal organisms, including humans. Besides, some authors show the high antioxidant and antiradical activity of coloured curded cauliflower landraces in comparison with the white curded commercial cultivars (Pizzocaro et al., 2000; Branca, 2006).

On the basis of glucosinolate activity and its presence in the pigmented cauliflower, breeding programs were started to add value to vegetable productions obtained from landraces (Branca, 1997; Branca, 1998; Branca, 2000; Branca et al., 2002, Acciarri et al., 2004). Also, genotypes and cultivars have been selected for maintenance of their nutritional value during one year of freezing (Acciarri et al., 1997; Maestrelli et al., 1999). For the coloured cauliflower curds, diversity among Italian landraces for maintaining some antioxidant compounds such as ascorbic acid, anthocyanin, etc. was studied (Genna et al., 2004).

Particular attention has also been paid to the aromatic profile of the produce which could affect consumer taste and which is also strictly related to the glucosinolate-isothiocynates system of Brassicaceae. Also, for these traits diversity among landraces and cultivars was found, but this remains to be exploited (Di Cesare et al., 2001; Di Cesare et al., 2005).

Concerning broccoli, in the last decade, efforts have concentrated on isothiocyanate enrichment of broccoli to increase concentration in healthy compounds (Mithen et al., 2003). Several epidemiological studies have shown that vegetable brassicas could reduce cancer risk, above all of the gastro-intestinal part (Block et al., 1992; Verhoeven et al., 1996). In particular, 3-methylsulfinylpropyland (3-MSP) and 4- methylsulfinylbutyl (4-MSB) isothiocyanate (ITCs) seem to derive from the same glucosinolate and they are potent inducers of phase II detoxification enzymes in rodents and human cell cultures (Zhang et al., 1992; Faulkner et al., 1998). The level of the above cited ITCs is low in broccoli florets, but increases significantly in the hybrids derived from crosses between broccoli and *B. villosa,* a *Brassica* wild relative widespread in the Mediterranean basin (Mithen et al., 2003).

On the basis of this evidence, new broccoli cultivars containing great amounts of the glucosinolate precursors which are mainly conversed in ITCs have been developed. The breeding program which permitted the introgression of *B. villosa* genome into the standard broccoli background is in progress, and this has allowed the identification of three genome segments on linkage groups 2, 5 and 9 which present alleles of the wild *Brassica* species that enhance the glucosinolate amount in broccoli heads (Mithen et al., 2003).

7 Breeding Methods and Techniques

Breeding methodologies are strictly related to the reproduction system of *Brassica* species. The inflorescence of brassicas is a typical corymbiform raceme, with bisexual and hypogynous flowers, with four free sepals and petals, the latter placed diagonally. Flowers have one pair of lateral stamens with shorter filaments and four median stamens with longer filaments.

The diversity of *B. oleracea* is a consequence of the reproduction characters, such as the characteristics of multiplication, the pollination system, the self-incompatibility and male sterility (Hervè, 1992). Pollination is made by insects (bees and *Diptera* spp.) because the pollen grains are usually aggregated and form structures that are too heavy to be transported by wind. The distance required to avoid cross pollination between *B. oleracea* crops is 1,000 m, so it is very important to plan isolation for seed production (Raymond, 1985). Most cauliflower and broccoli cvs. are cross-pollinated but some self-pollination can occur above all in the summer cvs. (Watts, 1980). Seed yield ranges from 5,000 to 30,000 seeds per plant, with a 1,000 seed weight of about 2.8 g. Both cauliflower and broccoli can be propagated vegetatively *in vivo* and *in vitro*. The range of organs utilized is very wide, such as shoots and curd portions, and regeneration by meristematic apices is used, especially for the breeding lines. *In vitro* propagation offers the possibility to regenerate clones affected by virus or used as parent lines.

For breeding it is important to know the techniques utilized for artificial hybridization such as flower emasculation. Flower buds, one-two days before anthesis, are opened to remove all stamens using forceps; emasculation is usually performed during afternoon or evening. Self-compatible cauliflower and broccoli types exist, but most of the types are usually self-incompatible. In the former case the buds are pollinated with the pollen of the same plant and they are protected from other pollens for about one week after pollination. All other flowers or florets are discarded and an identification tag is fixed to the flower pedicle. Pollination is done with the selected anthers picked with forceps for applying them to the stigma or brushing and dusting the collected pollen to the stigma. After pollination, the flowers are protected and the indication of plant donor is written on the tag.

For producing hybrid seeds of cabbage, cauliflower, broccoli, Brussels sprout and kale a self-incompatibility character is utilized which is controlled by the S-gene (King, 2003). The genotypes S-allele homozygous S_1S_1 and S_2S_2 are incompatible for genotype S_1S_2 for both stigma and pollen because the sporophyte controls pollen germination.

The exploitation of pollen self-incompatibility for genetic improvement of brassicas presents some advantages but also some disadvantages and difficulties. For hybrid production the parental lines have to be both self-incompatible; as a consequence also the hybrid obtained will be self-incompatible too. The expression of self-incompatibility is fluctuating and is greatly affected by temperature and humidity levels.

Inbred lines selected for strong expression of S-allele homozygosis show that few self pollinated. S_1S_1 and S_2S_2 genotypes, sown in alternate lines and crossed by insects, produce F_1 hybrid seeds identical in both inbred lines (Agrawal, 1998). F_1 hybrid seeds could be produced by "three way" crosses, utilizing F_1 hybrid genotypes originated by two S-allele homozygotic lines crossed with an S-allele homogozygous line, or the "four way" one, utilizing as parents two F_1 hybrid genotypes originated by two S-allele homozygotic lines. "Topcross" hybrids could be produced by crossing an open-pollinated cultivar, utilized as pollen donors, with a self-incompatible inbred line, utilized as a female parent (Agrawal, 1998). Three way top cross hybrids are produced utilizing a cultivar as male parent and F_1 hybrid genotypes originated by two S-allele homozygotic lines as female parent.

For self-incompatible types self-pollination could be achieved above all under high levels of humidity or by bud pollination. To apply this technique the largest unopened buds are not suitable because the incompatibility factor has already been biosynthesized (Dickson and Wallace, 1986). So some younger flower buds are opened 3-4 days before flowering at the same time with a toothpick or forceps and pollen from an older open flower transferred, with a brush or on a thumbnail, to the stigma.

However, for F1 hybrid production the genetic male sterility system presents advantages over self-incompatibility because no sibs are produced. The male sterility genes utilized are recessive and only one-half of plants of the female line will be male fertile (*Msms*) and have to be removed at flowering stage. These plants are identified with difficulty only at the appearance of the fist flowers but it is not simple to recognize them. To overcome this problem a marker gene has been inserted into the male sterile genotypes to facilitate their recognition but for the moment no reports exist on its use. In addition, the introgression of the recessive male in breeding lines requires some backcross generations and one generation of self-pollination to obtain the recessive homozygote for male sterility expression. The reproduction of the male sterile lines results in about 50% of male sterile plants and 50% of fertile ones; the latter have to be eliminate before flowering to avoid crossing with each other and a marker character expressed at plantlet stage which help to recognize the male fertile plants is usually utilized (Sampson, 1966). Male sterile plants could be propagated vegetatively for the production of F1 hybrids of cauliflower (Hervé, 1992). A dominant gene which controls male sterility was recently found and utilized in France for the production of F1 hybrids of cauliflower; the male sterile plants can be dominant homozygote S/S or heterozygote S/s whereas the fertile ones are recessive homozygote s/s. (Ruffio-Chable et al., 1997).

The cytoplasmatic male sterility found in radish was transferred to cauliflower by backcross (Ogura, 1968; Bannerot et al., 1974). The association of cauliflower and radish genome determines the reduction of chloroplast functionality, and hence of the chlorophyll in the leaves, above all at low temperature levels (Bannerot et al., 1977). The problems observed for cauliflower were solved for cabbage by protoplast fusion (cybridization) of a male sterile cabbage line and a sterile one (Pelletier et al., 1989).

The development of a new F1 hybrid is a way to reach improved yield and quality of the produce, to stabilize the cultivar, and to render the produce much more uniform. In cauliflower, heterosis was observed for yield, mainly as a consequence of the increasing curd weight and size, earliness, curd size and uniformity (Jones, 1932; Weiring, 1961; Swarup and Pal, 1966; Sandhu et al., 1977). Differences in the heterosis effect were observed between the mid-early F1 hybrids, for which the greatest improvements were reached, and the mid-late ones (Swarup and Chatterjee, 1974).

Interspecific and intergeneric hybridization may be used for the transfer of desirable genes from wild progenitor or related species, to exploit hybrid vigour, to produce new alloploid species and to determine the evolutionary relationship among species (Briggs and Knowles, 1967). The reproductive barriers, due to disharmonies between physiological and cytological systems of plants, prevent genetic interchange between two species.

In order to overcome the reproductive barriers, the plant breeder may employ several approaches directed to overcome pre-fertilization and post-fertilization barriers. Pre-fertilization barriers can be overcome by selecting the appropriate species for hybridization, the direction of the crosses, increasing the ploidy level, promoting indirect crosses, improving crossing techniques and *in vitro* pollination. The post-fertilization barriers could also be overcome by the use of growth hormones, mixed pollination, embryo culture, grafting and production of alloploids (Hadley and Openshaw, 1980).

Regarding cauliflower regeneration of inbred lines and maintenance of self-incompatible lines somatic embryoids were studied (Walkey and Woolfitt, 1970; Baroncelli et al., 1973; Crisp, 1974; Pareek and Chandra, 1978). Leaf callus cultures in MS medium with kinetin and IAA induced embryogenesis by firstly developing meristematic nodules and after, on them, embryoids after the reduction of IAA in the same medium (Pareek and Chandra, 1978). Leaf initial calli are obtained easily after two-three weeks of culture of curd explant tissue (Pow, 1969; Walkey and Woolfitt, 1970). LS medium was utilized for culturing marketable cauliflower curds (Linsmaier and Skoog, 1965). Friable callus was induced in a week utilizing MS medium supplemented with 0.5 – 1.0 mg l^{-1} kinetin and BA, whereas for shoot differentiation IBA and NAA were added (Singh and Mathur, 1985). LS medium supplemented with kinetin and IAA developed shoots from callus (Pow, 1969). Mesophyll protoplast of male sterile lines (CMS) were utilized for cauliflower regeneration.

For hybrid seed production of broccoli, regeneration of parental lines is carried out by culturing flower buds in MS medium supplemented with 2iP, IAA, adenine sulphate and sodium phosphate (Anderson and Carstens, 1977).

Concerning *B. oleracea*, several trials have been performed using inter and intra-specific crosses. It is well known that all botanical varieties of the species are interfertile, such as all the species of the Section *Brassica* which are grouping to *B. oleracea* cytodeme representing the primary gene pool (Harberd, 1976; Snogerup et al., 1990). The Mediterranean *Brassica* wild species (n=9) utilized in crossing experiments with *B. oleracea* in some cases determined a reduced hybrid fertility, the male fertility and fruit setting vary in F1 and F2, allowing the transfer of genes or gene blocks to different cultivated types and forms (Gustafsson, 1982).

Recently an extensive research was carried out to define the relative fertility of species belonging to the section *Brassica* including 10 wild taxa and 23 accessions representing six major cultivars of *B. oleracea,* and showed that the latter are closely related because they are interfertile (Bothmer et al., 1995). The fertility of crossing between cultivated forms and wild *B. oleracea* is reduced significantly compared to the other wild species (n=9); among the latter the lowest value in F1 and F2 were observed for *B. macrocarpa, B. montana* and *B. rupestris,* whereas the highest value with *B. cretica.* That shows that wild Mediterranean species of *Brassica* belong to the primary gene pool and represent a source of useful genes which easily flow towards *B. oleracea* (King, 1990).

Recently *B. villosa* has been utilized for crossing with broccoli breeding lines for obtaining high glucosinolate cvs. for use in the production of functional foods (Mithen et al., 1997; Mithen et al., 2000, 2003). This confirms the great interest in exploiting the Mediterranean wild *Brassica* (2n=18), especially for traits which can be useful in developing new cultivars for use in sustainable production systems and methods.

8 Integration of New Biotechnologies in Breeding Programs

During the last decades several biotechnologies have been tested and optimized for increasing the efficiency of the genetic improvement programs for several *Brassica* crops and in particular for cauliflower and broccoli.

There is great interest for using *Brassica* coenospecies to improve *Brassica* crops and enrich their gene pool by hybridization of naturally existing amphidiploids for obtaining artificial genome resynthesis (Namai et al., 1980; Downey and Robbelen, 1989; Buzza, 1995). Several wild *Brassica* species which are sources of interesting traits, such as resistance for biotic and abiotic stresses, often show fertilization barriers that limit obtaining new genomic combinations. Disturbance in chromosome pairing and/or in nuclear-chondriome of the new plastome combinations could determine several defects on flower biology and reproduction (Edwardson, 1970; Bannerot et al., 1977; Beillard et al., 1978; Rawat and Anand, 1979; Bonnet et al., 1991).

Several experiences started studying the possibility to propagate *in vitro* multicellular portions of several organs of the plant, such as single cells, protoplast or microspores, on different culture media, in aseptic environmental conditions, for obtaining new plants. Several protocols have been developed for optimizing the *in vitro* culture to regenerate organs or embryos for obtaining new functional plant, starting also from single cells (Figure 5).

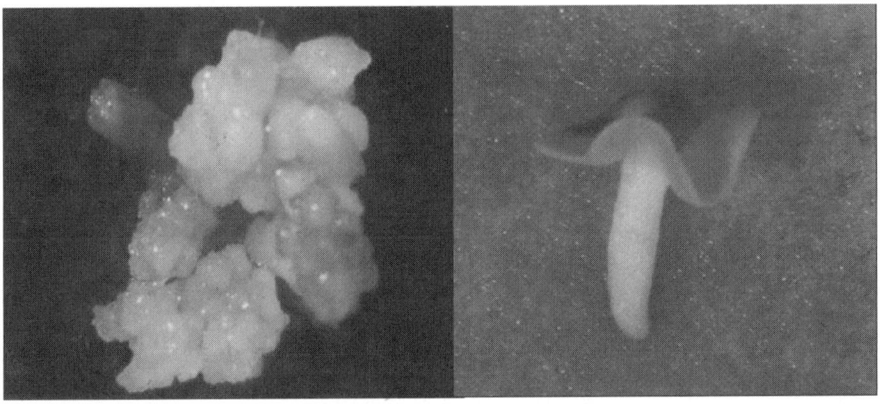

Fig. 5. Use of in vitro culture for cauliflower breeding: androgenic embryos (left) and germinating mature embryo (right).

The cellular totipotency, that is the ability of a single cell to express the full genome in the cells to which it gives rise by cell division, is the basic concept which has permitted to develop all in vitro biotechnologies and the related knowledge. Totipotent cells after division, and stimulated by environmental conditions and/or hormones, are specialized into pluripotent cells, that give rise to most of the tissues need for plant grown, which undergo further specialization into multipotent ones which are committed to give rise to cells that have a particular function. The totipotent cells must be able to differentiate not only into any cell in the organism but also into extraembrionic tissue associated with the organism.

On the basis of the new knowledge, sexual hybridization by protoplast fusion has been tested for overcoming genetic barriers among *Brassicaceae* species, such as chromosome pairing and new nuclear-cytoplasmatic combinations, obtaining somatic hybrids and cybrids (Glimelius, 1999).

Efficiency of mass fusions of *Brassica* protoplasts for plant regeneration has been obtained by using polyethylene glycol (Glimelius et al., 1986; Morgan and Maliga, 1987; Robertson et al., 1987). Several methods have been proposed for selecting the desired heterokaryons but the more interesting are based on flow cytometry to individuate the new genetic combinations cells which will be sorted, or on pretreating one parental with iodoacetate or iodoacetamide and irradiating the other parental (Sidorov et al., 1981; Gerdemann-Knörck et al., 1994; O'Neill et al., 1996).

Effective probes of the artificial hybridization of the different genomes are based on isoenzyme are RFLP analysis and lastly by repetitive species-specific sequences which estimate also the proportion of parental genomic material in asymmetric hydrids (Sundberg and Glimelius, 1986; Imamura et al., 1987; Piastuch and Bates, 1990; Fahleson et al., 1994; Fahleson et al., 1997).

Protoplast fusion techniques have been utilized not only for a functional integration of different nuclei, but also to integrate cytoplasmic traits by modifying organelle combination, such as chloroplasts and mitochondria, and organellar DNA (Pelletier, 1986). The Ogura rapeseed cultivar, characterized by cytoplasmic male sterility (CMS), has been largely utilized to introduce these traits to *B. napus* lines, such as the "CMS juncea", present in 'Anand' and in 'Tour' cvs., which was introduced from *B. tournefortii* (Pelletier et al., 1988; Renard et al., 1992; Rawat and Anand, 1979; Pradhan et al., 1991; Stiewe and Röbbelen, 1994; Liu et al., 1995). The mitochondria of the latter species were introduced for establishing alloplasmic CMS-lines which segregated into fertile and male-sterile lines; furthermore, it was found that the fertile lines restore the CMS ones.

The same results were obtained utilizing the hexaploid (AABBCC) somatic hybrid, by crossing *B. juncea* (AABB) containing *B. tounefortii* cytoplasm and *B. oleracea* (CC), and its derived fertile and male sterile lines (Arumugam et al., 1996). Following the same procedure CMS-*B. oleracea* lines were obtained by exploiting the mt-DNA recombinated lines derived from the cybrids obtained from "CMS-tour" and *B. rapa* (Cardi and Earle, 1997).

The traditional methods utilized for F1 hybrid production need about twelve cycles of self-pollination to obtain homozygous inbreed lines, whereas with anther and/or ovary culture it is possible obtain them in only one year. In *Brassica* spontaneous haploids occur at low frequency and can be produced by interspecific crosses (Prakash, 1973; Thompson, 1974; Prakash, 1974).

Haploids, defined as sporophytes with gametophytic chromosome constitution (Kimber and Rilley, 1963), rarely occur in nature by parthenogenesis process from an unfertilised egg which develops one or more embryos. Plant breeders started to direct their attention on androgenesis after the successful experiences on *Datura innoxia*, which offered numerous pollen plantlets by anther culture (Guha and Maheshwari 1964, 1966). In *Brassica*, it was found that homozygous plants can be obtained by the chromosome doubling of the haploids obtained from androgenesis.

Since the sixties, several researches were carried out with the aim to study the effect of several factors affecting androgenesis, such as nutritional requirements, stage of pollen development, temperature and light, or physiological status of the donor plant. All these factors have to be taken in account in order to obtain a successful protocol for producing haploids by *in vitro* culture. Generally, the culture of anthers is ineffective if these organs are at very young stage of development, such as when microspore mother cells are in meiosis, and at late ones, such as during the binucleate or trinucleate stages or when the pollen is filled by starch.

Several studies were carried out on *Brassica* species deal with the factors which affect haploids production with the aim to use them for the genetic improvement of several crops, such as cauliflower and broccoli. If we review them chronologically, the first studies dealt with the individuation of the best stage of pollen development, which as suggested by previous investigations on *Datura innoxia*, was just after pollen mitosis (Sunderland et al., 1974). For *Brassica*, it was found that the trinucleate pollen at the shedding offers the best response for androgenesis when mature anthers or isolated pollen are cultured (Kameya and Hinata, 1970). However, the optimal stage of development depends also on the genotype and ranges from the uninucleate pollen stage to the first mitotic division (Yang et al., 1992a).

Concerning with the photo-thermal conditions, it was found that for *B. campestris* that the thermal fluctuation of the temperature between 35°C and 25°C during the first day of anther culture improve the frequency of embryogenic anthers to 9.0% compared to a 0.5% when placed for all the time at 25°C (Keller and Armstrong, 1979). The comparison of the day/night temperatures 15/10°C increases by about 70% the embryogenic pollen in comparison with temperatures of 25/20°C for which only 27% of the pollen was embryogenic (Keller and Stringam, 1978). Also culture conditions of the donor plants are important and winter and spring season are the best for culture establishment (Yang et al., 1992a).

Several media have been tested for obtaining good embryogenic yield both for cauliflower and broccoli (Yang et al., 1992a, b). The effect of 2,4 D and of silver nitrate was tested on *B. oleracea* and good results using the former in absence of the latter were observed (Ockendon and McClenaghan, 1993). The addition into the medium of cytokinin (BAP) affects negatively the embryo production, whereas the use of liquid medium and of dark incubation after high temperature treatment is favourable for embryogenesis (Yang et al., 1992a).

Embryos obtained by anther culture obtained from an autumn type of cauliflower gave plants of which the 64% were of tetraploids, 26% diploids, 6% triploids and 0% haploids (Boucault et al., 1991). Good correlation was observed among the number of chloroplasts in the stomatic guard cells and some morphological characters and the ploidy level. The polymorphism for ACO and PGM enzyme loci allowed verifying the gametophytic origin of the plants obtained by anther culture and to detect the segregation of the correspondent alleles (Boucault et al., 1991).

The low level of embryogenic pollen grains of the cultured anthers and the interest to individuate more efficient methods for producing haploids has stimulated the opportunity to utilize isolated microspore for *in vitro* culture. In fact, the anther walls could limit qualitatively and quantitatively the production of embryos (Hoffmann et al., 1982). Mature pollen of *B. oleracea* and of the hybrid *B. oleracea* x *B. algoglabra* cultured *in vitro* allowed obtaining tissue formation and then haploids (Kameya and Hinata, 1970).

Although anther culture was successful for *Brassica species*, the culture of isolated microspores tested for other species showed great interest in most of the cases in which few cell divisions took place or that haploid callus or embryoids did not develop further (Wenzel et al., 1975; Sopory, 1977; Ono and Harashima, 1981).

For *Brassica* species microspore culture was positively tested also if in some new media cell clusters were formed by aggregation of microspores (Kameya and Hinata,

1970; Nitsch, 1977; Lichter, 1982). Several critical factors affect embryogenesis in *Brassica* species from isolated microspore culture and only by controlling them it is possible obtain good embryos yield, such as occurred for *B. napus* and *B. rapa* (Swanson, 1990; Ferrie and Keller, 1995) and for *B. oleracea*, for which there specific protocols have been developed (Lichter, 1989; Takahata and Keller, 1991; Duij et al., 1992).

Floral buds at the development stage, often characterized by specific bud or petal size, which allow obtaining the highest percentage of microspores from uninucleate to early binucleate stage are cultured, after filtration and centrifugation, in a reactive liquid medium and placed in dark conditions for about three-four weeks (Lichter, 1982; Fan et al., 1988; Swanson, 1990; Yang et al., 1992a, b; Telmer et al., 1993; Ferrie and Keller, 1995). The donor plant genotype is one of the main factors which affect embryogenesis frequency, embryos quality and plant regeneration (Arnison and Keller, 1990; Baille et al., 1992; Kieffer et al., 1993). Low temperatures, between 5°C and 20°C, during the growing cycle improve the plant physiology in terms to allow the production of microspores enough reactive for androgenesis (Keller *at al.,* 1987; Jain et al., 1989; Sorvari, 1985; Duijs et al., 1992).

Low temperatures probably vary endogenous levels of anther hormones and metabolites and could modify the frequency of symmetric cell division and control the characteristics of microspore (Lo and Pauls, 1992; Custers et al., 1994). High culture temperature is a key factor for the transition from gametophytic to sporophytic development of microspores by promotion of the vegetative nucleus in binucleate pollen (Custers et al., 1994).

For cauliflower and broccoli, androgenic embryo yield is related to the cultivar, the genotype and the culture medium; liquid medium containing silver nitrate gave the best results (Chauvin et al., 1993). Silver nitrate seems to inhibit the ethylene production, which is a limiting factor for microspore embryogenesis (Biddinton, 1992). Also, the carbohydrate levels play and important role for developing embryos and above all the sucrose and its high level in the culture medium could reduce the differences of response to embryogenesis among the cultivars (Dunwell and Thurling, 1985).

Regeneration rate from embryos to plantlets vary in accordance to the embryo developmental stage and to sucrose concentration into the medium. Mature pollen of *B. oleracea* and of the hybrid *B. oleracea* x *B. algoglabra,* cultured in hanging drop at low temperature, provide tissue formation and then haploid plants (Kameya and Hinata, 1970).

Ovaries excised from the plant four days after pollination between *B. oleracea* and *B. campestris* and between *B. campestris* and *B. cretica, B. montana* and *B. bourgeaui* cultured *in vitro*, produced hybrid plants by adding into the medium both casein hydrolysate and Nitsch and Nitsch's minerals (Inomata, 1985, 1986).

9 Seed Production

The specific flower biology of cauliflower makes seed production a difficult task, in particular due to the environmental conditions needed for the production of seed.

During the past, as said before, several genotypes were selected to improve seed production and to give a better cauliflower production. For this crop the sowing date for seed production is related to the biological materials utilized (population, breeding lines, etc.) and to their requirements according to environmental conditions which permit flower induction (vernalization is required for over-winter cultivars).

Several cauliflower and broccoli seed producers sow the seeds into containers and transplant the plantlets at the five-seven leaf stage in greenhouse, discarding off-types and plants affected by pathogens. Plant density has to be strictly related to the vigour of the plants of the cultivars or lines utilized for allowing plant inspection.

For the production of the basic seed, the mother plants are grown in their optimal season to facilitate selection on the basis of plant characters, mainly curd/head quality. For summer cultivars it is common to leave the plants *in situ* in open field, but for over-winter ones in continental climate conditions, the plants need to be transferred to greenhouse (but not in Mediterranean climates). The transfer of the selected plants is not always simple so several breeders produce seeds from propagules or by grafting pieces of the selected curd/head on selected cauliflower and broccoli rootstocks (Watts and George, 1963).

Vegetative propagation is often carried out using stumps of selected plants removed from open field and transferred to greenhouse or by cuttings which are inserted in a rooting substrate (Anonymous, 1980). Cauliflower and broccoli clones affected by virus (turnip mosaic virus and cauliflower mosaic virus) could be propagated vegetatively for production of virus-free materials by *in vitro* tissue culture (Walkey et al., 1974).

Flower induction is strictly related to temperature stimulus. The cold requirement varies among landraces and cultivars determine the biennial growing cycle for the over-winter types. The summer types do not require cold temperature for flowering and are considered annual.

In relation to pollination, cauliflower and broccoli genotypes are generally cross-pollinated, but self-pollination could occur especially for the summer types. Bees, blowflies and bumblebees often are responsible of the cross-pollination and are often utilized in greenhouse for seed production (Faulkner, 1962). Because of the pollination by insects, the recommended isolation distance is up 1,500 m between brassica fields (Raymond, 1985).

The two important roguing stages are one during the vegetative phase, in which precocious or button curds which develop too early are discarded, and the other by checking the uniformity in terms of leaf number, shape and crinkle according to the type; another rouging stage is at commercial harvesting, rejecting genotypes with inappropriate to curd colour, bractness, firmness, shape of the curd, and leaf protection.

For F_1 hybrid seed production the optimal pollinator/female ratio is 1:2 but this could vary in relation to the flowering ability of each parent. The parent lines utilized could show a high number of sibs produced.

At the beginning of the seed ripening stage the plant starts to dry and pods become light brown. Seed harvesting has to occur before pod shattering to avoid the loss of seeds. The plants could be cut and placed in windrows to continue drying before seed extraction. The dried plants are threshed. Care must be taken to use a slow cylinder speed not exceeding 700 rpm to avoid seed cracking (Raymond, 1985).

In relation to self-incompatibility which characterized flower biology of cauliflower and broccoli, it is very important that fields for the correspondent seed production be located at a distance of about 500-1,000 m. Alternatively, a net cabin can be used to guarantee genetic purity and to avoid genetic contamination. Crop density is about 3-6 plant m^{-2} depending on the plant vigour. After curd formation flower stems develop slowly and that is as consequence of hard selection for obtaining a firm curd which makes flowering more difficult. Often, if flowering occurs in autumn, the curd rots and only a part of the inflorescence reaches flowering and produces seed. Seed yield varies from 0.2 to 0.7 t ha^{-1}.

References

Acciarri N., Schiavi M., Vitelli G., Maestrelli A., Forni E., and Giovanessi L. 1997. *Breeding of green curded cauliflower for fresh market and freezing.* International Symposium on Brassicas, Acta Hort., 459: 403-410.

Acciarri N., Branca F., Sabatini E., Argento S., and Magnifico V. 2004. Miglioramento genetico dei cavolfiori a corimbo bianco e colorato. L'Informatore Agrario, LX, 25: 33-36.

Agrawal R.L. 1998. Fundamentals of plant breeding and hybrid seed production. Sscience Publishers Inc., Enfield, New Hampshire, USA.

Allegra G. 1934. Piante e Sementi, listino prezzi 1934-1935. Supplemento al Catalogo Generale, Stabilimento Tipografico Campione, Catania, 27-28.

Anderson W.C., and Carstens J.B. 1977. Tissue culture propagation of broccoli, *Brassica oleracea* (Italica Group), for use in F1 hybrid seed production. J. Am. Soc. Hortic. Sci., 102: 69.

Anonymous 1980. Top French seed starts from stumps. Grower, 93 (7) 16.

Anthony R.G., James P.E., and Jordan B.R. 1993. Cloning and sequence analysis of a FLO/LFY homologue isolated form cauliflower. Plant Ml. Biol., 22: 1163-1166.

Anthony R.G., James P.E., and Jordan B.R. 1996. Cauliflower curd development – the expression od meristem identity genes. J. Exp. Bot., 47: 181-188.

Arnison P.G., and Keller W.A. 1990. A survey on the anther culture response of *Brassica oleracea* L. cultivars grown under field conditions. Plant Breeding, 104: 125-133.

Arumugam N., Mukhopadhyay A., Gupta V., Pental D., and Praghan A. 1996. Synthesis of hexaploid (aabbcc) somatic hybrids: a bridging material for transfer of "tour" cytoplasmatic male sterility to different *Brassica* species. Theor. Appl. Genet., 56: 145-150.

Attia T., and Robbelen G. 1986. Cytogenetic relationship within cultivated *Brassica* analyzed in amphihaploids from the three diploid ancestors. Canadian Journal of Genetics and Cytology, 28: 232-329.

Baille A.M.R., Epp D.J., Jackson A., Semple C., and Keller W.A. 1990. Genotype-specific response of cultured broccoli (*Brassica oleracea* var. *italica*) anthers to cytokinins. Plant Cell Tissue Organ. Cult., 20: 217-222.

Bannerot H., Boulidard L., Cauderon Y., and Tempe J. 1974. Transfer of cytoplasmic male sterility from *Raphanus sativus* to *Brassica oleracea.* Proc. Eucarpia *Cruciferae* meeting. Dundee, 52-54.

Bannerot H., Boulidard L., and Chupeau Y. 1977. Unexpected difficulties met with the radish cytoplasm. Cruciferae Newsl., 2: 16

Baroncelli S., Buiatti M., and Bennici A. 1973. Genetics of growth and differentiation "*in vitro*" of *Brassica oleracea* var. *botrytis*. Z. Pflanzenzuecht., 70: 99.
Belliard G., Pelletier G., Vedel F., and Quétier F. 1978. Morphological characteristics and chloroplast DNA distribution in different cytoplasmatic parasexual hybrids of *Nicotiana tabacum*. Mol. Gen. Genet., 165: 231-237.
Bianco V. V., and Pimpini F. 1990. Orticoltura. Pàtron Editore, Bologna.
Biddinton N.L. 1992. The influence of ethylene in plant tissue culture. Plant Growth Regul., 11: 73-187.
Bonnett H.T., Kofer W., Hakansson G., and Glimelius K. 1991. Mitochondrial involvement in petal and stamen development studied by sexual and somatic hybridizations of *Nicotiana* species. Plant Sci., 80: 119-130.
Block G., Patterson B., and Subar A. 1992. Fruit, vegetables and cancer prevention: a review of the epidemiologicalm evidence. Nutr. Cancer, 18: 1-29.
Boukema I.W., and van Hintum Th. J.L. 1998. The European Brassica database. Acta Hort., 459: 249-254.
Boutelou C., and Boutelou E. 1801. Tratado de la Huerta. Imprenta Villalpando, Madrid.
Branca F. 1997. Morpho-biological traits of some italian local cultivars of broccoli. ISHS symposium "Brassica 97", Rennes 23-27 September 1997.
Branca F., and Iapichino G. 1997. Some wild and cultivated *Brassicaceae* exploited in Sicily as vegetables. FAO/IPGRI Plant Genetic Resources Newsletter, 110: 22-28.
Branca F. 1998. Caratterizzazione di tipi di cavolfiore violetto. Atti "IV Giornate Scientifiche Società Orticola Italiana, 79-80.
Branca F. 2000. Valutazione di cultivar locali di cavolfiore violetto. Workshop "Risultati del primo anno di attività del Piano Nazionale di Ricerca per l'Orticoltura del Mipa", Sirmione, 29 marzo 2000, 114-115.
Branca F., Li G., Goyal S., and Quiros C. 2002. Survey of aliphatic glucosinolates in Sicilian wild and cultivated *Brassicaceae*. Phytochemistry, 59: 717-724.
Branca F. 2002. Linee di attività per il miglioramento del cavolfiore violetto. Atti Vi Giornate Scientifiche Società Orticola Italiana, 461-462.
Branca F., Bahcevandziev K., Perticone V., and Monteiro A. 2005. Screening of Sicilian local cultivars of cauliflower and broccoli to *Peronospora parasitica*. Biodiversity and Conservation, 14: 841-848.
Branca F. 2006. Programmi e risultati per la valorizzazione della diversità specifica e genetica del genere *Brassica*. Italus Hortus, 13 (3): 204-210.
Branca F., and Candido V. 2007. Innovazione di processo e di prodotto della filiera del cavolfiore e del cavolo broccolo. In: Orticoltura di pien'aria in Italia: quali prospettive per il comparto (Miccolis V., Elia A., Candido V., eds.). Tipografia Vito Radio, Putignano: 165-193.
Branca F., Ferrari V., and Maestrelli A. 2007. Caratteristiche e nuove costituzioni di cavolfiore a corimbo pigmentato. Italus Hortus, 14 (2): 13-25.
Briggs F.N., and Knowles P.F. 1967. Introduction in plant breeding. Reinhold Publishing Corp., New York.
Buck P.A. 1956. Origin and taxonomy of broccoli. Econ. Bot., 10: 250-253.
Buzza G.C. 1995. Plant breeding. In: Kimber K., McGregor D.I. (eds.), *Brassica oilseeds, production* and utilization. Cambridge University Press, 153-175.
Cardi T., and Earle E. 1997. Production of new CMS *Brassica oleracea* by transfer of "Anand" cytoplasm from *B. rapa* through protoplast fusion. Theor. Appl. Genet., 94: 204-212.
Carr S.M., and Irish V.F. 1997. Floral homeotic gene expression defines developmental arrest stages in *Brassica oleracea* L. var. *botrytis* and *italica*. Planta, 201: 179-188.

Chiang M.S., and Crete R. 1970. Inheritance of clubroot resistance in cabbage (*Brassica oleracea* var. *capitata* L.). Can. J. Genet. Cytol., 12: 253.
Cole R.A. 1978. Epithiospecifier protein in turnip and changes in products of autolysis during ontogeny. Phytochemistry, 17: 1563.
Crisp P. 1974. Meristem culture and the breeding of cauliflowers. In Proc. Eucarpia Meeting on *Brassica,* Wills A.B., North C. (Eds.), Dundee, Scotland, 55.
Crisp P., and Walkey D.G.A. 1974. The use of aseptic meristem culture in cauliflower breeding. Euphytica, 23: 305-313.
Crisp P., Jewell P.A., and Gray A.R. 1975 a. Improved selection against the purple color defect of cauliflower. Euphytica, 24: 177-180.
Crisp P., Gray A.R., and Jewell P.A. 1975 b. Selection against the bracting defect of cauliflower. Euphytica, 24: 459-465.
Crisp P., Johnson A.G., Ellis P.R., and Hardman J.A. 1977. Genetical and environmental interaction affecting resistance in radish and to cabbage root fly. Heredity, 38: 209.
Crisp P. 1982. The use of an evolutionary scheme for cauliflowers in the screening of genetic resources. Euphytica, 31: 725-734.
Crisp P., and Gray A.R. 1984. Breeding old and new forms of purple heading broccoli. Cruciferae New, 9: 17-18.
Crisp P., and Tapsell C.R. 1993. Cauliflower, *Brassica oleracea* L., 157-178. In: Kaloo G., Bergh B.O. (eds) Genetic improvement of vegetable crops. Pergamon Press, Oxford.
Custers J.B.M., Cordewener J.H.G., Nöllen Y., Dons H.J.M., and Van-Lookeran-Champagne M.M. 1994. Temperature controls both gametophytic and sporophytic development in microspore cultures of *B. napus.* Plant Cell Rep., 13: 267-271.
David W.A.L., and Gardiner B.O.C. 1966. Mustard oil glucosides as feeding stimulants for *Pieris brassicae larvae in a semisynthetic diet.* Entomol. Exp. Appl., 9: 247.
De Candolle A.P. 1821. Regni vegetabilis systema naturale. II: 582-616. Paris.
Di Cesare L., Vitale, Acciarri N., and Branca F. 2001. Composizione volatile caratteristica di alcune cultivar di broccolo, cavolo e verza coltivate in Italia. Ortaggi-Industria alimentare, XL, 508-512.
Di Cesare L.F., Viscardi D., Genna A., Maestrelli A., Branca F., and Argento S. 2005. Indagini sulla composizione volatile e sul contenuto di isotiocianati di cultivar locali di cavolfiore violetto. Industrie Alimentari, XLIV, 262-271.
Dickson M.H., and Eckenrode C.J. 1975. Variation in *Brassica oleracea* resistance to cabbage looper and imported cabbage worm in the greenhouse of field. J. econ. Entomol., 68: 757.
Dickson M.H., and Wallace D.H. 1986. Cabbage breeding. In: Breeding vegetable crops, pp. 396-432. W.J. Bassett (ed.) AVI Publishing Company Inc., Westport, Connecticut, USA.
Dodoens R. 1578. A nieuve Herbal…Antverpiae, trasl H.lyte. London.
Downey R.K., and Röbbelen G. 1989. Brassica species. In: Röbbelen G., Downey R.K. and Ashri A. (eds.), Oil crops of the world. McGraw-Hill Inc. New York, 339-382.
Duij J.G., Visser D.L., and Custers J.B.M. 1992. Microspore culture is successful in most crop types of *Brassica oleracea* L. Euphytica, 60: 45-55.
Dunn J.A., and Kempton D.P.H. 1971. Differences in susceptibility to attack by *Brevicoryne brassicae* (L.) on Brussels sprouts. Ann. Appl. Biol., 68: 121.
Dunn J.A., and Kempton D.P.H. 1976. Varietal differences in the susceptibility of Bruxelles sprout to lepidopterous pests. Ann. Appl. Biol., 82: 11.
Dunwell J.M., and Thurling N. 1985. Role of sucrose in microspore embryo production in *Brassica napus* ssp. *oleifera.* J. Exptl. Bot., 36: 1478-1491.
Fahey J.W., Zhang Y., and Talalay, P. 1997. Broccoli sprouts: an exceptionally rich source of inducers of enzymes that protect against chemical carcinogens. *Proc. Natl. Acad. Sci. USA* 94: 10367-10372.

Fahey J.W., and Talalay P. 1999. Antioxidant Functions of Sulforaphane: a Potent Inducer of Phase II Detoxication Enzymes. Food and Chemical Toxicology, 37 (9-10): 973-979.

Faulkner G.J. 1962. Blowflies as pollinators of brassica crops. Commercial Grower, 3457: 807-809.

Fahleson J., Eriksson I., and Glimelius K. 1994. Intertribal somatic hybrids between *Brassica napus* and *Barbarea vulgaris* – production of *in vitro* plantlets. Plant Cell Rep., 13: 411-416.

Fahleson J., Lagercrantz U., Mouras A., and Glimelius K. 1997. Characterization of somatic hybrids between *Brassica napus* and *Eruca sativa* using species-specific repetitive sequences and genomic *in situ* hybridization. Plant Science, 123: 133-142.

Fan Z., Armstrong K.C., and Keller W.A. 1988. Development of microspores *in vitro* in *Brassica napus* L. Protoplasma, 147: 191-199.

Faulkner G.J. 1962. Blowflies as pollinators of brassica crops. Commercial Grower, 3457: 807-809.

Faulkner K., Mithen R., and Williamson G. 1998. Selective increase of the potential anticacinogen 4-methylsulphinylbutyl glucosinolate in broccoli. Carcinogenesis, 19: 605-609.

Ferrie A.M.R., and Keller W.A. 1995. Development of methodology and application of doubled haploids in *Brassica rapa.* Proc. 9th International Rapeseed Congress, Cambridge, U.K., 3: 807-809.

Finch S., Cole R.A., and Skinner G. 1974. Chemicals influencing cabbage root fly behaviour. Isolation attractants. Rep. Natl. Veg. Res. Stn. Wellesbourne, UK, p. 95.

Finley J.W. 2005. Proposed criteria for assessing the efficacy of cancer reduction by plant foods enriched in carotenoids, glucosinolates, polyphenols and selenocompounds. Ann. Bot., 95: 1075-1096.

Forti C. 1929. La coltivazione degli ortaggi. Unione Tipografico Editrice Torinese, 117-121.

Gamet-Payrastre L., Li P., Lumeau S., Cassar G., Dupont M.A., Chevolleau S., Gasc N., Tulliez J., and Tercé F. 2000. Sulforaphane, a naturally occurring isothiocyanate, induces cell cycle arrest and apoptosis in HT29 human colon cancer cells". Cancer Res. 60: 1426-1433.

Gass T., Gustafsson M., Astley D., and Frison E. A. 1995. Report of a working group on Brassica. 2nd meeting, 13-15 November 1994, Lisbon, Portugal. European Cooperative Programme for Crop Genetic Resources Networks (ECP/GR), International Plant Genetic Resources Institute, Rome, Italy. 87 pp.

Genna A., Avitabile Leva A., Acciarri N., Branca F., Di Cesare L.F., Lo Scalzo R., Bianchi G., and Maestrelli A. 2004. Caratteristiche qualitative di cavolfiori autoctoni da inserire nello scenario degli ortaggi surgelati. Italus Hortus, 11 (1): 37-40.

Gerdemann-Knörck M., Sacristán M.D., Braatz C., and Schieder O. 1994. Utilisation of asymmetric somatic hybridization fort he transfer of disease resistance frm *Brassica nigra* to *Brassica napus.* Plant Breeding, 113: 106-113.

Glimelius K. 1984. High growth rate and regeneration capacity of hypocotyl protoplast in some *Brassicaceae.* Physiol. Plant., 61: 38-44.

Glimelius K., Djupsjöbacka M., and Fellner F.H. 1986. Selection and enrichment of plant protoplast heterokaryons of Brassicaceae by flow sorting. Plant Sci. 45: 133-141.

Glimelius K. 1999. Somatic hybridization. In: Biology of Brassica coenospecies, Gómez-Campo (ed.). Elsevier Science B.V., Amsterdam.

Gómez-Campo C., and Gustafsson M. 1991. Germplasm of wild n=9 Mediterranean species of Brassica. Botanika Chronika: 10, 429-434.

Gray A.R. 1982. Taxonomy and evolution of broccoli (*Brassica oleracea* var. *italica*). Econ. Bot., 36 (4): 397-410.

Gray A.R. 1989. Majoring in minor brassicas. Science for growers, AFRC, Warwickshire: 2-3.

Guha S., and Maheshwari S.C. 1964. *In vitro* production of embryos from anther of *Datura.* Nature, 204: 497.

Guha S., and Maheshwari S.C. 1966. Cell division and differentiation of embryos pollen grains of *Datura in vitro.* Nature, 212: 97-98.

Gustafsson M. 1982. Germplasm conservation of wild (n=9) Mediterranean *Brassica* species. Sveriges Utsädesförenings Tidskrift, 92: 133-142.

Gustafson-Brown C., Savidge B., and Yanofsky M.F. 1994. Regulation of the *Arabidopsis* floral homeotic gene *APETALA 1.* Cell, 76: 131-143.

Hadley H.H., and Openshaw S.J. 1980. Interspecific and intergeneric hybridization. In: Hybridization of Crop Plants, pp. 133-159. W.R. Fehr and H.H. Hadley (eds.). ASA and CSSA, Madison Wisc., USA.

Harberd D.J. 1976. Cytotaxonomic studies of *Brassica* and related genera. In: Vaughan, J.G., A.J. McLeod and B.M.G. Jones, The biology and chemistry of the Cruciferae, Academic Press, London: 47-68.

Harlan J.R. 1975. Crops and man. American Society of Agronomy, Crop Science Society of America. Madison, Wisconsin.

Hervé Y. 1992. Les choux. In: Amélioration des espèces végétales cultivées (eds. Gallais A. and Bannerot H.). INRA editions, 435-447.

Hervé Y. 2003. Choux. In: Pitrat M., Four Y.C. (eds). Histories de legumes des origins è l'orée du XXI siecle INRA, 233-234.

Hintum, Th. J.L. Van and Boukema, I.W. 1993. The establishment of the European Database for *Brassica.* Plant Genetic Resources News., 11-13.

Hoffmann F., Thomas E., and Wenzel G. 1982. Anther culture as a breeding tool in rape II. Progeny analysis of androgenic lines and induced mutants from haploid culture. Theor. Appl. Genet., 61: 225-232.

Hosaka K., Kianian S.F., McGrath J.M., and Quiros C.F. 1990. Development and chromosomal localization of genome-specific DNA markers of *Brassica* and the evolution of amphidiploids and n=9 diploid species. Genome, 33: 131-142.

Hyams E. 1971. Cabbages and Kings. In: Plants in the service of man: 33-61, London.

Ibn al-'Awwâm, end of XII century). In Clément-Mullet J.J. (translator): Le livre de l'agriculture. Actes Sud, Arles, 1040 pages.

Imamura J., Saul M.W., and Potrykus I. 1987. X-ray irradiation promoted asymmetric somatic hybridization and molecula analysis of the products. Theor. Appl. Genet., 74: 445-450.

Inomata N. 1985a. A revised medium for *in vitro* cultures of *Brassica* ovaries. In Chapman G.P., Mantell S.H., Daniels R.W. (eds.), The experimental manipulation of ovule tissues. Longman, New York, London, 164-176.

Inomata N. 1985b. Interspecific hybrids between *Brassica campestris* and *B. cretica* by ovary culture *in vitro.* Cruciferae Newsletter, 10: 92-93.

Inomata N. 1986. Interspecific hybrids between *Brassica campestris* and *B. bourgeaui* by ovary culture *in vitro.* Cruciferae Newsletter, 11: 14-15.

Inomata N. 1987. Interspecific hybrids between *Brassica campestris* and *B. montana* by ovary culture *in vitro.* Cruciferae Newsletter, 12: 8-9.

Iori R., Bernardi R., Gueyrard D., Rollin P., and Palmieri S. 1999. Formation of glucoraphanin by chemoselective oxidation of natural glucoerucin: a chemoenzymatic route to sulforaphane. *Bioorg. Med. Chem. Lett.,* 9: 1047-1048.

Jain R.K., Brune U., and Friedt W. 1989. Plant regeneration from *in vitro* cultures of cotyledon explants and anthers of *Sinapis alba* and its implication on breeding crucifers. Euphytica, 43: 153-163.

Jeffery E.H., Brown A.F., Kurilich A.C., Keck A.S., Matusheski N., Klein B.P., and Juvik J.A. 2003. Variation in content of bioactive components in broccoli 16 (3): 323-330

Johnson I.T. 2002. Glucosinolates in human diet. Bioavailability and implication for health. Phytochemistry Reviews, 1, 2, 183-188.

Jones H.A. 1932. Vegetable breeding at the University of California. Proc. Am. Soc. Hortic. Sci., 29: 572.

Kameya T., and Hinata K. 1970. Induction of haploid plants from pollen grains of *Brassica*. Jpn. J. Breed., 20: 82-87.

Keller W.A., Arnison P.G., and Cardy B.J. 1987. Haploids from gametophytic cells – recent developments and future prospects. In: Plant Tissue and Cell Culture. Alan R. Liss, New York, 233-241.

Keller W.A., and Armstrong K.C. 1979. Stimulation of embryogenesis and haploid production in *Brassica campestris* anther culture by elevated temperature treatments. Thcor. Apll. Genet., 55: 65-67.

Keller W.A., and Stringam G.R. 1978. Production and utilisation of microspore-derived haploid plants. In: Thorpe T.A. (ed.): Frontieres of plant tissue culture, 113-122. Proc. LA.P.T.C., Calgary, 1978.

Kempin S.A., Savidge B., and Yanofsky M.F. 1995. Molecular basis of the cauliflower phenotype in *Arabidopsis*. Science, 267: 522-525.

Keum Y.S., Jeong W.S., and Tony Kong A.N. 2004. Chemoprevention by isothiocyanates and their underlying molecular signaling mechanisms. Mutation Research/Fundamental and Molecular Mechanisms of Mutagenesis 555 (1-2): 191-202.

Kieffer M., Fuller M.P., Chauvin J.E., and Schlesser A. 1993. Anther culture of kale (*Brassica oleracea* L. convar. *Acephala* (DC.) Alef.). Plant Cell Tissue Organ. Cult., 33: 303-313.

Kimber M., and Rilley R. 1963. Haploid angiosperms. Bot. Rev., 29: 480-531.

King G.J. 2003. Using molecular allelic variation to understand domestication processes and conserve diversity in *Brassica* crops. Acta Hort., 598: 181-186.

Kristal A.R., and Lampe J.W. Brassica vegetables and prostate cancer risk: a review of the epidemiological evidence. Nutr. Cancer, *42:* 1-9: 2002.

Labate J., Robertson L., and Björkman T. 2003. Genotypes at the *BoCAL-a* locus in *B. oleracea* do not predict broccoli, cauliflower and purple cauliflower phenotype. HortScience, 38: 736.

Lamarck J.B.A.1784. "Chou". Encyclopédie methodique botanique. I. Paris.

Lichtenstein E.P., Strong F.M., and Morgan D.G. 1962. Identification of 2-phenylethylisothiocyanate as an insecticide occuring naturally in the edible parts of turnips. J. Agric. Food Chem., 10: 30.

Licther R. 1982. Induction of haploid plants from isolated pollen of *Brassicsa napus*. Z. Pflanzenphysiol. Bd., 105: 427-434.

Lichter R. 1989. Efficient yield of embryoids by culture of isolated microspores of different *Brassicaceae* species. Plant Breeding, 103: 119-123.

Linnaeus C. 1753. Species Plantarum II: 561. Stockholm.

Linsmaier E.M., and Skoog F. 1965. Organic growth factor requirements of tobacco tissue cultures, Physiol. Plant., 18: 100.

Liu J-H., Dixelius C., Eriksson I., and Glimelius K. 1995. *Brassica napus* (+) *B. tournefortii,* a somatic hybrid containing traits of agronomic importance for rapeseed breeding. Plant Sci., 109: 75-86.

Lo K-H., and Pauls K.P. 1992. Plant growth environment effects on rapeseed microspore development and culture. Plant Physiol., 99: 468-472.

Maestrelli A., Acciarri N., Forni E., and Vitelli G. 1999. *The cauliflower cv Romanesco. An interesting green variety suitale for freezing*. 20th International Congress of Refrigeration, IIR/IIF, Sydney.

Maggioni L., Astley D., Gustafsson M., Gass T., and Lipman E., compilers. 1997. Report of a working group on *Brassica*. Third Meeting, 27-29 November 1996, International Plant Genetic Resources Institute, Rome, Italy, 92 pp.

Maggioni L. 1998. Activities and achievements of the ECP/GR Brassica Working Group. In: Thomas G. and Monteiro, A (eds), Proc. of an International Symposium on Brassicas, ISHS. Acta Hort. 459, 243-248.

Mandel M.A., Gustafson-Brown C., Savidge B., and Yanofsky M.F. 1992. Molecular characterization of the *Arabidopsis* floral homeotic gene *APETALA1*. Nature, 360: 273-277.

Massie I.H., Astley D., and King G.J. 1996. Patterns of genetic diversity and relationships between regional groups and population of Italian landraces cauliflower and broccoli (*Brassica oleracea* L. var. *botrytis* L. and var. *italica* Plenck). Acta Hort., 407: 45-53.

Massie I.H. 1998. Patterns of variation in the Italian landrace cauliflower and broccoli p. 223. PhD thesis, Univ. Lond., London.

McNaughton I.H. 1995a. Turnip and relatives. *Brassica napus* (Cruciferae). In: Smartt, J. and N.W. Simmonds (eds.), Evolution of Crop Plants, Chap. 17, pp. 62-68. Longman Group UK Limited.

McNaughton I.H. 1995b. Swedes and rapes. *Brassica napus* (Cruciferae). In: Smartt, J. and N.W. Simmonds (eds.), Evolution of Crop Plants, Chap. 18, pp. 68-75. Longman Group UK Limited.

Miller P. 1724. The gardener and floristic dictionary or a complete system of horticulture. Rivington, London.

Mithen R., Lewis, B.G., Heaney, R.K., and Fenwick, R. 1987. Glucosinolates of wild and cultivated *Brassica* species. Phytochemistry 26: 1969-1973.

Mithen R., Lewis, B.G., Heaney, R.K., and Fenwick, R. 1987. Glucosinolates of wild and cultivated *Brassica* species. Phytochemistry 26: 1969-1973.

Mithen R., Dekker M., Verkerk R., Rabot R., and Johnson I.T. 2000. The nutritional significance, biosynthesis and bioavailability of glucosinolates in human foods. J. Sci. Food Agric., 80: 967-984.

Mithen R., Faulkner K., Magrath R., Rose P., Williamson G., and Marquez J. 2003. Development of isothiocyanate-enriched broccoli, and its enhanced ability to induce phase 2 detoxification enzymes in mammalian cells. Theor. Appl. Genet., 106: 727-734.

Morgan A., and Maliga P. 1987. Interspecific T-DNA in *Brassica* cybrids. Mol. Gen. Genet., 209: 240-246.

O'Neill C.M., Murata T., Morgn C.L., and Mathias R.J. 1996. Expression of the C3-C4 intermediate character in somatic hybrids between *Brassica napus* and C_3-C_4 species *Moricandia arvensis*. Theor. Appl. Genet., 93: 1234-1241.

Namai H., Sarashima M., and Hosoda T. 1980. Interspecific and intergeneric hybridization breeding in Japan. In: Tsunoda S., Hinata K., Gomez-Campo C. (eds.), Brassica crops and wild allies. Biology and breeding. Japan Scientific Society, Tokio, 191-203.

Nitsch C. 1977. Culture of isolated microspores. In: Teinert J., Bajaj Y.P.S. (Eds.): Plant cell, tissue and organ culture, 268-278. Springer, Berlin, Heidelberg, New York.

Nitsch J.P., and Nitsch C. 1969. Haploid plants from pollen grains. Science, 163: 85-87.

Nonnecke I.L. 1989. Vegetable production. Van Nostrand Reinhold, New York, USA.

Nuez F., Gómez Campo C., Fernández de Córdova P., Soler S., and Valcárcel J.V. 1999. Colleccion de semillas de coliflor y broccoli. Instituto Nacional de Investigation y Tecnologia Agraria y Alimentaria, Madrid, pgg. 120

Ockendon D.J., and McClenaghan R. 1993. Effect of silver nitrate and 2,4D on anther culture of Brussels sprouts (*Brassica oleracea* var. *gemmifera*). Plant Cell, Tissue and Organ Culture, 32: 41-46.

Ogura H. 1968. Studies on the new male sterility in Japanese radish with special reference to the utilisation of this sterility towards the pratical raising of hybrid seeds. Mem. Fac. Agric. Kagoshima Univ., 6: 39-78.

Ono K., and Harashima S. 1981. Induction of haploid callus from isolated microspores of peony *in vitro.* Plant Cell Physiol., 22: 337-341.

Oost H., Brandenburg W.A., Reuling G.T.M., and Jarvis C.E. 1987. Lectotypification of *Brassica rapa* L., *B. campestris* L. and neotypification of *B. chinensis* L. (Cruciferae). Taxon, 36: 625-634.

Pareek L.K., and Chandra N. 1978. Somatic embryogenesis in leaf callus from cauliflower (*Brassica oleracea* var. *botrytis*). Plant Sci. Lett., 11: 311.

Pelletier G. 1986. Plant organelle genetics through somatic hybridization. Oxford surveys of plant molecular and cell biology, 3: 97-121.

Pelletier G., Primard C., Ferault M., Vendel F., Chetrit P., Renard M., and Delourme R. 1988. Use of protoplasts in plant breeding: cytoplasmic aspects. Plant cell, tissue and organ cult., 12: 173-180.

Pelletier G., Ferault M., Lancelin D., and Boulidard L. 1989. CMS *Brassica oleracea* cybrids and their potentiel for hybrid seed production. Proc. XIIth Eucarpia congress, 11-7.

Piastuch W.C., and Bates G.W. 1990. Chromosal analysis of *Nicotiana* asymmetric somatic hybrids by dot blotting and *in situ* hybridization. Mol. Gen. Genet., 222: 97-103.

Pizzocaro F., Ferrari V., Acciarri N., Morelli R., Russo-Volpe S., and Prinzivalli C. 2000. Antioxidant and antiradical activities in green and violet cauliflower ecotypes with different maturity stages. Workshop of VI Giornate Scientifiche SOI, Sirmione, 28 Marzo 2000: 34-35.

Pow J.J. 1969. Clonal propagation *in vitro* from cauliflower curd. Hortic. Res., 9: 151.

Poppel van G., Verhoeven D.T.H., Verhagen H., and Goldbohm R.A. 1999. Brassica Vegetables and Cancer Prevention. In: *Advances in Nutrition and Cancer 2*, (edit. Zappia et al.). Kluwer Academic/Plenium Publishers, New York.

Prakash S. 1973. Haploidy in *Brassica nigra* Koch. Euphytica, 22: 196-203.

Prakash S. 1974. Haploid meiosis and origin of *Brassica tournefortii* Gouan. Euphytica, 23: 591-595.

Prakash S., and Hinata K. 1980. Taxonomy, cytogenetics and origin of crop Brassicas, a review. Opera Botaniki, 55: 1-57.

Pradhan A.K., Mukhopadhyay A., and Pental D. 1991. Identification of the putative cytoplasmatic donor of a CMS system in *Brassica juncea.* Plant Breeding, 106: 204-208.

Purugganan M.D., Boyles A.L., and Suddith J. 2000. Variation and selection at the cauliflower floral homeotic gene accompanying the evolution f domesticated *Brassica oleracea.* Genetics, 155: 855-862.

Quiros C.F., Ochoa O., Kianian S.F., and Douches D. 1987. Analysis of the *Brassica oleracea* genome by the generation of *Brassica campestris-oleracea* addition lines: characterization by isozymes and rRNA genes. Theoretical and Applied Genetics, 74: 758-766.

Quiros C.F., Ochoa O., and Douches D. 1988. *Brassica* evolution: exploring the role of X=7 species in hybridisation with *B. nigra* and *B. oleracea.* Journal of Heredity, 79: 351-358.

Rawat D.S., and Anand I.J. 1979. Male sterility in Indian mustard. Indian J. Genet. Plant Breed., 39: 412-414.

Raymond A.T.G. 1985. Vegetable seed production. Longman, London and New York, 137-148.

Renard M., Delourme R., Mesquida J., Pelletier G., Primard C., Boulidard L., Dore C., Ruffio V., Herve Y., and Morice J. 1992. Male sterilities and fl hybrids in *Brassica.* In: Dattée Y., Dumas C., Gallais A. (eds), Reproductive Biology and Plant Breeding, Springer Verlag, Heidelberg, 107-109.

Röbbelen G. 1960. Beitrage zur Analyze des Brassica-Genomes. Chromosoma, 11: 205-228.

Robertson D., Palmer J.D., Earle E.D., and Mutschler M.A. 1987. Analysis of organelle genomes in a somatic hybrid derived from cytoplasmic male-sterile *Brassica oleracea* and arazine-resistant *B. campestris*. Theor. Appl. Genet., 76: 197-203.

Rosa E.A., Hearney R.K., Fenwick G.R., and Portas C.A. 1997. Glucosinolates in crop plants. Hort Rev. 19: 99-215.

Ruffio-Chable V., and Hervé Y. 1994. Chou-fleur et brocoli. In:Technologia des légumes (Tirilly Y., Bourgois C.M.eds), TEC & DOC Editions, 187-206.

Ruffio-Chable V., Hervé Y., Dumas C., and Gaude T. 1997. Distribution of S-haplotypes and its relationship with self-incompatibility in *Brassica oleracea*. 1. In inbred lines of cauliflower (*B. oleracea* var. *botrytis*). Theor. Appl. Genet., 94: 338-346.

Sampson D.R. 1966. Linkage of genetic male sterility with a seedling marker and its use in producing F1 hybrid seed of *Brassica oleracea*. Can. J. Plant. Sci., 46: 703.

Sandhu J.S., Thakur J.C., and Nandpuri K.S. 1977. Investigations on hibryd vigour in cauliflower (*Brassica oleracea* var. *botrytis* L.). Indian J. Hortic., 34: 430.

Sharma B.R., Swarup V., and Chatterjee S.S. 1972. Inheritance of resistance to black root in cauliflower. Can. J. Genet. Cytol, 14: 363.

Sidorov W.R., Menczel L., Nagy F., and Maliga P. 1981. Chloroplast transfer in *Nicotina* based n metabolic complementation between irradiated and iodoacetate treated protoplasts. Planta, 152: 341-345.

Singh S., and Mathur A. 1985. *In vitro* plantlet formation from cotyledons, leaf lamina and midrib of cauliflower. Curr. Sci., 54: 391.

Slocum M.K., Figdore S.S., Kennard W.C., Suzuki J.Y., and Osborn T.C. 1990. The linkage arrangement of restriction fragment length polymorphism loci in *Brassica oleracea*. Theoretical and Applied Genetics, 80: 57-64.

Smith L.B., and King G.J. 2000. The distribution of *BoCAL*-a alleles in *Brassica oleracea* is consistent with a genetic model for curd development and domestication of the cauliflower. Molecular Breeding, 6: 603-613.

Snogerup S., Gustafsson M., and Bothmer von R. 1990. Brassica sect. Brassica (Brassicaceae): 1. Taxonomy and variation. Willdenowia, 19: 271-365.

Song K., Osborn T.C., and Williams P.H. 1988. Brassica taxonomy based on nuclear restriction fragment length polymorfism (RFLPs). 2. Preliminary analysis of subspecies within *B. rapa* (syn. *campestris*) and *B. oleracea*. Theoretical and Applied Genetics, 76: 593-600.

Song K., Osborn T.C., and Williams P.H. 1990. Brassica taxonomy based on nuclear restriction fragment length polymorfism (RFLPs). 3. Genome relationships in Brassica and related genera and the origin of *B. oleracea* and *B. rapa* (syn. *campestris*). Theoretical and Applied Genetics, 79: 497-506.

Sopory S.K. 1977. Development of embryoids in isolated pollen culture of dihaploid *Solanum tuberosum*. Z. Pflanzenphysiol., 84: 453-457.

Sorvari S. 1985. Production of haploids from anther culture in agriculturally valuable *Brassica campestris* L. cultivars. Ann. Agric. Finniae, 24: 149-160.

Stiewe G., and Röbbelen G. 1994. Establishing cytoplasmatic male sterility in *Brassica napus* by mitochondrial recombination with *B. tournefortii,* Plant Breeding, 113: 294-304.

Sundberg E., and Glimelius K. 1986. A method for production of interspecific hybrids within *Brassicaceae* via somatic hybridization, using resynthesis of *Brassica napus* as a model. Pla. Sci., 43: 155-162.

Sunderland N. 1974. Anther culture as a means of haploid induction. In: Haploids in higher plants, Kasha K.J. (ed.), Proceeding 1st Int Symp, june 10-14, Univ Guelph, 91-22.

Swanson E.B. 1990. Microspore culture in *Brassica.* In: Pollard J.W. and Walker J.M. (eds.), Methods in Molec. Biol. Vol. 6. Plant cell and tissue culture. The Humana Press, 159-169.

Swarup V., and Pal A.B. 1966. Gene effects and heterosis in cauliflower. Indian J. Genet. Plant Breed., 26: 269.

Swarup V., and Chatterjee S.S. 1974. Heterosis in Indian cauliflower. Proc. 19th Int. Hortic. Cong. I. Section VII, Vegetables, 670.

Takahata Y., and Keller W.A. 1991. High frequency embryogenesis and plant regeneration in isolated microspore culture of *Brassica oleracea.* Plant. Sci., 74: 235-242.

Tamaro D. 1916. Orticoltura. Urico Hoepli, Milano, 186-191.

Telmer C.A., Newcomb W., and Simmonds D.H. 1993. Microspore development in *Brassica napus* and the effect of high temperature on division *in vivo* and *in vitro.* Protoplasm, 172: 154-165.

Thompson K.F. 1974. Homozygous diploid lines from naturally occurring haploids. 4th Internal. Rapskongr. Giessen, 119-124.

U N. 1935. Genome analysis in *Brassica* with special reference to the experimental formation of *B. napus* and peculiar mode of fertilization. Jpn. J. Bot., 7: 389-452.

Vang O., Mortensen J., and Andersen O. 2001. Biochemical effects of dietary intake of different broccoli samples. II. Multivariate analysis of contributions of specific glucosinolates in modulatine cytochrome P-450 and antioxidant defense enzyme activities. Metabolism, 50 (10): 1130-1135

Verhoeven D.T.H., Goldbohm R.A., Poppel van G., Verhagen H., and Brandt van den P.A. 1996. Epidemiological studies on *Brassica* vegetables and cancer risk. Cancer epidemiology, biomarkerms prevent, 5: 733-748.

Verhoeven D.T.H., Verhagen H., Goldbohm R.A., Brandt van den P.A., and Poppel van G. 1997. A review of mechanisms underlying anticarcinogenicity by brassica vegetables. *Chemico-Biological Interaction*, 103: 79-129.

Viani P. 1929. Trattato di Orticoltura. Battiato editore, Catania: 914-935.

Vriesenga J.D., and Honma S. 1971. Inheritance of seedlings resistance to club root in *Brassica oleracea* L. HortScience, 6: 295.

Walkey D., and Woolfitt J.M.C. 1970. Rapid clonal multiplication of cauliflower by shake culture. J. Hortic. Sci., 45: 205.

Walkey D.G.A., Cooper V.C., and Crisp P. 1974. The production of virus-free cauliflower by tissue culture. Journal Horticulture Science, 49: 273-275.

Watts L.E. 1980. Flower and Vegetable Plant Breeding. Grower Books, London.

Watts L.E., and George R.A.T. 1963. Vegetative propagation of autumn cauliflower. Euphytica, 12: 341-345.

Weiring D. 1961. The breeding of new cauliflower varieties for summer culture. Zaadhelangen, 15: 306.

Wenzel G., Hoffmann F., Potryeus I., and Thomas E. The separation of viable rye microspores from mixed population and their development in culture. Mol. Gen. Genet., 138: 293-297.

Williams P.H., Staub T., and Sutton J.C. 1972. Inheritance of resistance in cabbage to black root. Phytopatology, 62: 247.

Yang Q., Chauvin J.E., and Hervé Y. 1992a. A study of factors affecting anther culture of cauliflower (*Brassica oleracea* var. *botrytis*). Plant Cell, Tissue and Organ Culture, 28: 289-296.

Yang Q., Chauvin J.E., and Hervé Y. 1992b. Obtention d'embryons androgénétiques par culture *in vitro* de boutons floraux chez le broccoli (*Brassica oleracea* var. *italica*). C. R. Acad. Sci. Paris, 314 (III): 147-152.

Zhang Y., Talalay P., Cho C-G., and Posner G.H. 1992. A major inducer of anticarcinogenic protective enzme from broccoli: isolation and elucidation f structure. Proc. Natl. Acad, Sci, USA, 89: 2399-2403.

Zhang Y., Li J., and Tang L. 2005. Cancer-preventive isothiocyanates: dichotomous modulators of oxidative stress. Free Radical Biology and Medicine, 38 (1): 70-77.

Zhu C.Y., and Loft S. 2003. Effect of chemopreventive compounds from Brassica vegetables on NAD (P) H: quinone reductase and induction on DNA strand breaks in murine hepa1c1c7 cells. Food and Chemical Toxicology 41 (4): 455-462.

Family Chenopodiaceae

Spinach

Teddy E. Morelock[1] and James C. Correll[2]

[1] University of Arkansas, Department of Horticulture, morelock@uark.edu
[2] University of Arkansas, Department of Plant Pathology, jcorrell@uark.edu

1 Introduction

Spinach is arguably the #1 or # 2 most nutritious vegetable (broccoli being the other) that is consumed in the United States. It is very versatile since it is commonly used as a salad, a cooked vegetable or as a component of many other cooked meat and vegetable dishes. The recent development of baby leaf spinach coupled with an upswing of nutrition concerns has been responsible for increased consumption of spinach. It is widely stated that dark green leafy vegetables are the most lacking component of the American diet. Spinach is one of the most desirable dark green leafy vegetables because it is high in beta carotene (pro vitamin A) and folate, and is also a good source of vitamin C, calcium, iron phosphorous, sodium and potassium (Ryder, 1979; Nonnicke, 1989; Dicoteau, 2000). It is high in the carotenoid lutein which has been shown to prevent age related macular degeneration. Spinach is a good source of antioxidants and has one of the highest ORAC (oxygen radical absorbance capacity) values of any vegetable (Prior, 2003). While no attempts to breed for these traits have been reported in the literature there are reports of large differences between genotypes and it has been assumed that breeding for these traits is possible (Howard, 2001; Howard et al., 2002). Lutein values ranging from 10-25 mg/100g fresh weight have been reported (Murphy and Morelock, 2000). There also is a wide range of beta carotene levels between spinach genotypes (Murphy and Morelock, 2000). Moderate correlation between beta carotene content and lutein content has been reported (Murphy and Morelock, 2000; Murphy, 2001). Studies by Howard (2001) and Howard et al. (2002) have shown a wide variation in phenolic compounds in both amount and relative distribution between genotypes. The highest levels of these compounds occurred in genotypes that were white rust resistant (Howard, 2001; Howard et. al., 2002) and this is probably related to their disease resistance.

Spinach is widely known to be a good source of folate (Hine, 2003). The importance of folate in the human diet (particularly for women) can not be over emphasized. This is best illustrated by the fact that the United States RDA of folate for women was recently raised from 180 mcg to 400 mcg /day (Hine, 2003). This increase is primarily because of the effect of folate on the prevention of neural tube defects (Hine, 2003). It is also commonly known that folate can reduce homocystine levels in the blood which may impact coronary heart disease. The link is suspected but not proven. Difference in folate levels between spinach genotypes has not been reported probably due to the difficulty in folate analysis.

Spinach (*Spinacia oleracea* L) is a leafy cool season vegetable that produces a rosette during the vegetative growth phase. Leaves can be rounded to pointed and range from flat to (figures 1 and 2) fully savoy (crinkled). The original seed type for spinach was spiny but today round seed is the standard seed type in the United States and Europe. It is a diploid 2n = 12 that is native to the mid-east probably Iran. It is a member of the family Chenopodiaceae (the goosefoot family) which also contains beet sugar beet, chard and quinoa as well as some very successful weeds such as lambsquarter. Other common species include S. *turkestanica* Iljin and *S. tetranda* Stev. Spinach grows best in slightly acid to slightly basic soil (pH 6-7.5) (Nonnicke, 1989) but some very successful production occurs on soils above 8.0. The crop has a shallow root system and requires good levels of soil NPK. It also responds well to good levels of soil moisture but does not tolerate excess soil moisture.

Fig. 1. 'Wintergreen' semi-savoy spinach.

Fig. 2. 'F-415' flat spinach.

China produces the largest acreage of spinach followed by Indonesia, Japan, Turkey and the United States (Food and Agriculture Organization of the United Nations, 2006). China produces 76% of the world spinach and the United States ranks second with 4% of the world production (United States Department of Agriculture 2006a, 2006b). U.S. spinach consumption has actually increased recently (Food and Agriculture Organization of the United Nations, 2006) primarily due to health concerns of the general public and consumer acceptance of the prepacked triple washed baby leaf product. Baby leaf has also accounted for significant spinach acreage increases in California and Arizona (figures 3, 4, 5 and 6). In the past ten years California acreage has increased from 25,000 acres to 34,000 acres which Arizona has increased from 2500 acres to 6000 acres during the same period (United States Department of Agriculture, 2006a). Conversely during the same time period Texas spinach acreage has decreased by 6000 acres (United States Department of Agriculture, 2006a). These increases are due to increased demand for fresh spinach since during the same time period processed spinach has changed very little (United States Department of Agriculture, 2006a) with a slight increase in acreage for frozen spinach and a slight decrease for canned spinach.

2 Origin

Spinach is native to central Asia and widely thought to have originated in Persia (Iran) (Ryder, 1979; Nonnicke, 2000; Dicoteau, 2000; Swiader and Ware, 2002). It was first mentioned by the Chinese around 600 A.D. where it was referred to as the herb of Persia.

Fig. 3. Baby leaf spinach – newly planted.

Fig. 4. Baby leaf spinach about 1 week before harvest.

Fig. 5. Baby leaf spinach harvest in Salinas Valley, California.

Fig. 6. Band saw harvester cutting fresh market spinach.

It was first cultivated by the Arabs who introduced it to North Africa and it came to northern Europe around 1100 A.D. by way of Spain when it was introduced by the Moors (Ryder, 1979; Nonnicke, 2000; Dicoteau, 2000; Swiader and Ware, 2002). It was documented in Germany in 1280 and was a common garden vegetable by 1500 in England and France (Ryder, 1979; Nonnicke, 2000; Dicoteau, 2000; Swiader and Ware, 2002). It was brought to North America by the early Colonist and by 1806 three varieties were recognized and by 1828 the first savoy variety had been introduced (Nonnicke, 2000; Dicoteau, 2000). Prior to 1885 'Amsterdam Giant', 'Bloomsdale', 'Gaundry', 'Victoria' and 'Viroflay' had been introduced. Between 1900 and 1925 'Dark Green Bloomsdale', 'Long Standing Bloomsdale', 'Hollandia', 'King of Denmark', 'Juliana', 'Nobel' and 'Virginia Savoy' were introduced (Nonnicke, 2000). Between 1926 and 1950 'Canner King', 'Darkie', 'Del Monte', 'Domino', 'Old Dominion', 'Presto', 'Viking', 'Virginia Savoy Wilt Resistant', 'Winter Giant' and 'Winter King' were introduced (Magruder et al., 1938; Nonnicke, 2000). Hybrids were introduced in the 1950's and they have become the major type of spinach (Webb and Thomas, 1976).

3 Breeding and Genetics

3.1 Breeding Methods

Historically mass selection, recurrent selection and backcross have been the breeding techniques most commonly used in spinach but recent interest in more severe inbreeding to produce more uniform inbreds for use in hybrid production has spurred interest in methods such as paired crosses using individual female and male plants or use of monecious plants for selfing to develop inbreds.

3.2 Genetic Resources

Potential sources of genetic variation are always important to allow breeders to deal with changing condition such as new races of a pathogen, or a new disease or insect problem in addition changing environmental conditions, new production practices, new production areas, mineral content or phytonutrient content just are a few of the possible gene needs of the future. Even though large collections of genetic stocks do not exist for spinach the way they do for major crop species genetic resources are available. This point is probably best illustrated by the fact that in the past ten years 6 new downy mildew races have developed (Irish, 2004) and resistance to all of these new races has been discovered and incorporated into commercially available hybrids in a timely manner. Resistance for diseases such as white rust (Webb, 1969; Bowers, 1972) and Fusarium wilt (O'Brien and Winters, 1977) is available. It has also been demonstrated that genetic variation exists for such traits as aphid resistance (Mcleod et al., 1991), leafminer resistance (Mou, 2003, 2005), lutein content (Murphy and Morelock, 2000; Murphy, 2001), phenolic content (Howard, 2001; Howard et al., 2002; Pandjaitan, 2004; Pandjaitan et al., 2005), mineral content and curly top (Satou et al., 2002). It is also very likely that variation exists for many other traits that have not been investigated.

While large collections of genetic stocks do not exist, there are collections available to breeders. The USDA (USA) CGN (Holland) and IPK (Germany) all have larger collections of spinach germplasm but other countries such as Japan, Italy and Russia also have collections that are accessible to scientists (Irish, 2004).

Sneep (1958) studied inheritance of common traits in spinach and concluded the following traits to be single a gene or relatively simple inheritance: spiny vs. smooth seed, smooth vs. savoy leaves, light green vs. dark green leaves, and short vs. long petiole. Winter hardness and rate of growth were more complex in inheritance (Sneep, 1958).

3.3 Sex Determination

Since spinach is basically a dioecious species it has often been assumed that the mechanism of sex control is similar to the animal world in that it is controlled by a single pair of sex chromosomes (XY). Haga (1935) was unable to demonstrate the presence of a sex chromosome in a 1935 study. This was farther confirmed by Bemis and Wilson in 1953 (Bemis and Wilson, 1953). Janick and Stevenson proposed that X and Y could also be used to refer to alleles of a gene or a chromosome segment as well as whole chromosomes (Janick and Stevenson, 1954). They also examined crossing data and segregation patterns from a population of 'Long Standing Bloomdale' which indicated the heterogametic nature of spinach (Janick and Stevenson, 1955b). They proposed: XX-pistillate, XY staminate (may or may not produce seed) and YY (does not produce seed). They later proposed two possible genetic hypotheses to explain monoecious plants.

The first hypothesis proposed that in addition to XY that an independent incompletely dominate gene M^m accounted for the monoecious plants with XY being male and MM functioning only in the XX genotype which led to the following genotypes and phenotypes:

XXMM - true breeding monoecious
XXMm - monoecious, segregating more female than XXMM
XXmm - pistillate

Hypothesis number 2 - assume a third allelic X^m where

X^mX^m - true breeding monoecious
XX^m - segregating monoecious
XX - pistillate

The two hypotheses have similar expectations except when a true breeding monoecious is crossed with a staminate plant and the resulting staminate progeny are crossed with pistillate plants. The results of these crosses follow:

1.	a. XXMM monoecious	x	XYMM staminate	------- 1 XXMM monoecious	:	1 XYMM staminate
	b. XXMM pistillate	x	XYMM staminate	------- 1 XXMM monoecious	:	1 XXMM pistillate

2. a. X^mX^m x XY ------- $1\ X^mX$: $1\ X^mY$
monoecious staminate monoecious staminate

b. XX x X^mY ------- $1\ X^m$: $1\ XY$
pistillate staminate monoecious staminate

The appropriate crosses were made and segregation ratios proved the allelic model (#2) to be the correct scheme (Janick and Stevenson, 1955b) but these assumptions have been criticized by Dressler (1976).

Janick (1955a, 1955b) demonstrated with tetraploid spinach that the Y factor is necessary for maleness since XXXY is male but XXXX is female which indicates the presence of Y not the ratio of X and Y determines male or female. With XXXX x XXYY crosses a ratio of 5:1 staminate to pistillate was observed indicating chromosome segregation rather than chromatid segregation and that the sex genes are located near the centromere. Janick et al. (1959) also used trisomic analysis to locate the sex factor on the chromosome that produced the reflex trisomic.

The heterogametic nature of spinach has created considerable interest in cytogenetic investigation of spinach chromosomes (Ellis and Janick, 1960; Bose and Janick, 1961). Polyploidy has had limited investigation (Janick and Stevenson, 1954; Janick 1955 a, 1955b; Bose and Janick, 1961; Bragdo, 1962) much of which was prompted by interest in sex expression. This has also been considerable cytogenetic investigation aimed at synthesizing a heterogametic chromosome carrying the sex determination factor (Ellis and Janick, 1960; Iizuka and Janick, 1962, 1963, 1964).

4 Current Breeding Emphasis

Currently (2006) other than breeding for horticultural type disease resistance is the most common focus of breeding efforts with spinach. The primary emphasis is breeding for resistance to the downy mildew pathogen (*Peronospora farinosa* f.sp. *spinaciae*). This is due to the rapid rate at which new races of downy mildew are occurring.

Downy mildew was first described in 1824 and was reported in both the United States and Europe (Smith, 1885) (figure 7). Race 2 was reported in both California and Europe in 1958 (Zink and Smith, 1958). Race 3 was reported in the Netherlands in 1976 (Brandenberger, 1992), in California in 1978 (Brandenberger, 1992) and Texas in 1982 (Jones and Dainello, 1982). Race 4 was reported in California and Japan in 1990 (Brandenberger et al., 1991; Correll et al., 1990; Shimazaki, 1990), Texas in 1991 (Brandenberger, 1992) and Europe in 1994 (Lorenzini and Nali, 1994). Race 5 was reported in Colorado in 1996 (Correll et al., 1988), California in 1997 (Correll et al., 1988) and Europe in 1998 (Naili, 1998). Race 6 was reported in California in 1998 (Correll et al., 1988) and Japan in 2002 (Satou et al., 2002). Race 7 was reported in Europe in 1999 (Correll, 1998; Correll et al., 2001) and California in 2004 (Irish, 2004). Race 8 was reported in California in 2004 (Irish, 2004). Race 9 was reported in Europe in 2004 (Irish, 2004). Race 10 was reported in California in 2004 (Irish, 2004).

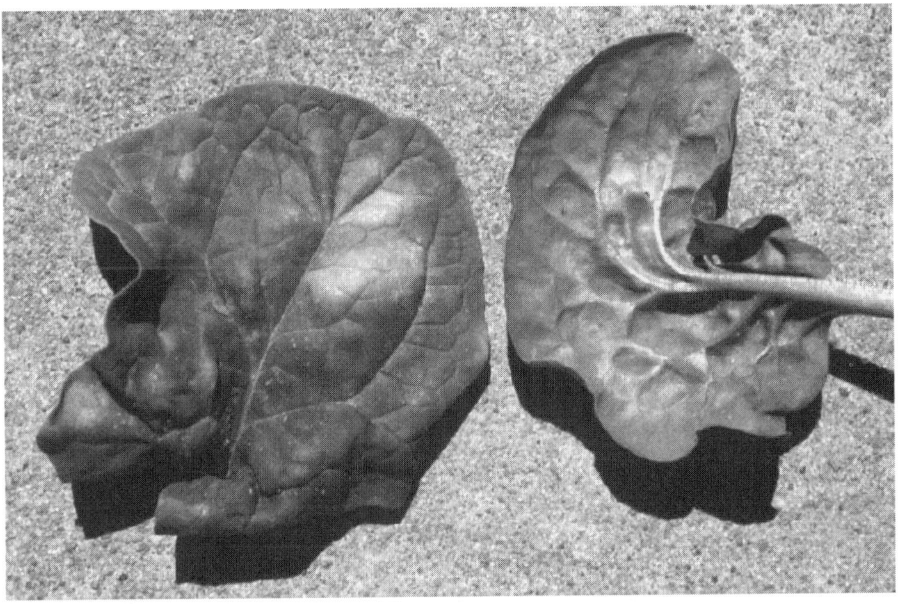

Fig. 7. Downy mildew of spinach.

The fact that seven new races of downy mildew developed between 1990 - 2004 (Koike et al., 2002) and six new races between 1996 - 2004 (Correll et al., 1990, 1998a, 1998b, 2001, 2003) has led to a tremendous effort aimed at developing hybrids that are resistant to the new races of the downy mildew pathogen. The major effort today (2006) is developing hybrids that carry resistance to all known races (1-10) of downy mildew. This philosophy coupled with year round production in California make it very likely that new races of this disease will continue to occur unless better methods of gene management are adopted.

White rust (*Albuigo occidentalis*) was first discovered on *Blitum capitatum* west of Ouray, Colorado in 1907 and was first found on commercial spinach in Virginia in 1910 (Brandenberger, 1992). White rust was first considered to be a serious problem in 1937 (Brandenberger et al., 1991; Brandenberger, 1992) when it was found on rail car lots of Texas spinach at the New York market. Since 1937 in Texas and 1944 in Arkansas it has occurred as epidemics repeatedly and today in the United States it can occur on commercial spinach in production areas east of the Rocky Mountains. In the United States white rust is second to downy mildew in the focus of plant breeding effort.

Fusarium wilt (*Fusarium oxysporum* f.sp. *spinaciae*) was first reported in 1923 (Bassi and Goode, 1978). It is presently a serious problem in Arkansas and Oklahoma in commercial production fields (Goode, 1969; Bassi and Goode, 1978) and much of the acreage has moved to the overwinter production season to avoid the problem by moving production to times where there are reduced soil temperatures (figures 8 and 9). It is also a problem in the seed production areas of the Puget Sound and Denmark. It has also been reported in Japan. The disease is potentially a problem for other production areas and seed producing areas because it can be seed borne disease (Bassi and Goode, 1978).

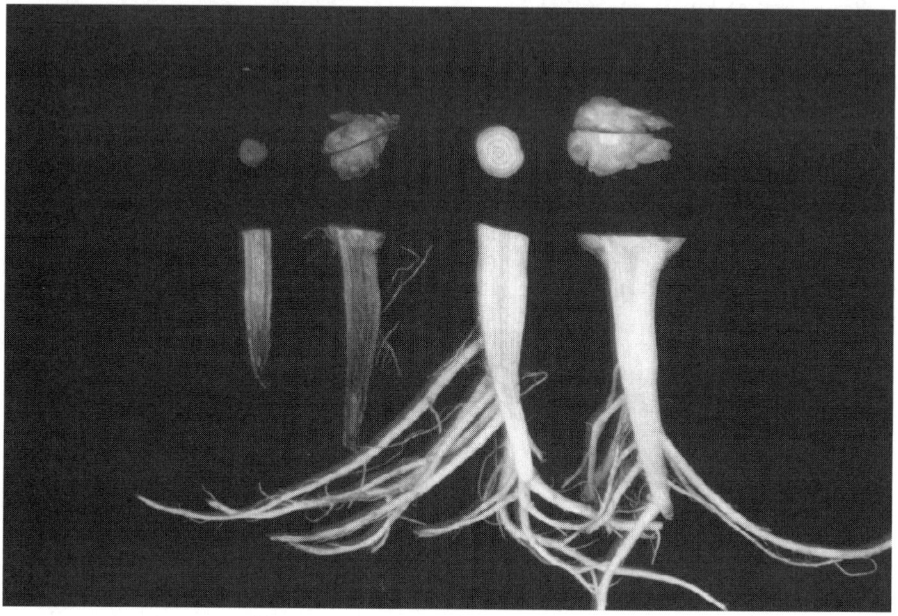

Fig. 8. Fusarium wilt: diseased roots (left); healthy roots (right).

4.1 Breeding for Downy Mildew Resistance

Downy mildew is the most important spinach disease worldwide. Race 1 first was first reported in 1824 (Brandenberger, 1992). In 1950 P.G. Smith identified resistance to race 1 downy mildew in PI 140467 and PI 140467 from Iran (Smith, 1950). The resistance from PI 140467 was isolated and found it to be due to a single dominate gene which was responsible for immunity to race 1 (Smith, 1950). This source of immunity was incorporated into 'Califlay' and 'Dixie market' and into the parents of 'Early Hybrid 7' and 'Early Hybrid 424' (Ryder, 1979). When race 2 occurred in 1958 it was found that some USDA material was resistant to race 2 (Smith et al., 1961, 1962). This was a single dominate gene which imparted immunity to race 2 and also to race 1 (Smith et al., 1962). Eenink (1974, 1976a, 1976b, 1976c) showed in 1976 that this resistance was actually due to two closely linked genes rather than a single dominate gene. When Race 3 occurred in 1976 resistance was incorporated into hybrids rather quickly and 'Mazurka', 'Polka' and 'Rhythm' were introduced in 1978 and several race 3 resistant hybrids were introduced in the next few years (Morelock, 1999). When race 4 occurred in 1990 (Brandenberger et al., 1991) and 'Bolero' and 'Bossanova' were introduced in 1991 (Morelock, 1999). Brandenberger et al. (1992) screened 707 spinach accessions and found that some resistance to race 4 resistance existed in 9 accessions (9-38%). Two accessions CGN09546 (60%) and SP1 82/87 (80%) exhibited a high level of resistance to race 4 (Brandenberger et al., 1992). Inheritance studies were not conducted but seed from the two accessions were increased and distributed to private sector breeders. The rapid development of races 5,6,7,8,9 and 10 (Correll et al., 1990, 2001, 2003; Correll, 1998; Handke et al., 2000; Irish, 2004; Irish et al., 2003, 2004) has sparked a race between private breeders with everyone trying to develop hybrids that carry resistance to all races that occur in specify production areas. Currently there are some hybrids such as 'El Dorado', 'El Palmar', 'Emilia', 'Lazio' and 'Lombardia' which carry resistance to all 10 races (Irish, 2004). This approach coupled with year round production in California make it highly probable that the rapid race development over the past 10 years will continue unless better gene management strategies are employed.

Fig. 9. Fusarium wilt: susceptible (left); resistant (right).

4.2 White Rust Breeding

Historically white rust is the most serious spinach disease in the Arkansas, Oklahoma and Texas production area (figure 10). It first was recognized as a serious problem in Texas in 1937 and in Arkansas in 1944 (Brandenberger, 1992) and it has been said that white rust first came to Arkansas on baskets of processing spinach that were shipped from Texas. Breeding for this problem was initiated by the USDA and Arkansas started a breeding program in the early 1970's when it became known that the USDA planned to terminate its spinach breeding program.

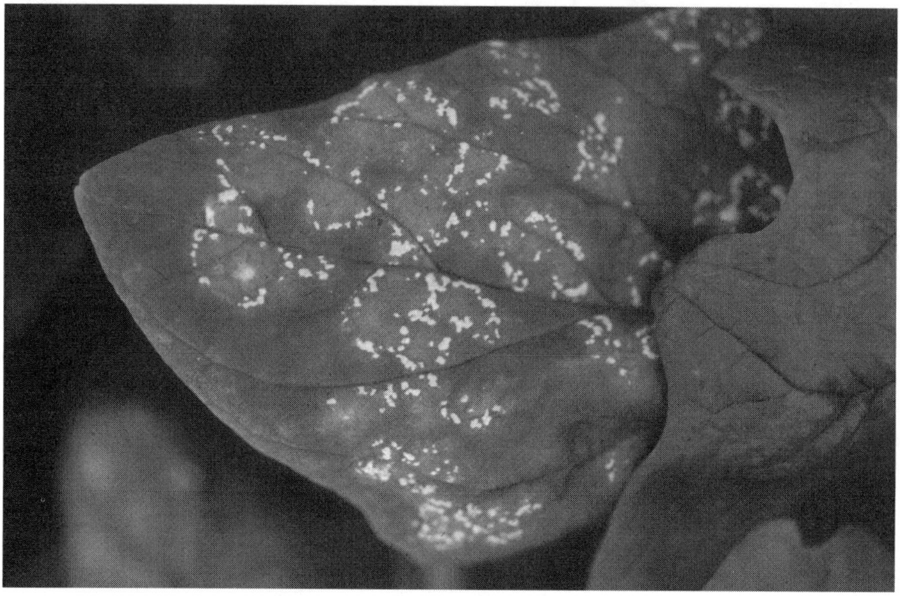

Fig. 10. White rust infected leaf.

Even through white rust and downy mildew are both obligate parasites breeding approaches to improve resistance are quite different. Unlike downy mildew, white rust does not have single gene immunity. The white rust resistance is a polygenic system (quantitative) and the exact number of genes responsible for resistance is unknown (Goode et al., 1988; Morelock et al., 2001, 2003, 2005). The first varieties with a degree of white rust resistance were 'Crystal', 'Jewel' and 'Wintergarden' which were released by the USDA in 1973 (Morelock, 1999), but these varieties were not widely grown. The University of Arkansas released 'Ozarka' and 'Greenvalley' in 1980 (Bowers and Goode, 1980a, 1980b; Morelock, 1999). 'Ozarka' was the more successful of the two varieties. These varieties utilized USDA material (WRG 70-5) and off-type plants from a commercial hybrid ('Hybrid 178') field as the sources of white rust resistance (Bowers, 1972; Bowers et al., 1974). In 1984 Alf Christianson Seed Company released 'Ozark II' and 'Greenvalley II' both of which were selections out of the original Arkansas releases (Morelock, 1999). In 1987 the University of Arkansas released 'Fallgreen' (figure 11) (Goode et al., 1988; Morelock, 1999) which was the first variety with a high level of white rust resistance and it comprised 90% or more of the fresh market acreage in the Texas wintergarden production area for the next 10 years (Goode et al., 1988; Morelock, 1999). 'Fallgreen' was replaced by 'Samish' hybrid because of 'Fallgreens' slow regrowth after the first cutting. Other hybrid semi-savoy types with some level of white rust resistance include 'Coho', 'Lessley', 'Padre', and 'Sassy' (Morelock, 1999).

Fig. 11. White rust: resistant Fallgreen (left); susceptible (right).

Flat leaf processing types with good levels of white rust resistance are widely grown in the Arkansas, Oklahoma and Texas production area. The first flat available to the public with high levels of white rust resistance was Arkansas 'F380' which is an open-pollinated variety released in 1992 (Morelock, 1999). 'F380' is currently being replaced by Arkansas 'F415' which is another open-pollinated variety released in 2005 (Morelock and Correll, 2005) with similar white rust resistance to 'F380' but a more upright plant. 'Del Monte 09' has a similar level of white rust resistance to 'F380' and 'F415' but this hybrid is not available to the general public.

One interesting aspect of varieties that have white rust resistance is that they also have a level of resistance or tolerance to some races of downy mildew (Brandenberger, 1992; Brandenberger et al., 1994). The exact relationship or mechanism of resistance is not fully understood but the phenomenon may very well be a useful tool in controlling downy mildew in the future if the number of downy mildew races continues to increase as rapidly as they have the past 10 years or if a new race of downy mildew develops for which an immunity gene can not be located.

The Arkansas white rust breeding program has utilized a modified recurrent selection program to generate open-pollinated varieties that have high levels of disease resistance. The original resistance came from USDA breeding line WRG 70-5 and off type plants from a field of 'Hybrid 178' processing spinach that was grown near Alma, Arkansas (Bowers, 1972). These plants and the USDA material were intercrossed and have been selected for several years under high levels of white rust pressure. This procedure has lead to development of the following open pollinated varieties: semi-savoy - 'Ozarka' and 'Greenvalley' 1984, 'Fallgreen' 1987, 'Wintergreen' 2001 and 'Evergreen' 2005; 'Flat F380' 1992 and 'F415' 2005. 'Fallgreen' and 'F380' have been successful varieties in their own right but both of these varieties and some of the other newer varieties have been used by private industry breeders to develop hybrids that have good levels of white rust resistance.

5 Breeding Program Mechanics

The fact that spinach is a wind pollinated dioecious species makes the logistics of breeding techniques somewhat different than would be typically used with other cross pollinated crops. These differences are also modified to some extent by the climate where the crop is being produced. For example it is possible to use small isolator cages (fine mesh cages) (figure 12) in areas that have cool summer climates for field crosses, sibbing, selfing or small masses. This is not possible in hotter climates such as Arkansas because of the high temperatures that would occur in these cages during late spring and early summer which is the period that spinach would flower in warmer climates.

Fig. 12. Small isolation cage.

Spinach breeding requires some method of effective isolation to successfully produce seed on plants that have been selected to use in crosses, sibbing or selfing. Various types of physical isolation structures are effectively used in different breeding programs. Isolation tents (cages) (figure 12) are one method that can be effectively used in cooler climates but other programs are forced to used air conditioned plastic isolation cages that have positive air pressure to prevent pollen movement between cages. For small increases or crosses where cages are used plants would simply be grown in pots or some other portable containers. It even makes it possible to transplant young plants into the field where they can be caged.

In other cases where field disease screening is to be done it may be necessary to dig plants that are moved to greenhouse cages for seed production (figure 13). For example, with the white rust screening in Arkansas selected plants are dug from the disease nursery (which has had at least one spinach crop per year for the last thirty years). Plants are removed from the field with a ball of soil and placed in a 3 gallon plastic pots that are partially filled with a peat base potting soil and potting soil is used to fill the sides of the pot around the soil ball. The pots are placed in a shady area for a few days to allow the plants to recover. After recovery the selected plants are transported to the greenhouse containing the isolation cages (4' x 2' cages hold 8 pots). The plants are placed on the floor in the greenhouse and day length is extended to 16 hours with fluorescent lights and the temperature is set at 65EF max 50EF min. When seed stalks start to appear the temperature is raised to a 75E-80EF max 65E min. As plants are far enough along in flowering that sex of the plants can be determined they are placed in the individual isolation cages. The female plants are always placed in the back of the cage and the male plants in the front (door) of the individual cages. This makes it easier to blow pollen onto the female plants and it also minimizes the chance of a stray foreign pollen grain getting into the cage and pollinating the female plants. After seed is set it is allowed to mature in the individual cages. The seed is then harvested, cleaned and planted back in the disease screening nursery and the cycle is repeated until lines are uniform enough for larger scale testing.

Fig. 13. Digging spinach Arkansas disease nursery.

The next step in the breeding program is for small scale seed production in field isolation. As small quantities of seed are produced in the field small scale testing is done in growers fields. If the results are favourable large quantities are produced for large scale tests in commercial fields and if warranted the breeding line is released as a variety.

For production of hybrids on a small scale field seed production would involve plantings of small crossing blocks where an individual male inbred would be planted with one or more female inbreds to produce small quantities of experimental hybrid seed which would be tested in the same manner as described for the open-pollinated variety. After superior hybrid combinations are determined large scale production of seed begins and the hybrid is sold to the general public as a named hybrid rather than under an experimental number designation.

6 Hybrid Production

Spinach is a dioecious species by nature and the following sex types are commonly recognized: vegetative male, extreme male, female and monoecious (Rosa, 1925). Although it is normally expected that male and female plants occur in a 1:1 ratio the relative numbers vary with genotypes and can be influenced by environmental conditions.

Interest in hybrid spinach has paralleled other crop species with yield increases of up to 40% more than open-pollinated varieties (Thompson, 1955, 1956; Webb and Thomas, 1976; Kalloo, 1988; Kalloo and Pandey, 1993). The USDA initiated a hybrid spinach breeding program in 1947 and the first hybrid, 'Early Hybrid 7', was released in 1955 (Jones et al., 1956; Ryder, 1979; Morelock, 1999). The hybrid production scheme used involved planting dioecious lines that were to be the female parent in the same field as the male parent which was also a dioeciously line. Plantings were arranged in the field with 6-8 adjacent rows of the female parent alternated with 1-2 rows of the male parent. Male plants in the female parent were removed by hand (rouged) prior to pollen being shed. After pollination the pollinator is normally rototilled or disc under to facilitate harvest of the hybrid seed and to prevent potential mixing of hybrid seed and the pollinator. Multiple plantings of the pollen parent could also be made to insure a proper nick (syncrony of flowering) when the male parent bolted (flowered) at different times than the female parent. While good quality hybrids could be produced using this method it was very labour intensive because of the large numbers of plants that had to be removed and the multiple trips through the fields were necessary to prevent self pollination in the female which would reduce the percentage of hybrid seed in the commercial product.

Excess labour cost and potential of inbred selfs in the commercial hybrid led to a modification of the technique. It was found that certain genotypes could be converted to all female lines by a process called sex reversion. This process involves planting a population of the dioecious inbred or open-pollinated variety and selecting the female plants by removing all the male plants before any pollen is shed. Female plants are allowed to flower. As flowering of the female plants continues with no pollination, the female plants often eventually produce a small amount of pollen late in the flowering cycle, thus allowing selfing and sibing of the female plants. A few cycles of this procedure allows the production of all female lines (actually gynomonoecious) (figure 14). The all female lines are used as the seed parent for production of hybrids using a procedure similar to the early USDA scheme but without the tremendous rouging costs (United States Department of Agriculture, 2004) (figure 15). These all female genotypes, when pollinated early tend to function as all female plants although they are weakly monoecious (gynomonoecious) (figure 14). This modification has improved the quality of commercial hybrids by reducing inbred selfs and has helped to hold down the cost of hybrid seed. This has probably contributed to the increased use of spinach hybrids by the U.S., Europe and Japan. Hybrids currently (2006) make up 85-90% of current production acreage.

Fig. 14. Sex reversed female (gynomonoecious).

7 Seed Production

Spinach seed production is favoured by areas that have long days and a cool maritime climate. Historically the Puget Sound area centred in Skagit County Washington USA (which still produces about 3000 acres of spinach seed) and the Netherlands were the major centres of spinach seed production but in recent years significant seed production has developed in Denmark and their production area continues to increase.

Spinach seed production in Denmark has increased from 5000 acres in 1996 to about 11,000 acres in 2005 which was about 70% of the world spinach seed production. About 90% of Denmark's production is hybrids. Production in Denmark also increased in 2006 but the extent of the increase is yet to be reported. In 2000 approximately 8700 acres in Denmark produced 7-8000 tons of spinach seed (Deleuran and Boelt, 2006).

Fig. 15. Production field spinach seed.

In Denmark to insure purity open-pollinated varieties are planted with 1200m between fields while hybrids are spaced 1500m between fields for the same groups (types) and 10,000m between groups (types). Open-pollinated varieties are grown in Northwestern Denmark and no hybrids are grown in this area. Denmark also recommends crop rotations of 5-6 years between spinach seed crops to reduce damage from root-pathogenic fungi such as Fusarium (Deleuran and Boelt, 2006).

In areas where spinach seed is grown production fields, they are planted in the early spring to give the plants size (frame) before they bolt (flower). For some hybrids it is necessary to plant the female and male parents at different times to insure a good nick (flower at the same time). It is also common to plant the pollinator 2-3 different times to insure nicking and to insure adequate pollen for the entire female flowering period. Since spinach is wind pollinated isolation is very important. Normally similar types i.e. savoy, semi-savoy or flat are planted in the same general area so if outcrossing occurs the outcross will be the same plant type and will be less obvious than if a semi-savoy or savoy were outcrossed to a flat. Asian types are usually well isolated from other types to insure the Asian types cannot possibly outcross to other types because the Asian types are so very different in plant type and tend to be very early flowering.

Fig. 16. Swathing spinach seed.

Fig. 17. Spinach seed harvest.

Fig. 18. Spinach seed harvest.

After the fields have been pollinated they are allowed to mature and in hybrid fields the pollinator is tilled under to eliminate the possibility of mixing hybrid seed with the male inbred. At the appropriate stage before the seed is fully dry the fields are cut with a swather (figure 16) and the seed is allowed to dry in a windrow (swath). After the seed is dry enough (about 12% moisture) it is harvested by a combine with a pickup header (figure 17). The combined seed is dumped into cardboard bulk bins (figure 18). The bins are transported to the seed cleaning facility where it is dried, cleaned and sized before it is bagged for sale. Presently spinach seed tends to be sold by seed count rather than by weight. Seed is normally sold by the thousand and a 1 million count bag is the standard package for larger seed lots.

8 Resistance to Diseases and Pests

8.1 Diseases

At least 35 biotic diseases have been reported on spinach (Goode, 1969) but fortunately less than one third of them consistently cause significant crop loss. Downy mildew (*Peronospora farinosa* f. sp. *spinaciae* Byford) and white rust (Albugo occidentalis G. W. Wils.) are serious foliar disease problems over large production areas (Correll et al., 1994). Other significant foliar diseases are anthracnose (*Colletotrichum dematium* (Preuss), Cercospora (*Cercospora beticola* sacc), Cladosporium (*Cladosporium macrcolarpum* G. Preuss), Alternaria sp. and

Stemphylium (*Stemphylium botrysum*) (Correll et al., 1994; du Toit and Dirie, 2001; du Toit, 2003, Mou et al., 2006).

Several soil borne pathogens have been shown to cause seedling disease. The disease causing pathogens include *Fusarium oxysporum* Schlechtend. :Fr., Fusarium sp., *Phytophthora aphanidermatum* (Edon) Fitzp., P. *irregulare* Buisman, Pythium sp., *Rhizoctonia solani* Kühn, and *Aphanomyces cochlioides* Drechs. (Correll et al., 1994) Furarium Wilt (*Fusarium oxysporum* f. sp. *spinaciae*) is a serious soil borne pest that can be spread from area to area on the seed (Goode, 1969; Bassi and Goode, 1978). Verticillium wilt and Cladosporium have recently been shown to be another pathogen that can be seed borne (Correll et al., 1998b).

There have been 14 naturally occurring virus diseases reported on spinach. Historically the most significant is cucumber mosaic virus (CMV) which is also called blight. CMV is normally controlled by genetic resistance. Other significant virus problems are Beet Western Yellows and Beet Curly Top. Several species of nematodes can also attack spinach and Koike et al has recently reported a bacterial disease (Koike et al., 2001).

8.2 Insects

It is widely recognized that at times insects can be a serious spinach pest. Historically the insect pest that is most often mentioned as a spinach pest is the green peach aphid (myzus persicae sulz) (Figure 19) (Mcleod, 2003). Recently Leafminer (*Liriomyza* spp) has become a serious pest of spinach in California and Arizona (Mou, 2003, 2005). Recently in Arkansas and Oklahoma three species of webworm (Garden webworm, *Achyra rantalis* (Guenee), Hawaiian beet webworm, Spolaclea (*Recurvalis fabricuis*) and Southern beet webworm, *Herpitogramma hepunctalis* (Fabricuis) have become a problem (Mcleod, 2006). The preferred host of these insects is pigweed located in soybean fields but as herbicides are applied later the insects move into spinach and other leafy greens (Mcleod, 2006). In recent years in Arkansas grasshoppers have become a problem for processing spinach. These large insects move into the succulent spinach fields from adjacent maturing Agronomic crops such as soybeans and corn as well as grass areas adjacent to spinach fields in the late summer and fall. In early spring overwintering grasshoppers move from resting sites into spinach and greens because these are some of the first green areas where these insects can feed. Other spinach insect pests include the seed corn maggot *Hylemyia cilicrura* (Rondlani), the cabbage looper *Trichoplusiani* (Huber), the cucumber beetle *Diabrotica duodecimpunctata* (Oliver) and leafhoppers which can carry the curly-top virus (Ryder, 1979).

8.3 Insect Resistance

Resistance to the green peach aphid has been reported in the Arkansas breeding line 86-70 (Mcleod et al., 1991). Greenhouse studies show fewer aphids per plant, reduced reproduction rates and shorter insect life (Mcleod et al., 1991). The exact mechanism was not determined and no inheritance studies were conducted.

Fig. 19. Heavy aphid infestation on spinach.

Fig. 20. Leafminer damage on spinach.

Mou conducted preliminary screening of leafminer resistance (figure 20). He screened 345 accessions from the USDA P.I. collection and 441 accessions from the CGN collections, (Holland and the IPK collection (Germany). Plants were screened in the field and in outdoor cages. Significant genotypic difference for leafminer stings per unit leaf area, mines per plant and mines per 100g plant weight were detected. Stings per unit of leaf area were highly correlated with cage stings. These studies indicate that breeding of leafminer resistance may be feasible (Mou, 2003, 2005).

9 Breeding for Improved Nutrition

The recent increased interest in human nutrition has put spinach in a very enviable position because of its excellent profile of vitamins, minerals and antioxidants. While spinach has long been recognized as a good source of vitamin A, folate, iron and magnesium the more recent increased interest is in lutein, phenolic compounds and ORAC.

Lutein has recently been a popular phytochemical that is currently being added to several types of multiple vitamins. It has been shown to be effective in the prevention of age related macular degeneration and as with many leafy greens spinach is an excellent source of lutein. Preliminary studies have shown considerable variation in spinach for lutein content ranges from 10mg / 100 gram fresh weight - 25 g/100 grams fresh weight (Murphy, 2001). Some 18 varieties, hybrids and breeding lines were assayed and the 2 highest genotypes were 'Fallgreen' and 'F380' (Murphy and Morelock, 2000). These results indicate that breeding for higher lutein content should be possible (82).

Phenolic compounds are good antioxidants and spinach has some very unique phenolic compounds i.e. patuletins and spinacetins (Howard et al., 2002). In a study of 26 spinach varieties, hybrids and breed lines showed that the breeding lines as a group were higher in phenolics that the commercial varieties and hybrids (Howard et al., 2002). Since all of the breeding lines were resistant to white rust it is very logical that they would probably be high in phenolic compounds. The phenolic content ranged from 2300-4800 mg/kg FW and the ORAC ranged from 10.7-25.0 TE/g FW (Howard et al., 2002). This variation for phenolics and ORAC indicated that it should be possible to breed for these traits (Howard et al., 2002). The limiting factor in breeding for these characteristics is the slow turn around on plant samples. It is possible to generate as many as 1000 samples in one afternoon and with only one HPLC it would take months to run the analysis. Once a spinach plant is dug and moved to greenhouses for seed production the breeder may only have 2-3 weeks before the plants need to be caged. This creates a significant problem if the breeder is tying to evaluate a large segregating population for lutein or phenolic content. Unless the turn around time can be significantly reduced it would be almost impossible to screen a large population and make the appropriate crosses before the plants are past the point that crosses could be made.

Spinach has long been considered to be high in calcium oxilate (Kitchen et al., 1964a). Kitchen et al. (1964b) found that savoy spinach tended to be lower in oxilate content than semi savoy types. Baker (1988) examined oxilate levels in ten varieties (7 semi savoy and 3 flat) and found that semi-savoy types tended to be higher than the flat types but significant difference were not observed between the spring and fall crop. Baker (1988) also found that mean oxilate values were higher for the spring crop than for the fall crop.

10 New Biotechnologies in Spinach Breeding

10.1 Genetic Transformation

In order to utilize *Agrobacterium* mediated genetic transformation it is necessary to develop tissue culture protocols for the crop that is to be transformed. These protocols were developed by Al-Khayri and these procedures are well documented (Al-Khayri, 1991; Al-Khayri et al., 1992). Al-Khayri (1995) also transformed spinach by placing the GUS (β-Glucoronidase) gene into spinach and successfully regenerated transformed plants. These protocols were used by other researchers to produce transgenic spinach that carried the coat protein gene for cucumber mosaic virus (Yang, 1996; Yang et al., 1997) and the gene for Glyphosate tolerance (Roundup Ready) (Wells, 1999; Bevitori, 2000; Burgos et al., 2001). Even though transgenic plants have been generated neither of these traits have been commercialized in spinach. These examples show that development of transgenic spinach is feasible should anyone want to devote the time and resources that are necessary to commercialize transgenic spinach. Maas Molina (2004) investigated the use of Glufosinate to develop transgenic herbicide tolerant spinach but did not successfully confirm presence of the gene in mature plants.

10.2 Marker Assisted Selection

The use of molecular markers that are linked to commercially important phenotypic traits is becoming more common in agricultural crops and is now a widely accepted approach to help expedite the development of improved varieties. However, few studies have utilized such an approach to improve spinach varieties. More recently, the University of Arkansas spinach program has initiated studies to develop molecular markers linked to major gene resistance loci for the downy mildew pathogen of spinach, *Peronospora farinosa f. sp. spinaciae*.

The marker assisted selection (MAS) approach that has been employed involves developing mapping populations by crossing commercial hybrids, with different downy mildew disease resistance profiles, with the universally susceptible variety Viroflay (which contains no none resistance genes to the downy mildew pathogen). The parental lines and the F_1 progeny are then screened for resistance to a given race of the pathogen as well as screened for their DNA fingerprints using amplified fragment length polymorphism (AFLPs). Any DNA bands that are unique to the resistant parent and also co-segregate with the resistance progeny are identified as "candidate" markers and are examined in greater detail. The candidate markers are extracted from a gel and the DNA of the fragment is sequenced. PCR based primers are then developed based on the sequence of the candidate markers in an approach that is known as SCAR (sequence characterized amplified region) marker development. The SCAR markers (Irish, 2004) are then used to determine how robust the marker is by examining the original parental lines and the F_1 progeny. Effective co-dominate SCAR markers are ones that can readily distinguish the genotype at a given resistance locus as either homozygous resistant (RR), heterozygous resistant (Rr), or homozygous susceptible (rr) for a given race. This approach has been used to develop a robust marker (Dm-1) that is linked to a major resistance gene locus (Pfs-1) in spinach that confers resistance to race 6 of the downy mildew pathogen (Irish, 2004; Irish et al, submitted). Efforts are currently underway to develop other SCAR markers, linked to other downy mildew resistance loci, which can be utilized in an MAS effort with spinach. We anticipate that such an approach can be effective to stack, or pyramid, major and minor genes for resistance to improve the durability of resistance to the downy mildew pathogen.

References

Al-Khayri, J.M, Huang, F.H., Morelock, T.E. and Busharar, T.A. 1992. In-vitro plant regeneration of spinach from mature seed-derive callus. In Vitro Cell 28: 64-66.

Al-Khayri, J.M. 1995. Genetic transformation in *Spinacia oleracea* L. (Spinach). Biotechnology Agric. For. 34:229-238.

Al-Khayri, J.M. 1991. In-vitro regeneration, sex alteration, genetic transformation and DNA isolation in spinach (*Spinacia oleracea* L.). Ph. D. Diss. University of Arkansas. Fayetteville, AR.

Baker, T.P. 1988. Analysis of oalic acid levels in spinach (*Spinacia oleracea* L.) using high-performance liquid chromatography MS. Thesis University of Arkansas

Bassi, A., Jr., and Goode, M.J. 1978. *Fusarium oxysporum* f.sp. *spinaciae* seedborn in spinach. Plant Dis. Rep. 62, 203-205.

Bemis, W.P. and Wilson, G.B. 1953. A new hypothesis explaining the genetics of sex determination. J. Hered. 44, 91-95.

Bevitori, R.A. 2000. Partial cDNA sequences coding for ACCASE in diclofop-resistant and susceptible ryegrass and glyphosate-tolerant spinach development. Ph.D. Diss. University of Arkansas. Fayetteville, AR.

Bose, S. and Janick, J. 1961. Karyo-races in *Spinacia oleracea*. Am. J. Bot. 48, 238-241.

Bowers, J.L. 1972. Spinach breeding program for disease resistance in Arkansas. Proc. Ark. State Hort. Soc. 93:53-54.

Bowers, J.L. and Goode, M.J. 1980a. Spinach varieties. Proc. Ark. State hort. Soc. 101:64.

Bowers, J.L. and Goode, J.J. 1980b. Ozarka and Greenvalley: new disease resistant spinach cultivars. Ark. Farm Res. 32(2):6.

Bowers, J.L. Goode, M.J. and Frankhauser, D.W. 1974. Program report on breeding for disease resistance in spinach. Proc. Ark. State Hort. Soc. 95:20-21.

Bragdo, M. 1962. Breeding of polyploid spinach. Euphytica 11, 143-148.

Brandenberger, L.P. 1992. Studies to Quantify Disease Resistance in spinach to the white rust (Albugo occidentalis) and downy mildew (*Peronospora farinosa* f sp. *spinaciae*) pathogens. Ph.D. Diss. University of Arkansas, Fayetteville, AR. USA

Brandenberger, L.P., Correll, J.C., and Morelock, T.E 1991. Nomenclature of the downy mildew fungus on spinach. Mycotaxon. 41:157-160.

Brandenberger, L.P., Correll, J.C., Morelock, T.E., and McNew, R.W. 1991. Identification of and cultivar reactions to a new race (race 4) of *Peronospora farinosa* f. sp. *spinaciae* on spinach in the United States. Plant Dis. 75:630-634.

Brandenberger, L.P., Correll, J.C., Morelock, T.E., and McNew, R.W. 1994. Characterization of resistance of spinach to the white rust (Albugo occidentalis) and downy mildew (Peronospora farinosa f.sp. Spinaciae). Phytopathology. 84:431-437.

Brandenberger, L.P., Morelock, T.E., and Correll, J.C. 1992. Evaluation of spinach germplasm for resistance to a new race (race 4) of *Peronospora farinosa* f. sp. *spinaciae*. HortScience. 27:1118-1119.

Burgos, N.R., Bevitori, R., Candole, B., Rajguru, S., Talbert, R.E. and Morelock, T.E. 2001. Roundup Ready Spinach (*Spinacia oleracea* L.) update. National Spinach Conference 14-15 November 2001. Fayetteville, AR USA abst p14.

Cheo, P.C., and Pound, G.S. 1952. Relation of air temperature, soil temperature, photoperiod and light intensity on the concentration of cucumber 1 virus in spinach. Phytopathology 42, 306-310.

Correll, J.C., Irish, B.M. Koike, S.T., Shafer, J. And Morelock T.E. 2003. Update on downey mildew of spinach. National Spinach Conference 20-21 November Fayetteville, AR USA Abst. p.18.

Correll, J.C., Irish, B.M., Koike, S.T. and Morelock, T.E. 2001. Update on downy mildew (*Peronospora farinosa* f.sp. *spinaciae*) of spinach in the United States. National Spinach Conference 14-15 November 2001, Fayetteville, AR USA abst p10.

Correll, J.C. 1998. Review of the biology of *Peronospora farinosa* f. sp. *spinaciae*. CAB International. Agriculture Biosciences. In: Crop protection Compendium. CAB International, Wallingford, U.K. - Electronic book chapter.

Correll, J.C., Koike, S.T., Brandenberger, L.P., Black, M.C., and Morelock, T.E. 1990. A new race of downy mildew threatens spinach. Calif. Agric. 44:14-15.

Correll, J.C., Morelock, T.E., and Koike, S.T. 1998a. Screening of USDA spinach germplasm to two new races of the downy mildew pathogen (*Peronospora farinosa* f. sp. *spinaciae*). Leafy Vegetable Conf. Atlantic City, N.J.

Correll, J.C., Koike, S.T., Schafer, Jl, Anders, J.M., Irish, B.M., and Morelock, T.E. 1998b. Two new races of the downy mildew pathogen (*Peronospora farinosa* f.sp. *spinaciae*) of spinach in the United States. Phytopathology. 88:S19.

Correll, J.C., Morelock, T.E., Black, M.C., Koike, S.T., Brandenberger, L. P., and Dainello, F.J. 1994. Economically important diseases of spinach.. Plant Dis. 78:653-660.

Deleuran, L.C. and Boelt, B. 2006. Spinach seed production in Denmark. 2006 International Spinach Conference 13-14 July 2006, LaConner, WA. USA abst. Pp13-15.

Dicoteau, D.R. 2000. Vegetable Crops. 221-237 Prentice Hall.

Dressler, O. 1976. Results of breeding monoecious spinach varieties. Proc. Eucarpia Meet. Leafy Vegetables, Wageningen, Holland, Mar. 15-18 pp. 67-77.

du Toit, L.J. 2003. Verticillium wilt of spinach. National Spinach Conference 20-21 November. Fayetteville, AR USA abst p20-25.

du Toit, L.J. 2003. Some epidemiological aspects of Stemphylium leaf spot and Cladosporium leaf spot in Spinach Seed Production. National Spinach Conference 20-21 November Fayetteville, AR USA abst p26-28.

du Toit, L.J. and Dirie, M.L. 2001. Stemphylium leafspot of spinach seed crops in Washington State. National Spinach Conference 14-15 November 2001, Fayetteville, AR USA abst p8.

du Toit, L.J. and Dirie, M.L. 2001. Further characterization of the pathogen that causes Stemphylium leaf spot of spinach in California. National Spinach Conference 14-15 November 2001. Fayetteville, AR USA abst. P7.

Eenink, A.H. 1974. Linkage in Spinacia oleracea L. Between the locus for resistance to *Peronospora spinaciae* Laub. and the locus for tolerance for cucumber virus 1. Euphytica 23, 485-487.

Eenink, A.H. 1976a. Linkage of *Spinacia oleracea* L. Of two race-specific genes for resistance to downy mildew *Peronospora farinosa* f.sp. *spinaciae*, Byford. Euphytica. 25:713-715.

Eenink, A.H. 1976b. Linkage in Spinacia oleracea L. Of two race-specific genes for resistance to downy mildew *Peronospora farinosa* f. sp. *spinaciae* Byford. Euphytica 25, 713-715.

Eenink, A.H. 1976c. Resistance in spinach to downy mildew. Proc. Eucarpia Meet. Leafy Vegetables, Wageningen, Holland, Mar. 15-18 pp. 53-54.

Ellis, J.R. and Janick, J. 1960. The chromosomes of *Spinacia oleracea*. Am. J. Bot. 47, 210-214.

Food and Agriculture Organization of the United Nations. 2006. FAO STAT data. http://faostat.fao.org/

Goode, M.J. 1969. Spinach Diseases in the Arkansas river valley. Proc. Ark. State Hort. Soc. 90:41-45.

Goode; M.J., Morelock, T.E. and Bowers, J.R. 1988. Fallgreen spinach. HortScience 23:931.

Haga, T. 1935. Sex and chromosomes in Spinacia oleracea L. Japan J. Genet. 10, 218-222.

Handke, S., Seehaus, H., and Radies, M. 2000. Detection of a linkage group of the four dominant mildew resistance genes "M1M2M3M4" in spinach from the wildtype *Spinacia turkestanica*. Gartenbauswissenschaft. 65:73-78.

Hine, J.R. 2003. Folic Acid: The Queen B? 2003. National Spinach Conference 20-21 November 2003. Fayetteville, AR USA abst p7.

Howard, L.R., Pandjaitan, N., Morelock, T. and M.I. Gil 2002. Antioxidant capacity and phenolic content of spinach as affected by genetics and growing season. J. Agric Food Chem. 50:5891-5896.

Howard, L. 2001. Antioxidant content of different spinach genotypes. National Spinach Conference. 14-15 November 2001, Fayetteville, AR USA abst p5

Iizuka, M. and Janick, J. 1962. Cytogenetic analysis of sex determination in *Spinacia oleracea*. Genetics 47, 1225-1241.

Iizuka, M. and Janick, J. 1963. Sex chromosome translocations in *Spinacia oleracea*. Genetics 48, 273-282.

Iizuka, M. and Janick, J. 1966. The synthesis of heteromorphic sex chromosomes in spinach. J. Hered. 57, 182-184.

Irish, B.M. 2004. New Races of the downy mildew pathogen of spinach, identification of molecular markers for disease resistance, and molecular diversity of spinach germplasm. Ph.D. Diss. University of Arkansas, Fayetteville, AR. USA

Irish, B.M., Correll, J.C., Raid, R.N. and Morelock, T.E. 2004. First report of Peronospora farinosa f. sp. spinaciae (race 5) of spinach in Florida. Plant Dis. 88:84.

Irish, B.M., Correll, J.C., Koike, S.T., Schafer, J., and Morelock, T.E. 2003. Identification and cultivar reaction to three new races of the spinach downy mildew pathogen (*Peronospora farinosa* f. sp. *spinaciae*), from the United States and Europe. Plant Dis. 87:567-572.

Janick, J. 1955a. Inheritance of sex in tetraploid spinach. Proc. Am. Soc. Hortic. Sci. 66, 361-363.

Janick, J. 1955b. The effects of polyploidy on sex expression in spinach. J. Hered. 46, 150-156.

Janick, J. Mahoney, D.L. and Pfahler, P.L. 1959. The trisomics of *Spinacia oleracea* L. J. Hered. 50, 46-50.

Janick, J. and Stevenson, E.C. 1954. A genetic study of the heterogametic nature of the staminate plant in spinach (*Spinacia oleracea* L.). proc. Am. Soc. Hortic. Sci. 63, 444-446.

Janick, J. and Stevenson, E.C. 1955a. Environmental influences on sex expression in monoecious lines of spinach. Proc. Am. Soc. Hortic. Sci. 65, 416-422.

Janick, J. and Stevenson., E.C. 1955b. Genetics of the monoecious character in spinach. Genetics 40, 429-437.

Jones, H.A., McLean, D.M. and Perry, B.A. 1956. Breeding hybrid spinach resistant to mosaic and downy mildew. Proc. Am. Soc. Hortic. Sci. 68, 304-308.

Jones, R.K., and Dainello, F.J. 1982. Occurrence of race 3 of *Peronospora effusa* on spinach in Texas and identification of sources of resistance. Plant Dis. 66:1078-1079.

Kalloo, G. 1988. Breeding methods in vegetable crops. CRC Press. Gainsville, FL.

Kalloo, G. and Pandey, S.C. 1993. Genetic improvement of vegetable crops, Pergamon Press: 325-339. Gainsville, FL.

Kitchen, J.W., Burns, E.E. and Langston, R. 1964a. The effects of light, temperature and ionic balance on oxalate formation in spinach. Proc. Am. Soc. Hortic. Sci. 85, 465-470.

Kitchen, J.W., Burns, E.E. and Perry, B.A. 1964b. Calcium oxalate content of spinach (spinacia oleracea L.). proc. Am. Soc. Hortic. Sci. 84, 441-445.

Koike, S.T., Azad, H.R. and Cooksey, D.C. 2001. New diseases of spinach: bacterial leaf spot caused by a *Pseudomonas syringae* pathogen. National Spinach Conference 14-15 November 2001. Fayetteville, AR USA abst p6.

Koike, S.T., Smith, R.F., and Schulbach, K.F. 1992. Resistant cultivars, fungicides combat downy mildew of spinach. Calif. Agric. 46:29-31.

Lorenzini, G. and Nali, C. 1994. A new race (race 4) of spinach downy mildew in Italy. Plant Dis. 78:208.

Magruder, R., Boswell, V.R., Scott, G.W., Work, P., and Hawthorn L.R. 1938. Description of types of principal american varieties of spinach. USDA Misc. Pub 316.

Maas Molina, L.F. 2004. Development of glufosinate - resistant spinach (*Spinacia oleracea* L.) M.S. Thesis University of Arkansas. Fayetteville, AR. USA

McLeod, P. 2006. Identification, Biology and Management of Insects Attacking Vegetables in Arkansas. Graf Lisina, Santa Cruz, Bolivia.

McLeod, 2003. Current and future insect management in spinach and greens. National Spinach Conference 20-21 November 2003 Fayetteville, AR USA abst p35

Mcleod, Paul, T.E. Morelock, and M.J. Goode. 1991. Performance developmental times, adult longevity and fecundity of green peach aphid (*Homoptera: aphidida*) on spinach. J. Ento. Sci. 28(1):95-98.

Morelock, T.E. and Correll, J.C. 2005. Spinach breeding in the mid-south. National Spinach Conference. 16-17 November 2005, Fayetteville, AR USA. abst. p14

Morelock, T.E. and Correll, J.C. 2003. Spinach breeding in the mid-south. National Spinach Conference 20-21 November Fayetteville, AR USA abst p39

Morelock, T.E. and Correll, J.C. 2001. Spinach breeding in Arkansas. National Spinach Conference 14-15 November 2001 Fayetteville, AR USA abst p27

Morelock, T.E. 1999. Spinach: variety test and description. Wehner, T., ed Hortsci 34(6):987-988

Mou, B., Koike, S., and duToit, L. 2006. Screening for resistance to Stemphylium leaf spot of spinach. 2006 International Spinach Conference 13-14 July 2006, La Conner, WA. USA abst. pp29-30

Mou, B. 2005. Leafminer resistance in spinach. National Spinach Conference. 16-17 November 2005, Fayetteville, AR. USA abst. p6

Mou, B. 2003. Screening and breeding for resistance to leafminer in spinach. National Spinach Conference 20-21 November. Fayetteville, AR USA abst p32.

Murphy, J.B. 2001. Lutein content of spinach cultivars and breeding lines. National Spinach Conference. 14-15 November 2001, Fayetteville, AR USA abst p4

Murphy, J.B. and Morelock, T.E. 2000. Spinach breeding program yields lines containing high levels of carotenoid antioxidants. Horticultural Studies id Richardson, M.D. and Clark, J.R. University of Arkansas, Research Series 475:36-39

Naili, C. 1998. A novel threat for spinach in Italy: a new race of downy mildew. Adv. Hort. Sci. 12:179-182.

Nonnicke, I.L. 1989. Vegetable Production 476-484 Van Nostrand Runhold

O'Brien, M.J. and Winters, H.F. 1977. Evaluation of spinach accessions and cultivars for resistance to Fusarium wilt. I. Greenhouse bench method. J. Am. Soc. Hortic. Sci. 102, 424-426.

Pandjaitan, N. Howard, L.R. Morelock, T. and Gil, M.I. 2005. Antioxidant capacity and phenolic content of spinach as affected by genetics and maturation. J. Agric. Food Chem. 53:8618-8623

Pandjaitan, N. 2004. Identification of carotenoid and flavonoids in spinach: effects of growing season, genotype, maturity and pressurized extraction technologies. Ph.D. Diss. University of Arkansas. Fayetteville, AR. USA

Pound, G.S. and Cheo, P.C. 1952. Studies on resistance to cucumber virus 1 in spinach. Phytopathology 42, 301-306.

Prior, R.L. 2003. Spinach as a source of antioxidant phytochemicals with potential health Effects.2003. National Spinach Conference 20-21 November 2003. Fayetteville, AR. USA Abst p3-4.

Rosa, J.T. 1925. Sex expression in spinach. Hilgardia 1:258.

Ryder, E.J. 1979. Leafy Salad Vegetables 195-227 AVI

Sams, D.W. and Bienz, D.R. 1974. Relative susceptibility of spinach plant introduction accessions to curly top. HortScience 9, 600-601.

Satou, M., Sugiura, T., Ohsaki, R., Honda, N., Horiuchi, S., and Yamauchi, N. 2002. A new race of spinach downy mildew in Japan. J. Gen. Plant Pathol. 68:49-51.

Shimazaki, Y. 1990. Appearance of a new race 4 of downy mildew on spinach. Ann. Phytopathol. Soc. Jpn. 56:95.

Smith, L.B. 1921. Breeding mosaic resistant spinach and notes on malnutrition. Virginia Truck. Exp. Stn. Bull. 31.

Smith, P.G. 1950. Downy mildew immunity in spinach. Phytopathology 40, 65-68.

Smith, P.G., Webb, R.E. and Luhn, C.H. 1962. Immunity to race 2 of spinach downy mildew. Phytopathology 52, 597-599.

Smith, P.G., Webb, R.E., Millett, A.M., and Luhn, C.H. 1961. Downy mildew on spinach. Calif. Agric. 15:5.

Smith, W.G. 1885. Disease of spinach, *Peronospora effusa* Grev. Gdnrs. Chron. 23:480.

Sneep, J. 1958. The present position of spinach breeding. Euphytica 7:1-8.

Swiader. J.M. and Ware, G.W. 2002. Producing Vegetable Crops 5th ed. 481-498 Interstate.

Thompson, A.E. 1954. The extent of natural crossing in inbred monecious spinach lines. Proc. Am. Soc. Hortic. Sci. 64, 405-409.

Thompson, A.E. 1955. Methods of producing first-generation hybrid seed in spinach. Cornell Agric. Exp. Stn. Mem. 336.

Thompson, A.E. 1956. The extent of hybrid vigor in spinach. Proc. Am. Soc. Hortic. Sci. 67, 440-444.

United States Department of Agriculture, Economic Research Service 2004. Vegetables and Melon yearbook. http://www.ers.usda.gov/

United States Department of Agriculture, National Agricultural Statistics Service. 2006a. Census of Agriculture. http://www.nass.usda.gov/Census_of_Agriculture

United States Department of Agriculture, Economic Research Service. 2006b. Commodity Highlight: Fresh market Spinach. http://www.ers.usda.gov/

Webb, R.E. and Thomas, C.E. 1976. Development of F_1 spinach hybrids. HortScience 11, 546.

Webb, R.E. 1969. Spinach improvement and resistance to white rust. Proc. Ark. State Hort. Soc. 90:36-40

Wells, J.J. 1999. Yellow nutsedge control in summer vegetables and development of glyphosate - tolerant spinach. Ph.D. Diss. University of Arkansas. Fayetteville, AR USA

Yang, Y., Al-Khayri, J.M. and Anderson, E.J. 1997. "Transgenic spinach plants expressing the coat protein of cucumber mosaic virus" In Vitro Cell 33: 200-204.

Yang, Y. 1996. Characterization of two seed-borne spinach isolates of cucumber mosaic virus and production transgenic spinach plants. Ph.D. Diss. University of Arkansas. Fayetteville, AR.

Zink, F.W. 1963. Effect of beet yellows virus on rate of growth and yield in spinach. Proc. Am. Soc. Hortic. Sci. 83, 675-679.

Zink, F.W., and Smith, P.G. 1958. A second physiologic race of spinach downy mildew. Plant Dis. Rept. 42:818.

Table Beet

Irwin L. Goldman[1] and John P. Navazio[2]

[1] University of Wisconsin, Department of Horticulture, ilgoldma@wisc.edu
[2] Abundant Life Seeds

1 Introduction

Table beet (*Beta vulgaris* subsp. *vulgaris* L) is a vegetable from the family Chenopodiaceae. Table beet is also known as garden beet or red beet in the U.S. scientific literature, and as beetroot in Europe and many other countries around the world. Table beet is a member of a crop complex from the genus *Beta* that includes Swiss chard, mangel, and sugarbeet. All three of these crops are derived from the same species, *vulgaris*, and are often represented with different subspecies designations. Table beet and Swiss chard are primarily used as vegetables, the mangel and its derivatives are used as animal feed, and the sugarbeet is used as a source of sucrose. Table beet breeding has been recently reviewed by Goldman and Navazio (2003) and we draw from sections of their work in this chapter.

Table beet is not one of the world's major vegetable crops in terms of acreage, production, or consumption; however, it occupies a unique niche in Europe, North America, the Middle East, and parts of Asia. Though typically portrayed as a red-rooted vegetable with bright green leaves, an array of root and leaf colours exist for this crop making it one of the most colourful of the vegetables. Table beet possesses a number of unique nutritional properties that make it valuable for human health, including an abundance of the B vitamin folic acid (Wang and Goldman, 1996) and a high concentration of the betalain pigments, which cause the colouring in the leaves and roots and have been shown to be powerful antioxidants.

The crop is consumed in a very large variety of styles around the world. Leaves are consumed as a baby leaf or salad crop, and roots are added to salads or consumed on their own. Beet root salads are popular the world over, and root slices are used on sandwiches in Australia and New Zealand. Roots can be pickled, roasted, boiled, or used to make soup. The pigments are extracted and used as a source of natural food dyes in a variety of food products. The earthy flavours associated with table beet, caused by compounds known as geosmins, are sought by chefs and consumers. Table beet was considered an "old fashioned" vegetable for many decades in the late 20th century, but has undergone renewed popularity in part due to its versatility, nutritional value, earthy flavours, and unique beauty among the root vegetables.

Table beet breeding is not widely practiced compared to the major vegetable crops, particularly because of the minor importance of this crop. Despite this, innovative breeding strategies have been developed for table beets that are instructive for crop breeding in general. In addition, as this vegetable crop continues to gain in acreage and importance around the world, interest in breeding will likewise increase. The goal of this chapter is to outline and review the key aspects of table beet breeding and genetics.

2 Origin and Domestication

Table beet has been cultivated and bred for millennia, but it is only in the last four or five centuries that the swollen rooted form we know today has been available. The modern table beet crop is, ironically, one of the progenitors of the sugarbeet, and thus it holds a unique place in the domestication of an important industrial and food crop as well as being an ancient vegetable crop in its own right. Later in this chapter, we discuss how the two crops have been used in modern times to improve one another.

The genus *Beta* is primarily found in Asia and Europe. Cultivated beets are derived from section Beta, one of the four sections in this genus (Ford-Lloyd, 1995). Section Beta includes six subspecies within *B. vulgaris*: *vulgaris* (sugar beet, table beet, mangel or fodder beet), *cicla* (Swiss chard, leaf beet, spinach beet), *maritima* (wild sea beet), *adanensis*, *trojana*, and *macrocarpa*. The section also includes the species *B. patula* and *B. atriplicifolia* (Ford-Lloyd, 1995). Beet has nine pairs of chromosomes and is a diploid, with 2n=2x=18. Tetraploid and triploid sugarbeet have been developed and are often used in commercial cultivars, but table beet cultivars are solely diploid.

The table beet was originally cultivated as a leaf vegetable in Asia and by the Romans (Ford-Lloyd and Williams, 1975). It is likely that this form of the table beet was represented by plant with an annual life cycle that did not possess a swollen root. A general theme in vegetable domestication is development of different plant organs to suit particular adaptive niches in human migration. One such adaptation is the colonization of northern latitudes and the need for food that could be stored through the winter months. Storage roots were particularly helpful in such cases, and thus vegetables with large storage roots were selected during these demographic shifts. Examples include the turnip and the kohlrabi among the *Brassicaceae*, and table beet in the *Chenopodiaceae*. Interestingly, domestication traits such as a shift from annual growth habit to biennial growth habit have received little attention from geneticists. This is an area of significant potential interest to scientists studying vegetable crops, and it may prove a fruitful avenue of research for understanding the biology of biennial crops such as table beet, turnip, carrot, and others. Of particular value here would be an understanding of the *B* allele in beet, which converts biennialism to annualism. Such research would be of benefit to our understanding of the domestication of table beet.

During the period of transition between Mediterranean zones to Northern Europe, or perhaps as this leaf crop moved into Northern Europe with human migration, selection for swollen-rooted forms resulted in the development of the swollen rooted vegetable that we associate with the modern table beet. Ford-Lloyd (1995) suggests that evidence of this change can be found first in the 16th century. Hybridization of these swollen rooted forms with leafy forms of *Beta vulgaris* likely resulted in a large array of types possessing varied colours and degrees of swollenness in the root and hypocotyl zone. An example of this hybridization is the description of a mangel, a swollen-rooted form of *B. vulgaris*, which arose from crosses between a red table beet and white leaf beet described in 1787 by the Abbe de Commerell (Ford-Lloyd, 1995). Clearly, intercrossing among cultivated *B. vulgaris* crops has played a role in the modern evolution of these subspecies (Figure 1).

By the 17th century, table beet was cultivated in Europe and it spread to many other regions of the world. The primary root shapes used since this period are: (1) round and globe shaped roots, which are the most common type; (2) flattened globe or Egyptian types; and (3) cylindrical types, which have specific value in the processing market. The primary root colours available since the 17th century have been red and yellow-rooted types, though great variation exists in root colour. Root colour and its biology and genetic control will be discussed later in this chapter, however it is important to note that the red colour present in most table beet roots is actually a representation of two different pigments; one reddish-purple and one yellow. The reddish-purple colour is due to betacyanin and the yellow to betaxanthin. When both are present, in approximate 3:1 ratios as is typical in table beet cultivars, the colour that appears is red. When betacyanin is absent, yellow root coloration can be found. The modern germplasm base of this crop is discussed in the next section.

As described earlier in this chapter, the progenitor of the table beet was originally selected for its use as a leaf vegetable in the Mediterranean region and then later for use as a fresh or stored root vegetable (Campbell, 1976). Early European herbals clearly point toward distinct uses for the leaf portion and the swollen red hypocotyl and root (Pink, 1992). Use of *B. vulgaris* as a leaf crop probably included the leaf beet, a vegetable form of beet grouped in the *vulgaris* subspecies. The leaf beet has fleshy petioles, similar in thickness to asparagus, although it does not possess a swollen root.

As the crop moved into Northern Europe, farmers would have faced a shorter growing season and a colder winter. These conditions may have resulted in the transition toward a biennial life cycle by creating selection pressure towards a swollen hypocotyl/root as an over-wintering propagule (Ford-Lloyd, 1995). Alternatively, selection for a swollen-rooted form may have taken place in the Mediterranean region prior to its movement into northern Europe. Some authors have suggested that swollen roots, which indicate the storage of carbohydrates for energy production during reproductive growth, may have been selected from leafy beets cultivated in Assyrian, Greek and Roman gardens (Ford-Lloyd and Williams,

1975; Williams and Ford-Lloyd, 1974). These leaf beets may have resembled the Swiss chard of today.

In the 18th century, the use of beet root was expanded to include animal feed. The fodder beet, as it came to be known, soon became an important component of European agriculture and served at least partially as the progenitor of the sugar beet (Pink, 1993). Fodder beet possessed edible roots and leaves, making it an excellent forage crop. Various names exist for the common fodder beet, including forage beet, mangels, mangolds, and mangel-wurzels. All of these possess very large swollen roots of various shapes and colours and were developed for animal feed (Ford-Lloyd, 1995).

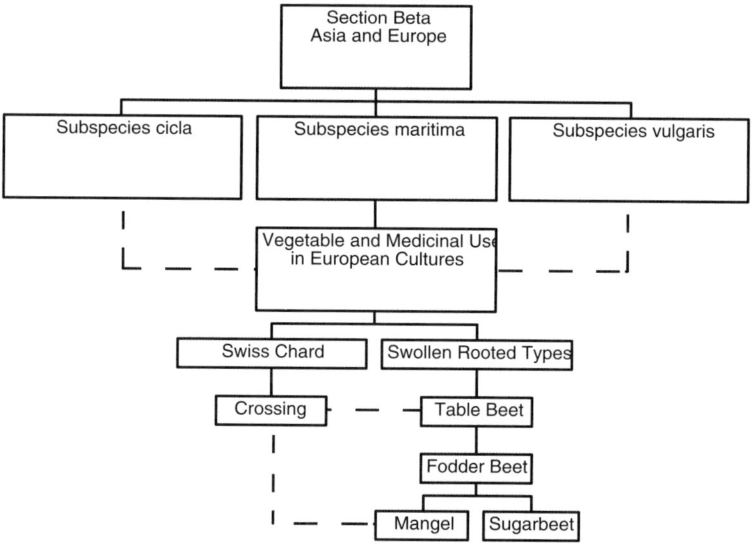

Fig. 1. Evolution of Beta Crops (adapted from Ford-Lloyd, 1995).

The sugar beet, a close relative of table beet, is designated as the same subspecies as table beet and is a crop of modern origin. Sugar beet was developed from a fodder beet population known as "White Silesian" (Fischer, 1989) during a search for alternative sources of sucrose when France was unable to obtain sugarcane sugar due to a British blockade during the Napoleonic wars (Winner, 1993). This work was predicated on the discovery of a sweet syrup from *B. vulgaris* by Olivier de Serres that was later identified as sucrose by Marggraf during the 18th century and discovered to be identical to the sugar from sugarcane (Winner, 1993). An alternative explanation for the development of a sugar source from *B. vulgaris* is that France wished to possess a domestic sugar industry that would hinder Britain's dominance of the sugar trade. Selection for high sucrose was practiced by a number of breeders during the 18th and 19th centuries, including one of Marggraf's students named Archard, and their methodologies were among the first to describe mass selection in a scientific manner (Goldman, 2000).

3 Varietal Groups

Goldman and Navazio (2003) described the key founding populations of table beet in the U.S. during the 20th century. These were originally described by Magruder et al. (1940) in a landmark publication on U.S. table beet production. These same populations form the basis for most modern table beet breeding in the U.S., although I will later describe how other sources of germplasm, such as sugar beet, were used to improve table beet. Among the key populations available in the first half of the 20th century, the most prominent were 'Flat Egyptian', 'Crosby's Egyptian', 'Light Red Crosby', 'Early Wonder', 'Detroit Dark Red', 'Morse Detroit', 'Ohio Canner', and 'Long Dark Red'. These populations comprised as much as 95% of the U.S. beet area in 1940 (Magruder et al., 1940). Because several of these populations appear to be synonymous, only Long 'Dark Red', 'Flat Egyptian', 'Crosby Egyptian', 'Early Wonder', and 'Detroit Dark Red' seem significantly separated from each other by origin. Among these, there are three clearly identifiable groups (Goldman and Navazio, 2003). These are the "Egyptian" group, which is comprised of Flat Egyptian, Crosby Egyptian, Light Red Crosby, and Early Wonder; the "Detroit" group, which is comprised of Detroit Dark Red and Ohio Canner; and the "Long" group, which is comprised of Long Dark Red and Cylindra. Goldman and Navazio (2003) considered these "founding" populations from the point of view of U.S. table beet breeding. The following is drawn from their descriptions of these materials, which were based on those of Magruder et al. (1940).

1. 'Flat Egyptian'. This population was introduced to the U.S. from Germany in 1868 by the Ernst Benary Company. It was listed in seed catalogues as 'Extra Early Egyptian' and 'Dark Red Egyptian'. The primary characteristics of this population were its earliness, small foliage, and flat root shape. 'Flat Egyptian' was popular as an early-maturing variety for northern climates and for the production of small-sized roots for canning. Root shape is the flattest among the founding populations.

2. 'Crosby Egyptian'. Developed by Josiah Crosby of Arlington, Massachusetts, from a population of 'Flat Egyptian'. This population was first listed in seed catalogues in 1885. The primary characteristics of 'Crosby Egyptian' were increased depth of the root and lessening of the rough root exterior. Thus, this population was much rounder and smoother than the 'Flat Egyptian' population from which it was derived.

3. 'Light Red Crosby'. This population was first widely described in 1904 by D.M. Ferry and Co. as a vermilion or light-red coloured table beet, although it was developed approximately ten years earlier by W.W. Tracy of the same company and first listed in 1896 in their wholesale catalogue. This population eventually became known as 'Ferry's Crosby', due to confusion generated from various other 'Crosby' populations. The root and exterior skin colour was lighter in colour than the original 'Crosby.'

4. 'Early Wonder.' This population was first listed in 1911 by the F.H. Woodruff and Sons and S.D. Woodruff and Sons catalogues. 'Early Wonder' was also listed in 1914 as the "Arlington strain of Crosby's Egyptian beet," and thus it is clearly a derivative of that population. The primary selection criterion used in development of this population was a root that was rounder in shape than 'Crosby's Egyptian'. The resulting 'Early Wonder' combined the early maturity of the original 'Flat Egyptian' with a very round root. This population has been and continues to be of great importance for fresh market table beet production, where its robust foliage is suitable for bunching.

5. 'Detroit Dark Red'. This population was originally selected from a population known as 'Early Blood Turnip' by a man named Reeves of Port Hope, Ontario, Canada. In 1892, this cultivar was listed as 'Detroit Dark Red Turnip' beet by D.M. Ferry and Co. This population is perhaps the most important and widely-adapted variety. 'Detroit Dark Red' is considered a multi-purpose cultivar and has been used for fresh market, processing, and market garden production. Roots are smooth and round and foliage is very dark green. 'Detroit Dark Red' has many synonyms that have been widely distributed throughout the world, most of which include the word "Detroit" in their names, such as 'Detroit Blood', 'Detroit Early Dark Red', and 'Early Detroit Dark Red'. The popular cultivar 'Morse Detroit' was selected from 'Detroit Dark Red' and offered first by C.C. Morse and Co. in 1928 as 'Morse's Improved Detroit.'

6. 'Ohio Canner.' This population was developed at the Ohio Agricultural Experiment Station and released in 1932. Its skin, flesh, and foliage colour is very similar to 'Detroit Dark Red'; however its primary attribute was reduced amount of differential colouring between cambial rings (zoning). This made 'Ohio Canner' an excellent choice for canning, because light coloured rings were not present in canned product.

7. 'Long Dark Blood'. This is perhaps the oldest founding population of table beet in the U.S. Magruder et al. (1940) suggest it may have been introduced by early settlers from France, where it was a very popular cultivar. 'Long Dark Blood', as the name suggests, is a long-rooted type with cylindrically-shaped roots. Maturity is later than for most other founding populations. The long, slender roots grow in part above-ground, and in some cases at slight angles to the below-ground portion of the root. This population has many synonyms, most of which include the words "Blood" and "Long." 'Long Dark Blood' was a primary founding population for cylindrical table beet germplasm in Europe and the U.S.

4 Genetic Resources and Major Breeding Achievements

There are three primary gene pools used by breeders for the improvement of table beet. These include wild species of *Beta*, close crop relatives such as sugarbeet and Swiss chard, and table beet populations and cultivars. Of these, the latter two have received the greatest attention from table beet breeders and will receive the most attention here.

Goldman and Navazio (2003) reviewed the use of wild species of the genus *Beta* to improve cultivated beet. The wild *Beta* species, including *Beta procumbens* and *Beta webbiana*, have been used to improve cultivated *Beta* crops. Gaskill (1954) reported viable hybrids from crosses of Swiss chard with both *Beta procumbens* and *Beta webbiana*. Both species may contain useful genes for crop improvement, including disease and pest resistance. Interestingly, neither species hybridized successfully with sugar beet, but both were successful in matings with Swiss chard. In Gaskill's program, Swiss chard served as a useful bridge species in introducing these desirable traits to sugar beet. *Beta procumbens* has been a source of root-knot nematode resistance genes for sugar beet. The $Hs1^{pro\text{-}1}$ gene carried on chromosome 1 of *B. procumbens* was transferred to sugar beet via interspecific hybridization and backcrossing (Jung and Wricke, 1987). Recently, Cai et al. (1997) used the technique of positional cloning to clone a nematode resistance gene from *B. procumbens* and transfer it to sugar beet. Resistance to root-knot nematode (*Meloidogyne* spp.) previously was introgressed into sugar beet (*Beta vulgaris* L.) from wild beet (*B. vulgaris* ssp. *maritima* (L.) and was demonstrated to be dominant and simply inherited. Since resistance conferred by this gene was effective against six different *Meloidogyne* spp. tested, the locus was designated R6m-1. Weiland and Yu (2003) found a CAPS marker designated NEM06 that co-segregated with resistance to the root knot nematode. Such a marker may make breeding for resistance to nematodes much easier than what could be accomplished through field or greenhouse screening.

Goldman and Navazio (2003) described the efforts made by Gabelman and students over many years to improve table beet using key traits from sugarbeet. The basis for the following section rests in personal communications between the author and Professor Gabelman as well as in an unpublished manuscript entitled "Table Beet Breeding" written by W.H. Gabelman, F.A. Bliss, and R.L. Engle in 1963. The following paragraphs are based on the discussion and interpretation in Goldman and Navazio (2003).

During the period from 1950-1970, the primary breeding objectives of Gabelman's table beet program focused on sterile and maintainer lines for the production of hybrids, disease resistance, round to globe shaped roots, improved colour and sweetness, multigerm and monogerm seed, and enhancement of combining ability. Gabelman took advantage of useful genes found in sugar beet in to improve table beet germplasm. Among the traits transferred from sugarbeet to table beet were self-fertility, annual growth habit, cytoplasmic male sterility, and the monogerm character. A description of this work is presented in the following paragraphs.

Cross-pollination in beet is obligate due to a self-incompatibility system with up to four loci (Lundqvist et al., 1973). However, a dominant gene for self-fertility, S^F (Savitsky 1954), overrides this system and allows for self-pollination. Gabelman knew that the development of table beet inbreds would be difficult if self-incompatibility was present, and for this reason sought to use the S^F allele in his breeding program. Inbreeding depression is one consequence of using S^F in a breeding program, but a number of inbred lines have been successfully developed.

The first gene to be introduced to table beet from sugar beet was the S^F allele. The S^F allele allowed for inbreeding individual plants, a technique that was not previously possible in beet because of its self-incompatibility. Inbreeding not only enabled the development of more homogenous populations that ultimately resulted in uniform inbred lines, but it made possible the maintenance of sterile inbred lines by allowing self-pollination of maintainer genotypes. In practice, the abundance of pollen present in plants carrying S^F made it difficult for foreign pollen to successfully pollinate and fertilize these plants. The value of this allele in maintaining fertile inbred maintainer lines without contamination from foreign windblown pollen was an important factor in successful table beet seed production in the Pacific Northwest, where a great majority of the U.S. table beet seed is currently produced.

Cytoplasmic male sterility (CMS) and the restoration of fertility was described in sugar beet by F.V. Owen (1945) and has been used since that time by sugarbeet and table beet breeders. A sterile cytoplasm will result in a male sterile plant if alleles at both the *x* and *z* loci are homozygous recessive. Dominant alleles at either or both loci will result in male fertility. Maintainer lines, referred to as "B" lines by many vegetable breeders and "O-types" by sugar beet breeders, possess normal, fertile cytoplasm and homozygous recessive alleles at the *x* and *z* loci. Gabelman introduced the *x* and *z* alleles, conditioning sterility at the nuclear restorer locus (in homozygous recessive condition) from sugar beet breeding lines obtained from F.V. Owen into table beet germplasm. These alleles in combination with the sterile cytoplasm obtained from Owen and the S^F allele obtained from Savitsky allowed for the development of male sterile breeding lines and their maintainer lines.

Recently, Hagihara et al. (2005) mapped the fertility restoration allele (*Rf1*) for O type cytoplasm in sugarbeet using bulked segregant analysis. Based on linkage position, these workers suggest that the *Rf1* locus corresponds to the *X* locus. Molecular markers closely linked to the *Rf1* allele should facilitate marker-based breeding strategies to identify maintainer lines in sugarbeet and table beet breeding programs. There is a distinct advantage to using a molecular marker system for identification of maintainer lines. The identification of maintainer genotypes is predicated on the existence of recessive alleles at the restorer loci. Some genetic backgrounds seem to be much more recalcitrant sources of maintainer lines, but this could be due to the low frequency of recessive alleles at restorer loci in those backgrounds. Havey and colleagues have found this to be true in onion, where certain open pollinated populations have not been sources of inbred lines specifically for this reason. With the use of a molecular marker that can tag the restorer alleles, a breeder could identify the frequency of that allele in a population and determine whether it is likely that maintainer lines could be extracted.

The *B* allele conditioning annual flowering habit was also obtained from sugar beet breeding material from Dr. V.F. Savitsky. Application of the *B* allele to sugar beet breeding has been discussed by Bosemark (1993) and to table beet breeding by Goldman and Navazio (2003). The *B* allele allows for efficient development of sterile inbred lines since spring-sown plants carrying *Bb* flower during the growing season. A cross of the constitution $SxxS^FS^FBb \times N\_\_s^Fs^F\_bb$ will give rise to 50% annual (*Bb*) progeny which, because they are flowering, can be classified for sterility in the field. These annual sterile plants can have their flower stalks removed, vernalised, and re-flowered in winter in the greenhouse nursery, assuring continuous inbreeding of the sterile line with its maintainer line. Thus far, the University of Wisconsin table beet breeding program is the only program to make use of this breeding scheme; though it offers promise for any program interested in gaining efficiency in inbred development or selecting monogerm types.

Hohmann et al. (2003) identified molecular markers very tightly linked to the *B* allele, and identified BACs that can be used to clone this gene. Gaafar et al. (2005) localized the *B* locus to a 360-kb contig in sugarbeet, as a prelude to positional cloning of this gene. If the *B* allele were cloned, it might be transformed into various *Beta* germplasm rather than backcrossed; though transgenic table beet poses some significant problems in gene flow as described later in this chapter. More interesting would be gaining an understanding of how this allele causes a switch from biennialism to annualism. It is possible that this genetic switch is associated with the domestication and early selection of table beet from leaf beet in Northern Europe many centuries ago.

Wild-type *Beta* possesses a trait known as multigerm seed, whereby each aggregate fruit may contain from one to five seeds, compressed into a single "seed ball." The seed ball is actually a lignified flower carcass with a corky appearance. Multigerm seed will obviously make precision seeding more difficult, as population density will be determined by the number of successful plants from each seed ball. The monogerm character, conditioned by recessive alleles at the *m* locus, was first identified as a mutant in a commercial sugar beet field (Savitsky, 1950). Many, if not most, sugar beet cultivars carry the monogerm trait. While monogerm table beet cultivars are still not widely accepted, their appearance in the marketplace is beginning to be of greater importance. Gabelman also incorporated the monogerm character into table beet germplasm, resulting in the development of the first table beet inbred lines carrying this trait (Goldman, 1996). The original sugar beet x table beet crosses required approximately 10 generations of backcrossing and selection before commercially-acceptable round, red roots were recovered. Recent molecular analyses of these materials suggest that the Wisconsin inbred lines have retained their intermediacy between sugar beet and table beet at the DNA level (Wang and Goldman, 1999). This finding suggests the Wisconsin material is unique in its genetic makeup among table beet germplasm sources.

5 Seed Production

Goldman et al. (1996) and Goldman and Navazio (2003) described the process of breeding a biennial crop such as table beet. Table beet has a biennial life cycle and requires specific environmental stimuli to promote a switch from vegetative growth to reproductive growth. Table beet is an obligate long-day plant. Vernalization, the exposure of the plant to a relatively short period of temperatures slightly above freezing, has the effect of increasing the competence of leaves to produce a flowering stimulus (Benjamin et al., 1997). While little work on the vernalization requirement in table beet has been conducted, some work has been reported in sugar beet. The minimum and maximum temperatures for vernalization in sugar beet were

0° and 15° C, respectively, with the fastest flowering response occurring at 12° C (Benjamin et al., 1997). Bolting, or the premature appearance of a flower stalk during the vegetative growth stage, is detrimental to crop production. Bolting is a fairly common occurrence in early-planted table beet crops, particularly in temperate environments. Significant yield losses can be expected where bolting has occurred, and selection against bolting is routinely practiced. Jaggard et al. (1983) reported that 50% of field-grown sugar beet plants bolted when temperatures were less than 12 C for 60 days during vegetative growth. Threshold temperature levels for bolting in table beet have not been reported. During a standard breeding cycle, vernalization typically takes place for 12 weeks at temperatures of approximately 2-5 C.

Production of beet seed for commercial purposes is usually accomplished over the course of two seasons in a climate with mild winters (average seasonal low – 8° C) and cool dry summers (average seasonal high 24° C). In North America, the "root-to-seed" method predominates. The majority of table beet seed production is the Puget Sound region of Washington, where significant damage to the roots can occur if they are left in the field over winter.

Stock seed is planted into a "plantbed" in late June or early July at densities approximating those of fresh market production. While plants are vigorously growing, between 6 and 10 weeks after planting, the crop can be rogued for off-types based on the foliage. By early October, when crop growth has slowed and roots have attained an adequate size, the crop is topped by mechanically cutting off the leaves and all but 2-5 cm of the petioles to prepare the crop for storage. Within a week to ten days of topping plants, roots are lifted and piled into "windrows," with each windrow representing four to six rows of beet roots (sometimes these windrows are pushed into a very shallow "pit" that is only 15 cm deep and 1 m wide). Enough soil is mounded over the roots to cover them with 30 cm of soil on all open sides. This is adequate thermal insulation to protect the roots from freezing damage in areas where annual low temperatures do not exceed –10° C and where the duration of cold periods near 0° C are not longer than 48 to 72 hr.

The roots are unearthed between the middle of March and early April. Roots for seed production are known as "stecklings." Stockseed roots are graded for shape, prominence of taproot, absence of disease, and trueness-to-type. As it is important to improve and maintain consistent internal colour, all stecklings used for stockseed production should be visually inspected for acceptable pigment levels. This is accomplished in two ways, both involving cutting the roots. The first method involves cutting approximately 1/10 to 1/15 of the root mass, diagonally, off the side of the root. The second method requires slicing the root diagonally, starting at the centre of the apical growing point, thereby cutting the root into two halves that are approximately the same size. Both methods allow the breeder to see both the overall intensity of colour as well as detecting the presence of "zoning" which is the differential colouring between cambial rings.

At the time of replanting the stecklings are mechanically dropped into furrows. The roots are placed upright and soil is firmly placed around each steckling. Rows are 76 to 91 cm apart and stecklings are 40 to 66 cm apart within the rows. An early irrigation, if necessary, should be made available after transplanting as adventitious roots that emerge in the second year are sensitive to drought during establishment.

Isolation of beet seed fields by at least 1.2 km from other beet seed production fields or at least 2.4 to 8.3 km from Swiss chard, mangel, or sugar beet seed fields is very important since *B. vulgaris* is a wind-pollinated species and its pollen can travel for 20 km (Poole, 1937). When producing stock seed, fields should also be planted at a minimum of 3 to 8 km apart. Harvest of beet seed is initiated when the seed balls at the base of each branch are mature and brown. It is not possible to combine standing beet seed in the field as it matures unevenly on the stalk. Plants are cut near the base or pulled by hand and laid in windrows to cure during a dry period in late summer. When the seed and plants are sufficiently dry they can be fed through a combine.

Goldman (1998) described the gene *ffs*, which conditions fasciation of the flower stalk. The fasciated character arose spontaneously in the inbred line W411. Its primary characteristics are a flattened flower stem with petioles coalesced into a twisted, ribbonlike appearance. Variable expression of the fasciated trait was noted in progenies produced from a variety of matings, however all plants carrying this recessive allele in the homozgous condition display a distinct flattening of the flower stem at the stem-hypocotyl junction. Fasciation due to *ffs* could be useful in seed production, as fasciated plants tend to exhibit heavy seed set with seed maturity occurring in a synchronous fashion. It is presently not known whether plants homozygous for *ffs* exhibit increased seed set under field conditions.

6 Breeding Methods

Goldman and Navazio (2003) described the trajectory of table beet breeding in the U.S. during the latter half of the 20th century. They compared the development of table beet inbred lines to that used in development of onion and carrot inbred lines via the inbred-hybrid method of breeding. Similar to onion and carrot, table beet hybrids are developed using cytoplasmic-genic sterility. This approach was based upon the sterile "O" type cytoplasm received from F.V. Owen and maintainer inbred lines extracted from open-pollinated cultivars. Crossing of this sterile table beet line with fertile plants from open-pollinated cultivars revealed the presence of maintainer genotypes (Bliss and Gabelman, 1965), although in certain matings it appeared that a single nuclear restorer locus was present. The first four maintainer lines developed were designated W32, W162, W163, and W187. Once sterile and maintainer pairs were available, the sterile lines could be used as females in various hybrid combinations with restorer lines, populations, or maintainer lines used as the pollen parent. Such early hybrids were promising, although they did not result in superior horticultural performance over open-pollinated cultivars. It was not until the second cycle of inbred lines was developed, including W218, W260, and W279, that heterosis for a variety of horticultural characteristics was apparent in various hybrid combinations (Goldman, 1996). Table beet inbred lines have been released as A and B pairs, where the A line refers to the sterile phenotype with genotype *Sxxzz*, and the B line refers to the maintainer of sterility (fertile) phenotype with genotype *Nxxzz*. All inbred lines released publicly are biennial and thus carry the *b* allele in the homozygous recessive condition.

During the initial phases of inbred development, open-pollinated cultivars and populations were chosen for desirable characteristics. Individual plants from these populations were mated in the greenhouse in small masses, usually containing two to four plants. Direct self pollination was not possible until the S^F allele was introgressed from sugarbeet; therefore initial small plant mass populations resulted in the production of hybrid seed. The small plant mass technique was originally developed by H.A. Jones for onion breeding (Jones, 1923; Jones and Emsweller, 1934). In this technique, vernalised roots were planted in clay pots in late fall and flowered during the winter months. Flowering plants were covered with paper bags, similar to those used in maize breeding, which were stapled shut at the bottom to prevent pollen release. The table beet flower is protandrous and pollination is by wind. Thus, a small amount of shaking of these pollination bags produces reliable seed set. Larger populations are grown in greenhouse isolations without the use of pollination bags (Figure 2).

Fig. 2. Clockwise from top left. Production of seed of a population undergoing recurrent selection on a greenhouse bench. Technique for seed production in a greenhouse using a clay pot. Technique for seed production of cylindrically-shaped roots in a greenhouse using aluminium cans. Technique of bagging individual plants using corn pollinating bags for isolation purposes.

With wild-type table beet germplasm, direct self-pollination was not possible. But with the introduction of the S^F allele, self-pollination became feasible. Gabelman and students found that inbreeding depression was not as great in table beet as expected, and despite the very high degree of self-pollination (>99%) with S^F many generations of direct selfing were possible. Inbred lines developed with S^F have since been self-pollinated more than 15 generations without declines in vigour that would preclude their continuation in a breeding program.

One of the unique features of inbred development in table beet has been the use of the annual gene, *B*, in assessing both sterility and degree of monogerm seed characteristics during first season of growth. Gabelman obtained this gene from V.F. Savitsky. Because standard table beet germplasm is biennial and of the genotype *bb*, germplasm carrying the *B* allele will flower (bolt) during the first season of growth. Although bolting is highly undesirable from the standpoint of crop production, it allows for the identification of floral characteristics prior to harvest. With a biennial life cycle, table beet roots must be harvested, stored, and vernalised prior to reproductive growth, a process that can take many months and a great degree of labour. The table beet breeding program at the University of Wisconsin relied upon a greenhouse environment from November – April in order to make crosses and produce seed for the next generation. Because this greenhouse space was limited, care was taken to identify useful genotypes in the field during the first season of growth. When plants carrying the *B* allele were flowering in the field during August, it became possible to identify these genotypes prior to harvest.

For example, a cross of the genotypes *SxxzzSfSfBb* X *NxxzzSfSfbb* yields progeny from the female parent which are 50% annual (*Bb*). These plants will flower during the first season of growth in the breeding nursery, where they can be classified for fertility or sterility. If, for example, one wishes to determine whether a fertile plant is carrying recessive alleles at the nuclear restorer locus, one can cross it with a sterile plant that is heterozygous for the *B* allele and the score the flowering progeny in the field the following season. If these progeny are sterile, the flowering plants can be harvested, and have their flower stalks separated from the roots. These roots of known sterile genotype can then be stored and planted again in the greenhouse, where they will again be crossed with maintainer genotypes. Roots of plants carrying the *B* allele obviously do not need to be vernalised in the same way *bb* plants. Using the technique described above, continuous inbreeding of the sterile line and continual crossing with its maintainer are managed in a relatively simple fashion. Use of the *B* allele has also facilitated the development of monogerm inbred lines, because flowering progeny in the nursery can also be checked for the number of flowers in the axil of each bract. Additionally, sterile plants carrying the *B* allele can be used to detect outcrossing during increases of fertile inbred lines under commercial seed production conditions.

When biennial plants carrying sterile cytoplasm are desired, such as during the latter stages of an inbred development program, the remaining 50% of the segregating progeny from the above-described cross that were not flowering can be chosen for appropriate testcrosses or commercial use. These are of the desired genotype *bb*. In practice, use of the *B* allele in table beet breeding allows for greater flexibility and precision in inbred development, because one can choose annual or biennial (or both) plants in the field and more accurately choose and plan the crosses to be made during winter months.

7 Current Goals in Breeding

Among the primary pests of table beet is *Cercospora beticola,* or Cercospora leaf blight (Figure 3). Delahaut and Stevenson (2004) have described the disease in a fact sheet used by the University of Wisconsin Extension Service. The fungus can over-winter on plant debris or other living hosts, and begins producing spores in the spring. Wet conditions favour this pathogen, and both high temperatures and high humidity favour its growth. Interestingly, much of the germplasm derived from the University of Wisconsin table beet breeding program has tolerance to *Cercospora*, but many popular cultivars are quite susceptible. Breeding for resistance to this fungus has largely been confined to selection in field environments where the pathogen is abundant, or in climates where infection has a high likelihood of causing damage. More breeding work is needed in this area because much of the table beet germplasm is susceptible to the pathogen. In addition, baby leaf production and requires the disease free leaf material. Recently, Setiawan et al. (2000) identified quantitative trait loci (QTL) associated with resistance to *Cercospora* in sugarbeet on chromosomes III, IV, VII, and IX. Molecular breeding for *Cercospora* resistance may increase the efficiency of selection and identification of resistant genotypes in table beet.

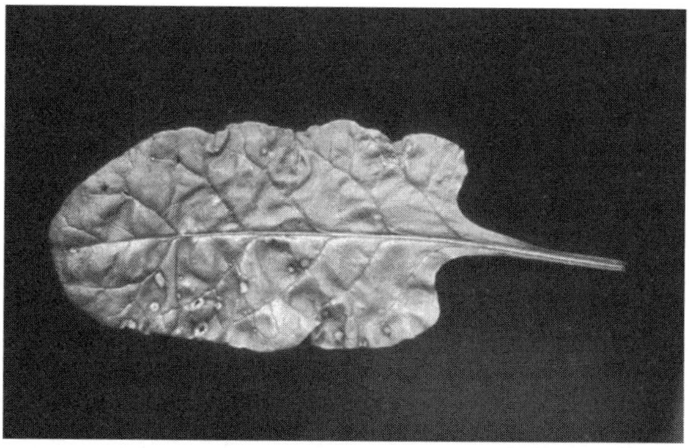

Fig. 3. Table beet leaf infected with *Cercospora beticola* (Delahaut and Stevenson, 2004).

Another key pest in table beet production is Rhizoctonia root rot caused by the fungus *Rhizoctonia solani.* Buttner et al. (2004) have developed sophisticated greenhouse and field screening techniques for developing sugarbeet germplasm resistant to this fungus. A suspension of *Rhizoctonia* mycelium (equivalent to 10 mg carbon per plant) is applied to the beet crown, and disease symptoms can be read in as early as three weeks. Symptomology was similar to that found in field situations. By 11 weeks, cultivars could easily be assessed for their ability to tolerate the inoculation. This test would be very helpful in the development of *Rhizoctonia* resistant table beet. At this time, no such materials are available, though there is a pressing need for table beet germplasm with this kind of tolerance or resistance in many regions of production.

The primary pigments in table beet are the betalains (Figure 4), a unique class of alkaloid pigments found primarily in the Caryophyllales and some fungi (Clement et al., 1992). Betalain pigments are comprised of the red-violet betacyanins and the yellow betaxanthins. Both are derived from betalamic acid following the cleavage of L-DOPA between the 4- and 5- positions, and differ from one another by conjugation of a substituted aromatic nucleus in the 1,7-diazaheptamethinium chromophore (Clement et al., 1992).

Fig. 4. Table beet roots expressing betaxanthin pigment only (left) and both betaxanthin and betacyanin pigments (right).

Betalain pigments extracted from red beet roots provide a unique natural alternative to synthetic red dyes. Betalains have been successfully used in commercial food colouring operations for a number of years (von Elbe et al., 1974), and continue to be an important source of red colour in the food industry, particularly in the U.S. and Europe. Red beet dye can be used in cosmetics, candy, ice cream, meat products, yogurt, and powdered drink mixes. The main limitation of beet concentrates in the food industry is the relatively low concentration of betacyanin in beet root juice, so large quantities of commercial product must be added to foods to obtain sufficient colouring. Concentrating betacyanin in the root juice is a time and energy-consuming process for the food processor, and it can lead to degraded pigment and reduced colour. Breeding red beet for increased betalain concentration has lowered the cost of red beet dye four-fold to a point where it is now only about two and one-half times as expensive as the petroleum-derived compounds in commercial use. At some point, it might be possible to reduce the cost of betalain-derived food colorants so that they are cost-competitive with synthetic dyes.

The presence of alleles at two linked loci (R and Y) condition betalain pigment production in beet (Keller, 1936). Sugarbeet workers also use the designation *G* as a synonym for the *Y* locus (Linde-Laursen, 1972). Wolyn and Gabelman (1990) demonstrated that three alleles at the R locus determine the ratio of betacyanin to betaxanthin in the beet root and shoot. Incomplete dominance for pigment ratio in R^t and *R* genotypes was observed. Colour patterning in the beet plant is affected by these *R* locus alleles as well as alleles at the *Y* locus. Red roots are observed only in the presence of dominant alleles at the *R* and *Y* loci, while white roots are conditioned by recessive alleles at both loci. A *yy* condition coupled with *rr*, which is characteristic of most sugar beet cultivars, produces no betacyanin and produces betaxanthin only in the hypocotyls. A third locus, *P*, appears to be required for pigment formation in *Beta*. Linde-Lauren (1972) suggested the *P* allele is indispensable for colour formation and demonstrated its close linkage with the *R* and *Y* loci. White-rooted beet plants likely carry the *p* allele in a homozygous recessive fashion along with dominant alleles at the *R* and *Y* loci.

Watson and Goldman (1997) described an allele known as *bl* conditioning a blotchy phenotype (irregular sectors of red and white root colour) in table beet. Austin and Goldman (2001) found transmission ratio distortion at the *bl* locus when *blbl* plants are used as both females and males in matings with wild type plants, but the degree of distortion was greater when *blbl* plants are used as females. The *Cl* allele interacts with the *R* locus and causes the formation of coloured leaves, most noticeably on the underside of the leaf. When *R-* is present, leaves are red, and when *rr* is present, leaves are yellow. However, the *Cl* allele also causes smaller than normal plants and its action appears to be highly variable. In many cases, *Cl* causes the appearance of blotches of pigment. *Cl* is linked to *Y*, *R*, and *B* (Owen and Ryser, 1942). The *Cl* allele is similar to the *Cv* allele, which causes the formation of very pronounced pigmentation near the veins of the leaf. The *Cv* allele is also linked to *Y, R,* and *B* (Owen and Ryser, 1942).

In 1978, a half-sib family recurrent selection program was initiated to increase pigment levels and decrease sugar levels in red beet. Decreasing sugar levels facilitates pigment extraction since sugar levels are very high in the root juice extract. During the first eight cycles of selection, total pigment increased approximately 200% (Goldman et al., 1996). To date more than 15 cycles have been completed and pigment levels have increased more than 450% from the starting materials. This experiment demonstrates pigment can be a target trait in breeding and that half-sib selection is an effective way to increase pigment levels in table beet.

Beet pigments also possess some unique bioactive properties with respect to human health, however the most promising results have been found for in vitro studies at this point. Lee et al. (2005) found that extracts of root tissue from table beet exhibited antioxidant and phase II enzyme-inducing activities. Two of the active fractions from these extracts were incorporated into rodent diets at 10-150 ppm over a two month period to assess bioavailability and in vivo activity. No statistically significant effect of diet was obtained, and wide ranges of tissue enzyme levels among individual animals were observed. More work is needed to determine if increased dosage may improve these results or if the bioavailability of betalain pigments is not high enough to be meaningful from a health standpoint.

One of the interesting elements of table beet consumption that has not been explored from a breeding point of view is modification of the geosmin compounds in the roots and leaves. The primary geosmin, *trans*-1, 10-dimethyl-*trans*-(9)-decalol, is responsible for the earthy flavour of beet and, as such, has supporters and detractors. A common comment on table beet flavour is that children don't like it because of the "earthiness." Were genetic variability to exist for geosmin concentration in table beet, it might be possible to make modification of geosmin concentration a breeding goal. Recently, Lu et al. (2003a) quantified geosmin in table beet using headspace solid-phase microextraction. Such a technique might make this a feasible trait for efficient screening in segregating populations. In addition, Lu et al. (2003b) showed that table beet is capable of endogenous geosmin production. Previously thought to occur only in association with soilborne microbes, geosmin production *in planta* suggests that a selection program designed to reduce or modify this compound or suite of compounds is possible.

Genetic engineering of sugarbeet through transgenic technology is now a reality, and transgenic sugarbeet cultivars have been developed for herbicide resistance. Little if any work has been accomplished in this arena in table beet; though it is likely that any genetic modification accomplished in sugarbeet could likewise be accomplished in table beet. One of the major areas of research needed here is in developing germplasm with high regenerative capacity. Ivic et al. (2005) reported significant genetic variability for regenerative ability from leaf disk callus in sugarbeet.

Sets of molecular markers are now available for use in beet genetics, possibly improving the ability of breeders to identify and select traits of interest. Single nucleotide polymorphism (SNP) markers were developed for each of the nine linkage groups of sugar beet (Mohring et al., 2005), and several workers have reported the availability of BAC clones and other genomic resources. It is only a matter of time before these tools become available for table beet workers; though very little molecular work has yet to be conducted on table beet. It would seem that the basic need to invest time and resources into classical breeding and cultivar development has dominated table beet work during the 20th and early 21st centuries. Were this crop to increase in popularity and acreage, and were more scientists to invest efforts in genetic improvement, they would find molecular tools from sugarbeet available for their use.

If transgenic table beet becomes a reality, it is certainly possible that gene flow could occur between wild or weedy beets and the cultivated crop. Desplanque et al. (2002) studied this phenomenon in sugarbeet in France. They found significant gene flow between cultivated sugarbeet and weedy beets in various ways in a field study. In cases where the weedy beets possessed the *B* allele and did not require vernalization for flowering, they were able to pollinate plants in sugarbeet seed production fields. Bolting plants in sugarbeet seed production fields could result in transgene escape into wild or weedy populations of plants. If the transgene is incorporated into a tetraploid pollinator plant in the seed production field, and transgene escape occurs, the resulting progeny will be triploid and therefore the gene will not immediately manifest itself. If, however, the pollinator is diploid, the possibility of transgene escape to wild or weedy populations is much higher. Table beet relies solely on diploid male and female plants in seed production, therefore the use of transgene technology will be particularly difficult wherever wild or weedy beet plants may exist. This problem is made even more significant when one considers the frequency of bolting in table beet crop and seed production fields, and could present a significant limitation to adoption of standard transgenic technology in this crop.

8 Acknowledgements

We thank D. Nicholas Breitbach, Warren Gabelman, and the graduate students of the University of Wisconsin Plant Breeding and Plant Genetics Program for many helpful discussions over the years. Portions of this chapter were published in or modified from Goldman and Navazio, 2003.

References

Austin, D., and Goldman, I.L. 2001. Transmission ratio distortion due to the *bl* gene in table beet. J. Amer. Soc. Hort Sci. 126:340-343.

Benjamin, L.R., A. McGarry, and Gray, D. 1997. The root vegetables: beet, carrot, parsnip and turnip. *In* Wein, H.C. (Ed). The physiology of vegetable crops. CAB International. pp. 553-581.

Bliss, F., and Gabelman, W.H. 1965. Inheritance of male sterility in beet. Crop Sci. 5:403-406.

Bosemark, N.O. 1993. Genetics and breeding. P. 67-119. In: The sugar beet crop: Science into practice. D.A. Cooke, R.K. Scott, R.K. (eds), Chapman and Hall, London.

Buttner, G., Pfahler, B., and Marlander, B. 2004. Greenhouse and field techniques for testing sugar beet for resistance to Rhizoctonia root and crown rot. Plant Breed. 123:158-166.

Cai, D., M. Kleine, S. Kifle, H-J. Haloff, N.N. Sandal, K. Marcker, R.M. Klein-Lankhorst, E.M.J. Salentijn, W. Lange, W.J. Stiekema, U. Wyss, F.M.W. Grundler, and Jung, C. 1997. Positional cloning of a gene for nematode resistance in sugar beet. Science 275:832834.

Campbell, G.K.G. 1976. Sugar beet. P. 25. In: Evolution of Crop Plants. N.W. Simmonds, ed. Longman, London.

Clement, J.S., T.J. Mabry, H. Wyler, and Dreiding, A.S. 1992. Chemical review and evolutionary significance of the betalains. P. 247-261. In: Evolution and Systematics of the Caryophyllales. H.D. Behnke and T.J. Mabry, eds. Springer Verlag.

Desplanque, B., Hautekeete, N., and van-Dijk, H. 2002. Transgenic weed beets: Possible, probable, avoidable? J. Appl. Ecol. 39:561-571.

Fischer, H.E. 1989. Origin of the 'Weisse Schlesische Rübe' (white Silesian beet) and resynthesis of sugar beet. Euphytica 41:75-80.

Ford-Lloyd, B.V. 1995. Sugarbeet and other cultivated beets. P. 35-40. n: Evolution of crop plants. 2nd ed. Smartt, J., Simmonds, N.W. (eds), Longman Scientific and Technical, Essex.

Ford-Lloyd, B.V., and Williams, J.T. 1975. A revision of *Beta* section *Vulgares* (Chenopodiaceae), with new light on the origin of cultivated beets. Botany J. Linn. Soc. 71:89-102.

Gaafar, R.M., Hohmann, U., and Jung, C. 2005. Bacterial artificial chromosome-derived molecular markers for early bolting in sugar beet. Theor. Appl. Genet. 110:1027-1037.

Gaskill, J.O. 1954. Viable hybrids from matings of chard with *Beta procumbens* and *B. webbiana.* Proc. Am. Soc. Sugar Beet Technol. 8:5.

Goldman, I.L. 1996. Inbred line and open-pollinated population releases from the University of Wisconsin beet breeding program. HortScience 31:880-881.

Goldman, I.L. 1998. Inheritance of *ffs*, a gene conditioning fasciated flower stem in red beet. J. Am. Soc. Hort Sci. 123:632-634.

Goldman, I.L. 2000. Prediction in plant breeding. Plant Breeding Rev. 19:15-40.

Goldman, I.L. and Austin, D. 2000. Linkage among the *R, Y,* and *bl* genes in table beet. Theor. Appl. Genet. 100:337-343.

Goldman, I.L., K.A. Eagen, D.N. Breitbach, and Gabelman, W.H. 1996. Simultaneous selection is effective in increasing betalain pigment concentration but not total dissolved solids in red beet (*Beta vulgaris* L.). J. Am. Soc. Hort Sci. 121:23-6.

Goldman, I.L., and Navazio, J.P. 2003. History and breeding of table beet in the United States. Plant Breed. Rev. 22:357-388.

Hagihara, E., Itchoda, N., Habu, Y., Iida, S., Mikami, T., and Kubo, T. 2005. Molecular mapping of a fertility restorer gene for Owen cytoplasmic male sterility in sugar beet. Theor. Appl. Genet. 111:250-255.

Hohmann, U., Jacobs, G., Telgmann, A., Gaafar, R.M., Alam, S., and Jung, C. 2003. A bacterial artificial chromosome (BAC) library of sugar beet and a physical map of the region encompassing the bolting gene B. Mol. Gen. Genomics. 269:126-136.

Ivic, H., Snezana, D., and Smigocki, Ann. 2005. Identification of highly regenerative plants within sugar beet (*Beta vulgaris* L.) breeding lines for molecular breeding, In Vitro Cell. Devel. Bio. Plant. 41:483-488.

Jaggard, J.W., R. Wickens, D.J. Webb, and Scott, R.K. 1983. Effects of sowing date on plant establishment and bolting and the influence of these factors on yields of sugar beet. J. Agr. Sci. Cambridge 101:147-161.

Jones, H.A. 1923. Pollination and self-fertility in the onion. Proc. Am. Soc. Hort. Sci. 20: 191-197.

Jones, H.A., and Emsweller, S.L. 1934. The use of flies as onion pollinators. Proc. Am. Soc. Hort. Sci. 31:160-164.

Jung, C., and Wricke, G. 1987. Plant Breed. 98:205.

Keller, W. 1936. Inheritance of some major colour types in beets. J. Agr. Res. 52:27-38.

Lee, C-H., Wettasinghe, M., Bolling, B.W., Ji, L.L., and Parkin, K.L. 2005. Betalains, phase II enzyme-inducing components from red beetroot (*Beta vulgaris* L.) extracts. 2005. Nutr. Cancer. 53:91-103.

Linde-Laursen, I. 1972. A new locus for colour formation in beet, *Beta vulgaris* L. Hereditas 70:105-112.

Lu, G., Fellman, J.K., Edwards, C.G., Mattinson, D.S., and Navazio, J. 2003a. Quantitative Determination of Geosmin in Red Beets (*Beta vulgaris* L.) Using Headspace Solid-Phase Microextraction. J Agric. Food Chem. 51:1021-1025.

Lu, G., Edwards, C.G., Fellman, J.K., Mattinson, D.S., and Navazio, J. 2003b. Biosynthetic origin of geosmin in red beets (*Beta vulgaris* L.). J Agric. Food Chem. 51:1026-1029.

Lundqvist, A., Østerbye, U., Larsen, K., and Linde-Laursen, I. 1973. Complex self-incompatibility systems in *Ranunculus acris* L. and *Beta vulgaris* L. Hereditas 74: 161-168.

Magruder, R., V.R. Boswell, H.A. Jones, J.C. Miller, J.F. Wood, L.R. Hawthorn, M.M. Parker, and Zimmerley, H.H. 1940. Descriptions of types of principal american varieties of red garden beets. USDA, Washington, DC.

Moehring, S., Salamini, F., and Schneider, K. 2004. Multiplexed, linkage group-specific SNP marker sets for rapid genetic mapping and fingerprinting of sugar beet (*Beta vulgaris* L.) Mol. Breed. 14:475-488.

Owen, F.V. 1945. Cytoplasmically inherited male-sterility in sugar beets. J. Ag. Res. 71: 423-440.

Owen, F.V., and Ryser, G.K. 1942. Some Mendelian characters in Beta vulgaris and linkages observed in the Y-R-B linkage group. J. Agr. Res. 65:155-171.

Pink, D.A.C. 1992. Beetroot. P. 473-477. In: Breeding Vegetable Crops. G. Kalloo (ed), AVI Press, Westport, CT.

Poole, C.F. 1937. Improving the root vegetables. USDA Yearbook. Washington, DC. Government Printing office. P. 300-325.

Savitsky, V.F. 1950. Monogerm sugar beets in the United States. Proc. Am. Soc. Sugar Beet Tech. 6:156-159.
Savitsky, H. 1954. Obtaining tetraploid monogerm self fertile, self fertile, and male sterile beets. Proc. Amer. Soc. Beet Technol. 8.
Setiawan, A., Koch, G., Barnes, S.R., and Jung, C. 2000. Mapping quantitative trait loci (QTLs) for resistance to Cercospora leaf spot disease (*Cercospora beticola* Sacc.) in sugar beet (*Beta vulgaris* L.) Theor. Appl. Genet. 100:1176-1182.
von Elbe J.H., J.H. Pasch, and Adams, J.P. 1974. Betalains as food colorants. Proc. IV Int. Congress Food Sci. Tech. 1:485-492
Wang, M., and Goldman, I.L. 1996. Phenotypic variation in free folic acid content among F_1 hybrids and open-pollinated cultivars of red beet. J. Am. Soc. Hort Sci. 121:1040-1042.
Wang, M., and Goldman, I.L. 1999. Genetic distance and diversity in table beet and subarbeet (*Beta vulgaris*) accessions measured by random amplified polymorphic DNA (RAPD). J. Am. Soc. Hort Sci. 124:630-635.
Watson, J.F., and Goldman, I.L. 1997. Inheritance of a recessive gene conditioning blotchy root colour patterning in *Beta vulgaris*. J. Hered. 88:540-543.
Weiland, J.J., and Yu, M.H. 2003. A cleaved amplified polymorphic sequence (CAPS) marker associated with root-knot nematode resistance in sugarbeet. Crop Science. 43:1814-1818.
Williams, J.T., and Ford-Lloyd, B.V. 1974. The systematics of the chenopodiaceae. Taxon 23:353-354.
Winner, C. 1993. History of the crop. P. 1-35. In: The sugar beet crop: Science into practice. Cooke, D.A., Scott, R.K. (eds), Chapman and Hall, London.
Wolyn, D.J., and Gabelman, W.H. 1990. Inheritance of root and petiole pigmentation in red table beet. J. Hered. 80: 33-38.

Family Cucurbitaceae

Cucumber

Jack E. Staub[1], Matthew D. Robbins[1], and Todd C. Wehner[2]

[1] USDA, ARS, University of Wisconsin, Department of Horticulture, jestaub@wisc.edu
[2] North Carolina State University, Department of Horticultural Science, todd_wehner@ncsu.edu

1 Introduction

Cucumber (*Cucumis sativus* var. *sativus* L.) is a member of the economically important family Cucurbitaceae which includes squash (*Cucurbita* ssp.), watermelon [*Citrullus lanatus* (Thunb.) Matsum. & Nakai], and melon (*Cucumis melo* L.). After tomato (*Solanum lycopersicum L*) and watermelon, cucumber and melon are cultivated more broadly than any other vegetable species (http://faostat.fao.org; Pitrat et al., 1999) where 2,479,728 hectares were harvested in 2005 producing 42,611,337 t under field and greenhouse culture. Production of cucumber is the second largest of all cucurbits, where China (26,559,600 t), Iran (1,400,000 t), Russia (1,414,700 t) Turkey (1,725,000 t) and the United States (981,660 t) represented 75 % of the world production in 2005.

The Cucurbitaceae consists of two subfamilies, Zanonioideae and the Cucurbitoideae (Jeffrey, 1980; Kirkbride 1993) (Figure 1). The Cucurbitoideae houses eight tribes, one of which (Melothrieae) includes the genus *Cucumis*, where the basic chromosome number is 2n = 2x = 24 (Dane and Tsuchiya 1976). *Cucumis* is partitioned into two subgenera designated as *Cucumis* (2n = 2x = 14 and 24) and *Melo* (2n = 2x = 24) that contain five cross-sterile species groups (Jeffrey 1980). The subgenus *Cucumis* comprises three or four Sino-Himalayan species, including *C. sativus* (2n = 2x = 14) and *C. hystrix* Chakr. (2n = 2x = 24). *C. sativus* houses several botanical varieties including var. *sativus*, the cultivated cucumber (hereafter referred to as *C. s.* var. *sativus*), and the wild, free-living var. *hardwickii* (R.) Alef. (hereafter referred to as *C. s.* var. *hardwickii*) (Kirkbride 1993).

Wild African *Cucumis* species (mostly 2n = 2x = 24) are cross incompatible with cucumber and melon, which are themselves cross-incompatible (Kroon et al., 1979). Likewise, the wild, free-living *C. hystrix* is only sparingly fertile with cucumber (Chen et al., 1995; 1997a and b). This species is found only in the Yunnan Province of Southern China, and has unique genetic attributes that make its taxonomic determination complex.

Cucurbitaceae (Family)
- Zanonioideae (Subfamily)
- Cucurbitoideae (Subfamily)
  - Melothrieae (Tribe)
    - *Cucumis* (Genus)
      - *Cucumis* (Subgenus)
        - *C. sativus* L. (Species)
          - var. *sativus*
          - var. *hardwickii*
        - *C. hystrix* Chakr. (Species)
      - *Melo* (Subgenus)
        - *C. melo* L. (Species)
          - subsp. *agrestis*
          - subsp. *melo*

Fig. 1. Taxonomic classification of cucumber (*C. sativus* L.) and melon (*C. melo*) L. in the family Cucurbitaceae according to Chung et al. (2006).

2 Origin and Domestication

The biosystematics and phylogeny of *Cucumis* species based on morphology, crossability, and protein analysis (Deakin et al., 1971; Staub et al., 1987 and 1992a; Perl-Treves and Galun, 1985) has led to an understanding of species relationships that have been largely confirmed by nuclear DNA analysis (Jobst et al., 1998; Zhuang et al., 2004). Most recently, Garcia-Mas et al. (2004) defined phylogenetic relationships among *Cucumis* species using ribosomal internal transcribed spacer sequences and microsatellite markers. Although their data did not agree with some of the previously described genetic relationships obtained using isozyme and restriction fragment markers (Staub et al., 1992; Perl-Treves et al., 1985; Jobst et al., 1998), their description of a clear separation between *C. sativus* and the rest of the *Cucumis* species supported these earlier studies.

The centre of origin for *Cucumis* species is likely Africa for the wild species. However, initial sites of domestication for melon and cucumber are probably the Middle East and Southern Asia, respectively, where genes from exotic sources have contributed extensively to plant improvement (Dane et al., 1980; McCreight et al., 1993; Staub et al., 1999). Cucumber (*C. s.* var. *sativus*) may have originated in Africa (Tapley et al., 1937), China, India, or in the Near East (Vavilov, 1926 and 1951; Harlan, 1975; De Candolle as cited by Hedrick, 1919), with domestication occurring later throughout Europe. It was domesticated about 3,000 years ago, and is indigenous to India (primary centre of diversity; Jeffrey, 1980; De Candolle as cited by Hedrick, 1919; Whitaker and Davis, 1962; Robinson and Decker-Walters, 1997).

Cucumber was brought to Greece and Italy by the Romans (2nd century BC; Mesopotamia), and it appeared in France in the 9th century, in England in the 14th century, and in North America by the mid-16th century. The Spanish brought cucumber to Haiti in 1494, and cucumber was reported in Montreal, Canada (by Cartier), in Florida, U.S. (by Desoto), and in Virginia, U.S. (by Amidas and Barlow) in 1535, 1539, and 1584, respectively (Whitaker and Davis, 1962; Robinson and Decker-Walters, 1997). Cucumber's dissemination westward from India is indicated by the profusion of ancient names that describe it. The English word "cucumber" comes from the Latin name *Cucumis*. Likewise, the Bohemian *agyrka*, German *gurke*, Greek *Aggouria*, and European *gherkin* trace back to an ancient Aryan word.

Cucumis s. var. *hardwickii* is a wild relative of *C. s.* var. *sativus* that grows in the foothills of the Himalayan mountains and is used by native peoples of Northern India as a laxative (Deakin et al., 1971). This botanical variety is sympatric and cross-compatible with *C. s.* var. *sativus* and possesses a multiple fruiting and branching habit that is not common in cucumber (Horst and Lower, 1978). *C. s.* var. *hardwickii*, therefore, represents the extreme in variation in *C. sativus* germplasm (Dijkhuizen et al., 1996), and, thus, has potential for increasing genetic diversity in commercial cucumber (Staub et al., 1992b).

The genetic variation in *C. s.* var. *sativus* accessions from India and China (the secondary centre of diversity for cucumber) has been assessed by protein and DNA marker analyses (Staub et al., 1997 and 1999; Horejsi and Staub, 1999). Data indicate that variation exists among accessions from different Indian states, the differences between Indian and Chinese accessions are distinct, and that Indian and Chinese accessions themselves are distinctly different from other *C. s.* var. *sativus* genotypes throughout the world. Genetic differences exist among cultivars grown in the same growing region [e.g., Shanghai and the Hunan region (Southern China) and Jiang Su, and Anhui (Northern China)]. These facts suggest that cultivar differences among Indian and Chinese cultivars can exist over a relatively limited geographical range.

It is has been hypothesized that the genetic variation present in cucumber germplasm in Southern China is endemic to that region and/or has been augmented by the infrequent immigration of germplasm from Northern India via ancient Himalayan trade routes (Staub et al., 1997 and 1999). This germplasm has subsequently been isolated by surrounding mountain ranges (i.e., Himalayas) and the region's political and social structure. In contrast, the genetic diversity of Northern Chinese cucumber germplasm is thought to be a direct beneficiary of the "Silk Road", where germplasm (genetic variation) has been continually introduced from India via Eastern Europe and the Middle East.

Two contrasting hypotheses regarding the origin of *C. s.* var. *sativus* and *C. melo* have been proffered; fragmentation of seven haploid chromosomes to form 12 (Bhaduri and Bose, 1947) and fusion of 12 haploid chromosomes to form seven (Trivedi and Roy, 1970). These species have been under reproductive and evolutionary isolation for a considerable amount of time (Garcia-Mas et al., 2004; Chung et al., 2006).

The phylogenetic relationship between *C. hystrix* (H) and *C. s.* var. *sativus* (S) is not substantially different since fertile amphidiploids have been synthesized between

C. hystix and *C. s.* var. *sativus* to create a synthetic species called *C. hystivus* (Chen et al., 1997a and b; Chen and Kirkbride, 2000; Chung et al., 2006). Furthermore, the development of the fully fertile *C. hystivus* (2n = 4x =38; HHSS; Chen and Kirkbride, 2000), partially fertile allotriploids (2n = 3x = 26; HSS), fertile diploids (2n = 2x = 14; SS), and partially fertile monosomics (2n = 15; SS + 1H) from *C. hystivus* and *C. s.* var. *sativus* matings indicates that chromosome reduction can occur through inbreeding and selection for chromosome number in this genus (Chen et al., 2004a and b). These facts, together with variation observed in the chloroplast genome among a broad array of *Cucumis* species, lend support to the hypothesis that *C. hystrix* is a progenitor species of *C. sativus*, or that they at least share a common ancestral lineage (Chung et al., 2006).

3 Market Classes or Cultivar Groups

There are two basic cucumber types; those eaten fresh (i.e., fresh or slicing market types; Wehner and Horton, 1986) and those consumed as a processed product (processing or pickling types; Staub and Bacher, 1997). The major fruit types are the American processing and fresh market types, the Dutch gherkin and greenhouse types, the German Schalgurken type, the Mideast Beit Alpha type, and the Oriental trellis (burpless) type.

Fresh market types are field or greenhouse grown, and are usually between 15 (i.e., U.S. and Mediterranean) to 40 (i.e., European) cm in length. Less common fresh market types include Sfran (compact fruit types marketed in the Persian Gulf), and "lemon" cucumber (shape similar to a lemon with pale, greenish-yellow skin; hermaphroditic). Processing types differ depending on cultural preferences (e.g., U.S. vs. Europe).

Europe

European market types include glasshouse cucumbers, Mediterranean or "Mini" types for glasshouse or poly-tunnel production, and processing cucumbers (i.e., gherkins). Mini cucumbers produce 14-17 cm long fruit. All of these types are primarily marketed for commercial production in Europe, North America, Greece, Spain, and Turkey.

European glasshouse types (32 to 40 cm long) are gynoecious, parthenocarpic (seedless), resistant to diseases such as powdery mildew (*Sphaerotheca fuliginea* (Schl. ex Fr.) Poll.), and cucumber mosaic virus (CMV), and will produce fruit under controlled climate conditions where commercial production is an exacting and costly enterprise. Most of the commercial cultivars are a product of Dutch breeding efforts. Disease resistance, high yield, and ability to grow at low temperatures are common breeding objectives. Some representative cultivars in Europe as well as in the U.S. and Canada include 'Jessica', 'Optima', 'Flamingo', 'Toska 70', 'Averyl', 'Niagara', 'Ladner', 'Sandra', 'Camaro', 'Dominica', 'Bella', 'Activa', and 'Sinaloa'. Some typical, disease resistant, gynoecious, mini cultivars are 'Jawell', 'Manar', 'Alamir', and 'Melita'.

Fig. 1. Varietal groups of endive and chicory: A escarole (photo ISI Sementi), B curled endive (photo ISI Sementi), C Witloof chicory, D pain de sucre, E radicchio Red of Chioggia (photo Veneto Agricoltura), F radicchio Early Red of Treviso (photo Veneto Agricoltura), G radicchio Late Red of Treviso (photo Veneto Agricoltura) , H radicchio Red of Verona (photo Veneto Agricoltura), I radicchio Variegated of Castelfranco (photo Veneto Agricoltura), L asparagus chicory (Catalogna) , M root chicory.

Fig. 2. Field plots with radicchio Early Red of Treviso hybrids: A and B inbred parents, C and D hybrid progenies.

Fig. 4. Head characteristics of the parents crossed for obtaining segregant F_1 population used for the construction of the genetic linkage maps: globe artichoke 'Romanesco clone C3' (A), globe artichoke 'Spinoso di Palermo' (B), cardoon (C) and wild cardoon (D). Examples of morphological variation of the heads observed in the F_1 segregant progenies.

Fig. 10. Drop of crisphead lettuce caused by *Sclerotinia minor* (Courtesy R. C. Grube).

Fig. 1. Different types of cauliflower curds.

Fig. 4. Colour diversity in cauliflower.

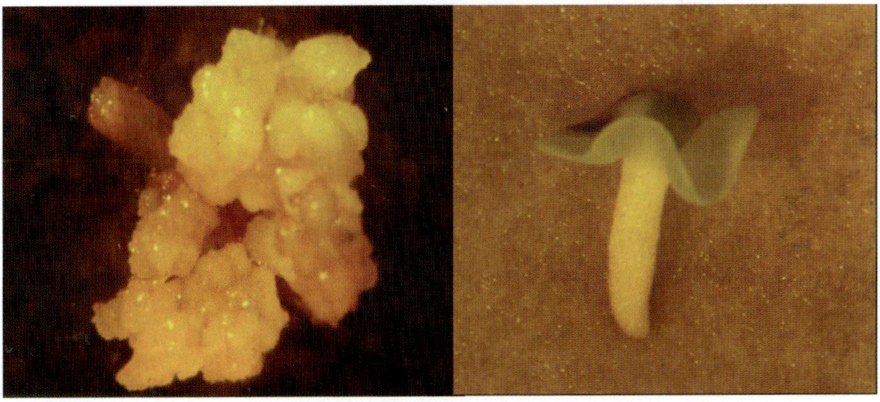

Fig. 5. Use of in vitro culture for cauliflower breeding: androgenic embryos (left) and germinating mature embryo (right).

Fig. 4. Table beet roots expressing betaxanthin pigment only (left) and both betaxanthin and betacyanin pigments (right).

Fig. 1. Some melon *varietas*. A = Wild melon plant in Sudan. B = Fruits of wild melons. C = var. *flexuosus*. D = var. *momordica*. E = var. *acidulus*. F = *var. tibish*.

Fig. 2. Some melon *varietas*. A = var. *cantalupensis* cultigroup Charentais. B = var. *dudaim*. C = var. *conomon*. D = var. *makuwa*. E = var. *chinensis*. F = var. *inodorus* cultigroup Piel de sapo.

Fig. 2. Young (summer squash, 2–5 days past anthesis) fruits of eight edible-fruited cultivars, one from each of the edible-fruited cultivar-groups of *Cucurbita pepo*. Left to right: subspecies *pepo*, Zucchini Group 'True French', Cocozelle Group 'Striato d'Italia', Vegetable Marrow Group 'Verte Petite d'Alger', Pumpkin Group 'Tondo Verde Scuro di Piacenza'; subspecies *texana*, Scallop Group 'Yellow Bush Scallop', Acorn Group 'Table Queen', Straightneck Group 'Early Prolific Straightneck', Crookneck Group 'Yellow Summer Crookneck'.

United States

The first cultivars used in the U.S. were brought from Europe (ca. 1700s), and included 'Early Short Prickley', 'Long Green Turkey', 'Smyrna', 'Roman', and 'White Spined' (Tapley et al., 1937). Additional cultivars sold in the U.S. were 'China Long' in 1862 and 'Chicago Pickling' in 1888 (Whitaker and Davis, 1962). Beginning in 1880, there was interest in cultivar development by and for American growers.

In the first few decades of the 1900's new cultivars were developed with improved fruit shape and colour (Anonymous, 1954-1958, 1960-1964; Barnes, 1969-1971; Minges, 1965-1968, Lower, 1973 and 1975). 'Model', introduced in 1946 with good fruit shape and adaptation to the southern production regions of the U.S., is a good example of that era. Then, beginning in 1937, emphasis was placed on disease resistance with the introduction of 'Shamrock', a cultivar resistant to CMV. After the introduction of cultivars with scab (causal agent: *Cladosporium cucumerinum* Ellis & Arthur) and downy mildew [causal agent: *Pseudoperonospora cubensis* (Berk. & Curt) Rostow] resistance, germplasm was developed with resistance to several other diseases. In 1955, resistance to scab and CMV were combined to produce the line Wis. SMR 12. Resistance to additional diseases was identified, and eventually combined to produce 'Sumter' (field resistances to seven diseases) and the gynoecious line Wis. 2757, with resistance to nine diseases [scab, CMV, bacterial wilt (causal agent: *Erwinia tracheiphila* {(E. F. Smith) Holland}, angular leaf spot (causal agent: *Pseudomonas lachrymans* {(E. F. Smith and Bryan) Carsner}, anthracnose (causal agent: *Colletotrichum lagenarium* {(Ross.) Ellis & Halst}, downy mildew, powdery mildew (causal agent: *Sphaerotheca fuliginea* {(Schl. ex Fr.) Poll.}, target leaf spot (causal agent: *Corynespora cassiicola* {(Berk. & Curt) Wei}, and Fusarium wilt (causal agent: *Fusarium oxysporum* {(Schlecht.) Snyd. & Hans f. sp. *cucumerinum* Owen}] (Table 1). Diseases where resistance needs to be uniformly incorporated into new cultivars include Rhizoctonia fruit rot [causal agent: *Rhizoctonia solani* Kuhn; telemorph *Thanatephorus cucumeris* {(A. B. Frank) Donk}], gummy stem blight [causal agent: *Didymelia bryoniae* (Auersw.) Rehm], watermelon mosaic virus (WMV) race 1, and zucchini yellow mosaic virus (ZYMV).

Relatively slow progress was made in the improvement of other traits such as sex expression and plant habit. 'Midget' was a dwarf-determinate cultivar introduced in 1940 (Table 1). The dwarf character was not used extensively, however, until later with the introduction of 'Castlepik', which was a semi-dwarf cultivar with determinate flowering habit. Monoecious hybrids were introduced in 1945, but seed was too expensive to permit wide commercial use. Development of gynoecious sex expression by Peterson and Anhder (1960) permitted hybrids to be produced economically. 'Spartan Dawn', developed from the gynoecious inbred 713-5, was the first (1962) gynoecious hybrid released for industry use. This initial USDA-ARS release was followed by the continued development of public gynoecious inbred lines between 1960-2004 (e.g., the Gy series, Gy-2, -3, -7, -8, and -14). Recent breeding efforts (1970-2000) have focused on improved fruit quality, yield, earliness, and adaptation to a broad array of U.S. production environments. Yield in once-over harvest systems may be improved by the introduction of dwarf plant types. However, research with the compact mutant (*cp*; Kaufman and Lower, 1976) for high-density cultivation, or multi-branching types such as littleleaf (Goode et al., 1980) that possesses simultaneous fruiting has not resulted in successful cultivars.

Table 1. Important steps in the genetic improvement of cucumber in the U.S.

Cultivar or line	Developer of seed source	Year introduced	Noteworthy trait(s)[1]
Improvement of disease resistance			
Shamrock	Iowa State Col., Ames	1937	CMV
Maine No. 2	Maine Agr. Exp. Sta.	1939	Scab
P.R. 39	Puerto Rico Agr. Exp. Sta.	1944	DM
Wis. SMR 12	Univ. of Wis., Madison	1955	Scab CMV
Ashe	N. C. Agr. Exp. Sta.	1959	Scab DM
Tablegreen	Cornell Univ., Ithaca	1960	CMV PM
Polaris	S. C. Agr. Exp. Sta.	1961	DM PM Anth
Poinsett	S. C. Agr. Exp. Sta.	1966	DM PM Anth ALS
Chipper	S. C. Agr. Exp. Sta.	1968	DM PM Anth ALS CMV
Sumter	S. C. Agr. Exp. Sta.	1973	DM PM Anth ALS CMV Scab WMV
Wis. 2757	U.S.D.A., Univ. Wis.	1982	DM PM Anth ALS CMV Scab TLS BW FW
Improvement of additional traits			
Midget	Minnesota Agr. Exp. Sta.	1940	Dwarf-determinate habit
Burpee Hybrid	W. Atlee Burpee Co.	1945	Mon-Hyb CMV DM
Model	Associated Seed Growers	1946	Fruit shape
MSU 713-5	Mich. Agr. Exp. Sta.	1960	Gyn
Spartan Dawn	Mich. Agr. Exp. Sta.	1962	Gyn-Hyb CMV Scab
Castlepik	A. L. Castle & Co.	unknown	Dwarf-determin., Gyn-Hyb
Littleleaf	Univ. Arkansas	1980	Multibranched habit

[1] CMV = cucumber mosaic virus resistance, DM = downy mildew resistance, Scab = scab resistance, PM = powdery mildew resistance, Anth = anthracnose resistance, ALS = angular leafspot resistance, WMV = watermelon mosaic race 2 resistance, TLS = target leafspot resistance, BW = bacterial wilt resistance, FW = Fusarium wilt resistance, Mon = monoecious sex expression, Gyn = gynoecious sex expression, and Hyb = hybrid.

4 Genetic Resources

The primary, secondary, and tertiary gene pools of *Cucumis* have been defined by Bates et al. (1995), den Nijs and Custers (1990), and Raamsdonk et al. (1989). Although the primary gene *Cucumis* pool includes *C. s.* var. *sativus* and var.

hardwickii, recent crossing and molecular analyses indicate that *C. hystrix* should perhaps be included in this gene pool (Chen and Kirkbride, 2000; Chen et al., 2004a). Wide hybridization in the *C. sativus* gene pool continues to be utilized for increasing the genetic diversity in cucumber (Nikolova et al., 2002). The secondary gene pool, however, includes wild African *Cucumis* species of varying ploidy levels, which are cross-incompatible with *C. sativus* (den Nijs and Custers, 1990).

The most recent compendium of cucurbit germplasm provides documentation of 68 world collections (Bettencourt and Konopka, 1990). It describes holdings in national genebanks, cites important breeding collections, and provides general information about these holdings, including their maintenance, availability, and evaluation. In Europe, the International Plant Genetic Resources Institute (IPGRI) coordinates institutional germplasm holdings. In the U.S., plant germplasm is maintained and evaluated by the U.S. National Plant Germplasm System (NPGS). The regional plant introduction (PI) station of NPGS at Ames, Iowa houses about 1,350 *C. sativus* accessions of worldwide origin. Molecular evaluation of this collection indicates that PI accessions are genetically diverse, not in Hardy-Weinberg equilibrium, and that they differ markedly from commercial germplasm in genetic structure (Meglic et al., 1996; Staub and Ivandic, 2000; Staub et al., 2002a). However, many of these accessions are as homozygous and homogeneous as elite inbred lines.

Plant introductions (i.e., PIs) have contributed significantly to cucumber improvement, and have been detailed by Tatlioglu (1993). Additional germplasm that has supplied traits for cucumber improvement include PI 183056 (India; large root size), PI 183967 (synom. LJ 90430; India; multiple lateral branching, sequential fruiting, nematode resistance), PI 197087 (India; downy mildew resistance), 200815 (Myanmar; powdery mildew and gummy stem blight resistance), PI 200818 (Myanmar; bacterial wilt resistance), PI 209065 (U.S.; high yield), PI 212233 (Japan; powdery mildew resistance), PI 220860 (South Korea; gynoecy), and PIs 418962, 419008, 419009, and 419135 from China [multiple disease resistances (Peterson et al., 1986a and b; Staub et al., 2002a)]. Other important germplasm used in cucumber improvement include 'Riesenschaal' (Germany), 'Zeppelin' (Germany), 'Chinese Long' (Japan), 'Tokyo Long Green' (Japan) 'Spotvrije' (The Netherlands), and ILG 58049 (The Netherlands; Peterson et al.,1986a and b).

5 Current Goals of Breeding

Cucumber improvement is a complex process involving the refinement of populations derived from intercrossing elite and/or exotic (unadapted) germplasm, the extraction of inbred lines from such populations, and the identification of commercially acceptable F_1 hybrids. Early genetic enhancement of cucumber (1850-1980) focused mainly on the incorporation of disease resistance and changes in plant architecture (e.g., sex expression, growth habit) that were augmented by improved cultural practices (Galun, 1961; McCollum, 1934; Sitterly, 1972; Peterson, 1975; de Ponti, 1975; George, 1970; Kubicki, 1980). General reviews of cucumber breeding (Lower and Edwards, 1986; Tatlioglu, 1993), and an examination of processing cucumber production (Staub and Bacher, 1997) have provided for rather complete treatments of cucumber improvement and culture. Therefore, the genetics and breeding information presented herein seeks to add to this early knowledge base, with an emphasis on new and emerging technologies as they relate to standard, commonly practiced breeding methods. Focus is placed on processing cucumber breeding since its breeding is relatively complex and the application of emerging technologies is well documented.

Yield and quality are a major focus of cucumber improvement and consist of many extensively reviewed, interrelated traits that are often the focus of the cucumber breeder (Lower and Edwards, 1986; Tatlioglu, 1993). These quantitatively and qualitatively inherited traits range from disease resistance to plant and fruit architecture and habit. Because of their diverse genetic nature, importance to plant improvement, and application in marker-assisted selection (MAS), the physiological interrelationships and genetics of a number of these traits are discussed below.

Yield

Yield has been a focus of cucumber breeders for over 50 years (Lower and Edwards, 1986; Wehner, 1989; Wehner et al., 1989). During the middle part of the 20th century, yield of U.S. processing cucumber maintained a steady increase from 4,685 kg/ha in 1949 to 11,455 kg/ha in 1979, an average annual increase of 226 kg/ha (Lower and Edwards, 1986). Similarly, yield trials of five popular gynoecious processing cultivars from the Southeastern United States released between 1969 and 1987 revealed an average annual yield increase of 400 kg/ha (Wehner, 1989). By 1980, the average yield of US processing cucumbers was 12,550 kg/ha, triple that of 4,076 kg/ha in 1920 (USDA, 1940, 1981). Most of the increase in yield during this period can be attributed to improved cultural practices and breeding for disease resistance (Lower and Edwards, 1986; Wehner 1989; Wehner et al., 1989). The introduction of the gynoecious flowering habit increased early yield, but did not affect total yield as measured over multiple harvests (Wehner et al., 1989).

As improved cucumber yield became increasingly important, it became the focus of many studies beginning in the late 1970's. Research conducted on many aspects of yield including breeding methodologies (e.g., selection methods and selection criteria), optimizing yield trials (e.g., methods to measure yield and optimal plot size), and the genetics of yield (e.g., heritability and genotype by environment interactions) from the late 1970's to the late 1980's are reviewed by Wehner (1989). Studies indicate that improvement by direct selection for yield is difficult. Yield is quantitatively inherited, has a low heritability [i.e., narrow-sense heritability (h^2) of 0.07 to 0.25], and is influenced mainly by genotype and environment, and to a lesser degree by genotype × environment interactions. Thus, selection for yield during population development should occur in intermediate stages of a recurrent selection scheme on a plot basis rather than on individual plants. Yield may effectively be evaluated in small (one row, single replication and harvest), multi-location (two to three) trials over seasons or years. The optimal time to harvest in trials depends upon a harvest index that is based on the number and weight of oversized fruit in check (control) plots.

Measurement of cucumber yield is often difficult because the fruits are harvested before they reach physiological maturity (yield measurement is reviewed by Wehner 1989). Cucumber growers usually measure yield by volume or weight per unit area, but the volume and weight of immature fruit can change rapidly from day to day, thus yield is dependent on the time of harvest. Converting yield to market value of processing cucumbers is further complicated because harvested fruit are graded by diameter where the smallest fruits have the greatest value, while oversized fruit have little or no commercial value. Although several methods for measuring yield (i.e., volume, mass, number, or dollar value) have been investigated, the most efficient measurement of yield in research studies is the total (marketable and oversize) number of fruits per plant, since it has a higher heritability, is more stable over time, and is easier to measure than other yield measurements. Furthermore, fruit number is highly correlated (genetic correlation = 0.87) with fruit weight (Wehner 1989).

Ironically, the increase in research on yield did not produce an increase in yield of U.S. processing cucumber, which has reached a plateau since the early 1980's (Shetty and Wehner, 2002; Fazio et al., 2003a; USDA, 2004). Mixed results have been obtained when selecting directly for yield, which may partially be explained by low heritability and environmental influence, combined with the difficulty in measuring yield (Wehner, 1989). The most effective approach to breeding for yield may be selecting for other traits correlated with yield that have a higher heritability (Wehner, 1989; Cramer and Wehner, 1998a and b; Cramer and Wehner, 2000a). Such traits correlated with yield are commonly referred to as yield components, and include number of harvests per plant, stem length, number of branches per plant, number of flowering nodes per branch, time to anthesis, percentage of pistillate flowers, and percentage of fruit set (Cramer and Wehner 1998a; Cramer and Wehner, 2000a). These traits can be manipulated to create various genotypes that possess an array of architectural habits. Recent studies suggest that MAS may be used to augment phenotypic selection for yield components (Fazio et al., 2003b; Fan et al., 2006).

Yield Components

Correlations among traits are important when manipulating plant architecture for yield improvement, since source/sink relationships provide practical constraints on fruit development. Correlations between yield component traits as well as with yield have been investigated in a variety of cucumber germplasm including slicing populations (Cramer and Wehner, 1998a and b), processing populations (Serquen et al., 1997a; Cramer and Wehner, 1998b; Cramer and Wehner, 2000a; Fazio, 2001), hybrids (Cramer and Wehner 1999a), and germplasm derived from *C. s.* var. *hardwickii* (Fredrick and Staub 1989). Correlative effects, a plant's reproductive biology (e.g., days to anthesis and sex expression), and gene action must be considered during breeding. Thus, studied attention to yield components has included the evaluation of the U.S. cucumber germplasm collection and elite lines for their combining ability for yield to create high-yielding wide- and narrow-based populations with acceptable fruit quality (Shetty and Wehner, 2002; Walters and Wehner, 1994; Wehner, 1997; Wehner, 1998; Wehner et al., 2000a and b).

Sex Expression

The type (e.g., gynoecious or monoecious) and intensity of sex expression is important to commercial cucumber production since differences in sex type and flowering can affect harvest date and relative yield. Genes that are hormonally controlled and influenced by growing environment affect both the type and intensity of sex expression (Lower and Edwards 1986; Tatlioglu 1993; Staub and Bacher, 1997).

Genetics of Sex Expression. Cucumber sex phenotypes are mainly monoecious (staminate and pistillate flowers) or gynoecious (pistillate flowers only), but androecious (staminate flowers only), hermaphroditic (perfect flowers), andromonoecious (staminate and perfect flowers), and trimonoecious (staminate, perfect, and pistillate flowers) types also exist. Plants possessing pistillate and perfect flowers have also been observed and used in hybrid production (El-Shawaf and Baker, 1981a). These sex types are determined by three major loci (*F*, *M*, and *A*; Shifriss, 1961; Galun, 1961 and Kubicki, 1969). The *F* locus influences the degree of femaleness (*FF* >*Ff*> *ff*), while the *M* locus determines whether flowers are unisexual (*M_*) or bisexual (*mm*). The *A* locus conditions increased male tendency if a plant is homozygous recessive *aa* and *ff*. Interactions between these loci yield the basic sex types found in cucumber.

While this three-gene model describes the basic regulation of sex types, a plant's phenotype is also influenced by modifying genes and environmental factors (Serquen et al., 1997a and b). The existence of sex modifying genes is supported by the observation that inbred gynoecious plants differ in their level of gynoecy and their capacity to confer femaleness in F_1 hybrids (Kubicki, 1969, Zhang et al., 1992). Monoecious plants also vary quantitatively in sex expression, ranging from predominately staminate to predominately pistillate. In fact, there are at least five genes that modify the expression of gynoecy in cucumber (Serquen et al., 1997b; Fazio et al., 2003a). Thus, hybrids between monoecious and gynoecious lines can show considerable variation in the frequency of female flowers depending upon the level of gynoecy in the parents (the *F* locus and the constitution of alleles at sex modifying loci). This variation in the level of gynoecy in gynoecious × gynoecious and gynoecious × monoecious hybrids remains a potential deficiency in many commercial cultivars.

Hormonal Factors Controlling Sex Expression. Genetic control and environmental variation of sex expression is mediated through changes in plant hormonal levels. Current theory holds that sex expression in cucumber is regulated by a balance between ethylene, auxins, absissic acid (ABA) and gibberellins (GA; Roy and Saran, 1990; Galun, 1959). While ethylene is considered the primary hormone affecting femaleness (Byers et al., 1972), gibberellins regulate male sex expression (Atsmon et al., 1968; Rudich et al., 1972a and b). Ethylene mediates primordial changes to determine gynoecy where the enzyme ACC (1-aminocyclopropane-1-carboxylic acid) synthase plays a critical regulatory role. Trebitsh et al. (1997)

isolated and mapped a partial sequence of the gene *CsACS1*, which co-segregates with *F* in cucumber. Another ACC synthase (*CSACS2*) gene was described by Kamachi et al. (1997; 2000), and subsequently the gene for femaleness (dominant *F* allele) in cucumber was characterized and isolated (Mibus and Tatlioglu, 2004). Sequencing of gene regions and assessment of their function will likely further elucidate the genetics of flower development including sex formation (Przybecki et al., 2004; Yamasaki et al., 2003).

Breeding For Improved Gynoecy. Associations between the number of female flowers per plant (sex expression) and fruit per plant (yield) have been identified in several studies. Selection for gynoecy has been successful in segregating progeny derived from European glasshouse by Chinese cultivar matings (Fang et al., 1995). In four U.S. slicing cucumber populations over several cycles of selection, Cramer and Wehner (1998a) found that the number of female flowers was positively correlated with yield in some population-season combinations. Highly significant, positive correlations (r) between percent pistillate nodes and yield were also identified in one of four pickling populations, with moderate, positive correlations in another (Cramer and Wehner, 2000b), suggesting sex expression has potential for increasing yield through indirect selection. In the other two populations, however, slight negative correlations between the two traits were identified. While Serquen et al. (1997b) found a slight negative phenotypic correlation ($r = -0.27$) between sex expression and the number of fruits per plant, Fazio (2001) found a positive correlation ($r = 0.24$) with the number of females nodes on lateral branches and total fruit per plant. Using similar germplasm, Fan et al. (2006) identified a positive correlation ($r = 0.40$) between gynoecy and fruit number. These data suggest that the association between yield and sex expression varies between populations and growing environments.

The most noticeable effect of sex expression on yield is not in total yield over multiple harvests, but on early yield. Gynoecious × gynoecious and gynoecious × monoecious hybrids produce significantly higher yields in the first harvest than monoecious × monoecious hybrids, but there is typically no significant difference among such hybrids for total yield over multiple harvests (Wehner and Miller, 1985). Because of their early, concentrated fruit set, gynoecious hybrids were instrumental in establishing a system for once-over mechanical harvesting of processing cucumber (Lower and Edwards 1986; Wehner 1989). Now almost all once-over mechanical harvest operations use exclusively gynoecious hybrids (Staub and Bacher, 1997).

Earliness

Earliness and stable gynoecious sex expression are important components of yield in processing cucumber, especially in once-over machine harvest operations. The introduction of early, gynoecious lines possessing a uniform, concentrated fruit set made once-over machine harvest systems economically practical (Lower and Edwards, 1986; Wehner, 1989). Earliness is often measured as days to anthesis or days to first harvest. Days to anthesis was found to be negatively correlated ($r = -0.23$) with the number of fruit per plant (i.e., fewer days to anthesis correlates to more fruit per plant; Serquen et al., 1997b). Fazio (2001) found a comparable result in a similar population in 2000 ($r = -0.31$), but these two characteristics were not significantly correlated in 1999. Additionally, a significant, positive correlation ($r = 0.26$) was identified between days to first harvest and number of fruit per plant in 1999. Interestingly, days to first harvest and days to anthesis were not correlated.

Multiple Lateral Branching

Evidence from several studies indicates that selection for multiple lateral branching (MLB) types can increase cucumber yield (i.e., fruit per plant). Number of lateral branches was found to be positively correlated ($r = 0.58$ to 0.42) with the number of fruit per plant in a processing cucumber population in two locations over two years (Fazio, 2001). Likewise, significant, positive correlations between yield and MLB were also detected in several diverse populations (Fredrick and Staub, 1989; Cramer and Wehner, 1998a; Cramer and Wehner, 1999a; Cramer and Wehner, 2000a).

Path analysis was employed in eight processing and slicing cucumber populations to determine the magnitude of correlations of yield component traits with each other as well as with yield (Cramer and Wehner, 2000b). Of the yield components tested (branches per plant, nodes per branch, pistillate nodes, and fruit set), only branches per plant were consistently correlated ($r > 0.7$) with yield (i.e., over populations, cycles of selection, and environments). Furthermore, the correlation between MLB and yield increased (from $r = 0.67$ to 0.82) with continued selection for yield (i.e., from early to later cycles). From their analyses, Cramer and Wehner (2000b) suggested that efforts to improve yield in cucumber should focus on increasing MLB. Multiple lateral branching is, in fact, quantitatively inherited (at least four genes; Wehner et al., 1989; Serquen et al., 1997a; Fazio et al., 2003a) with mostly additive genetic variance and a narrow sense heritability (h^2) of 0.00 to 0.61 depending on the population exploited, making it a candidate for use in plant improvement.

Fruit Size

Processing cucumbers in the U.S. are graded based on their size, with the smaller fruit usually bringing a higher price (Lower and Edwards, 1986; Tatlioglu, 1993). Thus, fruit length:diameter (L:D) is considered a yield component, since it determines marketable yield. For example, U.S. processing cucumbers must have an L:D of 2.9 to 3.3 to be commercially acceptable (Staub and Bacher, 1997). Although important for marketable yield, L:D is generally associated with lower fruit number per plant ($r = -0.98$, Serquen et al., 1997a; $r = -0.27$ to -0.36, Fazio, 2001).

Parthenocarpy

Parthenocarpy (seedless fruit) is an economically important yield- and quality-related trait in cucumber. Parthenocarpy is regulated by endogenous plant growth regulators (e.g., diffusible auxin, IAA), and their balance is dramatically influenced by environment (Kim et al., 1994). Phenotypic selection has, however, resulted in the development of parthenocarpic hybrids (More and Budgujar, 2002) and genetic stocks (Sztangret et al., 2004).

It is clear that parthenocarpy is genetically controlled, but there is little agreement regarding the number and type of gene action involved. Hawthorn and Wellington (1930) and Meshcherov and Juldasheva (1974) suggested that parthenocarpy is recessive and controlled by a single gene. Kvasnikov et al. (1970), however, proposed that many incompletely recessive genes control parthenocarpy. Pike and Peterson (1969) simultaneously proposed that a single dominate gene expressing incomplete dominance controls parthenocarpy in cucumber.

Results of de Ponti and Garretsen (1976) and El-Shawaf and Baker (1981a and b) indicate that parthenocarpy may be quantitatively inherited in this species. In fact, studies by Sun et al. (2006a and b) indicate that the genetics of parthenocarpy are complex. Generation means analyses in cross-progeny derived from elite processing cucumber lines indicated gene action generally could not be adequately explained by a simple additive-dominance model. Moreover, the analysis of F_3 families indicated that more than five genes control parthenocarpy, and that growing environment and epistatic interactions dramatically influence trait expression.

Fruit Quality

External Quality

External fruit quality differs for various market types (Lower and Edwards, 1986). For European glasshouse types, uniform green fruit must be fine-spined and possess a relatively high L:D (> 4), while dark green Asian greenhouse types tend to bear comparatively more warts. In contrast, medium green processing cucumbers in the U.S. possess a shorter L:D, and are typically blocky in shape. Such differences necessitate distinct breeding objectives, even though many external fruit quality characteristics are simply inherited (1-3 genes; Pierce and Wehner, 1990).

Internal Quality

The requirements for internal fruit quality differ dramatically between fresh market and processing types. For fresh market, breeding for traits such as keeping quality (e.g., no shrinkage) and internal taste (e.g., non-bitter) and colour characteristics are important (Wehner, 1996). Processing practices must be considered when breeding for improved pickling fruit quality (e.g., fruit storage characteristics; Wehner et al., 2000b; Shetty and Wehner, 2002). The U.S. cucumber processing industry produces a wide variety of products using three main processing methods: brine (fermented), fresh-pack (pasteurized), and cold-pack (refrigerated; Lower and Edwards, 1986; Miller and Wehner, 1989; Staub and Bacher, 1997). Brining, in general, involves preserving harvested cucumbers in a high salt solution (5-16% sodium chloride), which is allowed to ferment for several weeks. Breeding requires close scrutiny and testing for traits related to postharvest mesocarp disorders (Serce and Staub, 1999) and processing quality (seed cavity disorders; Staub and Bacher, 1997). In processing cucumber, particular attention is paid to the evaluation of seed cavity size and maturation, and fruit anomalies such as placental hollow and carpel separation (Lower and Edwards, 1986). Even though genetics of these quantitatively traits are not well documented and trait expression is dramatically affected by environment, recurrent selection has successfully improved fruit quality in U.S. processing cucumber (Wehner et al., 1996).

Disease and Insect Resistance Traits

The genetic control for resistance to scab (*Ccu*), downy mildew (*dm*), bacterial wilt (*Bw*), angular leaf spot (*psl*), anthracnose (*Ar*, *cla*), target leaf spot (*Cca*), Corynespora leaf spot, and Fusarium (*Foc*) is conditioned by few genes (Robinson et al., 1976; Pierce and Wehner, 1990). Seedling cotyledon tests have been developed to screen for resistance to the pathogens of each of these diseases allowing for the release of a wide array of resistant cucumber market types (Lower and Edwards, 1986). Seedling screening procedures are amendable to simple backcrossing and selfing strategies for line development.

In contrast, the genetics of resistance to viruses, such as CMV (*Cmv*), WMV (*Wmv*), Potyvirus, and ZYMV (*zymv*), or to powdery mildew (*pm-1*, *-2*, *-3*, *pm-h*), green mottle mosaic virus (GMMV), gummy stem blight, belly rot, cottony leak (causal agent: *Pythium* spp.), phytophthora rot (causal agent: *Phytophthora capsici* Leo.), and gray mold [causal agent: *Rhizopus stolonifer* (Ehrenb.: Fr) Vuill] is complex. Resistance to these diseases is quantitatively inherited and/or influenced dramatically by other pathogens (i.e., virus interactions) and by growing environment.

Breeding for resistance to these diseases requires exacting test protocols and extensive replicated testing (field and greenhouse) in multiple environments employing artificial and/or natural inoculation (Wehner and Shetty, 2000; St. Amand and Wehner, 1995; Uchneat and Wehner, 1998; Zijlstra et al., 1995). Usually accessions are screened for resistance, populations are developed through recurrent selection procedures, and then lines are extracted by backcrossing with subsequent selfing (St. Amand and Wehner, 2001a and b; Wehner et al., 2004). Lines and hybrids are then rigorously tested to determine their suitability for release (Wehner et al, 1996).

There is little genetic resistance to insect pests in cucumber (Dhillon and Wehner, 1991; Walters et al., 1993). One notable exception is resistance to root-knot nematode [causal agent: *Meloidogyne javanica* (Treub) Chitwood] for which resistance was found in *C. s.* var. *hardwickii* (PI 183967; Walters et al., 1991). In this case, resistance was conditioned by a single recessive gene (*mj*). Introgression of this gene required the development of screening protocols (Walters et al., 1992) and rigorous greenhouse selection in replicated "split-pot" tests (i.e., evaluating resistance for different nematode species) with subsequent field evaluation (Walters et al., 1999). Introgression breeding resulted in the release of resistant populations (Walters et al., 1996) and lines (Walters et al., 1997) of major importance to Southern U.S. growing regions.

Stress Resistance

Abiotic stresses (e.g., temperature extremes, water deficiencies) often depress yield, increase plant susceptibility to disease, and reduce fruit quality (Staub, 1996; Staub and Wehner, 1996). There is stress tolerance variability in cucumber germplasm (Chung et al., 2003; Smeets and Wehner, 1997; Staub and Krasowska, 1990; Staub et al., 1991; Walters and Wehner, 1994), and breeding has allowed gain from selection to produce germplasm with improved tolerance for some stresses (Staub et al., 1988; Staub et al., 1991). However, the genetics of stress resistance is largely not understood, and is, in most cases, likely complex and substantially influenced by growing and/or postharvest storage environment. For instance, the phenotypic expression of "pillowy," a fruit disorder caused by water deficiency, is directly influenced by fruit calcium concentration and is affected by temperature and relative humidity (Thomas and Staub, 1992; Staub and Navazio, 1993). Although the intensity of pillowy can be mitigated by appropriate postharvest handling (Navazio and Staub, 1994), the effect is genotype dependent (Serce and Staub, 1999). Large environmental effects make breeding for improved stress resistance expensive and laborious.

6 Breeding Methods and Techniques

Breeding objectives are determined by the requirements associated with cucumber market classes [e.g., U.S. processing (pickling), U.S. fresh market, European glasshouse, Mediterranean, Asian glasshouse]. Cucumber development for greenhouse and field growing environments involves specific cultural (e.g., chemically induced sex conversion) and market considerations (e.g., fruit type) that have been critically reviewed elsewhere (Lower and Edwards, 1986; Tatlioglu, 1993; Staub and Bacher, 1997). Only in rare cases are traits not inherited as Mendelian factors (Chung et al., 2003; Havey, 1997). Breeding plans are driven by historically proven procedures and emerging technologies. Often, genotypes with unique plant architecture (e.g., determinate, multiple lateral branching types) must be evaluated to determine cultural conditions to optimize their performance prior to their release (Staub et al., 1992b; Schultheis et al., 1998).

Breeding Plan

Program objectives (i.e., market type) determine the choice of parental types (plant introductions, accessions, cultivars, and breeding lines) that are selected based on the traits they possess. Typically, breeding follows a series of steps that consist of population development and improvement, line extraction, and hybrid evaluation.

Several breeding methods are usually employed in parallel to accomplish multiple objectives. That is, one program segment might use recurrent selection to develop a base population that possesses general adaptation, early yield, and appropriate fruit type. Pedigree selection might be used when crossing two parents to develop inbred lines with high, early yield borne on a unique plant habit (i.e., determinate) found in one parent, and high quality fruit (i.e., brine quality) along with other unique characteristics (i.e., high carotenes, disease resistance) that are typical of the other parent. A third program segment might use backcross breeding to make a disease resistant version of a parthenocarpic hybrid with top performance. Nevertheless, strategies that incorporate selection for disease resistance and improved yield require judicious implementation since selection for disease resistance can be negatively correlated with yield (Staub and Grumet, 1993). As molecular marker technologies become more efficient, effective, and affordable, they will be increasingly used to augment and enhance conventional phenotypic selection during population development and/or inbred line development.

Population Development

Recurrent Selection. Although cucumber is a cross-pollinated crop, population improvement methods that are popular in other cross-pollinated crops have not been frequently utilized. This is primarily due to the species' large plant size, and its low rate of natural outcrossing. In addition, the relatively few existing breeding programs (e.g., currently two public and three private breeders in the U.S.) often cannot bear the expense (i.e., additional years) of population development for quantitative traits during cultivar improvement.

The most effective method for the improvement of quantitative traits, such as yield in cucumber, may be recurrent selection. However, the initial populations must possess the necessary genetic diversity for selection (e.g., flesh colour, fruit size, and disease resistance; Wehner and Cramer, 1996). Due to the inherent characteristics of cucumber (i.e., large plant size and five-month generation time), recurrent selection methods (i.e., mass, full- and half-sib) are inherently limited to a few generations (2-3) per year (Wehner, 1989).

Intercrossing two to four superior, unrelated hybrids can create elite populations. Wide-based populations are created by manual intercrossing 20 or more elite cultivars for two or more generations, and then using bees for intermating in an isolation block for two or more generations before applying mild selection pressure for important quantitative traits such as yield and internal fruit quality. Simple recurrent selection can be utilized for selection among single-plant hills for a set of highly heritable traits. In contrast, reciprocal recurrent selection permits simultaneous improvement of two populations for traits with low heritability such as yield-associated combining ability (Cramer and Wehner, 1998a and c; Cramer and Wehner, 1999b). This is an expensive procedure, but produces two populations that are useful for male and female line development during elite hybrid construction.

Population development requires the identification of methods for yield testing that are efficient for large-scale yield trials (Wehner, 1989). Traditionally, recurrent selection procedures evaluate at least 200 individuals (or progenies of individuals) per population where 20 are intercrossed to create the next cycle of selection. Once a unique population is developed, the population can then be released and/or line extraction can proceed for hybrid evaluation and production (e.g., Wehner, 1998a and b).

Line Extraction

Pedigree Breeding. Selection based on pedigree is the most common cucumber breeding method. To initiate pedigree breeding, two or more adapted parents are chosen which complement each other in their traits. For instance, where the objective is to produce new lines with high yield, early maturity, high fruit quality, and good disease resistance, one parent might be generally acceptable (yield, earliness, fruit quality) except for disease resistance, and the other might be generally good (disease resistance, yield, earliness) except for fruit quality. Crossing the two parents results in a hybrid (F_1), which is then self- or sib-pollinated to produce a segregating (F_2) population, and subsequent selection for highly heritable traits produces the F_3 generation. If multiple progeny are tested from each selected F_2 plant (e.g., selection for anthracnose races 1 and 2), the best plants are typically chosen from each of the best F_3 families and are then used to produce the F_4 generation.

Beginning at the F_4 (or S_4) generation, selection emphasizes family-row performance for quantitative traits, and superior plants within family-rows are selected for the next generation. The F_6 (or S_6) are relatively uniform, and can then be handled as inbred lines. Selection typically involves the use of eight-plant plots for traits such as early flowering, number of pistillate flowers, and fruit number and quality. The number of plants or families selected typically in a cross might decrease from 54 F_2 plants to 36 F_3 families, 24 F_4 families, and then 18 F_5 lines during the selfing process.

Single-seed Descent. Single-seed-descent, a modification of pedigree breeding, is utilized to rapidly develop inbred lines by self-pollination in greenhouses and winter nurseries without selection until later generations (e.g., S_3 to S_6). This method can be employed to improve quantitative traits such as yield and earliness, rather than qualitative traits such as disease resistance. Selection for many qualitative traits (e.g., spine colour) can be performed in early generations (e.g., F_2, F_3/ S_3) by eliminating plants or families with unsuitable trait values.

Backcross Breeding. Backcross (BC) breeding is used to transfer one qualitative (highly-heritable) trait [e.g., determinate character (*de*), downy mildew resistance, (*dm*), nematode resistance (*mj*)] into an otherwise superior inbred, which is referred to as the recurrent parent. Often, six generations of selection and backcrossing to the recurrent parent are required to recover the desired genotype (recurrent parent with the additional trait) and eliminate the undesirable traits inherited from the non-recurrent (donor) parent.

Two versions of the backcross method are utilized depending on whether the gene of interest is recessive or dominant. For the transfer of a trait controlled by a recessive gene, the recurrent parent is crossed with the donor parent, and the F_1 is backcrossed to the recurrent parent. In one scheme, the F_1 is self-pollinated to produce the F_2, which segregates for the trait of interest. Individuals from the F_2 that possess the trait are backcrossed to the recurrent parent to produce the BC_1. The BC_1 generation is then self-pollinated to produce the BC_1S_1, which is evaluated for the trait. Individuals possessing the trait of interest are selected and backcrossed to the recurrent parent. This process is repeated until the BC_6 generation where the best individuals are self-pollinated and selected for the trait to produce the improved inbred.

For the transfer of a trait controlled by a dominant gene [e.g., anthracnose (*Ar*), bacterial wilt (*Bw*), or target leaf spot (*Cca*) resistance], the recurrent parent is crossed with the donor parent, and the F_1 is subsequently backcrossed to the recurrent parent. The BC_1 generation is then evaluated, and individuals possessing the trait are backcrossed to the recurrent parent. This process is repeated until the BC_6 generation where the best individuals are self-pollinated and selected for homozygous expression of the trait using progeny testing.

Hybrid Testing

Once developed, inbreds can be crossed in all possible combinations and evaluated to identify superior hybrid combinations. Hybrids are usually made as crosses between gynoecious and monoecious lines, or two monoecious inbred lines. In cases where many inbreds have been identified as potential parents, it may be necessary to limit the scope of the trialling [e.g., 20 inbreds could produce $(20 \times 19)/2 = 190$ different hybrids, without including reciprocal crosses]. Thus, hybrids for evaluation are usually made from pairs of inbreds having complementary traits. Consideration of potential combining ability is given when choosing lines for hybrid production.

Testing of experimental hybrids often progresses in stages, with fewer hybrids to test in later stages where more effort is spent on the evaluation of each hybrid. In the first trialling year, two replications are recommended in each of two locations. In the second year, the best hybrids should be evaluated under replication (2-4) in 8 to 12 diverse locations (i.e., grower fields, university experiment stations). In the third year, the hybrids are examined in grower trials (0.5-1.5 ha) in several production regions (~10-20). Information from the three years of trialling often leads to the release of the best one or two hybrids in the fourth year.

Even though publicly-released open-pollinated populations are often genetically broad-based and provide a source for further plant improvement, hybrids provide an avenue for proprietary protection of commercial inbred lines (Staub et al, 2005). Hybrid identification and production is, however, expensive, and thus cost/benefits are always critically assessed before initiating hybrid development.

Case Studies

Factors important to population improvement and inbred line extraction include the amount of genetic variation and gene action present, the heritability of the traits selected, and the degree of the linkage associations. These factors were considered in the development of populations and lines in the case studies given below. These studies highlight the use of the exotic *C. s.* var. *hardwickii* which possesses economically important genes not resident in *C. s.* var. *sativus*.

Architectural Habit. WI 6383 is a gynoecious, multiple disease resistant, white spined cucumber population produced by intermating elite USDA *C. s.* var. *sativus* processing lines and *C. s.* var. *hardwickii* accessions (PI 183967 and PI 215589; Staub et al., 1992c). This population was released to provide breeders with a source from which they could extract multiple disease resistant lines with a multiple lateral branching and sequential fruiting habit. The development of WI 6383 was supported by research on the inheritance of yield components in *C. s.* var. *sativus* × *C. s.* var. *hardwickii* derivatives (Kupper and Staub, 1988; Fredrick and Staub, 1989).

WI 6383 originated from a cross between four processing cucumber lines (WI 1606, WI 1589, WI 1983, and WI 1895) that also produced population WI 2843 (Peterson et al., 1985). These gynoecious, non-bitter lines are resistant to anthracnose, downy and powdery mildew, scab and Fusarium, and possess acceptable fruit quality. A selection from WI 2843 was crossed with the F_1 between PI 183967 and PI 215589 and subsequent pedigree selection produced disease resistant (via seedling screening tests), white spined, non-bitter, and gynoecious F_4 lines. About 100 F_4 plants were randomly mated to produce WI 6383, which is homogeneous for the traits selected. Seed of ~500 F_4 individuals were then subjected to three cycles (C) of recurrent, half-sib family selection for three-harvest yield. Self-pollination of selected C_3 plants led to F_7 families of which the highest yielding lines were designated WI 5098 and WI 5551. This population and attending lines are vigorous, indeterminate, produce between 4-6 primary lateral branches at the base (crown) of the plant (standard commercial types produce 1-3 primary lateral branches), and possess a sequential fruiting habit (no crown-set inhibition) not present in commercial cucumber.

Root-knot Nematode Resistance. Cultivars 'Lucia', 'Manteo', and 'Shelby' were developed with resistance to root-knot nematodes (*Meloidogyne* spp.; Walters and Wehner, 1997). Nematodes are important pests worldwide and cause about 11% crop loss annually in North Carolina (primary cucumber production state in U.S.). These elite lines possess resistance to *M. arenaria* races 1 and 2, *M. javanica*, and *M. hapla*. The development of these lines was supported by research that developed evaluation protocols (split-root technique; Walters et al., 1995) and identified the genetic control for resistance (Walters et al., 1993 and 1997).

'Lucia', 'Manteo', and 'Shelby' were developed from the NCH-1 population, which was created by intercrossing 12 cultivars, breeding lines, and plant introduction accessions with *C. s.* var. *hardwickii* PI 183967 (synom. LJ 90430). These F_1's were subjected to two cycles of bee-mediated intercrossing in open-field isolation. This resulted in a base population designated as NCH1 C_0 from which half-sib family recurrent selection (yield and fruit shape) was practiced to produce C_9. Random half-sib C_9 families were then self-pollinated and selected for nematode resistance using the split-root technique to produce indeterminate, monoecious S_7 lines from which 'Lucia', 'Manteo', and 'Shelby' were produced. Selection was initially applied for *M. javanica* (S_0-S_6), and then for *M. hapla* and races 1 and 2 of *M. arenaria* (S_6-S_7). In addition to nematode resistance, these three lines possess varying degrees of resistance to powdery mildew and anthracnose with acceptable fruit firmness, yield, and processing quality. The three cultivars differ mainly in their fruit L:D (i.e., short, medium-length or long).

7 Seed Production

Hybrid seed production is facilitated in either greenhouse or field environments (open-field or cage isolation) by hand- or insect-pollination as reviewed by Lower and Edwards (1986). Typically, breeder's seed of inbred lines is increased to produce enough seed for foundation and production seed that is then used to produce hybrids.

Breeder's and foundation seed of inbred lines is usually produced by hand-pollination under greenhouse or cage isolation. Isolation blocks or screen cages are often employed for large seed increases (inbred and hybrid). Open-field isolation blocks are separated from other cucumber fields by at least 1.5 km. Where the number of wild bees is insufficient to ensure adequate pollination, beehives are introduced into the isolation block or cage.

In the case of hybrid seed production for large-scale commercial use, open-field increases employ the strategic placement of rows such that cross-pollination can occur between lines of opposite sex types. Typically, one or two male rows are alternated with four to five female rows from which hybrid seed is harvested.

Hybrids commonly result from gynoecious × gynoecious, gynoecious × monoecious, monoecious × monoecious, and gynoecious × hermaphrodite line matings. In the case of gynoecious × gynoecious hybrids, the sex expression of one line is chemically altered by ethylene inhibitors such as silver nitrate, silver-thiosulfate, or aminoethoxyvinylglycine (Beyer, 1976; Lower and Edwards, 1986). The seed of gynoecious lines is also produced using such compounds. Likewise, staminate flowering lines (e.g., monoecious, hermaphrodite, androecious) can be converted to pistillate flowering by application of ethylene releasing compounds such as alpha-naphthalene acetic acid and ethephon (2-chloroethylphosphonic acid; Byers et al., 1972). Chemicals are usually applied at least three times, beginning at the first true-leaf stage, and then once a week thereafter to induce sex conversion.

8 Molecular Marker-Assisted Breeding

The application of genetic markers for MAS follows three major recurring cycles regardless of marker type (Figure 2). Markers are identified as potentially useful, and subsequently developed into efficient and effective genotyping systems. Polymorphic markers are then placed on a genetic map and associated with QTL through progeny analysis for their subsequent use in MAS.

8.1 Development of Molecular Markers

Marker development in cucumber has occurred in several marker systems [isozymes, restriction fragment length polymorphisms (RFLP), random amplified polymorphic DNA (RAPD), amplified fragment length polymorphisms (AFLP), simple sequence repeats (SSR), and single nucleotide polymorphisms (SNP)], where changes between marker systems have been driven by the steady progression of technological advances (Figure 3). In each case, the goal was the development of moderately saturated maps (i.e., ~150 to 200 markers to provide 90-95% coverage at 10-15 cM intervals).

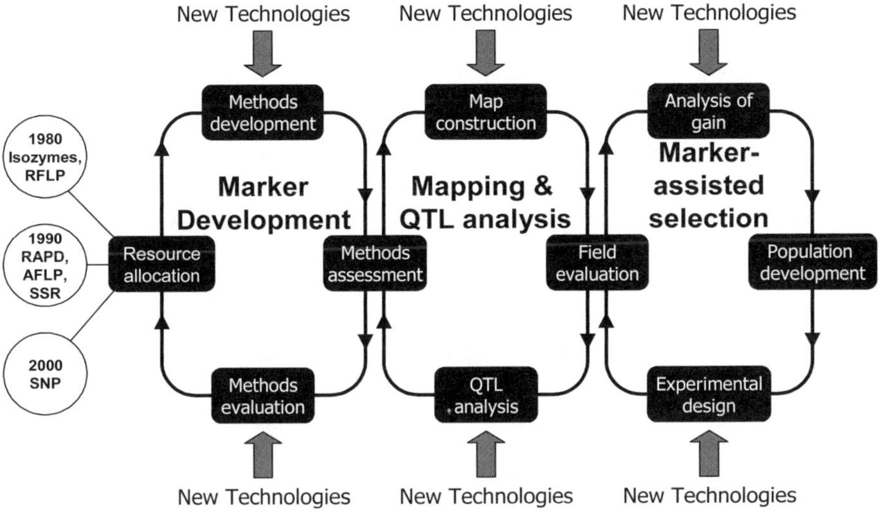

Fig. 2. Schematic of marker development and application in cucumber breeding.

During the period between 1984 and 1992, work in the U.S. progressed on the development of isozyme and RLFP markers leading to the construction of unsaturated maps (Knerr and Staub, 1992; Meglic and Staub, 1996; Kennard et al., 1994). The use of these codominant markers was assessed, and their development was terminated because of their utilization costs and the paucity of polymorphic markers when RAPD technologies were introduced (1992-2000; Figure 3). Although dominant in nature, RAPD and subsequent AFLP markers were attractive because of their comparatively low technological costs, methodological simplicity (RAPDs), and their potential to produce multiple, polymorphic markers from single assay (RAPDs and AFLPs). In the case of RAPD technology, putative polymorphism declaration (i.e., number of bands) was relatively high (10-15 bands per primer), but reproducibility and fit to 3:1 genetic ratios for many putative marker loci was also low (recovery rate = ~ 50 of 1,000 markers evaluated; Staub et al., 1996). This level of recovery is typical of many other marker systems in cucumber.

Dominant markers (RAPD and AFLP) were useful initially in the development of moderately saturated maps (Serquen et al., 1997b; Bradeen et al., 2001), but are not preferred in breeding programs. The mapped RAPD loci were, nevertheless, strategically important during early map construction (Serquen et al., 1997b; Figures 2 and 3), and were therefore subjected to conversion to more preferable sequence amplified characterized region (SCAR) markers by silver staining-mediated sequencing (Horejsi et al., 1999). Although 62 (83%) of the 75 RAPDs were successfully cloned, only 48 (64%) RAPD markers were successfully converted to

SCARs markers, and 11 (15%) of these reproduced the polymorphism observed with the original RAPD marker. The emergence of automated sequencing technologies made possible the development of codominant SSR and SNP technologies (Fazio et al., 2002 and 2003a), and the reassessment of RAPD to SCAR as well as SCAR to SNP marker conversion (Robbins, 2006). Two sources of sequence data [SCAR marker fragments and BAC library (Nam et al., 2006) clones] were employed to convert RAPD to SCAR and SNP markers for increased efficiency (multiplexing) and effectiveness (stable and codominant markers; Figure 4). A total of 39 new markers (SCAR and SNP) have recently (2006) been developed, seven of which have proven effective when multiplexed in MAS. The multiplexing potential of the remaining markers and those recently created from EST libraries (unpublished) has yet to be determined.

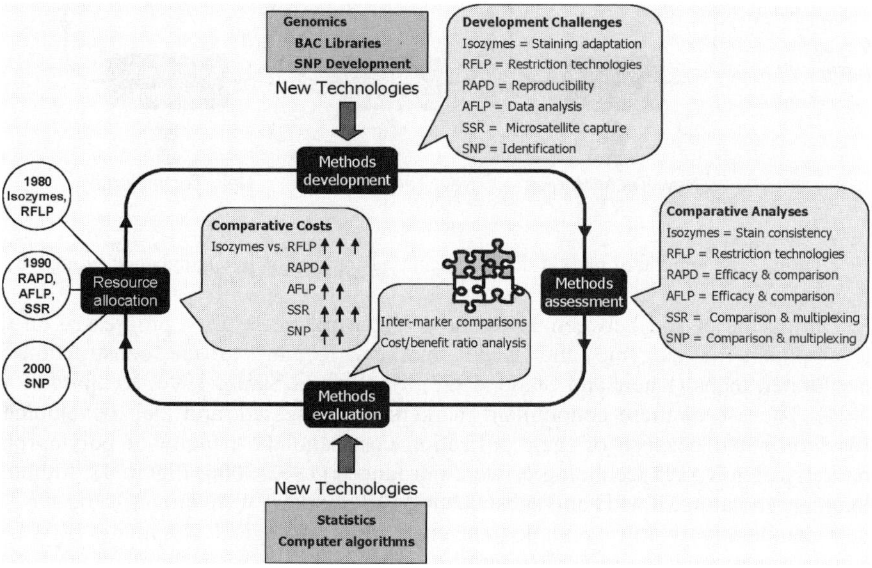

Fig. 3. Events surrounding marker development in cucumber where the puzzle icon indicates the critical element.

8.2 Development of Genetic Maps

The first genetic linkage maps in cucumber were reported almost 20 years ago and were based solely on phenotypic markers (Fanourakis and Simon, 1987; Vakalounakis, 1992; Pierce and Wehner, 1990). The first molecular markers mapped were isozymes (Knerr and Staub, 1992), which were subsequently combined with phenotypic markers for mapping purposes (Meglic and Staub 1996). As DNA-based molecular markers were developed (RFLP and RAPD), they were combined with existing marker types in linkage maps (Kennard et al., 1994) (Figure 5). More recent maps possess an array of phenotypic and DNA-based markers (RAPD, RFLP, AFLP, SCAR, SSR, and SNP; Serquen et al., 1997a; Park et al., 2000; Fazio et al., 2003a).

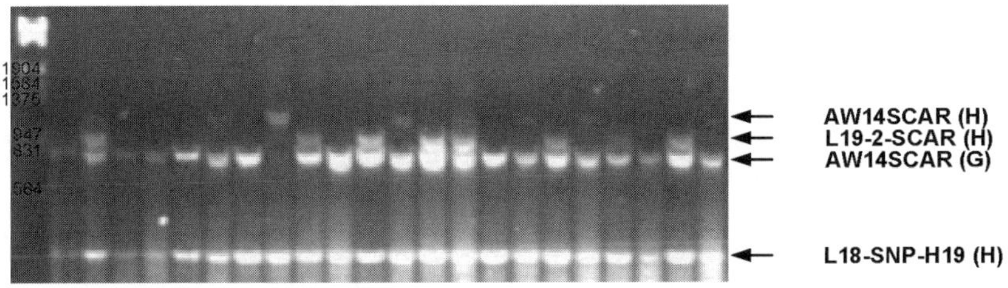

Fig. 4. Example of a multiplexing reaction in a cucumber population. The far left lane is a molecular weight marker (lambda DNA digested with *Eco*RI and *Hin*dIII) with the molecular weight of each band in base pairs. The four bands are (top to bottom) the H-19 (H) allele of AW14SCAR (a codominant marker), L19-2-SCAR (dominant H-19 marker), the Gy-7 (G) allele of AW14SCAR, and L18-SNP-H-19 (dominant H-19 marker).

As genetic maps continued to be refined and molecular markers were included, the total map distance generally expanded to approach the estimated genome size (750 to 1000 cM; Staub and Meglic, 1993). The total genetic distances of these maps spanned 166 (Fanourakis and Simon, 1987), 95 (Vakalounakis, 1992), 168 (Knerr and Staub, 1992), 766 (narrow-based), 480 (wide-based; Kennard et al., 1994), 584 (Meglic and Staub, 1996), 600 (Serquen et al., 1997b), 816 (Park et al., 2000), and 706 cM (Fazio et al., 2003a). These maps show varying degrees of colinearity (Table 2).

The map constructed by Park et al. (2000) employed 347 RAPD, RFLP, AFLP, and loci conditioning virus resistances, which were placed on 12 linkage groups with a mean marker interval of 4.2 cM. A map constructed by Serquen et al. (1997a) defined nine linkage groups with an average distance between markers of 8.4 cM (RAPD only). Information from this map was recently merged with other maps (Fanourakis and Simon, 1987; Knerr and Staub, 1992; Kennard et al., 1994; Meglic and Staub, 1996; Horejsi et al., 2000) to synthesize a consensus map containing 255 markers, including morphological traits, disease resistance loci, isozymes, RFLPs, RAPDs, and AFLPs on 10 linkage groups (Bradeen et al., 2001). The mean marker interval in this consensus map was 2.1 cM spanning a total length of 538 cM. More recently, Fazio et al. (2003a) constructed a map containing 14 SSR, 24 SCAR, 27 AFLP, 62 RAPD, one SNP, and three morphological markers (131 total markers) spanning seven linkage groups (the theoretical number based on the haploid chromosome number) using recombinant inbred lines (RIL). This map spanned 706 cM with a mean marker interval of 5.6 cM.

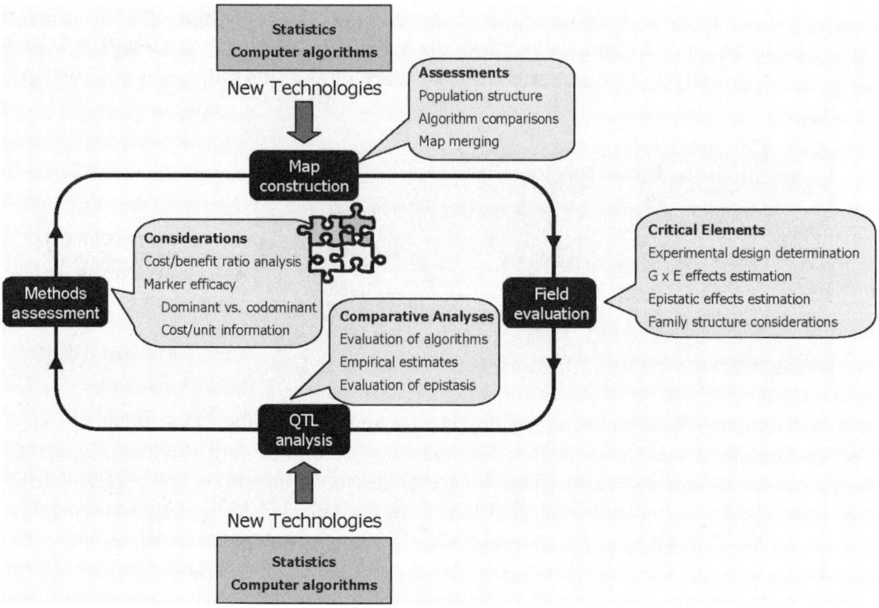

Fig. 5. Assessments during mapping and QTL analysis in cucumber where the puzzle icon indicates the critical element.

8.3 QTL Mapping

The development of genetic linkage maps has provided tools for the molecular analysis of important characteristics in cucumber including fruit quality (Wenzel et al., 1995), disease resistance, (Park et al., 2000), and yield components (Serquen et al., 1997b; Fazio et al., 2003a; Figure 5). The marker-QTL associations identified in these studies form the foundation for cucumber improvement through MAS.

Molecular mapping of economically important traits in cucumber has occurred using several inbred lines (Kennard et al., 1994; Horejsi et al., 2000; Park et al., 2000). These include lines GY-14, WI 1983, Zudm1, Straight-8, PI 183967 (*C. s.* var. *hardwickii*, India), and PI 432860 (China). These lines were chosen because of their disease resistance (e.g., downy mildew and virus resistance) or morphological (e.g., yield and quality) attributes.

Although extensive virus resistance mapping is still occurring (M. J. Havey, USDA, ARS), these maps have not been used extensively for QTL mapping (Figure 5). One notable exception involves the use of two inbred processing lines, Gy-7 (synom. G421; R.L. Lower, University of Wisconsin, Madison, Wisc.) and H-19 (synom. AR 7975; Goode et al., 1980), which have been exploited extensively as parents to create F_3 families for use in genetic analysis (Serquen et al., 1997a) and QTL mapping (Serquen et al. 1997b) of several yield components (Table 3).

Table 2. Common genetic markers across four linkage maps in cucumber [*Cucumis sativus* var. *sativus* and *C. sativus* var. *hardwickii* (R.) Alef.]

Sun, 2004 (2A x Gy8) F2 var. *sativus* x var. *sativus*	Revised Fazio et al. 2003 (Gy7 x H19) RIL var. *sativus* x var. *sativus*	Bradeen et al. 2001 Narrow-based Consensus F2/BC var. *sativus* x var. *sativus*	Bradeen et al. 2001 Broad-based Consensus F2/BC var. *sativus* x var. *hardwickii*
Linkage Group 1			
	F (LG1,0.0)		F (LGA,30.0)
CSWCT25-350 (LG1,6.5)*	CSWCT25-350 (LG1,9.4)		
	J5-SCAR (LG1,11.2)	**J5_1 (LGA,5.6)**	
	de (LG1,28.8)	de (LGA,15.6)	
	E14M62-214 (LG1,37.2)	*E14/M62-F-214--P2 (LGA,52.2)*	*E14/M62-F-214-P2 (LGA,77.9)*
	E14M62-112 (LG1,43.0)	*E14/M62-F-112-P1 (LGA,41.2)*	
E12M62-230 (LG1,56.2)	*E12M62-230 (LG1,49.0)*		
E18M48-188 (LG1-2A,64.6)	*E18M48-188 (LG1,54.7)*		
	I1B-SCAR (LG1,57.5)	**I1_1 (LGA,54.4)**	
	OP-AJ6 (LG1,59.5)	**AJ6 (LGA,52.2)**	
	E12M48-107 (LG1,62.2)	*E12/M48-F-107-P2 (LGA,53.0)*	
	BC523-SCAR (LG1,64.1)	**BC523 (LGA,52.2)**	
	OP-AD12-1 (LG1,68.4)	**AD12 (LGA,49.2)**	
	OP-W7-2 (LG1,76.2)	**W7_2 (LGA,71.8)**	
	E14M62-224 (LG1,76.9)	*E14/M62-F-224-P2 (LGA,48.2)*	
	ll (LG1,82.0)	ll (LGA,68.5)	
E18M48-303 (LG1-2A,68.5)	*E18M48-303 (LG1,84.0)*		
		BC551 (LGA,69.1)	**BC551 (LGA,92.1)**
	BC592-SCAR (LG1,100.1)	**BC592 (LGA,81.8)**	
	OP-AH14 (LG1,112.7)	**AH14 (LGA,96.2)**	
Linkage Group 2			
E18M58-101 (LG2-Gy8,0.0)	*E18M58-101 (LG2,8.9)*		
	OP-F4 (LG2,20.8)	**F4 (LGB,9.8)**	
	E11M60-114 (LG2,31.9)	*E11/M60-F-114-P1 (LGB,11.1)*	
	E11M60-125 (LG2,44.4)	*E11/M60-F-125-P1 (LGB,22.4)*	
	OP-AO7 (LG2,47.1)	**A07_1 (LGB,30.8)**	
E23M59-228 (LG2-Gy8,9.6)	*E23M59-228 (LG2,48.1)*		
	AW14-SCAR (LG3,0.9)	**AW14_1 (LGC,0.0)**	
	X15-SCAR (LG3,2.4)	**X15 (LGC,40.2)**	
	G14-SCAR (LG3,6.7)	**G14 (LGC,38.3)**	
	BC450-2 (LG3,7.8)	**BC450 (LGC,4.7)**	
	E11M60-342 (LG3,9.1)	*E11/M60-F-342-P2 (LGC,5.6)*	
		E14/M49-F-105-P1 (LGC,27.9)	*E14/M49-F-105-P2 (LGH,16.6)*
	AA9B-SCAR (LG4,18.1)	**AA9 (LGC,33.7)**	**OP_AA9 (LGH,14.7)**

	OP-H13 (LG4,24.5)	**H13 (LGC,38.3)**	
	OP-C1 (LG4,34.4)	**C1 (LGC,47.2)**	
	AJ18-SCAR (LG4,41.2)	**AJ18 (LGC,55.1)**	
		E14/M51-F-344-P1 (LGC,55.1)	*E14/M51-F-344-P2 (LGH,19.2)*
	OP-Y5 (LG4,44.5)	**Y5 (LGC,55.1)**	
	Y3-SCAR (LG4,48.2)	**Y3 (LGC,55.1)**	
	BC526-SCAR (LG4,48.8)		**BC_526 (LGH,15.8)**
	OP-L18-1 (LG4,52.9)	**L18_1 (LGC,46.5)**	
Linkage Group 4			
E14M52-85 (LG4-2A,13.8)	*E14M52-85 (LG4,74.0)*		
	OP-K7 (LG4,113.4)		**OP_K7-3 (LGH,14.7)**
		dm (LGC,55.1)	dm (LGH,27.7)
E23M50-210 (LG4,0.0)	*E23M50-210 (LG4,140.6)*		
E23M50-184 (LG4-2A,95.7)	*E23M50-184 (LG4,146.1)*		
OP-R13-580 (LG4-2A,86.7)	**OP-R13-580 (LG4,154.3)**		
E18M48-226 (LG4,12.9)	*E18M48-226 (LG4,193.7)*		
	E12M48-119 (LG5,6.0)	*E12M48-119 (LGE,15.8)*	
	BC503 (LG5,11.0)	**BC503 (LGE,7.2)**	
Linkage Group 5			
E23M50-181 (LG5-2A,0.0)	*E23M50-181 (LG5,14.5)*		
		<u>**CsC558/H3 (LGF,0.0)**</u>	<u>**CsC558/H3 (LGE,3.9)**</u>
		<u>**CsC137/H3 (LGF,2.4)**</u>	<u>**CsC137/H3 (LGE,5.6)**</u>
Linkage Group 6			
E26M54-345 (LG6,7.1)	*E26M54-345 (LG6,17.6)*		
	N6-A-SCAR (LG6,26.3)	**N6_2 (LGF,4.4)**	
	E11M60-332 (LG6,29.3)	*E11/M60-F-332-P2 (LGF,8.4)*	
	AK5-SCAR (LG6,33.5)	**AK5 (LGF,9.3)**	**OP_AK5-1 (LGE,13.0)**
E18M58-227 (LG6-Gy8,7.1)	*E18M58-227 (LG6,57.5)*		
		<u>**CsC362/E1 (LGF,19.2)**</u>	<u>**CsC362/E1 (LGE,23.0)**</u>
		<u>**CsP441/E1 (LGF,20.5)**</u>	<u>**CsP441/E1 (LGE,23.8)**</u>
		<u>**CsP280/H3 (LGF,22.1)**</u>	<u>**CsP280/H3 (LGE,25.4)**</u>
		BC_523 (LGF,28.1)	**BC_523 (LGE,30.6)**
		AP13 (LGF,32.4)	**AP13 (LGE,36.6)**
	BC605 (LG6,74.8)	**BC605 (LGG,0.0)**	
E18M17-227 (LG6-Gy8,11.1)	*E18M17-227 (LG6,85.1)*		
	E11M50-558 (LG6,91.7)	*E11M50-558 (LGG,15.7)*	
Linkage Group 7			
E13M50-277 (LG7,20.0)	*E13M50-277 (LG7,7.8)*		
	BC515 (LG7,15.8)	**BC515 (LGH,0.0)**	
		<u>**CsP308/E1 (LGH,4.1)**</u>	<u>**CsP308/E1 (LGI,5.1)**</u>

E25M60-545 (LG7-2A,7.5)	*E25M60-545 (LG7,21.7)*		
	L19-1-SCAR (LG7,27.0)	**L19_1 (LGH,11.9)**	
	OP-AT15-3 (LG7,28.2)	**AT15 (LGH,9.9)**	
	BC388-SCAR (LG7,28.4)	**BC388 (LGH,11.3)**	**BC388 (LGI,13.8)**
	BC231 (LG7,29.2)	**BC231 (LGH,11.9)**	
E23M49-237 (LG7,34.3)	*E23M49-237 (LG7,37.0)*		
E18M58-394 (LG7,68.4)	*E18M58-394 (LG7,56.3)*		
		CsP105/E1 (LGH,13.8)	CsP105/E1 (LGI,16.4)
		H5_4 (LGH,13.8)	**H5_4 (LGI,11.4)**
		CsC166/E1 (LGH,23.2)	CSC166/E1 (LGI,25.0)

Underline = single sequence repeat, italic = amplified fragment length polymorphism, bold = random amplified polymorphic DNA or sequence characterized region, and bold and underline = restriction fragment length polymorphism. Parenthesis indicates linkage group and position.

The traits mapped included multiple lateral branching, gynoecious sex expression, L:D, and earliness, and were further characterized by QTL analysis using RIL derived from the same parental lines (Fazio et al., 2003a). Subsequently, derivatives of these and other lines were used successfully in introgression of yield components by backcrossing using MAS (Fazio et al., 2003b; Fan et al., 2006), and are, therefore, employed herein for demonstration of specific accomplishments (Figure 6).

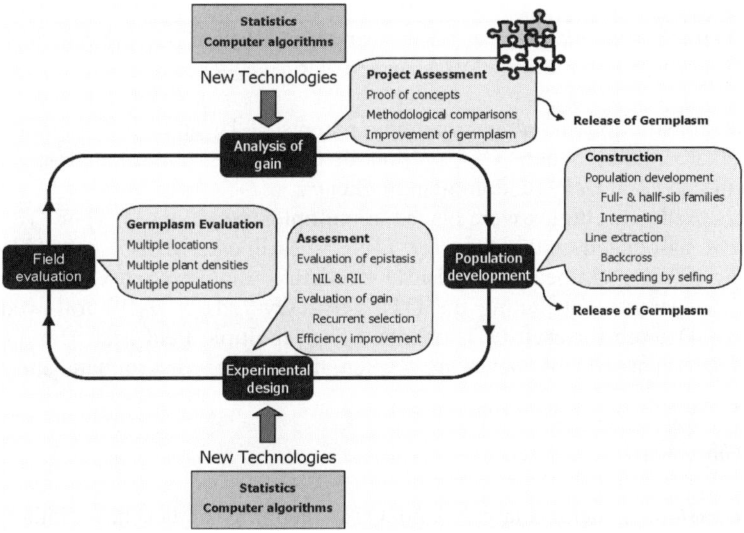

Fig. 6. The evaluation of marker-assisted selection in cucumber where the puzzle icon indicates the critical element.

Table 3. Characteristics of molecular markers defined in a genetic map of cucumber constructed by Fazio et al. (2003b) and used in marker-assisted selection for population improvement.

Marker	Type[a]	Linkage group	Map position (cM)	Parent[b]	Multiplex group[c]	Ideo-type	QTL (mapping parent and LOD score) and *gene* associations[d]
CSWCT28	SSR	1	5.0	G&H		G&H	EAR(G, 7.1), MLB(H, 10.4), GYN(G, 13.0), L:D(H, 5.7), *F*
L18-SNP-H19	SNP	1	7.4	H	1	H	EAR(G, 7.1), MLB(H, 10.4), GYN(G, 13.0), L:D(H, 5.7)
OP-AG1-1	RAPD	1	31.8	G		H	EAR(G, 6.4), MLB(H, 11.6), GYN(G, 7.3), *de*
AJ6SCAR	SCAR	1	61.4	G	3	H	MLB(H, 3.3)
BC523SCAR	SCAR	1	66.5	G	2	H	MLB(H, 3.3)
OP-AD12-1	RAPD	1	70.2	H		G	EAR(G, 4.1), MLB(H, 32.9), GYN(G, 3.7), L:D(G, 8.6), *ll*
AW14SCAR	SCAR	3	3.9	G&H	1	G	GYN(G, 5.1)
CSWTAAA01	SSR	4	34.1	G&H	2	H	MLB(H, 4.6)
OP-AI4	RAPD	5	101.0	G		G	GYN(G, 3.0)
OP-AO12	RAPD	5	117.3	G		G	GYN(G, 3.0)
OP-AI10	RAPD	6	22.5	H		G	L:D(G, 7.3)
AK5SCAR	SCAR	6	33.0	G	2	H	MLB(H, 3.0)
M8SCAR	SCAR	6	39.1	H		H	MLB(H, 3.0)
OP-W7-1	RAPD	6	83.4	H		G	GYN(G, 4.1)
L19-2-SCAR	SCAR	6	115.0	H	1	G	MLB(G, 4.2), GYN(G, 4.1)
NR60	SSR	6	137.4	G&H		G	MLB(G, 4.2)
BC515	RAPD	7	0.0	H		H	L:D(H, 4.2)
L19-1-SCAR	SCAR	7	9.9	H	3	H	L:D(H, 4.2)

[a] SSR = simple sequence repeat, SNP = single nucleotide polymorphism, RAPD = random amplified polymorphic DNA, and SCAR = sequence characterized amplified region.
[b] Allelic constitution based on mapping parents H-19 and Gy-7 (synom. G421) (Fazio et al. 2003b), where G = present in Gy-7, H = present in H-19, G&H = present in Gy-7 and H-19 (codominant marker).
[c] Markers used in multiplex were placed in multiplexing groups (1, 2, or 3).
[d] Markers associated with QTL for DTF = earliness, MLB = multiple lateral branching, GYN = gynoecious, and L:D = length to diameter ratio. The parentheses contain the parent contributing the QTL (G = Gy-7, H = H-19) followed by the highest LOD score for each QTL obtained from multiple field trials (Serquen et al. 1997a; Fazio et al. 2003b). Genes are *F* = femaleness, *de* = determinate, and *ll* = little leaf.

Yield Components

Sex expression. Genetic analyses and QTL mapping studies have indicated that several loci are involved in sex expression. In a population fixed for the *M* and *A* genes (i.e., segregating only at the *F* locus), Serquen et al. (1997a) estimated five effective factors involved in gynoecious sex expression in each of two locations.

Most of the gene action was attributed to dominance variance, with approximately a 1:3 ratio of additive to dominance variance. The narrow sense heritability (h^2) was estimated at 0.14 and 0.16 in two distinct environments, suggesting selection for sex expression would be difficult. In the same population, Serquen et al. (1997b) identified four QTL for sex expression common across two environments, plus a fifth QTL unique to one environment. These QTL accounted for over 85% of the observed variation in each environment with 67% and 74% of the variation attributed to a QTL near the *F* locus. In a QTL study of a RIL population derived from the same parents, three QTL were detected for the number of female nodes on the mainstem, accounting for 31% of the variation, 16% of which was attributed to a QTL at the *F* locus (Fazio et al., 2003a). Two of these QTL, including the one at the *F* locus, showed significant effects on the number female nodes on primary lateral branches. Although a large portion of the genetics of sex expression is controlled by the *F* locus, it is clear there are other regions of the genome involved in the expression of gynoecy.

Earliness. A QTL analysis of days to anthesis revealed a single QTL explaining 13% of the variation common in two environments, and a second QTL of smaller magnitude (R^2 = 8.1) in another environment (Serquen et al., 1997b). Fazio et al. (2003a) identified four QTL for days to anthesis, two of which were common in two environments tested. These two QTL accounted for 12% to 15% of the variation observed, with the environment specific QTL explaining an additional 4% and 15% of the variation. Fazio et al. (2003a) also identified four QTL in a single environment for days to first harvest. These QTL accounted for 21% of the variation observed, one of which mapped to the same genomic region as a QTL for days to anthesis common to two environments. Although a few earliness QTL (1-2) were identified in these studies, others likely remain undetected.

Multiple lateral branching. Four QTL affecting MLB have been identified by F_3 family analysis that explained 48% to 66% of the observed variation depending upon environment (Serquen et al., 1997b). Although a total of 13 QTL for MLB were subsequently identified by Fazio et al. (2003a) using RIL derived from the same parents, only five were detected in at least two locations with a combined R^2 of 37% to 55% depending on location. In both QTL studies, one major QTL was detected that accounted for 32% (Fazio et al., 2003a) to 40% (Serquen et al., 1997b) of the variation, which mapped near the little leaf locus (*ll*).

The number of lateral branches can be relatively stable across growing environments (Georgia and Wisconsin; Serquen et al., 1997b) and planting dates (early and late; Fredrick and Staub 1989). For instance, four QTL were found to be stable in diverse U.S. growing environments (Wisconsin in 1999 and 2000 and Utah in 1999; Fazio et al., 2003a). However, Fazio et al. (2003a) identified a QTL specific to Wisconsin (LOD 2.7-3.0 in both years), and seven other QTL (LOD 2.8-6.1) unique to a single environment. This result, coupled with the trait's moderate heritability and additive gene action (Serquen et al., 1997a), indicates that some QTL are affected by the environment [i.e., seasons (López-Sesé and Staub, 2002) and plant density (Staub et al., 1992b)]. Indeed, MLB has varied in *C. s.* var. *hardwickii* derived genotypes across years in another study in Wisconsin (Fredrick and Staub, 1989).

Fruit size. As with earliness, QTL analysis of fruit L:D suggests that a few stable QTL are involved with environmental factors playing a role in trait expression. In the QTL analysis of Serquen et al. (1997a), fruit length and fruit diameter were analyzed separately as well as L:D. One QTL was identified for fruit length in both environments tested (R^2 = 21% and 31%) and three QTL were identified for fruit diameter, one in both environments (R^2 = 15.7% and 9.6%) and one unique to each environment (R^2 = 21.9% and 9.6%). Two QTL were identified for L:D, but only in one environment (R^2 = 13.7% and 14.4%), both of which mapped to the same genomic regions as two QTL for fruit diameter, including the QTL identified in both environments. Although a total of 12 QTL for L:D were declared significant by Fazio et al. (2003b), only five were identified in both test locations with a combined R^2 of 31% and 30%. The total R^2 from all QTL was 36% and 57% in the two test environments. As with MLB number, L:D is effected by growing location (Serquen et al. 1997a and b; Fazio et al., 2003a and b) and plant density (Dijkhuizen and Staub, 2003). Efforts to isolate the specific genes regulating fruit growth in cucumber have resulted in the cloning of cDNAs for preferentially expressed genes (Suyama et al., 1999).

Disease Resistance

Horesji et al. (2000) identified RAPD markers linked to the downy mildew resistance gene (*dm*). Two F_3 family populations (WI 1983G × Straight 8 population and Zudm1 × Straight 8 population) were evaluated over five locations in North America and Europe to identify RAPD markers linked to *dm*. Five markers were identified 15 to 33 cM away from *dm*, which was subsequently mapped (0.1 and 1.9 cM away) and cosegrated with ten other markers (Bradeen et al., 2001). A scab resistance gene (*Ccu*) was also mapped by Bradeen et al. (2001). Park et al. (2000) found that resistances to papaya ringspot virus (PRV) and ZYMV were closely linked to each other (2.2 cM), and were also tightly linked (~5.2 cM) to three AFLP markers. Given their relatively closely linkage associations with resistance genes, markers from these studies will likely be exploitable in MAS.

8.4 Fruit Quality

Wenzel et al. (1995) used a wide cross [GY-14 (U.S. elite processing) × PI 432860 (China)] to identify QTL associated with fruit quality. The two-year, single-location study identified five, three, three, and two QTL for fruit length, diameter, seed-cavity size, and colour, respectively.

The fruits of parthenocarpic genotypes are typically of higher quality than their seeded counterparts. There have been 10 QTL detected for parthenocarpy in a narrow cross (2A × Gy8; Sun et al., 2006c), three of which map to the same genomic regions as QTL detected for fruit yield at first-harvest by Fazio et al. (2003a). Four of 10 QTL reside on Linkage Groups (LG) 1 and 4.

8.5 Use of Molecular Markers in Breeding

The pyramiding of simply inherited genes (e.g., disease resistance) during germplasm enhancement is common, and has proven useful in the improvement of many crop species. In cucumber, the pyramiding of disease resistance genes resulted in important inbred lines and populations [e.g., In the U.S. WI 2757 (Peterson et al., 1982), WI 1983 (Peterson et al., 1986a), WI 5207 (Peterson et al., 1986b), M-17 (Wehner et al., 1996), 'Lucia', 'Manteo', and 'Shelby' (Walters and Wehner, 1997), NCWBP, NCMBP, and NCEP1 (Wehner and Shetty, 1997), NCWBS, NCMBS, and NCES1 (Wehner, 1998a)]. Less well reported and understood are genetic approaches for the incorporation of quantitatively inherited traits. Molecular markers provide a tool for the dissection of quantitative variation, and thus are potentially important to cucumber improvement (Figure 6).

Cucumber possesses several characteristics that are favourable to MAS including a small genome size (~880 Mega base pairs; Staub and Meglic, 1993), low chromosome number, and rapid life cycle (three cycles per year). In addition, fairly saturated genetic linkage maps have been developed, and QTL analyses have identified several genomic locations involved with important traits (Serquen et al., 1997b; Fazio et al., 2003a; Table 3). Based on these associations, three experiments have been conducted to provide evidence for the potential benefits of MAS during population development and inbred line extraction in cucumber (Fazio et al., 2003b; Fan et al., 2006; Robbins, 2006).

Population Development

To evaluate the effectiveness of MAS, four genetically diverse processing cucumber inbred lines were intermated then bulked maternally to create four base populations (C_0; Robbins, 2006). Each of these populations underwent phenotypic selection (PHE), MAS (using marker-QTL associations identified by Fazio et al., 2003a; Table 3), and random mating without selection (RAN) for three cycles. The four traits under selection were MLB, gynoecy (GYN), earliness (EAR), and L:D (Table 3). Using the same C_0 populations and selection scheme allowed a direct comparison of the effectiveness of MAS and PHE. Since each C_0 population varied for any given trait, the response to MAS and PHE was not the same for each population. In general, C_0 populations that were inferior for a trait either responded favourably to selection or remained constant while those with superior trait values either did not change or decreased. Both MAS and PHE provided improvements in all traits under selection in at least one population with the exception of MAS for EAR. MAS and PHE were equally effective at improving MLB and L:D, but PHE was generally more effective than MAS for GYN and EAR. When considering all traits, responses to PHE were superior in three of the four populations. However, the population for which MAS was superior showed the only increase in yield (fruit/plant), which was not under direct selection. Thus, both MAS and PHE can be useful for multi-trait population improvement, but their effectiveness depends upon the traits and populations under selection.

Inbred Line Extraction

Fazio et al. (2003b) compared the response of MLB to PHE under open-field conditions, RAN, and MAS employing five markers (two SSRs, two RAPDs and one SNP) in two backcross generations (Table 3). No significant differences were detected in either backcross generation between the mean values of MLB from PHE and MAS, which were both significantly higher than the RAN control. Since two cycles of MAS required one year compared to three for PHE, MAS increased overall breeding efficiency.

The effect of MAS for four yield components (MLB, GYN, fruit L:D, and EAR) was evaluated in two backcross processing cucumber populations (line extraction) after two cycles of phenotypic recurrent selection (population improvement) for the same traits (Fan et al., 2006) (Table 3). Even after PHE provided gains in MLB and L:D, MAS continued to improve both these traits in one backcross population and L:D in the other. MAS also provided an increase in gynoecy (GYN) in both populations. Thus, MAS operated to fix favourable alleles that were not exploited by phenotypic selection.

The use of MAS requires the construction of robust markers (preferably codominant), the identification of marker-trait associations, and the development of strategies for their effective deployment in plant improvement programs (Figure 2). Although initial marker development efforts were largely ineffective, sequencing technologies and the availability of cucumber BAC libraries and expressed sequence tags (ESTs) will allow for the development of codominant SSR and SNP-based markers that will be extremely useful in MAS. RIL populations are now available which facilitate the identification of marker-trait associations. Phenotyping of individuals remains time consuming, but genotyping has been made more efficient through marker multiplexing during PCR (Staub et al., 2002b; Robbins, 2006). Recent MAS studies focusing on quantitatively inherited yield component traits are indicative of its potential for cucumber improvement as a tool to enhance selection efficiency. MAS will be most effective when it is used in conjunction with phenotypic selection, especially for quantitatively inherited traits where important genotype × environment interactions are known to exist.

9 Major Breeding Achievements

The early cucumber breeding achievements reviewed by Lower and Edwards (1986) include the: 1) development and use of disease screening technologies to develop resistant cultivars; 2) identification of biochemical pathways which regulate sex expression; 3) development and implementation of controlled pollination procedures, and; 4) characterization of genetics which stabilize gynoecious sex expression. Early selection for disease resistance was primarily performed in the open-field under conditions where the presence of economically important pathogens was unpredictable. In the 1970-1980's scientific collaborations between Drs. C. E. Peterson (cucumber breeder) and P. W. Williams (pathologist) at the University of Wisconsin resulted in the development of seedling screening methodologies that allowed for the highly controlled, high-throughput assessment of pathogen resistance in segregating progeny.

This led to the development of germplasm (i.e., lines, hybrids, and populations) with resistance to several important diseases, and the eventual transfer of this technology to the private sector by the late 1980's. Biochemical and comparative analyses of sex morphotypes in early 1960's led to the discovery of pathways that regulate sex expression in cucumber. This allowed for a better understanding of the biochemistry and physiology underlying sex expression that led to the ability to convert sex types for genetic manipulation. Chemical sex conversion allowed for more rapid cultivar improvement since plant types could be more predictably recovered from selection using more sophisticated selection techniques (e.g., reciprocal recurrent selection, tandem selection). The combination of the ability to manipulate sex expression, methodologies for accurate prediction of disease resistance, and sophisticated selection techniques allowed for the development of sex stable gynoecious lines that could be crossed to produce hybrids with distinctly improved attributes. Among those that provided such improvements were Drs. C. Barnes (Clemson University), B. Kubicki (Warsaw Agricultural University), H. Munger (Cornell University), C. E. Peterson (Michigan State University then USDA, ARS at the University of Wisconsin) and R. L. Lower (North Carolina State University and then the University of Wisconsin). Beginning in the early 1980s improved techniques for germplasm evaluation (e.g., improved field plot techniques) were documented and instituted for the systematic application of complex breeding systems resulting in improved germplasm (e.g., incorporation of exotic genes). These techniques and publicly released germplasm (gynoecy, disease resistance) have been used widely by the seed industry. It is likely that genes for parthenocarpy will be increasingly used to increase yield and fruit quality in the next decade.

More recently, the creation of sophisticated computer algorithms and the development of molecular marker technologies has allowed for an in-depth quantification of some economically important metric traits, the development of unique genetic stocks, and an improved understanding the cucumber genome (Figure 2). Much of the U.S. research on molecular marker development, map construction, and QTL analysis between 1980 and 2000 was partially funded by the seed industry. The use of new technologies (e.g., molecular markers) and genetic stocks [e.g., RIL and nearly-isogenic lines (NIL)] will likely increase in the future as they augment conventional breeding. Their wide-scale use will result from the availability of precise phenotypic data (i.e., cost and time), the development of a highly saturated map with attending marker-QTL associations (i.e., the identification of trait-linked SNP markers), and the ability to detect and characterize epistatic interactions (i.e., development of NIL and the availability of more sophisticated computer algorithms).

10 Acknowledgements

The authors are grateful for the assistance of their technicians and the many graduate students whose studies have led directly to the development of populations and lines, and to the cucumber seed and processing industry who provided support for their work. Mention of trade name, proprietary product, or specific equipment does not constitute a guarantee or warranty by the USDA and does not imply its approval to the exclusion of other products that may be suitable.

References

Anonymous. 1954. New vegetable varieties list I. Proc. Amer. Soc. Hort. Sci. 63: 503-525.

Anonymous. 1955. New vegetable varieties list II. Proc. Amer. Soc. Hort. Sci. 65: 493-511.

Anonymous. 1956. New vegetable varieties list III. Proc. Amer. Soc. Hort. Sci. 67: 587-609.

Anonymous. 1957. New vegetable varieties list IV. Proc. Amer. Soc. Hort. Sci. 69: 574-587.

Anonymous. 1958. New vegetable varieties list V. Proc. Amer. Soc. Hort. Sci. 71: 591-600.

Anonymous. 1960. New vegetable varieties list VI. Proc. Amer. Soc. Hort. Sci. 75: 842-850.

Anonymous. 1961. New vegetable varieties list VII. Proc. Amer. Soc. Hort. Sci. 77: 648-653.

Anonymous. 1963. New vegetable varieties list VIII. Proc. Amer. Soc. Hort. Sci. 82: 652-660.

Anonymous. 1964. New vegetable varieties list IX. Proc. Amer. Soc. Hort. Sci. 84: 665-673.

Atsmon, D., Lang, A., and Light, E. N. 1968. Contents and recovery of gibberellins in monoecious and gynoecious cucumber plants. Plant Physiol. 43: 806-810.

Barnes, W.C. 1969. New vegetable varieties list XVI. HortScience 4: 65-69.

Barnes, W.C. 1970. New vegetable varieties list XVII. HortScience 5: 146-149.

Barnes, W.C. 1971. New vegetable varieties list XVIII. HortScience 6: 124-127.

Bates, D. M., L.C. Merrick, and Robinson, R.W. 1995. Minor cucurbits. In: *Evolution of Crop Plants, 2nd ed.* J. Smartt and N.W. Simmonds, eds., Longman Scientific, Harlow, Essex, UK, pp. 105-111.

Bettencourt, E. and Konopka, J. 1990. *Directory of germplasm collections 4. Vegetables. Abelmoschus*, *Allium*, *Amaranthus*, Brassicaceae, *Capsicum*, Cucurbitaceae, *Lycopersicon*, *Solanum* and other vegetables. IBPGR, Rome, Italy.

Beyer, Jr., E. 1976. Silver: A potent antiethylene agent in cucumber and tomato. HortScience 11: 195-196.

Bhaduri, P.N., and Bose, P.C. 1947. Ctyo-genetical investigations in some common cucurbits, with special reference to fragmentation of chromosomes as physical basis of speciation. J. Genet. 48: 237-256.

Bradeen, J.M., J.E. Staub, C. Wyse, R. Antonise, and Peleman, J. 2001. Towards an expanded and integrated linkage map of cucumber (*Cucumis sativus* L.). Genome 44: 111-119.

Byers, R.E., L.R. Baker, H.M. Sell, R.C. Herner, and Dilley, D.R. 1972. Female flower induction on androecious cucumber, *Cucumis sativus* L. J. Amer. Soc. Hort. Sci. 98: 197-199.

Chen, J.F., S. Isshiki, Y. Tashiro, and Miyazaki, S. 1995. Studies on a wild cucumber from China (*Cucumis hystrix* Chakr.). I. Genetic distances between *C. hystrix* and two cultivated *Cucumis* species (*C. sativus* L. and *C. melo* L.) based on isozyme analysis. J. Jpn. Soc. Hort. Sci. 64: 264-265.

Chen, J.F., J.E. Staub, Y. Tashiro, S. Isshiki, Miyazaki, S. 1997a. Successful interspecific hybridization between *Cucumis sativus* L.and *Cucumis hystrix* Chakr. Euphytica 96: 413-419.

Chen, J.F., S. Isshiki, Y. Tashiro, and Miyazaki, S. 1997b. Biochemical affinities between *Cucumis hystrix* Chakr. and two cultivated *Cucumis* species (*C. sativus* L. and *C. melo.* L.) based on isozyme analysis. Euphytica 97: 139-141.

Chen, J.F. and Kirkbride, J.H. 2000. A new synthetic species *Cucumis* (Cucurbitaceae) from interspecific hybridization and chromosome doubling. Brittonia 52: 315–319.

Chen, J.F., F.Y. Zhuang., X.A. Liu, and Qian, C.T. 2004a. Reciprocal differences of morphological and DNA characters in interspecific hybridization in *Cucumis*. Can. J. Bot. 82: 16-21.

Chen, J.F., X.D. Luo, C.T. Qian, M.M. Jahn, J.E. Staub, F.Y. Zhuang, Q.F. Lou, and Ren, G. 2004b. *Cucumis* monosomic alien addition lines: morphological, cytological, and genotypic analyses. Theor. Appl. Genet. 108: 1343-1348.

Chung, S.M., J.E. Staub, and Fazio, G. 2003. Inheritance of chilling injury: A maternally inherited trait in cucumber (*Cucumis sativus* L.). J. Amer. Soc. Hort. Sci. 128: 526-530.

Chung, S.M., J.E. Staub, and Chen, J.F. 2006. Molecular phylogeny of *Cucumis* species as revealed by consensus chloroplast SSR marker length and sequence variation. *Genome* 49: 219-229.

Cramer, C.S. and Wehner, T.C. 1998a. Fruit yield and yield component means and correlations of four slicing cucumber populations improved through six to ten cycles of recurrent selection. J. Am. Soc. Hort. Sci. 123: 388-395.

Cramer, C.S. and Wehner, T.C. 1998b. Fruit yield and yield components of cucumber populations grown at low plant density, density. In: . J. D. McCreight, ed., *Cucurbitaceae '98: Evaluation and Enhancement of Cucurbit Germplasm.* ASHS Press, Alexandria, pp. 277-285.

Cramer, C.S. and Wehner, T.C. 1998c. Performance of three selection cycles for four slicing cucumber populations hybridized with a tester. J. Amer. Soc. Hort. Sci. 123: 396-400.

Cramer, C.S. and Wehner, T.C. 1999a. Little heterosis for yield and yield components in hybrids of six cucumber inbreds. Euphytica 110: 99-108.

Cramer, C.S. and Wehner, T.C. 1999b. Testcross performance of three selection cycles from four pickling cucumber populations. J. Amer. Soc. Hort. Sci. 124: 257-261.

Cramer, C.S. and Wehner, T.C. 2000a. Fruit yield and yield component correlations of four pickling cucumber populations. Cucurbit Genet. Coop. Rpt. 23: 12-15.

Cramer, C.S. and Wehner, T.C. 2000b. Path analysis of the correlation between fruit number and plant traits of cucumber populations. HortScience 35: 708-711.

Dane, F. and Tsuchiya, T. 1976. Chromosome studies in the genus *Cucumis*. Euphytica 25: 367-374.

Dane, F., D.W. Denna, and Tsuchiya, T. 1980. Evolutionary studies of wild species in the genus *Cucumis*. Z. Pflanzenzucht. 85: 89-109.

Deakin, J.R., G.W. Bohn, and Whitaker, T.W. 1971. Interspecific hybridization in *Cucumis*. Econ. Bot. 25: 195-211.

den Nijs, A.P.M. and Custers, J.B.M. 1990. Introducing resistances into the cucumber by interspecific hybridization. In: D. M. Bates, R. W. Robinson, C. Jeffrey, eds., *Biology and Utilization of the Cucurbitaceae.* Comstock Publishing Associates, Ithaca, New York and London, pp. 382–396.

de Ponti, O.M.B. 1975. Breeding parthenocarpic pickling cucumbers (*Cucumis sativus* L.): Necessity, genetical possibilities, environmental influences and selection criteria. Euphytica 25: 29-40.

Dhillon, N.P.S. and Wehner, T.C. 1991. Host-pathogen resistance to insect in cucurbits-germplasm resources, genetics, and breeding. Trop. Pest Manage. 37: 421-428.

Dijkhuizen, A., W.C. Kennard, M.J. Havey, and Staub, J.E. 1996. RFLP variability and genetic relationships in cultivated cucumber. Euphytica 90: 79-89.

Dijkhuizen, A. and Staub, J.E. 2003. Effects of environment and genetic background on QTL affecting yield and fruit quality traits in a wide cross in cucumber [*Cucumis sativus* L. x *Cucumis hardwickii* (R.) Alef.] J. New Seeds 4: 1-30.

El-Shawaf, I.I.S. and Baker, L.R. 1981a. Inheritance of parthenocarpic yield in gynoecious pickling cucumber for once-over mechanical harvest by diallel analysis of six gynoecious lines. J. Am. Soc. Hort. Sci. 106: 359-364.

El-Shawaf, I.I.S. and Baker, L.R. 1981b. Combining ability and genetic variances of G x H F_1 hybrids for parthenocarpic yield in gynoecious pickling cucumber for once-over mechanical harvest. J. Am. Soc. Hort. Sci. 106: 365-370.

Fan, Z., M.D. Robbins, and Staub, J.E. 2006. Population development by phenotypic selection with subsequent marker-assisted selection for line extraction in cucumber (*Cucumis sativus* L.) Theor. Appl. Genet. 112: 843-855.

Fang, X., Y. Yin, X. Han, and Gu, X. 1995. Selection of gynoecious lines and their hybrids with different ecotypes in cucumber (*Cucumis sativus*). Acta Hort. 402: 392-397.

Fanourakis, N.E. and Simon, P.W. 1987. Analysis of genetic linkage in cucumber. J. Hered. 78: 238-242.

Fazio, G. 2001. Comparative study of marker-assisted and phenotypic selection and genetic analysis of yield components in cucumber. PhD dissertation, University of Wisconsin, Madison.

Fazio, G., J.E. Staub, and Chung, S.M. 2002. Development and characterization of PCR markers in cucumber (*Cucumis sativus* L.). J. Am. Soc. Hort. Sci. 127: 545-557.

Fazio, G., J.E. Staub, and Stevens, M.R. 2003a. Genetic mapping and QTL analysis of horticultural traits in cucumber (*Cucumis sativus* L.) using recombinant inbred lines. Theor. Appl. Genet. 107: 864-874.

Fazio, G., S.M. Chung, and Staub, J.E. 2003b. Comparative analysis of response to phenotypic and marker-assisted selection for multiple lateral branching in cucumber (*Cucumis sativus* L.). Theor. Appl. Genet. 107: 875-883.

Fredrick, L.R. and Staub, J.E. 1989. Combining ability analysis and evaluation of nearly homozygous lines derived from *Cucumis sativus* var. *hardwickii* (R.) Alef. J. Amer. Soc. Hort. Sci. 114: 332-338.

Galun, E. 1959. Effects of gibberellic acid and naphthalene acetic acid on sex expression and some morphological characters in the cucumber plant. Phyton 13: 1-8.

Galun, E. 1961. Study of the inheritance of sex expression in the cucumber. The interaction of major genes with modifying genetic and non-genetic factors. Genetica 32: 134-163.

Garcia-Mas, J., A.J. Monforte, and Arus, P. 2004. Phylogenetic relationships among *Cucumis* species based on the ribosomal internal transcribed spacer sequence and microsatellite markers. Plant Syst. Evol. 248: 191-203.

George, W.L. Jr. 1970. Dioecism in cucumbers *Cucumis sativus* L. Genetics 64:23-28.

Goode, M.J., J.L. Bowers and Bass, A., Jr. 1980. Little-leaf, a new kind of pickling cucumber plant. *Arkansas Farm Research*, May/June, p. 4.

Harlan, J.R. 1975. Crops and man. American Society of Agronomy, Madison, Wis.

Havey, M.J. 1997. Predominant paternal transmission of the mitochondrial genome in cucumber. J. Heredity 88: 232-235.

Hawthorn, L.R. and Wellington, R. 1930. Geneva, a greenhouse cucumber that develops fruit without pollination. NY (Geneva) *Agr.* Exp. Stat. Bull. 580: 1-11.

Hedrick, U.P. 1919. Sturtevant's notes on edible plants. J. B. Lyon Co., Albany, New York.

Horejsi, T. and Staub, J.E. 1999. Genetic variation in cucumber (*Cucumis sativus* L.) as assessed by random amplified polymorphic DNA. Genet. Res. Crop Evol. 46: 337-350.

Horejsi, T., J.M. Box, and Staub, J.E. 1999. Efficiency of randomly amplified polymorphic DNA to sequence characterized amplified region marker conversion and their comparative polymerase chain reaction sensitivity in cucumber. J. Am. Soc. Hort. Sci. 124: 128-135.

Horejsi, T., J.E. Staub, and Thomas, C. 2000. Linkage of random amplified polymorphic DNA markers to downy mildew resistance in cucumber (*Cucumis sativus* L.) Euphytica 115: 105-113.

Horst, E.K., and Lower, R.L. 1978. *Cucumis hardwickii*, a source of germplasm for the cucumber breeder. Cucurbit Genet. Coop. Rpt. 1: 5.

Jeffrey, C. 1980. A review of the Cucurbitaceae. Bot. J. Linnean Soc. 81: 233–247.

Jobst, J., K. King, and Hemleben, V. 1998. Molecular evolution of the internal transcribed spacers (ITS1 and ITS2) and phylogenetic relationships among species of the family Cucurbitaceae. Mol. Phylogenet. Evol. 9: 204-219.

Kamachi, S., H. Sekimoto, N. Kondo, and Sakai, S. 1997. Cloning of a cDNA for a 1-aminocyclopropane-1-carboxylate synthase that is expressed during development of female flowers at the apices of *Cucumis sativus* L. Plant Cell Physiol. 38: 1197-1206.

Kamachi, S., H. Mizusawa, S. Mazuura, and Sakai, S. 2000. Expression of two 1-aminocyclopropane-1-carboxylate synthase genes, CS-ACS1 and CS-ASC2, correlated with sex phenotypes in *Cucumis* plants (*Cucumis sativus* L.). Plant Biotechnol. 17: 69-74.

Kaufman, D.S. and Lower, R.L. 1976. Inheritance of an extreme dwarf plant type in the cucumber. J. Amer. Soc. Hort. Sci. 101: 150-151.

Kennard, W.C., K. Poetter, A. Dijkhuizen, V. Meglic, J.E. Staub, and Havey, M.J. 1994. Linkages among RFLP, RAPD, isozyme, disease-resistance, and morphological markers in narrow and wide crosses of cucumber. Theor. Appl. Genet. 89: 42–48.

Knerr, L.D. and Staub, J.E. 1992. Inheritance and linkage relationships of isozyme loci in cucumber (*Cucumis sativus* L.). Theor. Appl. Genet. 84: 217–224.

Kroon, G.H., J.B.M. Custers, and Y.O. Kho, and den Nijs, A.M.P. 1979. Interspecific hybridization in *Cucumis* L. I. Need for genetic variation, biosystematic relations and possibilities to overcome crossing barriers. Euphytica 28: 723-728.

Kirkbride, J.H., Jr. 1993. *Biosystematic Monograph of the Genus Cucumis (Cucurbitaceae).* Parkway Publishers, Boone, North Carolina.

Kim, I.S., K.C. Yoo, K. Fujieda, and Okubo, H. 1994. Studies on parthenocarpy in *Cucumis sativus* L. V. Influence of exogenous plant growth regulators on growth and diffusible IAA level of cucumber ovaries. J. Kor. Soc. Hort. Sci. 35: 195-200.

Kubicki, B. 1969. Investigations on sex determination in cucumber (*Cucumis sativus* L.). Genet. Pol. 10: 3-143.

Kubicki, B. 1980. Investigations on sex determination in cucumbers *Cucumis sativus* L. IX. Induced mutant with recessive character in gynoecism. Genet. Pol. 21: 409-424.

Kupper, R.S. and Staub, J.E. 1988. Combining ability between lines of *Cucumis sativus* L. and *Cucumis sativus* var. *hardwickii* (R .) Alef. Euphytica 38: 197-210.

Kvasnikov, B.V., N.T. Rogova, S.I. Tarakonova, and Ignatova, I. 1970. Methods of breeding vegetable crops under the covered ground. Trudy-po-Prikladnoi-Botanike-Genetiki-I-Selektsii 42: 45-57.

López-Sesé, A. and Staub, J.E. 2002. Selection for early flowering, branching and gynoecy in cucumber (*Cucumis sativus* L.) Cucurbit Genet. Coop. Rpt. 25: 3-6.

Lower, R.L. 1973. New vegetable varieties list XIX. HortScience 8: 465-470.

Lower, R.L. 1975. New vegetable varieties list XX. HortScience 10: 467-470.

Lower, R.L. and Edwards, M.D. 1986. Cucumber breeding. In: M. J. Basset, ed., *Breeding Vegetable Crops*. AVI Publishing Co., Westport, Connecticut, pp. 173-207.

McCollum, J.P. 1934. Vegetative and reproductive responses associated with fruit development in the cucumber. Memo. N.Y. Agric. Exp. Stn. (Ithaca), 163.

McCreight, J.D., H. Nerson and Grumet, R. 1993. Melon, *Cucumis melo* L. In: *Genetic Improvement of Vegetable Crops,* G. Kalloo and B.O. Bergh, eds., Pergamon Press, New York, pp. 267-294.

Meglic, V. and Staub, J.E. 1996. Inheritance and linkage relationships of allozyme and morphological loci in cucumber (*Cucumis sativus* L.). Theor. Appl. Genet. 92: 865-872.

Meglic, V., F. Serquen, and Staub, J.E. 1996. Genetic diversity in cucumber (*Cucumis sativus* L.): I. A reevaluation of the U.S. germplasm collection. Genet. Resour. Crop Evol. 43: 533-546.

Meshcherov, E.T. and Juldasheva, L.W. 1974. Parthenocarpy in cucumber. Trudy-po-Prikladnoi-Botanike-Genetiki-I-Selektsii 51: 204-213.

More, T.A. and Budgujar, C.D. 2002. Isolation of parthenocarpic tropical gynoecious lines in cucumber (*Cucumis sativus* L.). Acta Horticulturae 588: 255-260.

Mibus, H. and Tatlioglu, T. 2004. Molecular characterization and isolation of the *F/f* gene for femaleness in cucumber (*Cucumis sativus* L.). Theor. Appl. Genet. 109: 1669-1676.

Miller, C.H. and Wehner, T.C. 1989. Cucumbers. In: *Quality and Preservation of Vegetables,* N. A. M. Eskin, ed., CRC Press, Inc., Boca Raton, Florida, pp. 245-264.

Minges, P.A. 1965. New vegetable varieties list X and XI. Proc. Amer. Soc. Hort. Sci. 86: 824-845.

Minges, P.A. 1966. New vegetable varieties list XII. Proc. Amer. Soc. Hort. Sci. 88: 718-726.

Minges, P.A. 1967. New vegetable varieties list XIV. Proc. Amer. Soc. Hort. Sci. 90: 567-569.

Minges, P.A. 1968. New vegetable varieties list XV. Proc. Amer. Soc. Hort. Sci. 92: 823-840.

Nam, Y.W., J.R. Lee, K. Song, M.K. Lee, M.D. Robbins, S.M. Chung, J.E. Staub, and Zhang, H.B. 2006. Construction of two BAC libraries from cucumber (*Cucumis sativus* L.) and identification of clones linked to yield component quantitative trait loci. Theor. Appl. Genet. 111: 150-161.

Navazio, J.P. and Staub, J.E. 1994. Effects of soil moisture and post-harvest handling on pillowy fruit disorder in cucumber. J. Amer. Soc. Hort. Sci. 119: 1234-1242.

Nikolova, V., M. Alexandrova, and Stoeva, V. 2002. Possibilities for the use of remote hybridization in the genus *Cucumis* for the development of genetic diversity. Acta Horticulturae 579: 39-43.

U.S. Dept. of Agriculture. 1940, 1981, and 1998. Agricultural Statistics. U.S. Government Printing Office, Washington, D.C.

USDA NASS. 2004. United States Department of Agriculture, National Agricultural Statistics Service, Vegetables: final estimates 1998-2003. 987: 1-136.

Park, Y.H., S. Senory, C. Wye, R. Antonise, J. Peleman, and Havey, M.J. 2000. A genetic map of cucumber composed of RAPDs, RFLPs, AFLPs, and loci conditioning resistance to papaya ringspot and zucchini yellow mosaic viruses. Genome 43: 1003-1010.

Perl-Treves, R., and Galun, E. 1985. The *Cucumis* plastome: physical map, intragenic variation, and phylogentic relationships. Theor. Appl. Genet. 71: 417-429.

Peterson, C.E. and Anhder, L.D. 1960. Induction of staminate flowers on gynoecious cucumbers with gibberellic A_3. Science 131: 1673-1674.

Peterson, C.E. 1975. Plant introduction in the improvement of vegetable cultivars. HortScience 10: 575-579.

Peterson, C.E., P.H. Williams, M. Palmer and Louward, P. 1982. Wisconsin 2757 cucumber. HortScience 17: 268.

Peterson, C.E., J.E. Staub, M.J. Palmer, and Crubaugh, L. 1985. Wisconsin 2843, a multiple disease resistant cucumber population. HortScience 20: 309-310.

Peterson, C.E., J.E. Staub, P.H. Williams, and Palmer, M.J. 1986a. Wisconsin 1983 cucumber. HortScience 21: 1082-1083.

Peterson, C.E., J.E. Staub and Palmer, M.J. 1986b. Wisconsin 5207, a multiple disease resistant population. HortScience 21: 335-33.

Pierce, L.K. and Wehner, T.C. 1990. Review of genes and linkage groups in cucumber. HortScience 25: 605-615.

Pike, L.M. and Peterson, C.E. 1969. Inheritance of parthenocarpy in the cucumber (*Cucumis sativus* L.). Euphytica 18: 101-105.

Pitrat, M., M. Chauvet, and Foury, C. 1999. Diversity, history, and production of cultivated cucurbits. Acta Horticulturae 492: 21-28.

Ponti, O.M.B. de. 1975. Breeding parthenocarpic pickling cucumbers (*Cucumis sativus* L.): Necessity, genetical possibilities, environmental influences and selection criteria. Euphytica 25: 29-40.

Przybecki, Z., M.E. Kowalczyk, J. Witkowicz, M. Filipecki, and Siedlecka, E. 2004. Polymorphom of sexually different cucumber (*Cucumis sativus* L.) NIL. Cell. Mol. Biol. Letters 9: 919-933.

Raamsdonk, L.W.D., A.P.M. den Nijs, and Jongerius, M.C. 1989. Meiotic analyses of *Cucumis* hybrids and an evolutionary evaluation of the genus *Cucumis* (Cucurbitaceae). Plant Syst. Evol. 163: 133-146.

Robbins, M.D. 2006. Molecular marker development, QTL pyramiding, and comparative analysis of phenotypic and marker-assisted selection in cucumber. PhD dissertation University of Wisconsin, Madison.

Robinson, R.W., H.M. Whitaker, and Bohn, G.W. 1976. Genes of the Cucurbitaceae. HortScience 11: 554-568.

Robinson, R.W., and Decker-Walters, D. 1997. *Cucurbits*. CAB International, Wallingford, England; 226 pp.

Roy, R.P. and Saran, S. 1990. Sex expression in the Cucurbitaceae, In: R. W. Robinson and C. Jeffery, eds., *Biology and Utilization of the Cucurbitaceae*. Comstock, Cornell University Press. Ithaca. pp. 251-268.

Rudich, J., A.H. Halevy, and Kedar, N. 1972a. Ethylene evolution from cucumber plants as related to sex expression. Plant Physiol. 49: 998-999.

Rudich, J., A.H. Halevy, and Kedar, N. 1972b. The level of phytohormones in monoecious and gynoecious cucumbers as affected by photoperiod and ethephon. Plant Physiol. 50: 585-590.

Schultheis, J.R., T.C. Wehner, and Walters, S.A. 1998. Optimum planting density and harvest stage for little-leaf and normal-leaf cucumbers for once-over harvest. Can. J. Plant Sci. 78: 333-340.

Shetty, N.V. and Wehner, T.C. 2002. Screening the cucumber germplasm collection for fruit yield and quality. Crop Sci. 42: 2174-2183.

Shifriss, O. 1961. Sex control in cucumbers. J. Hered. 52:5-12.

Serce, S. and Staub, J.E. 1999. Nearly-isogenic cucumber genotypes differing in leaf size and plant habit exhibit differential response to water stress. J. Amer. Soc. Hort. Sci. 124: 358-365.

Serquen, F.C., J. Bacher, and Staub, J.E. 1997a. Genetic analysis of yield components in cucumber (*Cucumis sativus* L.) at low plant density. J. Amer. Soc. Hort. Sci. 122: 522-528.

Serquen, F.C., J. Bacher, and Staub, J.E. 1997b. Mapping and QTL analysis of a narrow cross in cucumber (*Cucumis sativus* L.) using random amplified polymorphic DNA markers. Mol. Breeding 3: 257-268.

Shetty, N.V., and Wehner, T.C. 2002. Screening the cucumber germplasm collection for fruit yield and quality. Crop Sci. 42: 2174-2183.

Sitterly, W.R. 1972. Breeding of disease resistance in cucurbits. Annu. Rev. Phytopathol. 10: 471-490.

Smeets, L.and Wehner, T.C. 1997. Environmental effects on genetic variation of chilling resistance in cucumber. Euphytica 97: 217-225.

St. Amand, P.C. and Wehner, T.C. 1995. Greenhouse, detached-leaf, and field testing methods to determine cucumber resistance to gummy stem blight. J. Amer. Soc, Hort. Sci. 120: 673-680.

St. Amand, P.C. and Wehner, T.C. 2001a. Heritability and genetic variance estimates for leaf and stem resistance to gummy stem blight in two cucumber populations. J. Amer. Soc. Hort. Sci. 126: 90-94.

St. Amand, P.C. and Wehner, T.C. 2001b. Generation means analysis of leaf and stem resistance to gummy stem blight in cucumber. J. Amer. Soc. Hort. Sci. 126: 95-99.

Staub, J.E., L. Fredrick, and Marty, T. 1987. Electrophoretic variation in cross-compatible wild diploid species of *Cucumis*. Can. J. Bot. 65: 792-798.

Staub, J.E., R.L. Lower and Nienhuis, J. 1988. Correlated responses to selection for low temperature germination in cucumber. HortScience 23: 745-746.

Staub, J.E. and Krasowska, A. 1990. Screening of the U.S. germplasm collection for heat stress tolerance. Cucurbit Genet. Coop. Rpt. 13: 4-7.

Staub, J.E., L. Crubaugh, H. Baumgartner, and Hopen, H. 1991. Screening of the cucumber collection for tolerance to Clomazone herbicide. Cucurbit Genet. Coop. Rpt. 14: 23-24.

Staub, J.E., L.D. Knerr, and Weston, L.A. 1991. Evaluations and correlated responses for resistance to Chloramben herbicide in cucumber. HortScience 26: 905-908.

Staub, J.E., L.D. Knerr, D.J. Holder, and May, B. 1992a. Phylogenetic relationships among several African *Cucumis* species. Can. J. Bot. 70: 509-517.

Staub, J.E., L.D. Knerr, and Hopen, H.J. 1992b. Effects of plant density and herbicides on cucumber productivity. J. Amer. Soc. Hort. Sci. 117: 48-53.

Staub, J.E., C.E. Peterson, L.K. Crubaugh and Palmer, M.J. 1992c. Cucumber population WI 6383 and derived inbreds WI 5098 and WI 5551. HortScience 27: 1340-1341.

Staub, J.E. and Grumet, R. 1993. Selection for multiple disease resistance affects cucumber yield potential. Euphytica 67: 205-213.

Staub, J.E. and Meglic, V. 1993. Molecular genetic markers and their legal relevance for cultigen discrimination: A case study in cucumber. HortTechnology 3: 291-300.

Staub, J.E. and Navazio, J.P. 1993. Temperature and humidity affect pillowy fruit disorder in cucumber. HortScience 28: 822-823.

Staub, J.E. 1996. Noninfectious disorders: moisture stress (p. 65). In: T. A. Zitter, D. L. Hopkins, and C. E. Thomas, eds., *Compendium of cucurbit diseases Part II.* APS Press., St. Paul, MN. pp. 87.

Staub, J.E., J. Bacher, and Poetter, K. 1996. Factors affecting the application of random amplified polymorphic DNAs in cucumber (*Cucumis sativus* L.). HortScience 31: 262-266.

Staub, J.E. and Wehner, T.C. 1996. Temperature stress. In: T. A. Zitter, D. L. Hopkins, and C. E. Thomas. eds., *Compendium of cucurbit diseases Part II.* APS Press, St. Paul Minnesota, pp. 66-67.

Staub, J.E. and Bacher, J. 1997. Cucumber as a processed vegetable. In: D. S. Smith, J. N. Cash, W. Nip, and Y.H. Hui, eds., *Processing Vegetables: Science and Technology IV.* Technomic Publishing Co., Inc. Lancaster, PA., pp. 129-193.

Staub, J.E., F.C. Serquen, and McCreight, J.D. 1997. Genetic diversity in cucumber (*Cucumis sativus* L.): III. An evaluation of Indian germplasm. Genet. Res. Crop Evol. 44: 315-326.

Staub, J.E., F.C. Serquen, T. Horejsi, and Chen, J.F. 1999. Genetic diversity in cucumber (*Cucumis sativus* L.): IV. An evaluation of Chinese germplasm. Genet. Res. Crop Evol. 46: 297-310.

Staub, J.E. and Ivandic, V. 2000. Genetic assessment of the United States national cucumber collection. Acta Horticulturae 510: 113-121.

Staub, J.E., F. Dane, K. Reitsma, G. Fazio, and López-Sesé, A.I. 2002a. The formation of test arrays and a core collection in cucumber (*Cucumis sativus* L.) using phenotypic and molecular marker data. J. Am. Soc. Hort. Sci. 127: 558-567.

Staub, J.E., M.D. Robbins, and López-Sesé, A.I. 2002b. Molecular methodologies for improved genetic diversity assessment in cucumber and melon. In: J. D. Creight, ed., *Proceedings XXVI IRC,. Horticulture: Art and science for life- Advances in vegetable Breeding*. Acta Horticulturae 642: 41-47.

Staub, J.E., S.M. Chung, and Fazio, G. 2005. Conformity and genetic relatedness estimation in crop species having a narrow genetic base: The case of cucumber (*Cucumis sativus* L.). Plant Breed. 124: 44-53.

Stzangret, J., J. Wronka, T. Galecka, A. Korzeniewska, and Niemirowicz-Szczytt, K. 2004. Cucumber (*Cucumis sativus* L.) haploids developed from parthenocarpic hybrids. In: A. Lebeda and H.S. Paris, eds., *Progress in cucurbit genetics and breeding research, Proceedings of Cucurbitaceae 2004*, pp. 411-414.

Sun Z., R.L. Lower, and Staub, J.E. 2006a. Variance component analysis of parthenocarpy in elite U.S. processing type cucumber (*Cucumis sativus* L.) lines. Euphytica 138: 333-341.

Sun Z., R.L. Lower, and Staub, J.E. 2006b. Analysis of generation means and components of variance for parthenocarpy in cucumber (*Cucumis sativus* L.). Plant Breed. 125: 277-280.

Sun Z., R.L. Lower, S.M. Chung, and Staub, J.E. 2006c. Identification and comparative analysis of quantitative trait loci (QTL) associated with parthenocarpy in processing cucumber. Plant Breed. 125: 281-287.

Suyama, T., K. Yamada, H. Mori, K. Takeno, and Yamaki, S. 1999. Cloning cDNAs for gene preferentially expressed during fruit growth in cucumber. J. Amer. Soc. Hort. Sci. 124: 136-139.

Tapley, W.T., W.D. Enzie and van Eseltine, G.P. 1937. *The vegetables of New York. IV. The cucurbits*. Rpt. N. Y. Agr. Exp. Sta. J. B. Lyon Co., Albany, New York.

Tatlioglu, T. 1993. Cucumber *Cucumis sativus* L. In: G. Kalloo and B.O. Bergh, eds., *Genetic Improvement of Vegetable Crops*. Pergamon Press Ltd., Tarrytown, New York, pp. 197-234.

Thomas, R.S. and Staub, J.E. 1992. Effects of water stress and storage environment on pillowy fruit disorder in cucumber. J. Amer. Soc. Hort. Sci. 117: 394-399.

Trebitsh, T., J.E. Staub, and O'Neill, S.D. 1997. Identification of an 1-aminocyclopropane-1-carboxylate synthase gene linked to the *Female* gene (*F*) that determines female sex expression in cucumber (*Cucumis sativus* L.). Plant Physiol. 113: 987-995.

Trivedi, R.N. and Roy, R.P. 1970. Cytological studies in *Cucumis* and *Citrullus*. Cytologia 35: 561-569.

Uchneat, M.S. and Wehner, T.C. 1998. Resistance to belly rot in cucumber identified through field and detached-fruit evaluations. J. Amer. Soc. Hort. Sci. 123: 78-84.

USDA, 1940, 1981, and 1998. Agricultural Statistics. U.S. Government Printing Office, Washington, D.C.

USDA NASS. 2004. United States Department of Agriculture, National Agricultural Statistics Service, Vegetables: final estimates 1998-2003. 987: 1-136.

Vakalounakis, D.J. 1992. Heart leaf, a recessive leaf shape marker in cucumber: Linkage with disease resistance and other traits. J. Hered. 83: 217-221.

Vavilov, N.I. 1926. *Studies on the Origin of Cultivated Plants*. Institute of Applied Botany and Plant Breeding, Leningrad, USSR.

Vavilov, N.I. 1951. The origin, variation, immunity and breeding of cultivated plants. Chronica Botanica 13: 13-54.

Walters, S.A., T.C. Wehner, and Barker, K.R. 1991. Resistance to root-knot nematodes in cucumber and horned cucumber. J. Nematol. 23: 611-614.

Walters, S.A., T.C. Wehner, and Barker, K.R. 1992. Effects of root decay on the relationship between *Meloidogyne* spp. Gall index and egg mass number in cucumber and horned cucumber. J. Nematol. 24: 707-711.

Walters, S.A., T.C. Wehner, and Barker, K.R. 1993. Root-knot nematode resistance in cucumber and horned cucumber. HortScience 28: 151-154.

Walters, S.A. and Wehner, T.C. 1994. Evaluation of the U.S. cucumber germplasm collection for root size using a subjective rating technique. Euphytica 79: 39-43.

Walters, S.A., T.C. Wehner, and Barker, K.P. 1995. A split root technique for multiple nematode resistance in cucumber. Cucurbit Genet. Coop. Rpt. 18: 29-30.

Walters, S.A., T.C. Wehner, and Barker, K.R. 1996. NC-42 and NC-43: Root-knot nematode resistance cucumber germplasm. HortScience 31: 1246-1247.

Walters, S.A., and Wehner, T.C. 1997. 'Lucia', 'Manteo', and 'Shelby' root-knot nematode resistant cucumber inbred lines. HortScience 32: 1301-1303.

Walters, S.A., T.C. Wehner, and Barker, K.R. 1997. A single recessive gene for resistance to the root-knot nematode (*Meloidogyne javanica*) in *Cucumis sativus* var. *hardwickii*. J. Heredity 88: 66-69.

Walters, S.A., T.C. Wehner, and Barker, K.R. 1999. Greenhouse and field resistance in cucumber to root-knot nematodes. Nematology 1: 279-284.

Wehner, T.C. and Miller, C.H. 1985. Effect of gynoecious expression on yield and earliness of a fresh-market cucumber hybrid. J. Am. Soc. Hort. Sci. 110: 464-466.

Wehner, T.C. and Horton, R.R., Jr. 1986. Performance of cultivars of four different cucumber types for fresh-market use in North Carolina. Cucurbit Genet. Coop. Rpt. 9: 53-54.

Wehner, T.C. 1989. Breeding for improved yield in cucumber. Plant Breed Rev. 6: 323-359.

Wehner, T.C., R.L. Lower, J.E. Staub, and Tolla, G.E. 1989. Convergent-divergent selection for cucumber fruit yield. HortScience 24: 667-669.

Wehner, T.C. 1996. Bitter fruit. In: T. A. Zitter, D. L. Hopkins, and C. E. Thomas (eds.). *Compendium of cucurbit diseases*. APS Press, St. Paul, Minnesota, pp. 65.

Wehner, T.C. and Cramer, C.S. 1996. Gain for pickling cucumber yield and fruit shape using recurrent selection. Crop Sci. 36: 1538-1544.

Wehner, T.C., P.C. St. Amand, and Lower, R.L. 1996. 'M-17' gummy stem blight resistant pickling cucumber inbred. HortScience 31: 1248-1249.

Wehner, T.C. and Shetty, N.V. 1997. Three pickling cucumber populations: NCWBP, NCMBP, and NCEP1. HortScience 32: 941-944.

Wehner, T.C. 1998a. Three slicing cucumber populations: NCWBS, NCMBS, and NCES1. HortScience 33: 168-170.

Wehner, T.C. 1998b. Two special cucumber populations: NCH1 and NCBA1. HortScience 33: 766-768.

Wehner, T.C. and Shetty, N.V. 2000. Screening the cucumber germplasm collection for resistance to gummy stem blight in North Carolina field tests. HortScience 35: 1132-1140.

Wehner, T.C., N.V. Shetty, and Clark, R.L. 2000a. Screening the cucumber germplasm collection for combining ability for yield. HortScience 35: 1141-1150.

Wehner, T.C., N.V. Shetty, and Wilson, L.G. 2000b. Screening the cucumber germplasm collection for fruit storage ability. HortScience 35: 699-707.

Wehner, T.C., N.V. Shetty, and Sloane, J.T. 2004. Field and detached-fruit screening tests for resistance to belly rot in cucumber. HortScience 39: 149-152.

Wenzel, G., W.C. Kennard, and Havey, M.J. 1995. Quantitative trait analysis of fruit quality in cucumber: QTL detection, confirmation, and comparison with mating-design variation. Theor. Appl. Genet. 91: 53-61

Whitaker, T.W. and Davis, G.N. 1962. *Cucurbits: botany, cultivation, and utilization*. Interscience Publishers, Inc., New York.

Zhang, Q., A.C. Gabert, and Baggett, J.R. 1992. Parents and mating systems affect the transfer of gynoecious flowering to Chinese monoecious cucumbers. J. Amer. Soc. Hort. Sci. 117: 515-517.

Yamasaki, S., N. Fujii, and Takahashi, H. 2003. Characterization of ethylene effects on sex determination in cucumber plants. Sexual Plant Repro. 16: 103-111.

Zhuang, F.Y., Chen, J.F., Staub, J.E., and Qian, C.T. 2004. Assessment of genetic relationships in *Cucumis* species by SSR and RAPD analysis. Plant Breed. 123: 167-172.

Zijlstra, S., R.C. Jansen, and Groot, S.P.C. 1995. The relationship between powdery mildew (*Sphaerotheca fuliginea*) resistance and leaf chlorosis sensitivity in cucumber (*Cucumis sativus*) studied in single seed decent lines. Euphytica 81: 193-198.

Melon

Michel Pitrat[1]

[1] Institut National de la Recherche Agronomique, Unité de Génétique et Amélioration des Fruits et Légumes, Michel.Pitrat@avignon.inra.fr

1 Introduction

Melon (*Cucumis melo* L.) belongs to the *Cucurbitaceae* family, which includes several other vegetables of economic importance such as cucumber, watermelon, squash, pumpkin and gourds. The world production was estimated in 2005 at 28 millions tons (FAOSTAT Data, 2005), the main producing countries being China (15.1 millions tons) followed by Turkey, Iran, Spain and USA (between 1.7 and 1 millions tons). The world production increases regularly from 8.8 to 13.5 to 19.8 millions tons in 1980, 1990 and 2000 respectively.

Melon is a member of the *Cucumis* genus, which includes another important crop *Cucumis sativus* L. (cucumber). There have been propositions to separate the genus *Cucumis* into two genera: *Cucumis* including the cucumber and *Melo* with the melon (Ashurmetov, 1995). Even if some arguments are quite convincing, we will consider in this chapter that melon belongs to the genus *Cucumis* L. with more than 30 other species (Kirkbride, 1993).

Study of the genetic control of characters and "modern" breeding started at the beginning of the XXth century even if preliminary works have been published in the XIXth century (Sageret, 1825).

Previous articles on melon genetics include Robinson and Whitaker (1974), McCreight et al. (1993), Robinson and Decker-Walters (1997). Reviews on melon biotechnology include Guis et al. (1998), Pech et al. (2007). Complementary information can be found in this book in the "Cucumber" chapter.

Melon is mainly cultivated for the consumption of the fruits which can be harvested immature; in this case, the fruit is not sweet and can be eaten raw, cooked or pickled. In most cases, the fruit is harvested at maturity and high sugar content (mainly sucrose) is demanded. Fruit is mainly eaten raw; marginal uses are cubes canned in syrup, in "fruits confits", candies, ice-creams, biscuits and also in cosmetics. Locally, seeds or leaves can also be consumed.

2 Origin, Domestication and History

Melon is a diploid species with 2n = 2x = 24 chromosomes. All other species of *Cucumis* but *C. sativus* have also x = 12 chromosomes, some being tetraploid or hexaploid. The centre of origin is very probably East Africa. Wild melons, defined by fruits smaller than 50 grams, are commonly found in East and West Africa but also in Central Asia up to India. Melon could have been domesticated for its seeds rich in proteins and lipids of good nutritional value, like squash (*Cucurbita*) and watermelon [*Citrullus lanatus* (Thunb.) Matsum. & Nakai]. Indeed, in most cases, wild melon, squash and watermelon flesh is bitter due to the presence of cucurbitacins. Moreover, the flesh is very thin (Fig 1B). Domestication resulted in the development of the fruit mesocarp.

Most vegetables including melon are rich in water and very few archaeological records have been documented. The main difficulty with old representations or texts is the incertitude between cucumber and some botanical groups of melon like *conomon*, *chate* or *flexuosus* (see section 3). They have the same shape and both are not sweet and eaten in "salads". There is also an uncertainty between melon and watermelon when the fruit is ripe. Combining archaeological records (seeds), iconography and analysis of the available texts, it seems that melon was cultivated in China 3000 years BC, in India 2000 BC and in Egypt 1500 BC (Li, 1969; Watson, 1969; Vishnu-Mittre, 1974; Stol, 1987; Manniche, 1989; Walters, 1989; Decker-Walters, 1999).

The same uncertainty remains during the Greek and Roman periods (Andrews, 1956). The *cucumis* of Pliny and other authors could be a melon, a watermelon or a cucumber. Nevertheless Pliny mentions a *cucumis* with a strong aroma and dehiscence of the peduncle of the fruit which is clearly a melon. Paintings from Herculaneum and Pompeii show melon.

In their travels in Central Asia, Marco Polo (second half of XIIIth century) and Ibn Battuta (first half of XIVth century) described melon or watermelon of superior quality; some were dried and exported. In *Theatrum sanitatis* painted around 1400 in northern Italy, three types of melon including cantaloupes and casaba types are represented.

Melon was introduced in the new world by Columbus at his first travel and, like watermelon, was immediately adopted and grown by Amerindians.

3 Botanical Groups or Varietas

C. melo is the most variable species of the genus *Cucumis* and, even among vegetables, the great diversity of the fruit shape and size has always been recognized. C. von Linné himself described three species: *C. melo*, *C. flexuosus* and *C. dudaim*. Many other species such as *C. callosus* (Rittler) Cogniaux, *C. chate* Hasselquist, *C.*

conomon Thunberg, or *C. momordica* Roxburgh are now considered as synonyms of *C. melo*.

The centre of diversification is in Asia from the Mediterranean Sea to Eastern Asia. Melon has been divided in two subspecies (Jeffrey, 1980) according to the hypanthium's hairiness: subsp. *melo* with long hairs and subsp. *agrestis* with short hairs. Botanical groups belonging to the subsp. *agrestis* are found in eastern Asia from India to Japan and to the subsp. *melo* from India to Europe and the new world. The term *agrestis* is puzzling because wild melon defined by the small size of the fruits and of the seeds can have short or long hairs and so they can belong to the subsp. *melo* or to the subsp. *agrestis*. Independent domestication events may correspond to these two subspecies.

Naudin (1859) defines nine "tribes" of cultivated melons and one wild form (Naudin, 1859). Other researchers added groups or merged them (Cogniaux and Harms, 1924; Pangalo, 1928, 1933, 1958; Filov, 1960; Whitaker and Davis, 1962; Grebenšcikov, 1986; Munger and Robinson, 1991; Pitrat et al., 2000). A short description of the main botanical groups follows (Table 1, Figures 1 and 2) (Anonymous, 2006):

Table 1. Botanical groups of melon.

	Subsp *agrestis*	**Subsp *melo***
Non sweet	*acidulus, conomon, momordica*	*chate, flexuosus, tibish*
Sweet	*makuwa, chinensis*	*adana, ameri, cantalupensis, chandalak, reticulatus, inodorus*
Fragrance		*dudaim*

conomon (Thunberg) Makino

The foliage is dark green. Plants are andromonoecious with elongated fruit, a smooth thin skin and a white firm flesh. The fruit is not sweet, not aromatic and not climacteric. They are used like cucumber, eaten raw in salad or pickled. *Conomon* is cultivated only in Eastern Asia (China, Korea, Japan). Some instances are Freeman's cucumber, Shiro uri okayama, Shiro hagura uri, Aodaisimo uri, Wasada uri, Hyougo ao shima uri, Ko shiro uri, Kurona uri.

makuwa Thunberg

Leaves are dark green. The sex type is andromonoecious. Fruits are flat to round to oval with a smooth thin skin which can be white, yellow or light green, with or without sutures. The flesh is white, sweet with little aroma. The fruit is climacteric. They are grown in Eastern Asia and their importance is decreasing. The following cultivars belong to the varietas *makuwa*: Ginsen makuwa, Shirokawa nashi makuwa, Kanro makuwa, Showa kogane nashi makuwa, Kairvo ogate kogane sennari makuwa, Kinko makuwa, Ogon 9.

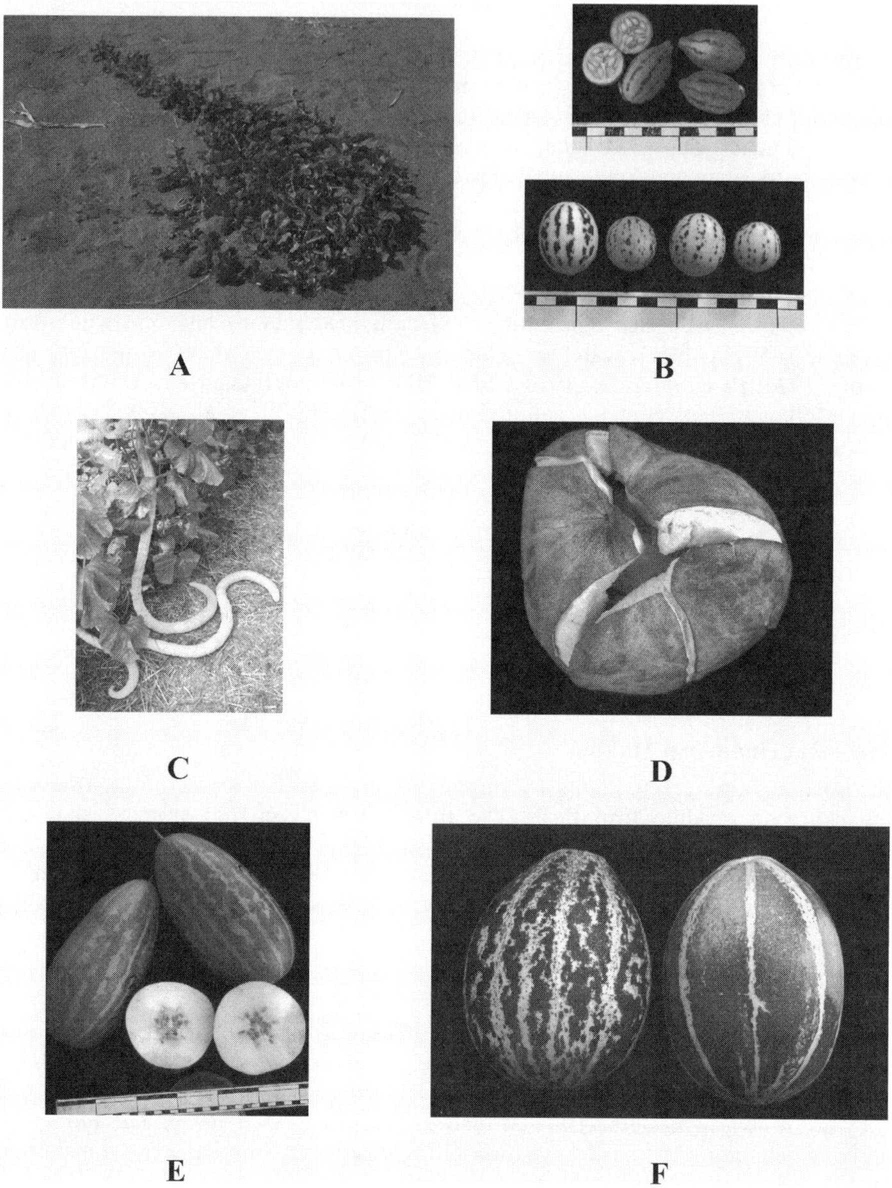

Fig. 1. Some melon *varietas*. A = Wild melon plant in Sudan. B = Fruits of wild melons. C = var. *flexuosus*. D = var. *momordica*. E = var. *acidulus*. F = *var. tibish*.

Fig. 2. Some melon *varietas*. A = var. *cantalupensis* cultigroup Charentais. B = var. *dudaim*. C = var. *conomon*. D = var. *makuwa*. E = var. *chinensis*. F = var. *inodorus* cultigroup Piel de sapo.

***chinensis* Pangalo**

The foliage is dark green. Plants are generally andromonoecious but a few hermaphrodite accessions have been described. The fruit is pear-shaped, with a light green or dark green skin with spots. The flesh is green or orange with medium sugar content and little or no aroma. The fruit can be climacteric or non climacteric. This varietas is found in China or Korea. PI 161375, PI 255478, PI 255479 belong to this type.

***momordica* (Roxburgh) Duthie & Fuller**

The foliage is light green. The sex type is monoecious. The fruit is flat to round to elongated. The thin fruit skin is smooth or slightly ribbed. A typical characteristic is the bursting of the fruit at maturity. The sugar content is low. The flesh is mealy and white at maturity with a low aroma. Fruit is climacteric. This varietas is grown only in India, for instance: PI 124111 (MR-1), PI 414723, Faizabadi phoont, PI 164343, Gill patti phut, PI 183307, PI 532841.

***acidulus* Naudin**

The plants are monoecious. The fruit is oval or elliptic, smooth, with a green or orange skin colour, uniform or with spots. The flesh is white, very firm and crisp. It has neither sugar nor aroma. *Acidulus* melons are found mainly in the southern part of India. Some instances are PI 164323, 90625 or Kekiri.

***tibish* Mohamed**

This is one of the most primitive forms of melon. The plants are andromonoecious. The fruit is oval-shaped. It has a small size (300 g) with a dark green skin with light green or yellow stripes. The fruit flesh is white, firm without sugar and aroma. It is harvested before maturity and eaten raw in salad like a cucumber. A very similar type (*seinat*) is cultivated for eating the seeds. *Tibish* and *seinat* are cultivated only in the Sudan.

***chate* Hasselquist**

The plants are monoecious, sometimes andromonoecious. The fruit is round to oval, with ribs and a light to dark green skin. The flesh is white to light orange, without either aroma or sugar. The fruit is climacteric. It is harvested before maturity and eaten raw in salad. It is cultivated in the Mediterranean basin and in western Asia. Carosello in Italy belongs to this varietas.

***flexuosus* (L.) Naudin**

The sex type is monoecious. The fruit is long to very long, up to 2 m. The skin is light green or dark green, ribbed or wrinkled. The flesh is white to light orange, mealy at maturity without either sugar or aroma. It is often called snakemelon, snakecucumber or Armenian cucumber. It is widely grown from Northern Africa to Turkey to Iraq to India. Some instances are: Fakouss, Fegous, Adjour, PI 222187, Alficoz, Silka, Kakri long green, Acur.

***cantalupensis* Naudin**

The plants are usually andromonoecious. The fruit is flat to oval, strongly to moderately ribbed with a smooth skin, sometimes with warts. The flesh is orange, sometimes green, aromatic and sweet. The fruit is climacteric. It is grown in Europe, Western Asia, North and South America. Charentais, Ogen, Ananas d'Amérique, Noir des Carmes, Prescott, Muscatello belong to the *cantalupensis* varietas.

***reticulatus* Séringe**

The sex type is andromonoecious. The fruit is round to oval, with a typical netted skin, with or without ribs. The flesh is usually orange, aromatic and sweet. The fruit is climacteric. This varietas is grown in Europe, Asia, North and South America. Instances are the American cantaloupes (Topmark, PMR 45, Hale's Best, Delicious 51), Sucrin de Tours, Earl's favourite, Netz marktgärtner, Galia.

***ameri* Gabaev**

The plants are andromonoecious. The fruit is elongated and oval shaped with a yellow to light green skin colour, usually without ribs. The skin can be slightly netted. The juicy flesh is white to light orange with a low aroma, and very good sugar content. The fruit is climacteric. It is grown mainly in Western and central Asia. Instances are Ananas, Altajskaja, Khatoni, Kzyl urum.

***inodorus* Jacquin**

The *inodorus* varietas is a very large group. Plants are andromonoecious. The fruit is round to elliptic, sometimes pointed at the peduncle end. The skin colour is white to yellow to dark green, uniform or with spots, often wrinkled, with or without ribs. The flesh is white, sweet with a low aroma, from which their name of *inodorus*. The fruit is non climacteric. They are grown in Central Asia, the Mediterranean basin, North and South America under the names Piel de sapo, Rochet, Amarillo (Canari), Casaba, Kirkagac, Yuva, Hasan bey, Tendral, Melon blanc d'Antibes d'hiver, Honeydew, Branco, Baskavas.

***dudaim* (L.) Naudin**

The sex type is andromonoecious. The fruit is round, small (orange size), yellow with ochre stripes and a velvety skin. It has a very strong typical aroma, a white thin flesh without any sugar. It is climacteric. This varietas is cultivated as ornamental or aromatic in Central Asia, from Turkey to Afghanistan. Queen Anne's pocket melon, Dastanbou, PI 177362 belongs to the *dudaim* varietas.

Within these botanical groups, cultivar-groups or cultigroups can be defined, for instance the Amarillo or the Piel de sapo within the *inodorus* or the Ogen or the Charentais within the *cantalupensis*. This large diversity is still living and cultivated in the world and not just maintained in gene banks as genetic resources. Chinese, Iranian, Spanish or French people think that the melon from Hami, the melons from Khorasan, the Piel de sapo or the Charentais are respectively the best in the world. This diversity obliges the plant breeders to a repetitive work by introducing the same characters in different cultigroups; for instance powdery mildew resistance in

Charentais, Piel de sapo, Honeydew, Yuva, *flexuosus*... Nevertheless it should be noticed that there is a diminution of the number of cultigroups and an increase of cultivars within some cultigroups. For instance in France, the Sucrin de Tours, Gris de Rennes, Ananas d'Amérique, Prescott, Noir des Carmes, Vert olive d'hiver, Blanc d'Antibes d'hiver... have disappeared from the commercial market for the benefit of the Charentais cultigroup with many cultivars and a diversification for disease resistance, fruit shelf life or sex expression within the Charentais.

4 Genetic Resources

Melon genetic resources are not threatened. Gene banks in different countries are active. Among the largest collections are Russia (2900 acc), USA (2300 acc), France (1800 acc), China (1200 acc). Collects have been made and are still conducted. Melon is not included in the international treaty for multilateral access to plant genetic resources for food and agriculture and this could reduce the exchange of accessions collected after 1993. Collaborations between gene banks is increasing for instance with the European Cooperative Programme for Crop Genetic Resources Networks (ECP/GR). Descriptors for melon have been edited by the International Plant Genetic Resources Institute (IPGRI). It seems that the wild melons are underrepresented in the gene bank. Up to now no original character has been observed only in wild melons; for instance wild melon can be resistant to powdery mildew, but the same trait is also found in cultivated landraces.

Genetic resources can be structured by geographical origin and by phenotypic traits (botanical groups). Other characters have also been utilized: Isozymes have been used as markers of the polymorphism since the 1980 (Esquinas Alcazar, 1981; Dane, 1983; Perl-Treves et al., 1985; Sujutha et al., 1991; Meglic et al., 1994; Kato et al., 1998; Akashi et al., 2001; McCreight et al., 2004). Polymorphism was also studied with different types of molecular markers (RAPD, RFLP, AFLP, SSR) (Neuhausen, 1992; Katzir et al., 1995; Garcia et al., 1998; Stepansky et al., 1999; Garcia-Mas et al., 2000; Staub et al., 2000; Akashi et al., 2001; Mliki et al., 2001; Baba et al., 2002; Liu et al., 2002; López-Sesé et al., 2002; Lee and Kim, 2003; Monforte et al., 2003; López-Sesé et al., 2004; Staub et al., 2004). There was usually a good agreement between the different types of markers (Staub et al., 1997; Garcia-Mas et al., 2000; Staub et al., 2000; López-Sesé et al., 2002). Relationship between the botanical groups and biochemical and molecular markers has been investigated. In general, accessions belonging to a botanical group fell in the same cluster determined by the molecular markers. But in many cases, accessions of different geographical origin belonging to one cultigroup were in different clusters (Mliki et al., 2001; López-Sesé et al., 2002). Molecular markers have not yet been used for the management of collections, for instance in the definition of core-collections.

Evaluations of the genetic resources for traits of horticultural interests are regularly conducted for yield and fruit quality or for pest and disease resistance.

5 Breeding for Pest and Disease Resistance

Disease resistance was among the first objectives for the development of new cultivars: PMR 45 resistant to powdery mildew was bred in the 1930 in California (Jagger and Scott, 1937). Three general comments should be done which are valid not only for melon but for all crops: (i) breeding for pest and disease resistance is much more successful when methods of artificial inoculations on plantlets have been developed; (ii) faced with the variability of the causal agent of the disease, durability of the resistance could be obtained by pyramiding genes for resistance and/or association of resistance with management practices and/or management in time and space of the resistances; (iii) multiple disease resistant cultivars should be developed. For instance, modern F_1 hybrids in the Charentais type can cumulate resistance to four races of *Fusarium* wilt, to several races of powdery mildew and to *Aphis gossypii*.

Powdery mildew

Powdery mildew is probably the most widespread disease of cucurbits, including melon. It can be caused by two pathogens: *Podosphaera xanthii* (Castagne) Braun & Shishkoff (formerly *Sphaerotheca fuliginea*) and *Golovinomyces cichoracearum* (DC) V.P. Heluta (formerly *Erysiphe cichoracearum*). The most prevalent species is *P. xanthii* but in some countries *G. cichoracearum* can also be found with a high frequency. Different races have been described mainly on *P. xanthii*; PMR 45 was released in 1937 and a new race overcoming its resistance was described in 1938 (Jagger et al., 1938). Since, many races have been described but the differential hosts were not always identical. A recent review has been published by McCreight (2006). Resistance to different races of *P. xanthii* is generally found in accessions from India, resistance to *G. cichoracearum* is quite frequent in *inodorus* botanical group (Amarillo, Piel de sapo). Inheritance studies concluded generally to monogenic dominant controls but recessive control has also been observed (Jahn et al., 2002; McCreight, 2003). Some genes can control several races of *P. xanthii*; some are specific of *P. xanthii* (*Pm-1* in PMR 45 for instance), some are specific of *G. cichoracearum* (*Pm-H* in Nantais oblong for instance) while some can control some races of both *P. xanthii* and *G. cichoracearum*. All the allelism tests have not been performed and the situation is quite confused. The differential interactions of some powdery mildew races and melon accessions are presented in Table 2, but other strains have been described using these melon lines or other accessions. The susceptibility of the accession Iran H to races 0 of *P. xanthii* and of *G. cichoracearum* should be considered as an exception as almost all melon accessions are resistant to races 0. The strains belonging to races 0 have been isolated from cucumber. This situation is similar to that described by Robinson and Providdenti (1975) where all watermelon accessions were resistant to the powdery mildew strain used in their test except PI 269677 which was susceptible.

Many lines with a high level of resistance to powdery mildew, for instance PMR 5, can present in some conditions (low light intensity, short day length, heavy fruit load) a severe leaf necrosis. These symptoms are not a hypersensitive reaction as they can be observed in the absence of the fungus. The "linkage" has not been broken and is more probably a pleiotropic effect than a true genetic linkage. The necrosis has a recessive genetic control while the powdery mildew resistance is dominant in most cases. F_1 hybrids between a resistant necrotic parent and a susceptible line are resistant without the risk of necrotic reactions.

Table 2. Reaction of powdery mildew *Podosphaera xanthii* (Px) and *Golovinomyces cichoracearum* (Gc) races on melon differential lines.

Accessions	**Px 0**	**Px 1**	**Px 2US**	**Px 2F**	**Px 3**	**Px 4**	**Px 5**	**Gc 0**	**Gc 1**
Iran H	S	S	S	S	S	S	S	S	S
Védrantais, Topmark	R	S	S	S	S	S	S	R	S
PMR 45	R	R	S	S	S	S	S	R	S
WMR 29	R	R	Het	R	R	S	S	R	S
EDISTO 47	R	R	S	R	R	R	S	R	S
PMR 5	R	R	R	R	S	R	R	R	R
Nantais oblong	R	S	S	S	S	S	S	R	R

S = Susceptible, R = Resistant, Het = Heterogeneous

Genetic control of powdery mildew resistance is generally quite simple with major dominant genes. But modifier genes are also commonly observed. For instance F_1 hybrids between one resistant parental line and several susceptible lines have not the same level of resistance.

Since PMR 45, many powdery mildew resistant cultivars have been released with resistance to one or several races.

Downy mildew

Downy mildew, caused by *Pseudoperonospora cubensis* (Berk. & Curtis) Rostovzev, can be found in temperate and tropical areas with high relative humidity. Different pathotypes have been described and good level of resistance have been found in Indian accessions: PI 124111 and its derivative MR-1 (Thomas, 1986; Kenigsbuch and Cohen, 1989) or PI 124112 (Kenigsbuch and Cohen, 1992). Major genes have been described (Kenigsbuch and Cohen, 1989, 1992) but also a polygenic control (Epinat and Pitrat, 1994a, b; Perchepied et al., 2005a). Very few improved cultivars have been released with a high level of resistance to downy mildew.

Fusarium wilt

Fusarium oxysporum Schltdl. f. sp. *melonis* Snyder & Hansen is a soil-borne pathogen causing Fusarium wilt. Four races have been described (Risser et al., 1976) (Table 3); strains belonging to race 1.2 can induce yellowing symptoms followed by the death of the plant (race 1.2 yellowing) or wilting without yellowing (race 1.2 wilting). A first major gene (*Fom*-1) has been described segregating in Charentais populations in France and open pollinated cultivars homogeneous for resistance have been selected such as Doublon or Védrantais. These cultivars were observed susceptible in some fields where race 1 was present. Screening of the genetic resources led to the discovery of an independent dominant gene (*Fom*-2) in accessions from Far-East (CM 17187) (Risser, 1973). A third gene (*Fom-3*) controlling resistance to races 0 and 2 like *Fom-1* has also been described (Zink and Gubler, 1985). A polygenic recessive resistance has been observed in few accessions, for instance Ogon 9, from Far-East (Perchepied et al., 2005b); this partial resistance is effective against all races, including race 1.2. A non commercial breeding line, Isabelle, cumulating *Fom-1*, *Fom-2* and the polygenic recessive resistance of Ogon 9 has been released.

Table 3. Genetic control of resistance to races of *Fusarium oxysporum* fsp. *melonis*.

Accessions	Genotype	Race 0	Race 1	Race 2	Race 1.2
Charentais T		S	S	S	S
Doublon, Charentais Fom1,	*Fom-1*	R	S	R	S
CM 17187, Charentais Fom-2	*Fom-2*	R	R	S	S
Perlita FR	*Fom-3*	R	S	R	S
	Polygenic recessive	r	r	r	r
Isabelle	*Fom-1, Fom-2, polygenic recessive*	R	R	R	r

S = Susceptible, R = Resistant, r = partially resistant

Almost all the F_1 hybrids in the Charentais cultigroup have *Fom-1* and *Fom-2*. A few have also a good level of resistance to race 1.2 coming from Isabelle but it is difficult to cumulate a high level of resistance and good horticultural characteristics.

Control of Fusarium wilt includes also grafting on resistant rootstocks. Resistant melon accessions can be used but usually other cucurbits are preferred. *Benincasa hispida* (Thunb.) Cogn. has first be proposed but today *Cucurbita* rootstocks are more commonly used (see chapter "Pumpkins and winter squash" in this book).

Gummy stem blight

Didymella bryoniae (Fuckel) Rehm [syn. *Mycosphaerella melonis* (C.O. Sm.) Grossenb.] affects leaves, fruits and stems. The disease is observed mainly in hot and humid conditions (tropical and subtropical areas or greenhouses). Five genes (four dominant *Gsb-1* to *Gsb-4*, and one recessive *gsb-5*) have been described in accessions from China, Mexico and Zimbabwe (Frantz and Jahn, 2004). Up to now, no study on the interactions between these resistant accessions and *D. bryoniae* strains has been conducted and no races have been clearly defined.

Viruses

More than 30 viruses can infect cucurbits and melon (Lecoq, 2003). Some of them can induce very severe losses. They are very difficult to control. Symptoms are not always easy to recognize, particularly in the case of mixed infections which are common.

Table 4. Resistance to some viruses infecting melon.

Genus: Species	**Source of resistance (genetic control)**	**References**
Polerovirus: *Cucurbit aphid borne yellows virus* (CABYV)	PI 414723, 90625, PI 124112 (2 recessive genes, *cab-1 cab-2*)	Dogimont et al., 1997
Carmovirus: *Melon necrotic spot virus* (MNSV)	PI 161375, Gulfstream, Planters Jumbo, PMR 5 (*nsv*) Doublon (*Mnr1 Mnr2*)	Coudriet et al., 1981 Mallor et al., 2003
Cucumovirus: *Cucumber mosaic virus* (CMV)	Freeman's cucumber PI 161375, Ginsen makuwa (Oligogenic recessive)	Karchi et al., 1975 Risser et al., 1977; Dogimont et al., 2000
Tobamovirus: *Cucumber green mottle mosaic virus* (CGMMV)	Phoot, VRM5-10, VRM29-1… (inheritance?)	More, 2001
Tobamovirus: *Kyuri green mottle mosaic virus* (KGMMV)	Mawatauri, PI 161375 (inheritance?)	Daryono et al., 2005
Crinivirus: *Cucurbit yellow stunting disorder virus* (CYSDV)	TGR 1551 (*Cys*)	López-Sesé and Gómez-Guillamón, 2000
Crinivirus: *Lettuce infectious yellows virus* (LIYV)	PI 313970 (*Liy*)	McCreight, 2000
Crinivirus *Beet pseudo yellows virus* (BPYV)	Nagata Kin makuwa (*My* partially dominant)	Nuez et al., 1999
Potyvirus: *Zucchini yellow mosaic virus* (ZYMV)	PI 414723 (*Zym*, *Zym-2*, *Zym-3*)	Pitrat and Lecoq, 1984; Danin-Poleg et al., 1997
Potyvirus: *Papaya ringspot virus* (PRSV-W)	PI 180280 (Prv^1), PI 180283 (Prv^2) PI 124112 (*Prv-2*)	Kaan, 1973; Webb, 1979; Pitrat and Lecoq, 1983; McCreight and Fashing-Burdette, 1996
Potyvirus: *Watermelon mosaic virus* (WMV)	PI 414723 (*Wmr*), TGR 1551 (one recessive and one dominant)	Gilbert et al., 1994 Diaz-Pendon et al., 2005

The prevalence of the different viruses varies according to the agrosystem: vectors, plastic mulches, greenhouses, weeds and other crops as vectors and virus reservoirs… For instance in southern USA, the substitution of *Bemisia tabaci* Gennadius by the silverleaf whitefly *Bemisia argentifolii* Bellows & Perring (= biotype B of *B. tabaci*) led to the near disappearance of *Lettuce infectious yellows virus* (LiYV). In southern Europe, the development of *B. tabaci* has been rapidly followed by the whiteflies transmitted viruses *Cucurbit yellow stunting disorder virus* (CYSDV) and *Cucumber vein yellowing virus* (CVYV). In France, it seems that the incidence of *Cucumber mosaic virus* (CMV) is decreasing and that the incidence of *Watermelon mosaic virus* (WMV) is increasing. In the beginning of the 1980s, *Zucchini yellow mosaic virus* (ZYMV) is a good example of emerging disease.

One of the best approaches to control the viruses is breeding for resistance, starting with the evaluation of genetic resources. Sources of resistance to several of them have been described and their inheritance has been studied in many cases (Table 4). Sources of resistance are mainly found in accessions from India (PI 414723, PI 180280, PI 180283, PI 124112, PI 313970…) but also from Far-East (PI 161375, Freeman's cucumber) or Zimbabwe (TGR 1551). Novel strategies have also been developed using the virus coat protein or ribozymes in transgenic plants (see section 8 Integration of new biotechnologies in breeding programmes).

Up to now, few improved cultivars with virus resistance have been released. Techniques for screening resistant plants in segregating progenies are available, at least with mechanically transmitted viruses. This screening is more difficult when the vectors, aphids or whiteflies for instance, must be used. There are often mixed virus infections in a melon crop and resistance to one of the viruses of the complex would not bring any horticultural interest. Seed company strategy could be to have a cultivar with resistance to several viruses which would bring a commercial advantage over competitors and not to release cultivars with resistance to only one of the viruses.

Pests

Resistance to pests has received less attention than resistance to fungi or viruses (Robinson, 1992). Nevertheless resistances to the aphid *Aphis gossypii* Glover (Kishaba et al., 1971; Bohn et al., 1973; Lecoq et al., 1979; Soria et al., 2003), to the whitefly *Bemisia tabaci* Gennadius (Boissot et al., 2003), and to leafminers *Liriomyza sativae* Blanchard (Kennedy et al., 1978) and *L. trifolii* Burgess (Dogimont et al., 1999) have been described. Resistance to *A. gossypii* is particularly interesting because resistance to colonization of the plant by the aphid is linked with resistance to virus transmission by *A. gossypii*. This resistance to virus transmission is specific of the aphid species but not of the virus (Lecoq et al., 1979; Lecoq et al., 1980; Soria et al., 2003). It brings a few days delay in the development of the epidemics which could be cumulated with partial virus resistance and technical practices such as weeding and using repulsive plastic mulches.

Resistance to many other diseases has been described including among others *Alternaria cucumerina* (Ellis & Everh.) Elliott (Thomas et al., 1990), *Monosporascus cannonballus* Pollack & Uecker (Crosby et al., 2000; Dias et al., 2004), *Diaphania hyalinata* L. (Guillaume and Boissot, 2001).

Finally, there are many diseases against which no resistance has been found: for instance root-knot nematodes (*Meloidogyne sp*) or bacterial wilt [*Erwinia tracheiphila* (Smith) Bergey et al.].

Sources of resistance to pests and diseases are mainly found in accessions from India and Far-East belonging to the *momordica*, *acidulus*, *conomon,* and *makuwa* varietas. Some accessions are particularly interesting as they cumulate resistance to several diseases. For instance PI 414723 from India is resistant to Fusarium wilt, powdery mildew, *A. gossypii*, ZYMV, PRSV, and CABYV; MR-1 from India is resistant to Fusarium wilt, powdery and downy mildews, and *Alternaria*; PI 161375 from Korea is resistant to CMV, *A. gossypii*, MNSV, and Fusarium wilt. Other geographical origins could also be interesting, for instance TGR 1551 from Zimbabwe which is resistant to powdery mildew, CYSDV, *A. gossypii* and WMV. All these accessions are cultivated and even if wild melons could be resistant to some disease, no resistance has been found only in non cultivated melons. This could be also due to the fact that wild melons are under represented in collections.

6 Plant Growth, Sex Expression and Fruit Quality

Melon has usually a long main stem with several lateral branches. The repartition of the flowers at the different nodes is more regular in melon than in other cucurbits. At nodes of the main stem are located multiflowered inflorescences of male flowers. At the first and second nodes of the lateral branches are one-flowered, sometimes two-flowered inflorescences of female or perfect flowers, in monoecious or andromonoecious plants respectively. On the following nodes of the lateral branches are again male inflorescences. The more the plant is ramified, the more female flowers are produced. But not every flower gives a fruit as there is a strong competition between young fruits. Plants with 20-30 female flowers bear only 4-5 fruits. If the growing (fertilizers, temperature) and phytosanitary conditions are good enough, a second fruit set can be obtained. After harvesting the first fruits, the plants grow again, produce new female flowers and a second wave of fruits can be harvested. According to the temperature, a typical timetable is one month between sowing and the first male flowers, one month more to the female flowers and one to two months between pollination and fruit maturity depending of the fruit size and type. Melon is very susceptible to temperature and to light intensity and day length. Contrary to cucumber, there is no all-year-round production of melon under glasshouses in temperate climate, northern or central Europe for instance. For breeders, it is difficult in temperate countries to make three generations per year. During the winter, a third generation can be produced in tropical countries or in the southern hemisphere.

Genes with effect on the plant habit have been identified. Several cultivars with very short internodes (gene *si*-1) have been released (Denna, 1962). A very high density plantation with one fruit per plant maturing more or less simultaneously is attractive; up to now, it seems that this idea is not commercially successful. Plants with the gene *si-2* (Paris et al., 1984) have short internodes when the plant is young and normal internode length when the plant is older; fruits are grouped at the centre of the plant giving a “birdnest” phenotype. Intermediate internode length is controlled by the gene *si-3* (Knavel, 1990). As female or perfect flowers are mainly at the first and second nodes of the lateral branches, increasing the ramifications could increase the pistillate flower number and the fruit number. The number of ramifications or lateral branches is under a polygenic control (Zalapa et al., 2004).

Melon is particularly interesting for its flower biology. The Cucurbitaceae family has a very high proportion of dioecious species but not *C. melo*. Three types of flowers can be observed: male, female and hermaphrodite or perfect (Fig. 3). Wild types of melon, like most species of the genus *Cucumis*, are monoecious (male and female flowers on the same plant). About 2/3 of accessions in collections of cultivars or land races are andromonoecious (male and hermaphrodite flowers on the same plant) and 1/3 are monoecious. Few accessions are hermaphrodite (all plants have only perfect flowers) (Poole and Grimball, 1939; Kubicki, 1969). From these hermaphrodite accessions, gynoecious lines have been developed. Two genes are involved in the genetic control: the locus *a* controls the presence/absence of stamens in female flowers, the locus *g* controls the presence/absence of two types of flowers on the same plant (Table 5). The phenotype can be temporarily altered by external factors, particularly ethylene. Spraying ethrel (a precursor of ethylene) on the leaves of a plant mimics the effect of the allele *g*, by suppressing the male flowers; spraying silver nitrate (an inhibitor of ethylene) mimics the effect of the allele *a*, by inducing the presence of stamens in female flowers.

Fig. 3. Melon flowers: A = female, B = perfect, C = male.

In order to produce more easily F_1 commercial hybrids (see section 9 Seed production), seed companies have tried to introduce the allele a^+ in traditionally andromonoecious cultivars, for instance in the *cantalupensis* or the *reticulatus* groups. The locus a/a^+ has pleiotropic effect on fruit shape. The absence of stamens in female flowers induces larger and more elongated fruits. It has been suggested that there was a linkage between the locus *a* and another locus named *Oval* fruit shape (symbol *O*) (Wall, 1967). But according to other experiments made by inducing stamens in female flowers by spraying silver nitrate (Risser, 1984), it is more likely a pleiotropic effect *i.e.* interactions between a^+ and the genetic background. There are some flat monoecious accessions, for instance MR-1; by backcrossing the allele a^+ by a round andromonoecious, an oval-shaped monoecious line is obtained. Breeder's experience shows also that once a round monoecious line is obtained, in new crosses with round andromonoecious, oval monoecious fruits are observed: the "linkage" has not been broken. In addition to the locus *a*, QTLs involved in the fruit shape have been detected (Périn et al., 2002c; Monforte et al., 2004).

Table 5. Genetic control of sex expression in melon and effect of ethylene (Ethrel and silver nitrate $AgNO_3$ are respectively a precursor and an inhibitor of ethylene).

	Locus a		
Locus g	**Absence of stamens in pistillate flowers (allele a^+)**	**Presence of stamens in pistillate flowers (allele a)**	
Two types of flower on a plant (allele g^+)	Monoecious	Andromonoecious	**+ Ethrel** ↓
One type of flower on a plant (allele g)	Gynoecious	Hermaphrodite	
	+ $AgNO_3$ →		

Sugar content, volatile compounds and flesh texture and colour are the main components of fruit quality along with external aspect (fruit shape, skin colour, netting, sutures…). Many studies have been conducted on the aroma composition, sugar accumulation, or micronutriment content, in different cultivars or according to the fruit development, growing and storage environments.

Melon fruit flesh composition for minerals and vitamins is given in Table 6, in comparison with other fruits and vegetables of temperate climate. Concerning minerals, melon is under the median for calcium, iron and phosphorus, and above the median for magnesium, potassium and sodium. For vitamins, melon is under the median for vitamins B2 and E, close to the median for vitamins B1, B6 and pantothenic acid and above the median for vitamins B3, C, β-carotene (orange-fleshed melons) and folic acid. There is a growing interest in developing "neutraceutical foods", for instance fruits and vegetables rich in beneficial phytochemicals such as vitamins, antioxidants, minerals… Using the natural variability, cultivars have been developed with higher concentration in vitamins and melon breeders could be interested in such objectives. It should be mentioned that concentration of a particular molecule in a fruit is not always correlated with its final availability for human health because of complex interactions which are not fully understood, for instance between carotenoids and vitamin C. For a melon breeder interested in developing a new cultivar rich in micronutriments, there is generally a good variability in the genetic resources, but studies on the genetic control of these traits are scarce and probably polygenic in most cases.

Table 6. Composition in minerals and vitamins (mg for 100 g of fresh weight) of melon and temperate fruits and vegetables (F&V), excluding seeds (almonds, nuts, hazelnuts, peas, beans, chickpeas,…).

	Melon	Range in F&V	F&V with low value	F&V with high value
Minerals				
Calcium	15	5 - 160	Grape, apple, watermelon	Spinach
Iron	0.2	0.2 – 2.5	Apple, pear, watermelon, melon	Spinach, swiss chard
Magnesium	14	4 - 80	Grape, apple	Swiss chard, spinach
Phosphorus	17	9 - 70	Apple, watermelon	Broccoli, celery, asparagus
Potassium	300	100 - 500	Cabbage, watermelon	Potato, swiss chard
Sodium	18	1 - 130	Squash, peach, apricot	Swiss chard, Celery
Vitamins				
β-carotene (pro vitamin A)	4[a]	0 - 10	Potato, onion, grape	Carrot
C	25	2 - 200	Eggplant, carrot	Black currant, pepper
B1: Thiamine	0.04	0 – 0.9	Pumpkin, pepper	Brussels sprouts
B2: Riboflavin	0.02	0.02 – 0.2	Pepper, apple, cucumber, grape	Spinach
B3: Niacin	0.5	0.1 - 1.5	Apple, watermelon, onion	Potato, asparagus, peach
Pantothenic acid	0.2	0.04 - 1	Grape, apple	Chicory, broccoli
B6: Pyridoxine	0.09	0.02 – 0.3	Peach, pear, root beet	Pepper, potato
B9: Folic acid	0.1	0.004 – 0.2	Watermelon, apricot	Spinach, lamb's lettuce
E	0.1	0 - 2	Radish, squash, eggplant	Spinach, asparagus

[a] Orange-fleshed melon

A two genes model is classically admitted for the genetic control of flesh colour, the orange-fleshed cultivars having a much higher level of β-carotene; *white flesh* (symbol *wf*) and *green flesh* (symbol *gf*) are epistatic with the following genotypes and phenotypes: wf^{+} / - - is orange, *wf wf* / *gf gf* is green and *wf wf* / gf^{+} - is white (Hughes, 1948; Iman et al., 1972; Clayberg, 1992). However other genes are involved in the genetic control of fruit colour and the intensity of the colour (Monforte et al., 2004).

Fruit bitterness usually disappears when fruits approach maturity but in the case of fruits harvested immature (var. *flexuosus* for instance), one must be careful to eliminate bitterness due to the presence of the triterpene cucurbitacins. Non bitter plants at the seedling stage produce non bitter fruits but other genes can be involved (Lee and Janick, 1978; Elawed and El Jack, 1992; Ma et al., 1997).

In a cross between a non-sweet melon (var. *flexuosus*) and a sweet melon, sucrose accumulation was controlled by one recessive gene (Burger et al., 2002). QTLs were detected in crosses between medium-sweet and sweet melons (Park et al., 2003; Monforte et al., 2004; Sinclair et al., 2006). Usually sweet melons have a high pH (low organic acid content). Combination of high sucrose and high organic acid, mainly citric acid, controlled by the gene *Sour* (symbol *So*) leads to new flavours.

Genes involved in the biosynthesis of esters, the main volatile components, of the Alcohol Deshydrogenase (ADH) and Alcohol Acyl-Transferase (AAT) families have been isolated but their phenotypic effect has not yet been determined (El-Sharkawy et al., 2005).

Genetic control of fruit skin colour has been studied and some genes with strong effect have been described: *Mottled* rind pattern (symbol *Mt* and *Mt-2*), *ridged* fruit surface (symbol *ri*), *striped* epicarp (symbol *st* and *st-2*), *sutures* (symbol *s* and *s-2*), *white* colour of mature fruit (symbol *w*), *White colour of immature* fruit (symbol *Wi*), or *Yellow* colour of the fruit (symbol *Y*),

Breeders have tried to improve the shelf-life or the keeping quality after harvest. Melon is one of the few species where climacteric and non-climacteric fruits have been selected. Peach, apricot, banana, tomato, melons belonging to var. *cantalupensis*, *dudaim*, *makuwa*, or *momordica* are climacteric fruits with a burst of respiration and autocatalytic ethylene production at fruit maturity. Strawberry, citrus, grape, melons belonging to var. *inodorus* with the cultigroups piel de sapo, honeydew, tendral, yuva, and casaba are non climacteric. In melon, some traits are ethylene independent like sugar and carotenoid accumulation; while other traits like production of volatile compounds, peduncle abscission, change of skin colour, flesh softening are ethylene dependent (Guis et al., 1997a). Decreasing the intensity of the climacteric crisis enables to increase the shelf-life.

Melon seeds are rich in lipids and proteins which represent a great amount of energy. Parthenocarpic melons could theoretically produce more fruits for the same photosynthetic activity. Parthenocarpic fruits can be obtained by spraying growth substance, for instance auxins. Up to now, no natural parthenocarpy has been described in melon as it is known in cucumber.

7 Breeding Methods and Techniques

Melon is a semi-allogamous species. Presence of nectar in male and female flowers attracts bees and other insects. Cross pollination rate is higher in monoecious plants than in andromonoecious. Usually there is no vigour or fertility depression (=inbreeding effect) due to the homozygous state and pure line can be cultivated. Heterosis or hybrid vigour is clearly observed in F_1 hybrids between two different parental lines. But when the parents belong to the same cultigroup, heterosis is not very important; for instance, both parents of a commercial F_1 Charentais type belong to the Charentais type. This allows to cumulate easily dominant traits, for instance resistance to Fusarium wilt (alleles *Fom-1* and *Fom-2*), to *A. gossypii* (allele *Vat*) and to powdery mildew without the risk of necrotic reactions (see section 5). Heterosis is more important when parents are more divergent; for instance the F_1 Galia is a hybrid between an Ogen line from central Europe and a netted Asiatic parent. Most of the modern cultivars are F_1 hybrids which allow also a protection of breeder's investments.

Flowers are quite large and hand pollination is easy to handle. Male flowers (in inflorescences) appear before female or bisexual flowers at the nodes of the main stem and after the second or third node of the lateral branches, while the solitary pistillate flowers are present at the first and second nodes of the lateral branches. In the case of gynoecious or hermaphrodite plants, female or perfect flowers respectively appear at all the nodes of primary or secondary branches. Controlled pollinations can be achieved by working in insect-proof greenhouses or by closing the flowers before and after pollination with bags or other means. The typical number of stamens is five but partial fusion leads generally to two large bilocular and one small unilocular stamens. The stamens open and release the pollen the day the corolla opens. To perform crosses, perfect flowers must be emasculated before stamens open. The stigma is receptive one day before and one day after the corolla opens. Pollination can be achieved at the same time as emasculation. It is usually more efficient to make pollinations in the morning, as in the afternoon nectar can wet the pollen. One successful pollination will produce 300 to 500 seeds.

Classical breeding methods using crosses and selfing such as pedigree or backcrosses can be used. Usually backcrosses are used to introduce a new character in a cultigroup, for instance disease resistance from an exotic accession in a Charentais or Piel de sapo type. Within a cultigroup, pedigree and recurrent selection are used to cumulate favourable traits. For characters with a simple inheritance, selection is efficient in the first generations (F_2 or F_3). For instance, tests can be performed on individual F_2 plants for many disease resistance; moreover techniques have been developed allowing multiple tests on a single plant, for instance by using leaf disks or detached cotyledons. For characters with a more complex inheritance, such as fruit quality, earliness or yield, selection is more efficient if applied in more advanced generations (F_4 or F_5).

For developing populations, pollen mixture can be achieved by spreading the pollen in n-pentane which evaporates very quickly; pollen can then be stored for a few days.

Vegetative propagation of selected plants is easy by cuttings. The tip (about 5 cm) of a young shoot is cut and the leaves are removed, except the growing point. The basis is dipped in a solution of indole-3-butyric acid and placed in a soil/sand mixture. The shoots are placed in conditions with medium temperature and saturated humidity for about a week, for instance under a plastic sheet; then relative humidity is progressively decreased for another week. Roots are usually observed within two weeks.

Haploid plants can be obtained by *in vivo* gynogenesis induced by irradiated pollen followed by *in vitro* embryo rescue (Sauton and Dumas de Vaulx, 1987). Male flowers are irradiated with γ-rays from Co^{60} at 300 Gy and used to pollinate female flowers (or emasculated perfect flowers). About twenty days after pollination, the fruits are surface sterilized, seeds are opened and small embryos are *in vitro* cultured. The detection of seeds with small embryos can be enhanced by X-rays radiography of the seeds. Haploid plantlets can be *in vitro* propagated by cuttings. Spontaneous chromosome number doubling can be observed; chromosome number can be doubled by *in vitro* colchicine treatment (0.2 % for 2 hrs) (Yetisir and Sari, 2003). As for most species, the efficiency of haploid production is genotype- and season-dependent.

Tetraploid melons are of superior quality for their sugar content and longer shelf-life. Tetraploid plants can be obtained by colchicine treatment or by spontaneous *in vitro* regeneration (Ezura et al., 1992). No commercial cultivar has been released due to defaults such as fruit cracking or late maturity (Dumas de Vaulx, 1974). Triploid melons obtained by crossing a tetraploid and a diploid line have also been studied (Adelberg et al., 1995).

Numerous attempts of interspecific crosses between melon and other *Cucumis* species such as *C. metuliferus* or *C. sativus* have been made (Fassuliotis, 1977; van der Knaap and de Ruiter, 1978; Custers and Bergervoet, 1984; Lebeda et al., 1996). For instance, resistance to root-knot nematodes, *Meloidogyne* sp., has not been observed in melon but is present in wild species such as *C. metuliferus* or *C. anguria*. Up to now, no character of horticultural interest has been introduced from another species in the melon genome (Norton and Granberry, 1980).

8 Integration of New Biotechnologies in Breeding Programmes

Melon has a small genome size, estimated at 450-500 Mbp (1C) (Arumanagathan and Earle, 1991), a chloroplast genome estimated at 150 kbp (Ward et al., 1981; Palmer, 1982; Perl-Treves and Galun, 1985) and a huge mitochondrial genome of 2400 kbp (Ward et al., 1981). The maternal transmission of the chloroplast and the paternal transmission of the mitochondrial genome have been demonstrated (Havey et al., 1998).

An International Cucurbit Genomics Initiative started in 2005 with the aim of the better knowledge of cucurbits genomics: genetic and physical maps, EST (Expressed Sequence Tags) and BAC (Bacterial Artificial Chromosome) libraries, genome sequencing. Melon is the most advanced species for genomics among the cucurbits and will be used as a model for the family.

A gene list is regularly published by the Cucurbit Genetics Cooperative and can be consulted at http://cuke.hort.ncsu.edu/cgc/index.html.

Different types of molecular markers have been used to study the genetic diversity (see section 4) and to develop genetic maps by crossing distant parents. Several types of population have been used: F_2, BC, doubled haploid (DH) and recombinant inbred lines (RILs). Genes and QTLs controlling phenotypic traits have been localized (Pitrat, 1991; Baudracco-Arnas and Pitrat, 1996; Wang et al., 1997;

Brotman et al., 2000; Danin-Poleg et al., 2000; Oliver et al., 2001; Danin-Poleg et al., 2002; Périn et al., 2002a; Périn et al., 2002b; Périn et al., 2002c; Silberstein et al., 2003; Brotman et al., 2004; Fukino et al., 2004; Gonzalo et al., 2005; Perchepied et al., 2005a; Perchepied et al., 2005b) (Fig. 4). BAC and EST libraries are available and provide new sequences which can be used as markers. The next step will be the merging of these different maps in a consensus one using anchor points such as microsatellites (SSR).

The melon genetic map is the most advanced among the cucurbits. Using SSR as common points, first results indicate that linkage group B of cucumber could correspond to linkage groups E and 2 of melon (Danin-Poleg et al., 2000). No synteny was observed between the regions controlling ZYMV resistance in melon and cucumber (Park et al., 2004). Microsynteny was observed between a melon BAC clone and chromosomes 3 and 5 of *Arabidopsis thaliana* (van Leeuwen et al., 2003).

The development of genetic maps allows the use of marker assisted selection (MAS). Breeders can use markers closely linked with genes or QTLs of horticultural interest such as disease resistance, fruit quality or flower biology. One of the main problems in MAS is the recombination event between the gene and the marker. Markers should be developed in different elite material genetic background. When a gene has been cloned, polymorphic sequences between the alleles allow to define markers within the allele. For instance a single nucleotide difference has been found between the alleles *nsv* and nsv^+ controlling resistance and susceptibility to MNSV (Morales et al., 2005).

In vitro regeneration by organogenesis or somatic embryogenesis is not easy; one of the main problem is the great number of tetraploid plants which are regenerated. Shoot formation from cotyledons, hypocotyls, roots, or leaf explants has been obtained (Moreno et al., 1985; Kathal et al., 1994; Guis et al., 2000; Curuk et al., 2002). Somatic embryogenesis has also been successful (Oridate and Oosawa, 1986; Guis et al., 1997b; Akasaka-Kennedy et al., 2004). Regeneration is genotype dependent and sexually transmissible; the genotype BU-21/3 has superior competence for regeneration by organogenesis (Molina and Nuez, 1996, 1997; Galperin et al., 2003a; Galperin et al., 2003b). Genetic transformations of melon with genes involved in disease resistance, salt tolerance, or long shelf-life have been successful (Fang and Grumet, 1990; Yoshioka et al., 1993; Clough and Hamm, 1995; Ayub et al., 1996; Bordas et al., 1997; Fuchs et al., 1997; Plagès, 1997; Clendennen et al., 1999; Taler et al., 2004). For instance, transgenic Charentais-type melon plants with antisense ACC oxidase, the last enzyme in the biosynthetic pathway of ethylene, have been extensively studied for shelf-life and ethylene-dependent or independent traits. Sugar and β-carotene accumulation are ethylene independent whereas fruit peduncle abscission, skin colour change, fruit flesh softening and emission of volatile compounds are ethylene dependent.

Agrobacterium tumefaciens has generally been used as the vector. An attenuated non-aphid transmissible strain of ZYMV has also been used as vector for herbicide resistance or for producing antiviral and antitumor proteins (Shiboleth et al., 2001; Arazi et al., 2002).

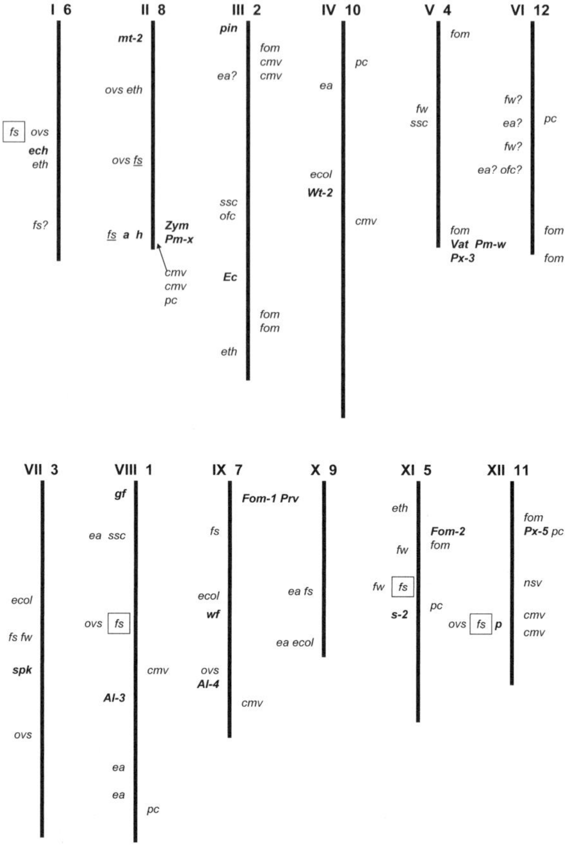

Fig. 4. A tentative synthetic genetic map of melon. Linkage groups are labelled with roman numbers (Périn et al. 2002a) or with Arabic numbers (Oliver et al. 2001). Genes and QTLs are indicated in bold and normal characters respectively. As there are some doubts on the alignment of the two maps (top and bottom), QTLs with a question-mark on LG I, III and VI could be in other positions. Genes or QTLs for disease resistance are on the right side of each linkage group and genes or QTLs for other traits (flower biology, fruit characters) are on the left side. Genes for disease resistance: *Fom* = *Fusarium* wilt, *nsv* = *Melon necrotic spot virus*, *Pm* = Powdery mildew, *Prv* = *Papaya ringspot virus*, *Px* = *Podosphaera xanthii* (Powdery mildew), *Vat* = Virus aphid transmission resistance, *Zym* = *Zucchini yellow mosaic virus*. QTLs for disease resistance: *cmv* = *Cucumber mosaic virus*, *fom* = Fusarium wilt race 1.2, *pc* = *Pseudoperonospora cubensis* (downy mildew). Genes for other traits: *a* = andromonoecious, *Al* = Abscission layer, *Ec* = empty cavity, *ech* = exaggerated curvature of the hook, *gf* = green flesh, *h* = halo cotyledons, *mt* = mottled (spots) fruit, *p* = pentamerous, *pin* = pine seed shape, *s* = suture, *spk* = speckled fruit epidermis, *wf* = white flesh, *Wt* = white seed. QTLs for other traits: *ea* = earliness, *ecol* = external colour of the fruit, *eth* = ethylene, *fs* = fruit shape, *fw* = fruit weight, *ofc* = orange fruit colour, *os* = ovary shape, *ssc* = soluble solid contents. QTLs for *fruit shape* described by Périn et al. (2002c) are underlined, those described by Monforte et al. (2004) are in normal characters and those which could be in common are in boxes.

9 Seed Production

Conditions for high quality melon seed production are not well defined and, usually, the production of good quality fruit allows the production of good seeds. For F_1 hybrid seed production, different methods are available. If the female parental line is andromonoecious, the perfect flowers must be emasculated before hand pollination (see section 6) with the pollen of male parental line. If the female parental line is monoecious, hand pollination can be performed without emasculation in insect-proof conditions *i.e.* under greenhouses or by closing the flowers. Alternatively, spraying ethrel on a monoecious line suppresses temporarily the male flowers and bees can make the pollination between the male and the female parental lines. Up to now, no gynoecious genotype has been used as female parental line but this would be an interesting possibility. Neither cytoplasmic male sterility nor self-incompatibility have been described in melon, but five recessive genes for male-sterility have been described and at least one (*ms-5*) has been used for production of commercial F_1 hybrids. The main difficulty is that the female parental line is segregating for male sterility; 50 % of the plants are fertile and must be removed. Male sterile plants can be vegetatively propagated by cuttings or *in vitro* but this is quite difficult to manage on a practical point of view.

Seeds must be separated from the placentas, washed and dried. Removing the placentas can be done by fermenting the mixture in water for 1 or 2 days. Instead of natural fermentation, a better method is to add a pectinolytic enzyme in the mixture. The seeds sink at the bottom and the placentas and empty seeds float. Seeds are then washed in running water and dried. Seeds, stored in the dark at 5°C and low relative humidity, are still viable for 10 years or more.

Seed-borne diseases include viruses, bacteria and fungi. *Squash mosaic virus* (SqMV), *Melon necrotic spot virus* (MNSV), *Cucumber green mottle mosaic virus* (CGMMV) and *Kyuri green mottle mosaic virus* (KGMMV) are the most frequent seed-borne viruses; the transmission of ZYMV is still controversial. The bacteria *Acidovorax avenae* subsp. *citrulli* was first described on watermelon but is now observed on melon in several countries. Some fungi like *Fusarium*, *Colletotrichum*, *Didymella* or *Cladosporium* can also be seed transmitted. Detection methods include serological tests such as ELISA and PCR-based techniques. Seed production in disease free areas must be recommended as seed disinfection is not always possible and usually decreases the seed germination rate.

For quality controls, seed companies can use biochemical or molecular markers for fingerprinting the cultivars and to check their purity, particularly to detect the inbreds in F_1 seed production.

Seed coating include usually treatments with fungicides. Priming *i.e.* pregerminating the seeds and then drying them again allows a more uniform germination.

References

Adelberg, J., P.E. Nugent, B. Rhodes, X.-P. Zhang, and Skorupska, H. 1995. Fertility and fruit characters of hybrid triploid melon. Breeding Science 45:37-43.

Akasaka-Kennedy, Y., K. Tomita, and Ezura, H. 2004. Efficient plant regeneration and Agrobacterium-mediated transformation via somatic embryogenesis in melon (*Cucumis melo* L.). Plant Science 166:763-769.

Akashi, Y., H. Ezura, Y. Kubo, M. Masuda, and Kato, K. 2001. Varietal variation in microsatellite and CAPS for ethylene-related genes and its possible association with agronomic characters in melon (*Cucumis melo*). 313-316 *In* Proceedings of the Second International Symposium on Cucurbits (S. Nishimura, H. Ezura, T. Matsuda and A. Tazuke, eds). Tsukuba (JP)

Andrews, A.C. 1956. Melons and watermelons in the classical era. Osiris 12:368-375.

Anonymous. 2006. Commercial types of melons. (International standardisation of fruit and vegetables OECD/OCDE), Paris (FR). 81 pp.

Arazi, T., L.P. Huang, L. Zhang, Y.M. Shiboleth, A. Gal-On, and Lee-Huang, S. 2002. Production of antiviral and antitumor proteins MAP3O and GAP31 in cucurbits using the plant virus vector ZYMV-AGII. Biochemical and Biophysical Research Communications 292:441-448.

Arumanagathan, K., and Earle, E.D. 1991. Nuclear DNA content of some important plant species. Plant Molecular Biology Reporter 9:208-209.

Ashurmetov, O.A. 1995. On morphology and taxonomy of the genera *Cucumis* L. and *Melo* Mill. Feddes Repertorium 106:155-159.

Ayub, R., M. Guis, M. Ben Amor, L. Gillot, J.P. Roustan, A. Latché, M. Bouzayen, and Pech, J.C. 1996. Expression of an antisense ACC oxydase gene inhibits ripening in cantaloupe melons fruits. Nat Biotech 14:862-866.

Baba, E., V. Zarka, T. Deak, A. Pedryc, I. Velich, and Bisztray, G.D. 2002. Molecular diversity of Hungarian melon varieties revealed by RAPD markers. International Journal of Horticultural Science 8:11-13.

Baudracco-Arnas, S., and Pitrat, M. 1996. A genetic map of melon (*Cucumis melo* L.) with RFLP, RAPD, isozyme, disease resistance and morphological markers. Theor. Appl. Genet. 93:57-64.

Bohn, G.W., A.N. Kishaba, J.A. Principe, and Toba, H.H. 1973. Tolerance to melon aphid in *Cucumis melo* L. J Amer Soc Hort Sci 98:37-40.

Boissot, N., D. Lafortune, C. Pavis, and Sauvion, N. 2003. Field resistance to *Bemisia tabaci* in *Cucumis melo*. HortScience 77:77-80.

Bordas, M., C. Montesinos, M. Dabauza, A. Salvador, L.A. Roig, R. Serrano, and Moreno, V. 1997. Transfer of the yeast salt tolerance gene HAL1 to *Cucumis melo* L. cultivars and *in vitro* evaluation of salt tolerance. Transgenic Research 6:41-50.

Brotman, Y., L. Silberstein, I. Kovalski, J. Klingler, G. Thompson, N. Katzir, and Perl-Treves, R. 2000. Linkage groups of *Cucumis melo*, including resistance gene homologues and known genes. 441-448 *In* Proceedings of Cucurbitaceae 2000 (N. Katzir and H. S. Paris, eds). Ma'ale Ha Hamisha (IL), 19-23/03/2000

Brotman, Y., I. Kovalski, C. Dogimont, M. Pitrat, N. Katzir, and Perl-Treves, R. 2004. Molecular mapping of the melon *Fom-1/Prv* locus. 485-489 *In* Progress in cucurbit genetics and breeding research. Proceedings of Cucurbitaceae 2004, the 8th EUCARPIA Meeting on Cucurbit Genetics and Breeding (A. Lebeda and H. S. Paris, eds). Olomouc (CZ), 12-17/07/2004

Burger, Y., U. Saar, N. Katzir, H.S. Paris, Y. Yeselson, I. Levin, and Schaffer, A.A. 2002. A single recessive gene for sucrose accumulation in *Cucumis melo* fruit. J Amer Soc Hort Sci 127:938-943.

Clayberg, C.D. 1992. Interaction and linkage tests of flesh color genes in *Cucumis melo* L. Cucurbit Genet Coop Rep 15:53.

Clendennen, S.K., J.A. Kellogg, K.A. Wolff, W. Matsumura, S. Peters, J.E. Vanwinkle, B. Copes, M. Pieper, and Kramer, M.G. 1999. Genetic engineering of cantaloupe to reduce ethylene biosynthesis and control ripening. 371-379 *In* Biology and biotechnology of the plant hormone ethylene II. Proceedings of the EU-TMR-Euroconference (A. K. Kanellis, C. Chang, H. Klee, A. B. Bleecker, J. C. Pech and D. Grierson, eds). Thira (Santorini, GR), 5-8/09/1998

Clough, G.H., and Hamm, P.B. 1995. Coat protein transgenic resistance to watermelon mosaic and zucchini yellows mosaic virus in squash and cantaloupe. Plant Dis. 79:1107-1109.

Cogniaux, A., and Harms, H. 1924. Cucurbitaceae - Cucurbiteae - Cucumerineae, p. 116-157. *In* Das Pflanzenreich. Regni vegetabilis conspectus (A. Engler ed.). Vol: 88 (IV.275.II). Wilhelm Engelmann, Leipzig (DE).

Coudriet, D.L., A.N. Kishaba, and Bohn, G.W. 1981. Inheritance of resistance to muskmelon necrotic spot virus in a melon aphid resistant breeding lines of muskmelon. J Amer Soc Hort Sci 106:789-791.

Crosby, K., D. Wolff, and Miller, M. 2000. Comparisons of root morphology in susceptible and tolerant melon cultivars before and after infection by *Monosporascus cannonballus*. HortScience 35:681-683.

Curuk, S., C. Elman, E. Schlarman, O. Sagee, I. Shomer, S. Cetiner, D.J. Gray, and Gaba, V. 2002. A novel pathway for rapid shoot regeneration from the proximal zone of the hypocotyl of melon (*Cucumis melo* L). In Vitro Cellular & Developmental Biology - Plant 38:260-267.

Custers, J.B.M., and Bergervoet, J.H.W. 1984. Embryo size in *Cucumis sativus* x *C. melo* as affected by irradiation of the pollen and genotype of the female parent. Cucurbit Genet Coop Rep 7:94-95.

Dane, F. 1983. Cucurbit, p. 369-390. *In* Isozymes in plant genetics and breeding, part B (S. D. Tanksley and T. J. Orton eds). Elsevier Science Publication, Amsterdam (NL).

Danin-Poleg, Y., H.S. Paris, S. Cohen, H.D. Rabinowitch, and Karchi, Z. 1997. Oligogenic inheritance of resistance to zucchini yellow mosaic virus in melons. Euphytica 93:331-337.

Danin-Poleg, Y., N. Reis, S. Baudracco-Arnas, M. Pitrat, J.E. Staub, M. Oliver, P. Arús, C.M. de Vicente, and Katzir, N. 2000. Simple Sequence Repeats in *Cucumis* mapping and map merging. Genome 43:963-974.

Danin-Poleg, Y., Y. Tadmor, G. Tzuri, N. Reis, J. Hirschberg, and Katzir, N. 2002. Construction of a genetic map of melon with molecular markers and horticultural traits, and localization of genes associated with ZYMV resistance. Euphytica 125:373-384.

Daryono, B.S., S. Somowiyarjo, and Natsuaki, K.T. 2005. Screening for resistance to Kyuri green mottle mosaic virus in various melons. Plant Breeding 124:487-490.

Decker-Walters, D.S. 1999. Cucurbits, sanskrit, and the Indo-Aryas. Economic Botany 53:98-112.

Denna, D.W. 1962. A study of the genetic, morphological and physiological basis for the bush and vine habit of several cucurbits. PhD Ithaca (NY, US).

Dias, R.D.S., B. Pico, A. Espinos, and Nuez, F. 2004. Resistance to melon vine decline derived from *Cucumis melo* ssp *agrestis*: genetic analysis of root structure and root response. Plant Breeding 123:66-72.

Diaz-Pendon, J.A., R. Fernandez-Munoz, M.L. Gomez-Guillamon, and Moriones, E. 2005. Inheritance of resistance to Watermelon mosaic virus in *Cucumis melo* that impairs virus accumulation, symptom expression, and aphid transmission. Phytopathology 95:840-846.

Dogimont, C., A. Bussemakers, J. Martin, S. Slama, H. Lecoq, and Pitrat, M. 1997. Two complementary recessive genes conferring resistance to Cucurbit Aphid Borne Yellows Luteovirus in an Indian melon line (*Cucumis melo* L.). Euphytica 96:391-395.

Dogimont, C., D. Bordat, C. Pages, N. Boissot, and Pitrat, M. 1999. One dominant gene conferring the resistance to the leafminer *Liriomyza trifolii* (Burgess) Diptera: Agromyzidae in melon (*Cucumis melo* L.). Euphytica 105:63-67.

Dogimont, C., L. Lecomte, C. Périn, A. Thabuis, H. Lecoq, and Pitrat, M. 2000. Identification of QTLs contributing to resistance to different strains of cucumber mosaic cucumovirus in melon. 391-398 *In* Cucurbitaceae 2000, VIIth EUCARPIA Meeting on Cucurbit Genetics and Breeding (N. Katzir and H. Paris, eds). Ma'ale Hahamisha (IL), 19-23/03/2000

Dumas de Vaulx, R. 1974. Etude des possibilités d'utilisation de la polyploïdie dans l'amélioration du melon (*Cucumis melo* L.). Ann Amélior Plantes 24:389-403.

Elawed, H.S., and El Jack, A.E. 1992. Bitterness in snake cucumber *Cucumis melo* var. *flexuosus* Naud. Cucurbit Genet Coop Rep 15:54.

El-Sharkawy, I., D. Manriquez, F.B. Flores, F. Regad, M. Bouzayen, A. Latché, and Pech, J.C. 2005. Functional characterization of a melon alcohol acyl-transferase gene family involved in the biosynthesis of ester volatiles. Identification of the crucial role of a threonine residue for enzyme activity. Plant Mol Biol 59:345-362.

Epinat, C., and Pitrat, M. 1994a. Inheritance of resistance to downy mildew (*Pseudoperonospora cubensis*) in muskmelon (*Cucumis melo*). I Analysis of a 8x8 diallel table. Agronomie 14:239-248.

Epinat, C., and Pitrat, M. 1994b. Inheritance of resistance to downy mildew (*Pseudoperonospora cubensis*) in muskmelon (*Cucumis melo*). II Generation means analysis of 5 genitors. Agronomie 14:249-257.

Esquinas Alcazar, J.T. 1981. Allozyme variation and relationships among Spanish land races of *Cucumis melo* L. Kulturpflanze 29:337-352.

Ezura, H., H. Amagai, K. Yoshioka, and Oosawa, K. 1992. Highly frequent appearance of tetraploidy in regenerated plants, a universal phenomenon, in tissue cultures of melon (*Cucumis melo* L.). Pl Science 85:209-213.

Fang, G.W., and Grumet, R. 1990. *Agrobacterium tumefaciens* mediated transformation and regeneration of muskmelon plants. Plant Cell Reports 9:160-164.

FAOSTAT Data. 2005. http://faostat.fao.org

Fassuliotis, G. 1977. Self fertilization of *Cucumis metuliferus* Naud. and its cross compatibility with *C. melo* L. J Amer Soc Hort Sci 102:336-339.

Filov, A.I. 1960. [The problem of melon systematics]. Vestnik sel'skochozjajstvennoj nauki 1:126-132.

Frantz, J.D., and Jahn, M.M. 2004. Five independent loci each control monogenic resistance to gummy stem blight in melon (*Cucumis melo* L.). Theor. Appl. Genet. 108:1033-1038.

Fuchs, M., J.R. McFerson, D.M. Tricoli, J.R. McMaster, R.Z. Deng, M.L. Boeshore, J.F. Reynolds, P.F. Russell, H.D. Quemada, and Gonzalves, D. 1997. Cantaloupe line CZW-30 containing coat protein genes of cucumber mosaic virus, zucchini yellow mosaic virus, and watermelon mosaic virus-2 is resistant to these three viruses in the field. Molecular Breeding 3:279-290.

Fukino, N., M. Kunihisa, and Matsumoto, S. 2004. Characterization of recombinant inbred lines derived from crosses in melon (*Cucumis melo* L.), 'PMAR No. 5' x 'Harukei No. 3. Breeding Science 54:141-145.

Galperin, M., L. Patlis, A. Ovadia, D. Wolf, A. Zelcer, and Kenigsbuch, D. 2003a. A melon genotype with superior competence for regeneration and transformation. Plant Breeding 122:66-69.

Galperin, M., A. Zelcer, and Kenigsbuch, D. 2003b. High competence for adventitious regeneration in the BU-21/3 melon genotype is controlled by a single dominant locus. HortScience 38:1167-1168.

Garcia, E., M. Jamilena, J.I. Alvarez, T. Arnedo, J.L. Oliver, and Lozano, R. 1998. Genetic relationships among melon breeding lines revealed by RAPD markers and agronomic traits. Theor. Appl. Genet. 96:878-885.

Garcia-Mas, J., M. Oliver, H. Gómez-Paniagua, and de Vicente, M.C. 2000. Comparing AFLP, RAPD and RFLP markers for measuring genetic diversity in melon. Theor. Appl. Genet. 101:860-864.

Gilbert, R.Z., M.M. Kyle, H.M. Munger, and Gray, S.M. 1994. Inheritance of resistance to watermelon mosaic virus in *Cucumis melo* L. HortScience 29:107-110.

Gonzalo, M.J., M. Oliver, J. Garcia-Mas, A. Monfort, R. Dolcet-Sanjuan, N. Katzir, P. Arús, and Monforte, A.J. 2005. Simple-sequence repeat markers used in merging linkage maps of melon (*Cucumis melo* L.). Theor. Appl. Genet. 110:802-811.

Grebenšcikov, I. 1986. Cucurbitaceae, p. 914-951. *In* Rudolf Mansfelds Verzeichnis landwirtschaftlicher und gärtnerischer Kulturpflanzen (J. Schultze-Motel ed.). Vol: 2. Akademie Verlag, Berlin (DE).

Guillaume, R., and Boissot, N. 2001. Resistance to *Diaphania hyalinata* (Lepidoptera: Crambidae) in *Cucumis* species. J. Econ. Entom. 94:719-723.

Guis, M., R. Botondi, M. Ben-Amor, R. Ayub, M. Bouzayen, J.C. Pech, and Latché, A. 1997a. Ripening-associated biochemical traits of cantaloupe Charentais melons expressing an antisense ACC oxidase transgene. J Amer Soc Hort Sci 122:748-751.

Guis, M., A. Latché, J.C. Pech, and Roustan, J.P. 1997b. An efficient method for production of diploid Cantaloupe Charentais Melon (*Cucumis melo* L. var. *cantalupensis*) by somatic embryogenesis. Scientia Hort. 69:199-206.

Guis, M., J.P. Roustan, C. Dogimont, M. Pitrat, and Pech, J.C. 1998. Melon biotechnology. Biotech Genet Engin Rev 15:289-311.

Guis, M., M.B. Ben Amor, A. Latché, J.C. Pech, and Roustan, J.P. 2000. A reliable system for the transformation of cantaloupe charentais melon (*Cucumis melo* L. var. *cantalupensis*) leading to a majority of diploid regenerants. Scientia Hort. 84:91-99.

Havey, M.J., J.D. McCreight, B. Rhodes, and Taurick, G. 1998. Differential transmission of the *Cucumis* organellar genomes. Theor. Appl. Genet. 97:122-128.

Hughes, M.B. 1948. The inheritance of two characters of *Cucumis melo* and their interrelationship. Proc Amer Soc Hortic Sci 52:399-402.

Iman, M.K., M.A. Abo-Bakr, and Hanna, H.Y. 1972. Inheritance of some economic characters in crosses between sweet melon and snake cucumber. I. Inheritance of qualitative characters. Assiut J Agricult Sci 3:363-380.

Jagger, I.C., and Scott, G.W. 1937. Development of powdery mildew resistant cantaloupe No. 45. USDA Circul. 441:1-5.

Jagger, I.C., T.W. Whitaker, and Porter, D.R. 1938. A new biotic form of powdery mildew on muskmelon in the Imperial Valley of California. Plant Disease Reporter 22:275-276.

Jahn, M., H.M. Munger, and McCreight, J.D. 2002. Breeding cucurbit crops for powdery mildew resistance, p. 239-242. *In* The Powdery Mildews, A Comprehensive Treatise (R. R. Belanger, W. R. Bushnell, A. J. Dik and L. W. Carver eds). APS Press, St Paul (MN, US).

Jeffrey, C. 1980. A review of the Cucurbitaceae. Botanical Journal Linnean Society 81:233-247.

Kaan, J.F. 1973. Recherches sur la résistance du melon aux maladies, notamment à la mosaïque de la pastèque et au *Pseudoperonospora*, appliquées au type variétal "Cantaloup Charentais". 41-49 *In* EUCARPIA meeting on melon (G. Risser, ed.). Avignon (FR), June 19-22, 1973

Karchi, Z., S. Cohen, and Govers, A. 1975. Inheritance of resistance to Cucumber Mosaic Virus in melons. Phytopathology 65:479-481.

Kathal, R., S.P. Bhatnagar, and Bhojwani, S.S. 1994. Plant regeneration from the callus derived from roots explant of *Cucumis melo* L. cultivar 'Pusa Sharbati'. Pl Science 96:137-142.

Kato, K., Y. Akashi, A. Okamoto, S. Kadota, and Masuda, M. 1998. Isozyme polymorphism in melon (*Cucumis melo* L.) and application to seed purity test of F1 cultivars. Breeding Science 48:237-242.

Katzir, N., Y. Danin-Poleg, G. Tzuri, Z. Karchi, U. Lavi, and Cregan, P.B. 1995. Application of RAPD and SSR analyses to the identification and mapping of melon (*Cucumis melo* L.) varieties. 196 *In* 'Cucurbitaceae '94 Evaluation and Enhancement of Cucurbits Germplasm' (G. E. Lester and J. R. Dunlap, eds). South Padre Island (TX, US), 01-04/11/1994

Kenigsbuch, D., and Cohen, Y. 1989. Inheritance of resistance to downy mildew in a gynoecious muskmelon. Plant Dis. 73:994-996.

Kenigsbuch, D., and Cohen, Y. 1992. Inheritance of resistance to downy mildew in *Cucumis melo* PI 124112 and commonality of resistance genes with PI 124111F. Plant Dis. 76:615-617.

Kennedy, G.G., G.W. Bohn, A.K. Stoner, and Webb, R.E. 1978. Leafminer resistance in muskmelon. J Amer Soc Hort Sci 103:571-574.

Kirkbride, J.H. 1993. Biosystematic Monograph of the Genus *Cucumis* (Cucurbitaceae). Parkway publishers, Boone (NC, US)

Kishaba, A.N., G.W. Bohn, and Toba, H.H. 1971. Resistance to *Aphis gossypii* in muskmelon. J. Econ. Entom. 64:935-937.

Knavel, D.E. 1990. Inheritance of a short internode mutant of 'Mainstream' muskmelon. HortScience 25:1274-1275.

Kubicki, B. 1969. Sex determination in muskmelon (*Cucumis melo* L.). Genet Polonica 10:145-165.

Lebeda, A., E. Kristkova, and Kubalakova, M. 1996. Interspecific hybridization of *Cucumis sativus* x *Cucumis melo* as a potential way to transfer resistance to *Pseudoperonospora cubensis*. 31-37 *In* Cucurbits toward 2000. VIth EUCARPIA meeting on Cucurbit Genetics and Breeding (M. L. Gómez-Guillamón, C. Soria, J. Cuartero, J. A. Torès and R. Fernandez-Munoz, eds). Málaga (ES), 28-30/05/1996

Lecoq, H., S. Cohen, M. Pitrat, and Labonne, G. 1979. Resistance to cucumber mosaic virus transmission by aphids in *Cucumis melo*. Phytopathology 69:1223-1225.

Lecoq, H., G. Labonne, and Pitrat, M. 1980. Specificity of resistance to virus transmission by aphids in *Cucumis melo*. Ann. Phytopathol. 12:139-144.

Lecoq, H. 2003. Cucurbits, p. 665-688. *In* Virus and virus-like diseases of major crops in developing countries (G. Loebenstein and G. Thottappilly eds). Kluwer Academic Publishers, Dordrecht (NL).

Lee, C.W., and Janick, J. 1978. Inheritance of seedling bitterness in *Cucumis melo*. HortScience 13:193-194.

Lee, S., and Kim, Z. 2003. Genetic relationship analysis of melons (*Cucumis melo*) germplasm by RAPD method. J Korean Soc Horticult Sci 44:307-313.

Li, H.L. 1969. The vegetables of ancient China. Economic Botany 23:253-260.

Liu, W., M. Song, F. Liu, and Wang, H. 2002. Assessment of genetic diversity of melon (*Cucumis melo*) germplasm based on RAPD and ISSR. Journal of Agricultural Biotechnology 10:231-236.

López-Sesé, A.I., and Gómez-Guillamón, M.L. 2000. Resistance to Cucurbit Yellowing Stunting Disorder Virus (CYSDV) in *Cucumis melo* L. HortScience 35:110-113.

López-Sesé, A.I., J. Staub, N. Katzir, and Gómez-Guillamón, M.L. 2002. Estimation of between and within accession variation in selected Spanish melon germplasm using RAPD and SSR markers to assess strategies for large collection evaluation. Euphytica 127:41-51.

López-Sesé, A.I., J.E. Staub, and Gómez-Guillamón, M.L. 2004. Genetic analysis of Spanish melon (*Cucumis melo* L.) germplasm using a standardized molecular-marker array and geographically diverse reference accessions. Theor. Appl. Genet. 108:41-52.

Ma, D., L. Sun, Y.H. Liu, Y. Zhang, and Liu, H. 1997. A genetic model of bitter taste in young fruits of melon. Cucurbit Genet Coop Rep 20:27-29.

Mallor, C., J.M. Álvarez, and Luis-Arteaga, M. 2003. Inheritance of resistance to systemic symptom expression of melon necrotic spot virus (MNSV) in *Cucumis melo* L. 'Doublon'. Euphytica 134:319-324.

Manniche, L. 1989. An ancient egyptian herbal. University of Texas Press, Austin (USA)

McCreight, J.D., H. Nerson, and Grumet, R. 1993. Melon *Cucumis melo* L., p. 267-294. *In* Genetic Improvement of Vegetable Crops (G. Kalloo and B. O. Bergh eds). Vol: Pergamon Press, Oxford (GB).

McCreight, J.D., and Fashing-Burdette, P. 1996. Resistance of PI 124112 and 'Eldorado-300' melons (*Cucumis melo* L.) to papaya ringspot virus watermelon strain. 298-301 *In* Cucurbits toward 2000. VIth EUCARPIA meeting on Cucurbit Genetics and Breeding (M. L. Gómez-Guillamón, C. Soria, J. Cuartero, J. A. Torès and R. Fernandez-Munoz, eds). Málaga (ES), 28-30/05/1996

McCreight, J.D. 2000. Inheritance of resistance to Lettuce Infectious Yellows virus in melon. HortScience 35:1118-1120.

McCreight, J.D. 2003. Genes for resistance to powdery mildew races 1 and 2U.S. in melon PI 313970. HortScience 38:591-594.

McCreight, J.D., J.E. Staub, A. López Sesé, and Chung, S.M. 2004. Isozyme variation in indian and chinese melon (*Cucumis melo* L.) germplasm collections. J Amer Soc Hort Sci 129:811-818.

McCreight, J.D. 2006. Melon-powdery mildew interactions reveal variation in melon cultigens and *Podosphaera xanthii* races 1 and 2. J Amer Soc Hort Sci 131:59-65.

Meglic, V., T.F. Horejsi, J.D. McCreight, and Staub, J.E. 1994. Genetic diversity and inheritance and linkage of isozyme loci in melon (*Cucumis melo* L.). HortScience 29:449.

Mliki, A., J.E. Staub, Z.Y. Sun, and Ghorbel, A. 2001. Genetic diversity in melon (*Cucumis melo* L.): an evaluation of African germplasm. Genetic Resources and Crop Evolution 48:587-597.

Molina, R.V., and Nuez, F. 1996. The inheritance of organogenic response in melon. Plant Cell Tissue Organ Cult 46:251-256.

Molina, R.V., and Nuez, F. 1997. Sexual transmission of the *in vitro* regeneration capacity via caulogenesis of *Cucumis melo* L. in a medium with a high auxin/cytokinin ratio. Scientia Hort. 70:237-241.

Monforte, A.J., J. Garcia-Mas, and Arus, P. 2003. Genetic variability in melon based on microsatellite variation. Plant Breeding 122:153-157.

Monforte, A.J., M. Oliver, M.J. Gonzalo, J.M. Alvarez, R. Dolcet-Sanjuan, and Arus, P. 2004. Identification of quantitative trait loci involved in fruit quality traits in melon (*Cucumis melo* L.). Theor. Appl. Genet. 108:750-758.

Morales, M., G. Orjeda, C. Nieto, H. van Leeuwen, A. Monfort, M. Charpentier, M. Caboche, P. Arus, P. Puigdomenech, M.A. Aranda, C. Dogimont, A. Bendahmane, and Garcia-Mas, J. 2005. A physical map covering the *nsv* locus that confers resistance to Melon necrotic spot virus in melon (*Cucumis melo* L.). Theor. Appl. Genet. 111:914-922.

More, T.A. 2001. Enhancement of muskmelon resistance to disease via breeding and transformation. 205-211 *In* Proceedings of the Second International Symposium on

Cucurbits (S. Nishimura, H. Ezura, T. Matsuda and A. Tazuke, eds). Tsukuba (JP), 28/09-01/10/2001

Moreno, V., M. Garcia-Sogo, I. Granell, B. Garcia-Sogo, and Roig, L.A. 1985. Plant regeneration from calli of melon (*Cucumis melo* L. cv Amarillo Oro). Plant Cell, Tissue and Organ Culture 5:139-146.

Munger, H.M., and Robinson, R.W. 1991. Nomenclature of *Cucumis melo* L. Cucurbit Genet Coop Rep 14:43-44.

Naudin, C. 1859. Essais d'une monographie des espèces et des variétés du genre *Cucumis*. Ann Sci Nat 11:5-87.

Neuhausen, S.L. 1992. Evaluation of restriction fragment length polymorphism in *Cucumis melo*. Theor. Appl. Genet. 83:379-384.

Norton, J.D., and Granberry, D.M. 1980. Characteristics of progeny from interspecific cross of *Cucumis melo* L with *C. metuliferus* E. Mey. J Amer Soc Hort Sci 105:174-180.

Nuez, F., B. Picó, A. Iglesias, J. Esteva, and Juarez, M. 1999. Genetics of melon yellows virus resistance derived from *Cucumis melo* spp. *agrestis*. European J. Plant Pathol. 105:453-464.

Oliver, J.L., J. Garcia-Mas, M. Cardús, N. Pueyo, A.I. López-Sesé, M. Arroyo, H. Gómez-Paniagua, P. Arús, and de Vicente, C.M. 2001. Construction of a reference linkage map of melon. Genome 44:836-845.

Oridate, T., and Oosawa, K. 1986. Somatic embryogenesis and plant regeneration from suspension callus culture in melon. Japanese Journal of Breeding 36:424-428.

Palmer, J. 1982. Physical and gene mapping of chloroplast DNA from *Atriplex triangularis* and *Cucumis sativus*. Nucleic Acids Research 10:1593-1605.

Pangalo, K.I. 1928. Melons. Leningrad

Pangalo, K.I. 1933. Cucurbitacées, p. 518-559,873-882. *In* La Turquie Agricole (P. Zhukovsky ed.). Les éditions de l'Etat section agricole "Selkhozghiz", Moscou (SUN).

Pangalo, K.I. 1958. [Dyni]. Gosudarstvennoe izdatel'stvo, Kisinev (MD)

Paris, H.S., H. Nerson, and Karchi, Z. 1984. Genetics of internode length in melons. J Hered 75:403-406.

Park, S.O., J.W. Sinclair, K.S. Yoo, L.M. Pike, and Crosby, K.M. 2003. Detection of QTL for sugar-related traits in ananas and cantaloupe melons. HortScience 38:676.

Park, Y., N. Katzir, Y. Brotman, J. King, F. Bertrand, and Havey, M. 2004. Comparative mapping of ZYMV resistances in cucumber (*Cucumis sativus* L.) and melon (*Cucumis melo* L.). Theor. Appl. Genet. 109:707-712.

Pech, J.C., A. Bernadac, M. Bouzayen, A. Latché, C. Dogimont, and Pitrat, M. 2007. Melon, p. 209-240. *In* Transgenic Crops V (Biotechnology in Agriculture and Forestry) (E. C. Pua amd M.R. Davey, eds). Springer, Berlin-Heidelberg (DE),

Perchepied, L., M. Bardin, C. Dogimont, and Pitrat, M. 2005a. Relationship between loci conferring downy mildew and powdery mildew resistance in melon assessed by QTL mapping. Phytopathology 95:556-565.

Perchepied, L., C. Dogimont, and Pitrat, M. 2005b. Strain-specific and recessive QTLs involved in control of partial resistance to *Fusarium oxysporum* f. sp. *melonis* race 1.2 in a recombinant inbred line population of melon. Theor. Appl. Genet. 111:65-74.

Périn, C., M.C. Gomez-Jimenez, L. Hagen, C. Dogimont, J.C. Pech, A. Latché, M. Pitrat, and Lelièvre, J.M. 2002a. Molecular and genetic characterisation of a non-climacteric phenotype in melon reveals two loci conferring altered ethylene response in fruit. Plant Physiol 129:300-309.

Périn, C., L.S. Hagen, V. de Conto, N. Katzir, Y. Danin-Poleg, V. Portnoy, S. Baudracco-Arnas, J. Chadoeuf, C. Dogimont, and Pitrat, M. 2002b. A reference map of *Cucumis melo* based on two recombinant inbred line populations. Theor. Appl. Genet. 104:1017-1034.

Périn, C., L.S. Hagen, N. Giovinazzo, D. Besombes, C. Dogimont, and Pitrat, M. 2002c. Genetic control of fruit shape acts prior to anthesis in melon (*Cucumis melo* L.). Mol Genet Genom 266:933-941.

Perl-Treves, R., and Galun, E. 1985. The *Cucumis* plastome: physical map, intrageneric variation and phylogenetic relationships. Theor. Appl. Genet. 71:417-429.

Perl-Treves, R., D. Zamir, N. Navot, and Galun, E. 1985. Phylogeny of *Cucumis* based on isozyme variability and its comparison with plastome phylogeny. Theor. Appl. Genet. 71:430-436.

Pitrat, M., and Lecoq, H. 1983. Two alleles for Watermelon Mosaic Virus 1 resistance in melon. Cucurbit Genet Coop Rep 6:52-53.

Pitrat, M., and Lecoq, H. 1984. Inheritance of Zucchini Yellow Mosaic Virus resistance in *Cucumis melo* L. Euphytica 33:57-61.

Pitrat, M. 1991. Linkage groups in *Cucumis melo* L. J Hered 82:406-411.

Pitrat, M., P. Hanelt, and Hammer, K. 2000. Some comments on infraspecific classification of cultivars of melon. 29-36 *In* Cucurbitaceae 2000, VIIth EUCARPIA Meeting on Cucurbit Genetics and Breeding (N. Katzir and H. Paris, eds). Ma'ale Hahamisha (IL), 19-23/03/2000

Plagès, J.N. 1997. L'avenir des variétés génétiquement modifiées pour la résistance aux virus (un exemple développé par Limagrain). C R Acad Agric France 83:161-164.

Poole, C.F., and Grimball, P.C. 1939. Inheritance of new sex forms in *Cucumis melo* L. J Hered 30:21-25.

Risser, G. 1973. Étude de l'hérédité de la résistance du melon (*Cucumis melo*) aux races 1 et 2 de *Fusarium oxysporum* f.sp. *melonis*. Ann Amélior Plantes 23:259-263.

Risser, G., Z. Banihashemi, and Davis, D.W. 1976. A proposed nomenclature of *Fusarium oxysporum* f.sp. *melonis* races and resistance genes in *Cucumis melo*. Phytopathology 66:1105-1106.

Risser, G., M. Pitrat, and Rode, J.C. 1977. Etude de la résistance du melon (*Cucumis melo* L.) au virus de la mosaïque du concombre. Ann Amélior Plantes 27:509-522.

Risser, G. 1984. Correlation between sex expression and fruit shape in muskmelon (*Cucumis melo* L.). 100-103 *In* Cucumis and Melon's 84. 3rd meeting EUCARPIA Plovdiv (BG), 2-5/07/1984.

Robinson, R.W., and Whitaker, T.W. 1974. Cucumis, p. 145-150. *In* Handbook of Genetics (R. C. King ed.). Vol: 2. Plenum, New York (US).

Robinson, R.W. 1992. Genetic resistance in the Cucurbitaceae to insects and spider mites, p. 309-360. *In* Plant Breeding Reviews (J. Janick ed.). Vol: 10. John Wiley & Sons, New York (US).

Robinson, R.W., and Decker-Walters, D.S. 1997. Cucurbits. CAB International, Oxon (GB)

Robinson, R.W., and Provvidenti, R. 1975. Susceptibility to powdery mildew in *Citrullus lanatus* (Thunb.) Matsum. & Nakai. J Amer Soc Hort Sci 100:328-330.

Sageret. 1825. Mémoire sur les Cucurbitacées, et principalement sur le melon avec des considérations sur la production des hybrides, des variétés, etc... *In* Extrait des Mémoires de la Société royale et centrale d'Agriculture. Imprimerie de Madame Huzard (née Vallat La Chapelle), Paris (FR). 60 pp.

Sauton, A., and Dumas de Vaulx, R. 1987. Obtention de plantes haploïdes chez le melon (*Cucumis melo* L.) par gynogenèse induite par du pollen irradié. agronomie 7:141-147.

Shiboleth, Y.M., T. Arazi, Y.Z. Wang, and Gal-On, A. 2001. A new approach for weed control in a cucurbit field employing an attenuated potyvirus-vector for herbicide resistance. Journal of Biotechnology 92:37-46.

Silberstein, L., I. Kovalski, Y. Brotman, C. Périn, C. Dogimont, M. Pitrat, J. Klingler, G. Thompson, V. Portnoy, N. Katzir, and Perl-Treves, R. 2003. Linkage map of *Cucumis melo* including phenotypic traits and sequence-characterized genes. Genome 46:761-773.

Sinclair, J.W., S.O. Park, G. Lester, K.S. Yoo, and Crosby, K. 2006. Identification and confirmation of RAPD markers and andromonoecious associated with quantitative trait loci for sugars in melon. J Amer Soc Hort Sci 131:360-371.

Soria, C., E. Moriones, A. Fereres, E. Garzo, and Gómez-Guillamón, M.L. 2003. New source of resistance to mosaic virus transmission by *Aphis gossypii* in melon. Euphytica 133:313-318.

Staub, J.E., J. Box, V. Meglic, T.F. Horejsi, and McCreight, J.D. 1997. Comparison of isozyme and random amplified polymorphic DNA data for determining intraspecific variation in *Cucumis*. Genetic Resources and Crop Evolution 44:257-269.

Staub, J.E., Y. Danin-Poleg, G. Fazio, T. Horejsi, N. Reis, and Katzir, N. 2000. Comparative analysis of cultivated melon groups (*Cucumis melo* L.) using random amplified polymorphic DNA and simple sequence repeat markers. Euphytica 115:225-241.

Staub, J.E., A.I. López-Sesé, and Fanourakis, N. 2004. Diversity among melon landraces (*Cucumis melo* L.) from Greece and their genetic relationships with other melon germplasm of diverse origins. Euphytica 136:151-166.

Stepansky, A., I. Kovalski, and Perl-Treves, R. 1999. Intraspecific classification of melons (*Cucumis melo* L.) in view of their phenotypic and molecular variation. Plant System Evol 217:313-332.

Stol, M. 1987. The Cucurbitaceae in the cuneiform texts. Bulletin Sumerian Agriculture 3:81-92.

Sujutha, V.S., V.S. Seshadri, K.N. Srivastava, and More, T.A. 1991. Isozyme variation in muskmelon (*Cucumis melo* L.). Indian J Genet Pl Breed 51:438-444.

Taler, D., M. Galperin, I. Benjamin, Y. Cohen, and Kenigsbuch, D. 2004. Plant eR genes that encode photorespiratory enzymes confer resistance against disease. The Plant Cell 16:172-184.

Thomas, C.E. 1986. Downy and powdery mildew resistant muskmelon breeding line MR-1. HortScience 21:329.

Thomas, C.E., J.D. McCreight, and Jourdain, E.L. 1990. Inheritance of resistance to *Alternaria cucumerina* in *Cucumis melo* line MR-1. Plant Dis. 74:868-870.

van der Knaap, B.J., and de Ruiter, A.C. 1978. An interspecific cross between cucumber (*Cucumis sativus*) and muskmelon (*Cucumis melo*). Cucurbit Genet Coop Rep 1:6-8.

van Leeuwen, H., A. Monfort , H.B. Zhang, and Puigdomènech P. 2003. Identification and characterisation of a melon genomic region containing a resistance gene cluster from a constructed BAC library. Microcolinearity between *Cucumis melo* and *Arabidopsis thaliana*. Plant Mol Biol 51:703-718.

Vishnu-Mittre. 1974. Palaeobotanical evidence in India, p. 3-30. *In* Evolutionary studies in world crops (J. Hutchinson ed.). Cambridge University Press, Cambridge (GB).

Wall, J.R. 1967. Correlated inheritance of sex expression and fruit shape in *Cucumis*. Euphytica 16:199-208.

Walters, T.W. 1989. Historical overview on domesticated plants in China with special emphasis on the Cucurbitaceae. Economic Botany 43:297-313.

Wang, Y.H., C.E. Thomas, and Dean, R.A. 1997. A genetic map of melon (*Cucumis melo* L.) based on amplified fragment length polymorphism (AFLP) markers. Theor. Appl. Genet. 95:791-797.

Ward, B.L., R.S. Anderson, and Bendich, A.J. 1981. The mitochondrial genome is large and variable in a family of plants (*Cucurbitaceae*). Cell 25:793-803.

Watson, W. 1969. Early cereal cultivation in China, p. 397-402. *In* The domestication and exploitation of plants and animals (P. J. Ucko and G. W. Dimbledy eds). Gerald Duckworth & Co, London (GB).

Webb, R.E. 1979. Inheritance of resistance to watermelon mosaic virus in *Cucumis melo* L. HortScience 14:265-266.

Whitaker, T.W., and Davis, G.N. 1962. Cucurbit, botany, cultivation and utilization. Interscience Publisher, New York (US)

Yetisir, H., and Sari, N. 2003. A new method for haploid muskmelon (*Cucumis melo* L.) dihaploidization. Scientia Hort. 98:277-283.

Yoshioka, K., K. Hanada, T. Harada, Y. Minobe, and Oosawa, K. 1993. Virus resistance in transgenic melon plants that express the cucumber mosaic virus coat protein gene and their progeny. Japanese Journal of Breeding 43:629-634.

Zalapa, J.E., J.E. Staub, and McCreight, J.D. 2004. Genetic analysis of branching in melon (*Cucumis melo*). 373-379 *In* Progress in cucurbit genetics and breeding research. Proceedings of Cucurbitaceae 2004, the 8th EUCARPIA Meeting on Cucurbit Genetics and Breeding (A. Lebeda and H. S. Paris, eds). Olomouc (CZ), 12-17/07/2004

Zink, F.W. and Gubler, W.D. 1985. Inheritance of resistance in muskmelon to Fusarium wilt. J Amer Soc Hort Sci 110:600-604.

Pumpkin and Winter Squash

María Ferriol[1] and Belén Picó[2]

[1] Universidad Politécnica de Valencia, Institute of Mediterranean Agroforestry, mafermo@upvnet.upv.es

[2] Universidad Politécnica de Valencia, Institute of Conservation and Improvement of Agrodiversity, mpicosi@btc.upv.es

1 Introduction

The common terms "pumpkin", "squash", "gourd", "cushaw", "ayote", "zapallo", "calabaza", etc. are often applied indiscriminately to different cultivated species of the New World genus *Cucurbita* L. (Cucurbitaceae): *C. pepo* L., *C. maxima* Duchesne, *C. moschata* Duchesne, *C. argyrosperma* C. Huber and *C. ficifolia* Bouché. These species are mainly grown for their fruits (botanically a *pepo*) which are a significant source of carbohydrates and vitamins (Whitaker and Davis, 1962). The fruits can be picked either when immature or fully mature, and this type of use conditions the culture techniques, cultivar selection and breeding objectives.

"Pumpkin" is mostly used to refer to cultivars with round fruits which are used when mature for baking or for feeding livestock. "Squash", by contrast, is differentially applied to cultivars grown for their edible immature fruits (often referred to collectively as "summer squash") and to cultivars grown for their mature fruits that store well and are not usually round (often referred to as "winter squash") (Decker-Walters and Walters, 2000). Despite the supposedly differential characteristics that make "summer" and "winter squashes" two different crops, these common terms overlap, generating confusion which also affects production statistics. For example, only one FAO category includes "pumpkins", "squash" and "gourds" together (FAOSTAT data, 2006) (see the chapter on "Summer Squash" for production data). *Cucurbita* species rank collectively among the 10 leading vegetable crops worldwide. China and India lead the world production. Other major producers are U.S., Egypt, Mexico, Ukraine, Cuba, Italy, Iran and Turkey.

C. pepo is today the most economically important species distributed worldwide. Although its great economic value is based mainly on the culinary use of immature fruits ("Zucchini", "Vegetable Marrow", "Cocozelle", "Croockneck", "Straightneck" and "Scallop" types, described in the chapter on "Summer squash"), there are two horticultural groups, "Pumpkin" and "Acorn", which display a major use as "winter squashes" (Paris, 1989). A different situation can be found in the other two economically significant species, *C. maxima* and *C. moschata,* which mainly include cultivars grown as "winter squashes" in developing countries under low-input agricultural systems. However, in southern Latin America, *C. maxima* has been largely bred for immature fruit consumption (zapallito varieties) and some *C. moschata* cultivars are also valued as "summer squashes". The other two species, *C. argyrosperma* and *C. ficifolia,* have less economic importance and a narrower distribution. In both cases, the mature fruits are the most valued, but some varieties are eaten as a vegetable as well.

In addition to the use of the mature edible fruits, pumpkins and winter squashes are also grown for ornamental purposes, as in the popular use of pumpkins of *C. pepo* for Halloween, or as containers. Pumpkin seeds are also important as snacks, as a source of edible oil and protein for human and animal consumption, and in the pharmaceutical industry. In Latin American countries, the flowers, leaves, and vine tips of *Cucurbita* are also consumed.

2 Origin and Domestication

Prehistoric *Cucurbita* species distribution and usage have been determined through archaeological remains that include dried rinds, as well as peduncles and seeds. These have persisted for centuries as the fruits of the wild species and some of the first domesticates had lignified rinds, trait which has been linked to the phytolith formation in *Cucurbita* fruits (Piperno et al., 2002).

The five cultivated species of the genus *Cucurbita* were domesticated in different places, ranging from North America to southern South America. Each species probably represents an independent domestication event from different wild ancestor populations. All were cultivated during the pre-Columbian era, and some were members of the earliest crop plant complexes known to the New World, along with maize, beans, etc. *C. pepo* is the earliest documented domesticate in Mesoamerica (dating back to 7,920 calibrated years B.P., after reassessing Coxcatlan Cave, Puebla, Mexico) (Smith, 2005). Phytoliths of domesticated *Cucurbita* from southwestern Ecuador have recently been dated to about 12,000 to 10,000 calendar years ago, providing evidence for an independent use of *Cucurbita* crops in lowland South America and in highland Mesoamerica (Piperno and Stothert, 2003).

Wild *Cucurbita* species have a flesh which is bitter due to its high content of cucurbitacins. Prehistoric cultures could have used these species primarily as containers or for their seeds, which are edible and nutritious. Initially, young fruits could have been consumed after multiple boilings as supported by the high frequency of immature peduncles among the archaeological remains (Decker-Walters and Walters, 2000). The cucurbitacin content is genetically controlled by only a few genes (Paris and Brown, 2005). Mutations in these genes, leading to non-bitter flesh, could have been selected by natives. A later selection for non-lignified rinds and thicker, more starchy and less fibrous flesh would have allowed for the consumption of the mature fruit, leading to its use today as a winter squash. In the process of *Cucurbita* domestication, larger seeds and fruits, relatively uniform germination, loss of seed dormancy, adaptation to shorter growing seasons, and a reduction in size and abundance of trichomes could also have been favoured (Lira-Saade and Montes-Hernández, 1994).

The origin and domestication of *C. pepo* is discussed in detail in chapter 11 (Summer squash). For this species, the data indicate two separate domestications. The potential zone of domestication for the cultivated *C. pepo* subsp. *texana* (Scheele) Filov (syn. subsp. *ovifera* (L.) Decker), which includes the acorn types, ranges from eastern North America to northeastern Mexico. Wild plants of this subspecies can be found growing in the U.S.A. The wild populations of *C. pepo* subsp. *fraterna* (L.H. Bailey) Lira, Andres & Nee from northern Mexico are closely related to subsp. *texana* (Sanjur et al., 2002), but many specimens are also completely interfertile with the other domesticated subspecies, *C. pepo* subsp. *pepo,* which includes the pumpkin type (Nee, 1990). The wild ancestor of the subsp. *pepo* is unknown. Sanjur et al. (2002) indicated that wild populations of the subsp. *fraterna* more closely related to subsp. *pepo*, may still exist in central or southern Mexico, meaning that additional collections in these areas are needed. Teppner (2004) reported an additional subspecies, *C. pepo* subsp. *gumala* Teppner, cultivated in Guatemala and Mexico, closely related to wild species which could have been the starting point of the domestication of subsp. *pepo.*

The earliest archaeological remains indicative of the domestication of *C. moschata* were discovered in Mexico (about 5,000 B.C.), which was initially proposed as the domestication centre (Cutler and Whitaker, 1967). However, when reassessing some Mexican caves this dating was not confirmed (the oldest *C. moschata* remains date back to 800 B.C.), which suggests that these older specimens were not correctly identified in the first analysis or have disappeared (Smith, 2005). More recent *C. moschata* remains have been found in coastal Peru (3,000 B.C.) and Guatemala (2,000 B.C.). Some authors have suggested two independent domestications, in Mexico and northern South America, as *C. moschata* has native names in both areas (Lira-Saade, 1995; Robinson and Decker-Walters, 1997). Nowadays, the archaeological records and the diversity of the South American landraces point to this area as the domestication centre or as a secondary site of early diversification (Nee, 1990; Sanjur et al., 2002). In addition, some morphological traits considered as primitive have been observed in landraces from Colombia, Panama and Bolivia (Wessel-Beaver, 2000). Some bitter fruits have also been reported in Colombia, supporting the existence of hybridization events between *C. moschata* and a wild local species. The wild ancestor of *C. moschata* is still unknown. *C. lundelliana* L.H.Bailey, distributed across the Yucatan peninsula, was initially proposed. However, several studies suggest that *C. argyrosperma* is much more closely related to *C. moschata* than *C. lundelliana* is (Merrick, 1990). Nonetheless, the distinctive electrophoretic patterns and the presence of reproductive barriers between *C. moschata* and wild *C. argyrosperma* support the idea that this wild taxon is not the progenitor of *C. moschata* (Sanjur *et al.*, 2002). The current data point to a single origin for *C. moschata*, somewhere in northern lowland South America, from a wild ancestor closely related to wild *C. argyrosperma* that has yet to be identified.

C. argyrosperma and *C. moschata* are closely related. Traditionally, they were considered to be a single species until Pangalo (1930) described *C. argyrosperma* as a distinct species (named *C. mixta* Pang.). The reassessment of some archaeological remains in southwestern and central Mexico suggests that this species could have been domesticated in this region between 3,085 B.C. and 115 B.C. (Merrick, 1990; Smith, 2005). This initial appearance of *C. argyrosperma* in the Mexican cultural deposits of Coxcatlan Cave is more recent than originally estimated (Cutler and Whitaker, 1967). The wild *C. argyrosperma* subsp. *sororia* (L.H. Bailey) Merrick & Bates, distributed nowadays from Mexico to Central America, is probably the wild progenitor of the cultivated group (belonging to subsp. *argyrosperma*) due to their reproductive compatibility, morphological similarity, geographical distribution, and phylogenetic relationships (Sanjur et al., 2002).

The poor archaeological data available for *C. maxima* indicate that this species was domesticated on the Peruvian coast by around 2,000 B.C. In pre-Columbian times, landraces of *C. maxima* were being cultivated in northeastern Argentina and Paraguay by the Guarani indigenes as well as in the Andean valleys. Archaeobotanical macroremains of *C. maxima* have been recently reported from northwest Argentina, dated between ca. 1,750 and 1,450 B.P (Oliszewski, 2005). Nowadays, it is assumed that the wild ancestor of the cultivated form is *C. maxima* subsp. *andreana* (Naud.) A.I. Filov, endemic to South America. Fruits of this subspecies have been found in warmer temperate areas of Argentina, Uruguay, and, more recently, in humid lowland regions of Bolivia, extending the potential zone of domestication into this region (Nee, 1990).

Archaeological remnants of domesticated *C. ficifolia* have been found only in coastal Peru, dating back to between 3,000 and 6,000 B.C. This leads to the hypothesis of an Andean origin and domestication (Nee, 1990; Sanjur et al., 2002). However, *C. ficifolia* is often though to be of Mesoamerican origin because it has a *Nahuatl* name and there are some reports which describe Aztec religious practices using the fruits. Nonetheless, some *Quechua* and *Aymara* names have also been reported for the *C. ficifolia* fruits, and intensive searches for a wild progenitor in Mexico have been unsuccessful (Andres, 1990). Hybridization studies have shown that *C. ficifolia* partially cosses with *C. lundelliana*, *C. foetidissima* H.B.K, and *C. pedatifolia* L.H. Bailey. However, these species bear little resemblance to *C. ficifolia* in morphology and ecology. On the other hand, the leaf lobes of *C. ecuadorensis* Cutler & Whitaker are similar to those of *C. ficifolia*, but both species differ greatly in their ecological and morphological traits. Presently, there are reports of weedy plants in Guatemala and Bolivia which may be revelatory in this respect (Decker-Walters and Walters, 2000).

3 Varietal Groups

Cultivated *Cucurbita* species appear frequently in 16th century reports by European priests and explorers that describe Native American agriculture. Squashes were some of the first vegetables that arrived in Europe soon after the contact, as demonstrated by the abundant representations of different *Cucurbita* species in paintings and illustrations in the European herbals of the Renaissance (Paris, 1989). In some species, like *C. ficifolia* and cultivars of *C. maxima*, the diffusion outside America was also *via* other continents. This early expansion caused a great diversification reflected, in part, in the high variability of the current cultivars.

Most of the current winter squash production occurs in developing countries, oriented toward self-consumption or sale in local markets. In many regions farmers save seeds, as commercially produced cultivars are not readily available. For example, in a recent study on traditional cropping systems of Mexico (milpas), Montes-Hernández et al. (2005) found that the cultivated varieties of *C. argyrosperma* and *C. moschata* were exclusively locally adapted landraces that had not been replaced by modern cultivars. Some of these farmer-selected landraces have been the basis for the development of commercial cultivars, at first open-pollinated and more recently hybrids, by Seed Companies or Agricultural Experiment Stations. Different proposals for grouping the commercial cultivars exist. Market groups are mainly based on fruit shape and secondarily on a number of interesting phenotypic characteristics (rind and flesh colour, seed traits, etc.), meeting taxonomic infraspecific classifications only in some species. Even though more and more traditional varieties are being introduced by seed companies in international commercial trade, as awareness of preserving and using biodiversity is increasing, these informal classifications of market types no longer satisfactorily accommodate the extant diversity.

In pre-Columbian times, *C. moschata* spread to the Caribbean, where some indigene cultivars were developed, and to the eastern and southwestern U.S.A (Robinson and Decker-Walters, 1997; Andres, 2004a). Further spreading, especially in North America, may have been facilitated by early Spanish explorers. In Florida, a distinctive landrace developed by the Seminoles with small, oval to pyriform fruits, is still grown with the same name, "Seminole Pumpkin". In Europe, no illustration or description gives evidence of the cultivation of *C. moschata* before the late 17th century. This could be due to the poor adaptation of this species to the temperate climates of middle to high latitudes. By the end of the 19th century, *C. moschata* had spread worldwide.

C. moschata is highly diverse and mostly adapted to climates that are hot, humid, and at low elevations. It displays fruits whose size and shape are very variable, are frequently furrowed and sometimes warty, have a different rind colour and high quality flesh, ranging from deep yellow to orange. The proposed infraspecific taxonomic classifications have failed to accurately describe the genetic relationships between the extant forms. The market classifications are based on fruit morphology. Castetter (1925) and Whitaker and Davis (1962) proposed three horticultural groups, which are those established in the North American commercial trade (Robinson and Decker-Walters, 1997) (Table 1). "Cheese pumpkins" include usually oblate fruits with a buff-coloured rind. The fruits of "Crooknecks" are round at the blossom end and necked. "Bell squashes" have bell-shaped to almost cylindrical fruits. This group includes "Butternut", the first and one of the most popular cultivars for its high quality flesh. This classical commercial grouping does not encompass all the fruit types that we can find in tropical landraces.

Table 1. Commercial cultivars of *C. moschata* from recent seed catalogues classified according to the horticultural groups proposed for this species. Other cultivar names can be found in Andres (2004b).

Varietal group	Cultivars
Cheese	Calhoun, Chirimen, Fairytale*, Futtsu Black, Kentucky Field, Large Cheese, Large Sweet Cheese, Long Island Cheese, Magdalena Big Cheese, Musquée de Provence, Quaker Pie, Tan Cheese.
Crookneck	Argonaut*, Bugle Gramma, Canada Crookneck, Longue de Nice, Lunga di Napoli, Neck Pumpkin, Pennsylvania Dutch crookneck, Tromba d´Albenga, Winter Crookneck.
Butternut-Bell	Alagold op, Atlas*, Autumn Glow, Avalon*, Butterboy, Burpee's Butterbush, Butternut, Canesi*, Early Butternut*, Estribo*, Li'l Abner*, Menina Rajada Seca, Metro PMR*, Nicklow's Delight, Pilgrim*, Ponca Butternut, Puritan Butternut, Really Big*, Rebenque*, Sucrine du Berry, Supreme*, Tahitian Butternut, Ultra Butternut*, Violina, Waltham, Zenith*
Others	Buckskin*, Cuban Pumpkin (Zapallo), Dickinson, Early Buckskin, Fordhook Acorn, Golden Cushaw, Seminole Pumpkin

*indicates F1. The absence of the asterisk indicates open-pollinated, breeding line, or lack of information about this cultivar.

C. argyrosperma was poorly diffused in pre-Columbian times, arriving in the southwestern and eastern U.S.A by 900 and 1,400 A.D., respectively (Fritz, 1994). The first documents which describe its cultivation in the Old World, especially China, are recent, from the second half of the 80s (Lira-Saade, 1995). However, the information about this species is frequently confusing as it is difficult to distinguish it from *C. moschata*. Generally, fruits of *C. argyrosperma* have enlarged, corky peduncles and rarely display orange rind colour, unlike *C. moschata*. Fruits of *C. argyrosperma* range from globose to long-necked, are usually unfurrowed with green or white rinds, which sometimes mature to yellow, and have pale yellow to orange flesh. The flesh has poor culinary quality and many cultivars and landraces are grown for their edible seeds or as forage.

The diversity of *C. argyrosperma* is low when compared to the other *Cucurbita* species and only a few commercial cultivars exits (Decker-Walters and Walters, 2000). They are grouped according to the botanical infraspecific classification. All commercial cultivars belong to 3 varieties of subsp. *argyrosperma*: var. *argyrosperma*, var. *callicarpa* Merrick and Bates, and var. *stenosperma* (Pang.) Merrick and Bates (Table 2). The fruits of var. *argyrosperma*, considered the more primitive, are generally striped, with bright colour, smooth skin, and large seeds. The var. *callicarpa*, the most recent and variable, includes most of the commercial cultivars and landraces, with solid, striped or blotchy-coloured fruits; some of these cultivars have a good flesh quality. The fruits of var. *stenosperma* are mostly striped, and the cultivars are mainly grown for their edible seeds. However, some authors think that these 3 varieties are morphologically difficult to distinguish and include all the cultivars in the subsp *argyrosperma* (Lira Saade, 1995).

Table 2. Infraspecific classification of *C. argyrosperma* subsp. *argyrosperma* and list of commercial cultivars from recent seed catalogues. None of these cultivars appeared as hybrids, and many derive from native landraces of southern North America, Mexico, and Central America.

Varietal group	Cultivars
Var. *argyrosperma*	Silverseed Gourd
Var. *callicarpa*	Allneck Cushaw, Black Tennessee Sweet Potato, Campeche, Chompa, Cushaw Crookneck Green Striped, Green Striped Cushaw, Hopi, Japanese Pie, Mayo Arrote, Navajo Cushaw, Prima Bajo Sequalca, Tennessee Sweet Potato, Tricolor Cushaw, Veracruz Pepita, White Cushaw, Zebra Mystery
Var. *stenosperma*	Elfrida Taos

Horticultural grouping in *C. pepo*, also based on fruit shape, meets infraspecific taxonomic classifications (see Chapter on "Summer Squash"). Only two horticultural groups, acorn (subsp. *texana*) and pumpkin (subsp. *pepo*), are almost exclusively consumed as winter squashes (Table 3). Both horticultural types seem to have been developed under the guidance of Native Americans in pre-Columbian times. Cultivars of "Pumpkin" and "Acorn" were among the first to be described in European depictions and paintings of the 16th century. Unlike *C. pepo* summer squashes, where F1 hybrids have replaced the ancient cultivars, some old cultivars are still popular as winter squashes. In addition to the flesh consumption, "Pumpkins" have the largest seeds in the species, which are commonly consumed. Some forms are hull-less seeded and are grown for the extraction of seed oil. They are also grown as a snack in the U.S.A and elsewhere. Acorn cultivars are less variable, small, turbinate, furrowed, and usually dark in colour. Cultivars of other *C. pepo* horticultural groups are also consumed when mature. The traditional "Vegetable Spaghetti" is considered to belong to the "Vegetable Marrow" horticultural group in the subsp. *pepo* (Maynard et al., 2001). It is typically characterized by its flesh that breaks into strands after cooking. "Delicata" is also an early introduced cultivar that belongs to subsp. *texana*, which cannot be assigned to any known horticultural group (Paris, 2001). Apart from these, the ornamental gourds, which also reach maturity, usually include small, bitter fruits with hard and coloured rinds. The round ones that have smooth or warted forms are included in the subsp. *pepo*, while the oviform or pear-shaped ones with smooth rind forms are in the subsp. *texana*.

Table 3. Commercial cultivars from recent seed catalogues of horticultural groups of *C. pepo* used as winter squashes.

Varietal group	Cultivars
Acorn	Autumn Prince op, Autumn Queen, Celebration*, Carnival, Cream of the Crop, Early acorn, Festival, Harlequin*, Heart of Gold, Mardi Gras*, Mammoth Table Queen, Mesa Queen*, Royal Acorn, Swan White Table Queen, Sweet Dumpling, Table Ace, Table Gold, Table King, Table Queen, Table Queen Ebony, Table Star, Taybelle*, Tuffy, White Acorn
Pumpkin	Aladdin*, Appalachian, Aspen*, Autumn King, Big Autumn, Connecticut Field, Early Autumn*, Early Harvest*, Gladiator*, Gold Fever, Gold Keeper, Gold Rush, Gold Standard, Gold Strike*, Howden*, Howdy Doody*, Jack O' Lantern, Jack of All Trades*, Li'l Goblin*, Li'l Ironsides*, Longface, Magic Lantern*, Merlin*, Mother Lode, Mystic*, New England Pie, Peek a Boo*, Phantom, Pick-a-Pie, Racer*, Rocket, Schooltime*, Small Sugar, Sorcerer*, Spirit*, Spookie, Spooktacular, Tallman, Tours, Trickster*, Wee Be little, Winter Luxury Pie, Wizard
Hull or semihulless	Baby Bear, Eat All, Gleisdorfer Öilkurbis, Hull-Less, Lady Godiva, Snackjack*, Streaker, Styrian Pumpkin, Trick or Treat, Triple Treat.
Spaghetti	Goldetti*, Hasta la Pasta, Heaven*, Orangetti*, Pasta*, Small Wonder*, Spaghetti Tivoli, Stripetti*, Trifetti*, Vegetable Spaghetti, Vermicelli*.
Ornamental	Autumn Wings, Baby Boo, Baby Pam, Bicolor Pear, Flat Stripped, Galeuses, Gremlin, Jack B Little, Li'l Pum-Ke-Mon, Little Boo, Miniature Ball, Munchkin, Orange, Orange Small, Orange Warted, Pam*, Pear, Spoon, Spoon Bicolor, Striped Crown of Thorns, Striped Pear, Wee-B-Little op, White Ball, White Egg
Others	Bush Delicata, Camäleon, Delicata

*indicates F1. The absence of the asterisk indicates open pollinated, breeding line, or lack of information about this cultivar.

Apparently, *C. maxima* never left its continent of origin during the pre-Columbian era, but a great diversity of forms were developed by South American aborigines during this time (Decker-Walters and Walters, 2000; Sanjur *et al.*, 2002). Different types of this species were introduced into Western Europe from America. Other cultivars reached Europe via Asia, Australia and Africa, where local landraces evolved. As with *C. moschata* the evolutionary relationships among the extant cultivars and landraces have not been clarified.

C. maxima is the most diverse *Cucurbita* species after *C. pepo* and is grown throughout tropical to temperate regions. It exhibits fruits very variable in size, shape, and coloration and has a flesh (yellow to orange) of the highest quality among squashes. Early in the 19th century, American sailors introduced various cultivars into the U.S., where new varieties were developed. Most of these cultivars entered the commercial trade at the beginning of the 20th century, which traditionally includes 6 horticultural groups as proposed by Castetter (1925). This classification has persisted until the modern day (Whitaker and Davis, 1962; Robinson and Decker-Walters, 1997; Decker-Walters and Walters, 2000) (Table 4). The "Banana" squashes are long, pointed at both ends, and have a soft rind and brown seeds. They were introduced into the U.S. from Mexico at the end of the 19th century. "Delicious" squashes were also introduced directly from South America (Brazil) into the U.S. The fruits are turbinate and shallowly ribbed, with a hard rind, white seeds, and a high-quality flesh. "Marrows" probably originated in Chile and include oval to pyriform fruits with white seeds. "Hubbard" squashes are oval, with tapering to curved ends, with a very hard rind which is variable in colour and white seeds. Old and very popular commercial cultivars are included in this group, like "Arikara" and "Marblehead", which were cultivated by North American aboriginal tribes. "Show" includes large, orange fruits, with a soft rind and white seeds. Some cultivars develop large fruits used for shows and contests. "Turban" squashes are turban-shaped. The fruits of this group were among the first to be brought to and cultivated in Europe, and many cultivars were also selected in secondary centres of diversity. The fruits are grown mostly for their decorative value, but they can also be cooked and stuffed. The most known cultivar is "Turk's Turban", but other varieties exist that vary in colour or size. Cultivars that evolved from the popular "Buttercup" with a high quality flesh are also included in this group. As with other species, not all the commercial cultivars can be classified within these groups (Table 4).

Recently, some cultivars of the high-quality "Kabocha" type, developed in Japan, have been introduced in the commercial world trade and are widely grown in Japan, Australia, and New Zealand (Table 4). Kabocha cultivars are similar to Buttercup. Presumably, Japanese breeders utilized crosses between open pollinated Buttercup varieties and related germplasm for developing it. Although similar in size, kabocha squash lacks the protruding ovary and large blossom scar common to Buttercup, have a more mottled green colour, and are usually more oblate with rounded shoulders. Some "Kabocha" cultivars have been hybridised with *C. moschata* cultivars originating in Japan. These hybrids appear as "Japanese squashes" or "Oriental squashes" in the seed catalogues, and are sometimes confused with "Kabocha" cultivars. One of the best-known interspecific hybrids is "Tetsukabuto", also called "Late Potkin" in Australia (Morgan and Midmore, 2003).

After its domestication, *C. ficifolia* gradually spread to the north and arrived in Mexico, where the Aztecs cultivated it in pre-Columbian times. This species is cold-tolerant, but due to its short-day requirement, it has been poorly diffused outside the tropics. *C. ficifolia* fruits reached the Malabar Coast of India in the 16th and 17th centuries by way of the Portuguese and Dutch trade routes, and arrived in Europe from there as late as the 1800s. Due to this, *C. ficifolia* is also known as Malabar melon, Angora squash, Siam squash, Thai marrow, etc. (Andres, 1990).

Table 4. Commercial cultivars of *C. maxima* from recent seed catalogues classified according to the horticultural groups proposed for this species.

Varietal group	Cultivars
Banana	Blue Banana, Orange Banana, Pink Banana, Pink Banana Jumbo, Mammoth Jumbo, Plymouth Rock.
Delicious	Delicious, Golden Delicious, Green delicious, Faxon, Quality.
Marrow	Autumnal Marrow, Boston Marrow, Golden Bronze, Ohio, Valparaiso, Wilder.
Hubbard	Arikara, Baby Blue, Baby Green, Baby Red (Red Kuri), Blue Ballet, Blue Hubbard, Blue Magic*, Brighton, Chicago Warted Hubbard, Golden Hubbard, Green Hubbard, Green Warted Hubbard, Kitchenette, Little Gem, Marble Head, Orange Magic*, Red Hubbard, True Hubbard, Uchiki kuri (Red kuri), Umatilla Marble Head, Warted Hubbard, Yakima Marble Head.
Show	Atlantic Giant, Big Max, Big Moon, Cinderella (Rouge Vif d´Estampes), Dill's Atlantic Giant, Etampes, First Prize*, Mammoth Chili, Mammoth Gold, Prizewinner*, Virginia Mammoth, Wyatt's Wonder.
Turban	Aladdin's Turban, Autumn Cup*, Bonbon*, Burgess Buttercup, Bush Buttercup, Buttercup, Crown, Crown Prince*, Essex, Mini-Red Turban, Mooregold, New Zealand Blue, Orange Dawn*, Queensland Blue, Red China, Red Warren, Sweetmeat, Tiny Turk, Turk's Turban, Warren, Zapallito de Tronco.
Other	Casper, Confection*, Dulce de Horno (Buen Gusto), Galeuse d'Eysines, Gold Nugget, Lakota, Lumina, One Too Many, Red Warty Thing, Triamble, Valenciano, Zapallo Plomo.
Kabocha	Aijehei*, Ambercup*, Black Forest*, Cha-Cha*, Delica* (Ebisu), Eclipse*, Emiguri*, Gatton*, Golden Debut*, Golden Orbit*, Hokkori*, Hokkori*, Honey Delight*, Japan Cup*, Jarrahdale*, Kurijiman*, Nutty Delica*, Pacifica*, Sweet Mama* (Tsurunashi Yakko), Sunshine*.

*indicates F1. The absence of the asterisk indicates open pollinated, breeding line, or lack of information about this cultivar.

Thus, although cultivated for a long time, *C. ficifolia* displays little diversity. Fruits of this species are very uniform, varying mostly in the rind colour pattern: white and green, solid, reticulated or striped. It normally exhibits a coarsely fibrous white flesh used in candies or for forage. Only a few cultivars are used in commercial trade, and most of them are used as rootstock. Recent European and American seed catalogues include cultivars such as "Cabello de Ángel", "Chilacayote", "Chilacayote White", "Courge de Siam", and "Siamkürbis".

4 Genetic Resources

In addition to the five domesticated *Cucurbita*, there are about 10 wild species in the genus (ranging from mesophytic to xerophytic and from annual to perennial). These species are naturally distributed from the central United States to central Argentina, with the greatest diversity in Mexico. They include the wild ancestors of the cultivated species as well as others. The domesticates are partially crossable, so the possibility of transferring genes from one to another exists. Many wild species are also cross-compatible with the domesticates and increase their genetic diversity through natural crossings (Lira Saade, 1995; Montes-Hernández and Eguiarte, 2002).

Both domesticated and wild *Cucurbita* have a high potential value for breeding, although many landraces with differential traits have not yet entered the commercial trade and some wild populations are not well known. Firstly, several interesting characteristics have been reported, such as resistance to different stresses, agronomic and quality traits (Provvidenti, 1990; Loy, 2004). Secondly, *Cucurbita* species are still used by local people in different ways, many of which have ancient origins (Andres, 2000). This ethnobotanical knowledge is fundamental in promoting conservation strategies and for using them as sources for new cultivars in order to attend the market demands of new natural products and new uses.

By far the greatest diversity in *C. moschata* occurs among the innumerable landraces that are grown in the American tropics (Andres, 2004a). Some examples of this variability are the Colombian varieties, whose diversity has only begun to be described (Wessel-Beaver, 2000; Montes et al., 2004), the Cuban landraces, a source of genetic material for tolerance to marginal growth conditions, or the landraces from the north of Peru, including the special type called "Loche" which typically has warty fruits and deep, orange flesh highly esteemed locally as a flavouring for stews (Andres, 2004a). There also exists significant diversity in landraces from warmer regions of Asia or Africa (the warty and wrinkled Japanese fruits; the Indian landraces with large, soft-skinned fruits; the abundance of barbell-shaped fruits in Asia Minor: the Nigerian landraces resistant to diseases, etc.) (Lira-Saade, 1995; Decker-Walters and Walters, 2000). *C. moschata* increase its variability by crossing with wild species. For example, some authors think that the genetic diversity in some northwestern Mexican landraces of *C. moschata* may be the result of introgression from wild subsp. *sosoria* (Decker-Walters and Walters, 2000; Montes-Hernández and Eguiarte, 2002). It is also partially compatible with *C. lundelliana* which is reportedly resistant to diseases. The possibilities of hybridization that *C. moschata* has shown with other cultivated species (for example, *C. maxima*) also increases its genetic pool.

C. argyrosperma has remained limited in its geography and genetic variability. There are still wild populations of the wild subsp. *sosoria* in low-elevations of

Central America and Mexico where they are used by local peoples (by repeated boiling in sugar water, eaten by wild animals, as a soap substitute to wash clothing, and the seeds for medicinal purposes). The landraces of subsp. *argyrosperma* from southern U.S.A and Central America are apparently variable only in fruit and seed size and colour, being those of the var. *callicarpa* the most variable. There is a fourth weedy variety (var. *palmeri* (Bailey) Merrick and Bates) that has mixed characteristics between subsp. *sosoria* and var. *callicarpa* (intermediate fruit size, bitter and non-bitter fruits) and may represent escapes of *callicarpa* that have hybridised with *sosoria*. Like *C. moschata*, high levels of gene flow have been found between cultivated and wild taxa in Mexico (Montes-Hernández and Eguiarte, 2002). Another wild species, *C. okeechobeensis* subsp. *martinezii* L.H (Bailey), endemic to Mexico, can be found growing sympatrically with subsp. *sororia* and is used by local peoples in a similar way. This species, partially crossable with some of the domesticates, has been reported as a source of viral and fungal resistance.

Wild *C. pepo* populations still cross with cultivated forms of *C. pepo* and other species in their distribution range. In addition to these wild resources, primitive landrace pumpkins are widely grown in Mexico and northern Central America that constitute an extraordinary genetic stock. As it was the earliest and most diffused species outside of America, a significant diversification occurred in different places. Mexican pumpkins are often grey–green or black–green striped with thick, lignified rinds; those from U.S.A. are often orange, grooved and not lignified; those of Europe are slightly ribbed, often black-green and orange striped, with thin lignified rinds (Paris, 2001). In Central Europe also a great amount of landraces of the hull-less mutant have originated. A great diversity of types can also be found in Asia Minor (very high in Turkey). However, diversity in acorn cultivars can hardly be found outside of northern America. In this area many landraces used in pre-Columbian times have been lost. Among the domesticates, *C. pepo* is closely related to *C. moschata* and *C. argyrosperma*, which extends its genetic pool.

The centre of diversity of *C. maxima* lies in South American temperate zones, where landraces exhibit interesting traits (such as the large fruits from Chile or those from Bolivia with woody skin suitable for long storage). Many landraces of this species are also in North America, Australia and different countries of Africa (Zambia, Zimbabwe, Nigeria), Asia (China, India, Iran, Afghanistan), and Europe (Spain, Turkey). *C. maxima* is closely related to different wild species of South America. Its wild progenitor, subsp. *andreana*, with small, bitter fruits, is found in Uruguay, Argentina, Bolivia and Paraguay, where spontaneous hybridisation with cultivars increases its genetic variability. This is so frequent that local squash farmers claim this weed causes their cultivars to become bitter. Another closely related species is *C. ecuadorensis*, endemic to Ecuador (also related to *C. moschata*). It has been reported to have been semi-domesticated in pre-Columbian times and then nearly lost from use. There exist free-living populations with wild traits and others with non-bitter fruits of intermediate size that are consumed by local people especially in dry years. This species is easily crossed with *C. maxima* and is being used as a source of disease resistance in breeding programs and is also reported to be well-adapted to drought.

There is a lack of genetic diversity in *C. ficifolia* as compared to the other domesticates (Andres, 1990). It is widely cultivated in small gardens at high altitudes from Mexico to central Chile, where it is used to prepare candied squash. The mature flesh is also eaten raw in salads, the flowers eaten fried with cheese in quesadillas, and the seeds are toasted. The plant is also used in traditional medicine (Andrade-Cetto and Heinrich, 2005). It has not been very diffused outside America, but landraces are grown in the U.S.A, Europe and in developing countries of the Old World. The main variation occurs in the size and colour of its fruits and seeds. Some authors have reported variability in other agronomic characteristics as well, but these are yet to be studied (Lira-Saade, 1995). This species has been reported to be resistant to diseases, and tolerant to low temperatures and to long storage periods. Despite its supposed South American origin, it is partially compatible with xerophytic perennial wild species endemic to Mexico and the southern U.S.A. These include *C. foetidissima*, suitable for arid lands, that has oil and protein-rich seeds (Gathman and Bemis. 1990), and others.

The need for the conservation of these resources is widely recognized. In some places, landraces and wild species are threatened by genetic erosion. Landraces are displaced by modern commercial cultivars and by new cash crops, and the wild species disappear by the alteration of their habitats. In Panama, elderly farmers remember that subsp. *sosoria* was common prior to the practice of using herbicides and the introduction of invasive African grasses (Andres, 2000). In Mexico, most of the farmers that cultivate winter squash landraces in Mexican milpas are elderly, over 60, indicating that genetic erosion of *Cucurbita* is likely in the region in the near future (Montes-Hernández et al., 2005).

There exist many *ex-situ* collections of these genetic resources distributed both in their areas of origin and in different centres of diversity. Many belong to national genebanks, but there are also private organizations dedicated to preserving genetic heritage. A general overview, although not updated, of the most important collections can be found at the IPGRI Directory of Germplasm Collections (http://www.ipgri.cgiar.org/ germplasm/dbintro.htm). Some of the most interesting germplasm collections are in major genebanks in Mexico (INIFAP, http://www.inifap.gob.mx), and Costa Rica (CATIE, http://www.catie.ac.cr), with accessions coming not only from these countries, but also from Central America and the West Indies (Andres, 2004b). Other significant collections are those from Brazil (EMBRAPA, http://www.cenargen.embrapa.br), Colombia (CORPOICA, http://www.corpoica.org.co), Bolivia (Centro de Investigaciones Fitoecogenéticas de Pairumani, http://www.fundacionpatino.org/fitoeco.htm), etc. However, more information is needed about the availability and characteristics of *Cucurbita* collections held in these countries. In addition, some gaps in these collections exist. For landraces and wild species, Mesoamerican countries other than Mexico are not well represented. Also, in South America, Argentina is best represented than Uruguay, Paraguay, Chile, and Peru.

In the U.S.A. the most relevant collection is the Plant Introductions in the U.S. National Germplasm System (http://www.ars-grin.gov) which includes wild species, landraces from different countries and modern cultivars. The different accessions have been collected by different institutions or provided by different contributors (such as some relevant USDA researchers in cucurbits, T.W. Whitaker, R.L. Knight, R.L Clark, H.F. Winters, L. Merrick, T. Andres, etc.). In Europe there also exist large collections. Currently, an international cooperation of European genebanks, promoted by IPGRI within the ECP/GR program (Díez *et al.*, 2002), has been established. This network includes genebanks from Bulgaria, the Czech Republic, Germany, Hungary, Portugal, the Russian Federation, Spain, the Netherlands and Turkey (http://www.comav.upv.es). The most relevant collection is that of the N.I. Vavilov Research Institute of Plant Industry, in St. Petersburg (http://www.vir.nw.ru). During the middle of the twentieth century, many plant-collecting expeditions were organized from Russia (N.I. Vavilov and co-workers) to search for new landraces. During these expeditions, even some *Cucurbita* species were described for the first time (Pangalo, 1930). Other important collections exist in different countries of Asia and Africa such as the collections of *C. moschata* landraces in India, the Philippines, Zambia, Kenya, Ethiopia, Nigeria, Zimbabwe, and that of the Institute of Crop Science (CAAS) in China (Gwanama et al., 2002; Pandey et al., 2003).

The *ex situ* conservation of these resources has some risks. Many institutions use open pollination which threatens genetic integrity of the collections. Regarding its characterization, many of these genebanks have only partially characterized these collections using the descriptors lists published by IPGRI (Esquinas–Alcázar and Gulick, 1983), the UPOV guidelines, and different descriptor lists published by national programs (Vinter et al., 2004). Thus, one of the most important tasks of the Cucurbitaceae working group of the ECP/GR program is the creation of a minimum descriptor list to facilitate a uniform characterization of genetic resources of *Cucurbita*. Despite this characterization, there still exists a great deal of taxonomic misclassification. Some genebanks also perform evaluation assays. Resistance to major viruses has been reported mainly in wild species (Provvidenti, 1990). However, collections are seldom used as sources of quantitatively inherited traits that can contribute to overall performance of the crop and many landraces have not been adequately evaluated, so further screenings are necessary.

Additionally to this *ex situ* conservation programs, programs on *in situ* conservation and participative breeding are gaining importance, mainly in traditional systems of Mexico (Chávez et al., 2002; Montes-Hernández et al., 2005). These programs include the collection of the existing diversity and its evaluation in different environments, but also the socioeconomic study about the management of native seeds and the participative breeding of the outstanding material.

5 Major Breeding Achievements

Squash breeding started during its domestication in America, was under way in Europe by the 1800s and, by the twentieth century, breeding programs were well established in Europe, North America and Asia.

5.1 Genetics of Traits Selected During Domestication

The genetics of the traits selected during domestication have been largely researched (reviewed in Paris and Brown, 2005). These include the fruit size, which is highly polygenic, but also the rind texture and hardness and the fruit cucurbitacin content, which are controlled by single genes (*Hr, Hard rind* and *Wt, Warty* fruit in *C. pepo; Hi, Hard* rind *inhibitor* in *C. maxima; Bi, Bitter* fruit, in *C. maxima* and *C. pepo*). Knowledge of the genetics of these traits facilitates their use in breeding programs. Non-lignified rinds are preferred in fruits consumed when ripe, such as many *C. pepo* winter squashes, in contrast with ornamental gourds and summer squashes. In *C. maxima* and *C. moschata,* the Butternut and Buttercup varieties are the most popular, in part due to their lack of a hard shell, as they can be cut easily with a kitchen knife.

Many fruit variants were also selected since pre-Columbian times. Selection for shape and colour is still a relevant objective for current breeding programs. Fruit shape is polygenic, but some major genes have been reported in *C. pepo* (*Di*, *Disc*) and *C. moschata* (*Bn*, *Butternut*). Many genes have been reported to control rind colour (some affecting flesh colour), most in summer squash forms of *C. pepo*, but other allelic and non-allellic genes have also been reported in *C. maxima* and *C. moschata*. Some of these genes are multiple-allelic and interact with each other (reviewed in Paris and Brown, 2005). For example, the alleles *l-1 (light), l-2 (light), d (D: Dark), b (B: Bicolour), Qi (q: quiescent intense), mo-1 (mature orange 1),* and *mo-2 (mature orange 2)* are responsible for the light orange colouration in "Vegetable Spaghetti" fruits; *qi* for the intense bright orange colour of some Halloween pumpkins; and *Mo1* and *Mo2* for the green colour of "Table Queen" Acorn (Tadmor et al., 2005). In *C. maxima* B^{max} (*Bicolour*), derived from the subsp. *andreana*, is responsible for precocious yellow fruit pigmentation, and *Rd* (*Red* skin) for the red colour of "Turk's Caps", dominant to other colours.

5.2 Crop Productivity: Variation in Plant Architecture. Bush or Semi-Bush Growth Habit

Considering the worldwide importance of *Cucurbita* as a food source, crop productivity has received relatively little attention. A review of the physiological basis of vegetative growth, flowering pattern and fruit set, and of the latest advances in *Cucurbita* breeding for yield components has recently been published by Loy (2004).

Some of the most spectacular increases in plant productivity achieved through plant breeding have been accomplished through genetic changes in plant architecture or growth habit. Wild *Cucurbita* species have generally viny plants. The bushy habit due to short internodes is conditioned by the *Bu* allele (*Bush* habit) in *C. pepo* (dominant to vine habit in young plant stage, but recessive at maturity). Bush-vine heterozygotes produce a uniform bush phenotype early in the season and then a more trailing phenotype beginning in midseason (Denna and Munger, 1963). The same gene has been proposed to explain the bushy habit in *C. maxima,* but the conversion from compact to a more spreading growth occurs much faster in bush-vine heterozygotes of *C. maxima* than those of *C. pepo*. A different gene confers determinant plant growth in *C. moschata* (*de, determinate*) (Kwack, 1995).

The development of *Cucurbita* cultivars with a more compact growth habit has increased substantially during the past two decades. The bushy phenotype, common in summer squashes of *C. pepo*, has been transferred into the Pumpkin and Acorn forms of the species. Many of the Halloween Pumpkins that are offered are either semibush or have restricted vine growth. The first Acorn cultivars, such as "Table Queen", had a vine habit, but bush cultivars have already been bred ("Table Ace"). Similarly, the bushy habit has also been introduced into some cultivars of *C. maxima* by developing productive short- x long-vined hybrids. The introduction of the bushy habit in *C. moschata* is more recent and was first transferred to temperate types by inter-specific hybridization (between *C. pepo* and *C. moschata*) followed by backcrosses to *C. moschata*. More recently, some compact hybrids have been derived from these bushy *C. moschata* cultivars and long-vined types with tropical fruit characteristics (adapted to tropical conditions and resistant to diseases) (Maynard et al., 2002). These hybrids showed higher yields, but smaller fruits than the long-vine landraces (Chesney et al., 2004). The wide acceptance of the bush growth habit by breeders and vegetable growers can be attributed to different factors. Bush plants are characterized by earlier flowering, a higher ratio of pistillate to staminate flowers, earlier maturation, a higher ratio of fruit to vegetative biomass, amenability to high-density planting, rapid leaf canopy closure, and more sustainable weed control.

5.3 Adaptation to Growing Cycles and Modification of the Male:Female Flower Ratio

As stated before, earliness has been achieved by introducing a bushy or semibushy habit in many varieties of *C. pepo, C. maxima* and *C. moschata*. Early types are much easier to mature in cooler areas and can be marketed earlier. However, many *Cucurbita* landraces are of tropical origin and are unadapted to temperate areas, specially many cultigens of *C. moschata* which require long growing seasons. Variation in the production period has been reported in local varieties. For example, the short-cycle variety "Xmejem Cum", commonly grown in the typical Mayan maize-bean-squash milpa system (Lira and Montes-Hernández, 1994; Graephe, 2003). Other species with adaptation problems are *C. argyrosperma* and *C. ficifolia that* often fail to flower when day lengths are too long (Ferriol et al., 2005). Earlier-flowering, day-neutral forms of *C. ficifolia* have been selected to be used as rootstocks, but even these selections are late to mature (Robinson and Decker-Walters, 1997). Furthermore, the higher ratio of female:male flowers of the bushy plants may result in greater fruit setting and yield. This trait is controlled by both environmental and genetic factors. Intra-specific variation of this trait has been reported in several species and transferred between them. High female lines of *C. moschata* and *C. maxima* have been obtained from selected *C. pepo* varieties (Robinson et al., 1978; Kwack and Fujieda, 1985; Gwanama et al., 2001; Loy, 2004).

5.4 Pest and Disease Resistance

5.4.1 Resistance to Pests

Leaf-silvering disorder in *Cucurbita* is a response to the feeding of the immature stage of the silverleaf whitefly, *Bemisia argentifolii*. This is economically significant since silvered plants are less vigorous and silvered fruits are commercially unacceptable. Large collections of landraces of the five cultivated species and some wild taxa (subsp. *sosoria*, *C. ecuadorensis*, *C. foetidissima*, *C. digitata*, *C. okeechobeensis* subsp. *martinezii*) have been screened to identify silverleaf resistance (Maynard, 2001). This screening is sometimes difficult as silvering syndrome is very similar to silver-mottling leaf character (controlled by the dominant gene *M, Mottled leaves,* in *C. pepo*, *C. moschata* and *C. maxima*, and modifiers). Differences in the responses of the different horticultural groups of *C. pepo* have been reported (McAuslane et al., 1996), along with the high resistance of different accessions of *C. moschata* from different origins. The resistance found in *C. moschata* (the Paraguayan landrace PI-162889 and Butternut types) is controlled by a single recessive gene (*sl*, *silverleaf* resistance) (González-Roman and Wessel-Beaver, 2002).

Resistance to other pests have been reported in different species: resistance to the fruit fly (*Dacus cucurbitae*) in *C. maxima* (controlled by the *Fr* gene, *Fruit fly*), resistance to pickle worm in introductions of *C. pepo*, *C. moschata* and *C. maxima*, and tolerance to squash vine borer in *C. moschata* (Whitaker and Bemis, 1976; Paris and Brown, 2005).

5.4.2 Resistance to Fungi

Powdery mildew is one of the major diseases affecting field and glasshouse cucurbit production worldwide. The main causal agents are *Golovonomyces cichoracearum* (formerly *Erysiphe cichoracearum*) and *Podosphaera xanthii* (formerly *Sphaerotheca fuliginea*). Screening of commercial cultivars of *C. pepo* has revealed differential responses according to the horticultural groups (Lebeda et al., 2000). Field resistance has also been reported in *C. maxima* landraces. Adittionally, resistance to *G. cichoracearum* has been reported in *C. lundelliana* (controlled by a dominant gene, *Pm*) and in *C. moschata* (controlled by two genes, *pm-1*, from the tropical cultivar "La Primera", and *pm-2*, from "Seminole") (reviewed in Paris and Brown, 2005). Resistance to *P. xanthii,* derived from *C. okeechobensis* subsp. *martinezzi,* is governed by a single dominant gene with modifiers. This resistance has been transferred to *C. pepo* (Cohen et al., 2003), using *C. moschata* as a bridge. Also, this resistance has been used to breed resistant oriental varieties of *C. moschata* (Cho et al., 2003). Today, powdery mildew-resistant varieties of winter squash (*Cucurbita* spp.) are available in seed catalogues.

Downy mildew caused by *Pseudoperonospora cubensis* is another major disease of cucurbits in temperate and humid areas. Screening of commercial cultivars of *C. pepo* reveals a negative correlation between resistance to downy and to powdery mildew (Lebeda et al., 2000). Resistance from *C. ecuadorensis* and *C. okeechobeensis* was introgressesd into *C. pepo, C. maxima* and *C. moschata* producing germplasm useful for breeding mildew-resistant squashes (Robinson and John, 1987)

5.4.3 Resistance to Viruses

The aphid-transmitted *Zucchini Yellow Mosaic Virus* (ZYMV) remains one of the most widespread and destructive pathogens of cucurbits (Tobias and Palkovic, 2003). Resistance to different strains of ZYMV has been reported in *C. ecuadorensis,* and is controlled by a recessive gene (*zym*^{ecu}), with modifiers, that has been introgressed into *C. maxima*. Different genes of resistance have been reported in *C. moschata,* such as the *zym*^{mos} gene, derived from the tropical cultivar "Soler", and oligogenic resistance, which is controlled by 3 genes, *Zmy-1, Zmy-2 and Zmy-3,* from the "Menina" landrace from Portugal (reviewed in Paris and Brown, 2005). The most promising resistance is that found in a landrace of *C. moschata* from Nigeria referred to as "Nigerian Local". It is controlled by a single, incompletely dominant, major gene (*Zym-0*) with modifiers. This resistance was transferred to different *C. pepo* cultivars (Robinson and Provvidenti, 1997). "Nigerian Local" has also been used in breeding both *C. moschata* and *C. pepo* as a source of resistance to *Watermelon mosaic virus* (WMV), controlled by the dominant *Wmv*, *Papaya ringspot virus*-W (PRSV-W), controlled by the recessive *prv*, and *Cucumber mosaic virus* (CMV), controlled by the dominant *Cmv* gene (Brown *et al*., 2003).

C. ecuadorensis also offers resistance to WMV (*Wmv*^{ecu}), which has been transferred to *C. maxima,* along with a low level of tolerance to *Squash mosaic virus* (SqMV) (Provvidenti, 1997). Resistance to *Squash leaf curl virus* (SqLCV) (*Slc*) has been reported in *C. moschata*, and has been transferred from *C. ecuadoresis* and *C. lundelliana* to *C. maxima*. Resistance to these and other important viruses has been reported in some *C. maxima* PIs (Kristkova and Lebeda, 2000).

5.5 Seed-Pumpkins: the Hull-Less Trait and High Seed Yielding Cultivars

Pumpkin seeds are a high-energy source (mainly lipids and proteins) and are also a reasonable good source of K, P, Fe, and *B-carotene*, derived from the inner seed coat. The main uses of pumpkin seeds are as oil sources and as snacks.

The production of high-quality vegetable oil from seeds is a very traditional practice in Austria, Hungary and Slovenia, where mostly *C. pepo* pumpkins are used. There, pumpkin oil is today a high-value product recognized by the European community ("Sytirian Pumpkin-Seed Oil" is recognized as a Stamp of origin) (Winkler, 2000). *C. maxima, C. moschata* and even *C. ficifolia* are also used as oil sources in other regions (Lira-Saade, 1995; Castro et al., 2006). *C. pepo* pumpkin seeds normally possess a thick, leathery seed coat (hull) that comprises about 20% of the seed weight. At first, the seeds were dehulled by hand and then pressed to extract the oil. A spontaneous mutation of "naked" (hull-less) seeds occurred about 100 years ago in Central Europe. This trait is controlled by the recessive allele of a single major gene (*n, naked* seeds) with modifiers (Zraidi et al., 2003). A recessive gene with a similar effect to *n* has been reported in *C. moschata* (Xianglin, 1987).

The hull-less mutant of *C. pepo* was used as the basis of the breeding work in oil-seed pumpkins which began in Central Europe by the mid 20th century. The first oil-seed pumpkin cultivar in Austria "Gleisdorfer Ölkürbis", was developed from hull-less landraces. Further breeding of the hull-less cultivars included a compact bushy growth habit and higher seed yield. High yields were firstly obtained by selecting large seed size, which is highly correlated to fruit size. Thus, the first oil seed pumpkin cultivars had rather large fruit with thick flesh (fruits of Gleisdorfer Olkurbis range from 3 to 7 kg). These cultivars are suitable for a combined use, as fruit flesh can be used to feed the cattle after the removal of the seeds. The more recent use of seed yield indices that evaluate the ratio of seed/fruit yield and energy has been more efficient at selecting high seed-yielding cultivars. Furthermore, by focusing on genetic variability in seed thickness, which is less correlated to fruit size than seed length and width, it has been possible to develop small-fruited cultivars (with fruit size in the 0.5 to 1.5 kg range) with good seed yield (Winkler, 2000; Cui and Loy, 2002; Loy 2004). The male:female flower ratio is another factor to be considered when breeding high seed yielding cultivars. For example, a high ratio of pistillate to staminate flowers early in the season may lead to limited pollen availability which can decrease seed yield per fruit.

Hull-less seed pumpkins are also being bred for the snackseed trade by producing cultivars with high seed yields and larger seeds. The recently released 'Snackjack' variety meets these characteristics (Loy, 2004). Other hulled forms of *C. pepo*, large white-seeded genotypes of *C. maxima* (such us "Golden Delicious"), and Central American cultivars of *C. argyrosperma* are also used throughout the world as snacks in baked products, and in various confectionary items and trail mixes.

6 Current Goals of Breeding

Breeding for improved fruit colour and morphology, for increased productivity and for pest and disease resistance are still major objectives of current winter squash breeding. Many sources of pest resistance have been identified, although relatively few have been deployed in commercial cultivars. Furthermore, in any given region, the seriousness of infection of different pests and diseases often varies from year to year and destructive, new viral or fungal races can appear suddenly (Desbiez et al., 2003). Therefore, the pyramiding of genes of resistance is also a major goal of future breeding (Ahmed et al., 2001). In addition to biotic stress, breeding for resistance to abiotic stress is also necessary. The basis of the resistance mechanisms of *C. ficifolia* to soil stresses, such as low temperatures and saline conditions is being studied, and it will facilitate the use of this source in breeding other winter squashes and other cucurbits (Lee and Chung, 2005; Wang *et al.*, 2006).

6.1 Fruit Quality

Fruit quality is a major concern of current squash breeders. Usually, cultivars with marginal quality, but high yields, were widely adopted by growers and distributors. However, the consumers′ demands for quality are increasing and there is a critical need to provide a consistently good product. In some markets, such as those for Kabocha squash in Japan, imported squash of inadequate quality may be rejected (Wright and Grant, 1999). The standards for quality vary by the species and cultivar group and according to the final use of the product (fresh market, canning, baby food, seeds, etc.) (Loy, 2004).

6.1.1 Fruit Quality of Fresh Market Winter Squash

Commercial standards for fresh market squash usually refer to cultivar purity and lack of fruit defects, and do not reflect quality traits appreciated by consumers. These are different in summer and winter squash. In the latter, consumers appreciate fruit size, shape, and colour. However, eating acceptability is most often related to flesh colour, consistency, flavour, and sweetness, traits that may play a minor role in summer squashes (Daniel et al., 1995).

One of the parameters most directly related to flesh quality is the dry matter content of mesocarp (DM), responsible for flesh dryness. The main component of squash mesocarp is starch, which is gradually converted into sugar. DM varies according to the balance of starch and sugar content (Culpepper and Moon, 1945; Stevenson and Yoo, 2005). Flesh sweetness increases as the fruit matures, both on the vine or during storage, but flesh dryness increases with maturity in the field, while decreasing during storage. If the DM is initially too low (due to early harvest or environmental stress) or is reduced to unacceptable levels during storage, the cooked texture of the squash will be unsuitably moist and fibrous. By contrast, a DM which is too high may lead to unacceptably dry flesh and low sweetness. Both genetic and environmental factors affect DM, just as different values are preferred for the different types of winter squash. For example, the DM in Acorn and Butternut squash is usually in the range of 12 to 20%, considerably below than that of Kabocha and Buttercup cultivars, which are considered to be of the highest quality (DM 20 to 30% and soluble solids 11 to 13%) (Harvey et al., 1997). Tropical cultigens of *C. moschata* that are frequently used in sauces, soups, stews, cakes and puddings do not require high DM, and a smooth, pasty texture is popular (Daniel et al., 1995).

Nutritional value is another important quality factor in winter squashes. These crops are a source of antioxidants, especially carotenoids, and also tocopherol and ascorbic acid, which have a major role in nutrition in the form of provitamin A, vitamin E and vitamin C (Murkovic et al., 2002). Winter squashes play an important role in local diets and could provide a viable solution to diseases in developing countries. The total carotenoid content has a positive association with the intensity of fruit-flesh colour (Hidaka et al., 1987; Tadmor et al., 2005). Many winter squashes have dark-orange or dark-yellow flesh in contrast with the white or light colours of summer squashes. The relationship between the genetic configuration of major genes *B* and *L-2* and the carotenoid content of fruits flesh has been studied in detail in

C. pepo. The dominant *B* allele promotes carotenoid accumulation, and prevents tocopherol accumulation. The dominant *L-2* allele doubles carotenoid content and, in combination with *B*, significantly increases carotenoid content as compared to the recessive *l-2* allele. Additionally, large collections of this and other species (*C. pepo, C. moschata* and *C. maxima* and crosses between *C .maxima* and *C .moschata*) are being evaluated for these traits. Varieties with a high content of carotenes have an orange appearance, whereas varieties with a high lutein content and a low carotene content show a bright yellow colour. Generally, varieties of *C. maxima* show a much higher content of lutein than those of the remainder species. Promising genotypes are being selected in *C. moschata* (Gwanama et al., 2002; Pandey et al., 2003). However, the inheritance of carotenoid content has not been studied in these species.

6.1.2 Fruit Quality of Processed Squash

C. maxima and *C. moschata* cultivars with the tastiest and deepest-coloured fruits are those mainly used in the pumpkin canning industry. Specific cultivar improvement has been minimal during the past years. Growers have selected the high-yielding cultivars available for fresh market with the proper flesh colour and consistency for this use. Processing varieties must have a soft and, preferably, bright orange rind. Only moderate-to-low levels of DM (9 to 11%) are needed to provide enough consistency. High sweetness is not necessary for producing squash pure for pies, as it can be artificially flavoured, but it is preferred for baby food. A reasonable consistency with a high enough sugar level is necessary for the frozen food industry (Loy, 2004).

6.2 Seed Quality and Diversification of Uses

The economic importance of the pumpkin oil production in Central Europe has enhanced the interest for increasing its nutritional and pharmaceutical value. One current breeding objective is the selection for pumpkin cultivars with a good seed chemical composition, like increased unsaturated fatty acid as well as tocopherol contents (Berenji, 2000; Bavec et al., 2002). Tocopherols are antioxidants and have the potential to reduce the risk of cancer and cardiovascular diseases. The use of other species as sources of pumpkin oil is also being studied.

Many studies on the chemical composition of pumpkin seeds reveal new compounds which suggest new uses or which explain the molecular basis for ancient uses (Castro et al., 2006). For example, different *C. moschata* seed products (raw, roasted autoclaved, germinated, fermented, pumpkin protein concentrate and pumpkin protein isolate) are assayed as a way to improve the nutritional quality of bread. It has been proven that the incorporation of pumpkin products in wheat flour increases its protein, lysine and mineral content (Soukkary, 2001). Pumpkin seeds are also important in traditional medicine in China and Latin America (Diaz et al., 2004; Koike et al., 2005). Pumpkin seeds are also a rich source of phytosterol (compounds that have been shown to reduce the levels of blood cholesterol, decrease the risk of certain types of cancer and enhance the immune system) (Phillips et al. 2005).

6.3 Other Breeding Objectives: Use of Squashes as Rootstocks

Due to the success of certain winter squashes and pumpkins under marginal conditions, they are being tested as rootstocks for other cucurbits, mainly melon, watermelon and cucumber (Eldestein et al., 2004). The vigorous root system of *Cucurbita* species increases the efficiency of water and nutrient absorption, resistance against biotic stress (mainly soil-borne pathogens) and may also serve as a source of endogenous hormones, thus leading to increased yield. Some commercial types, many local landraces of *C. moschata*, *C. maxima* and *C. argyrosperma*, and intra- or inter-specific hybrids, like Tetsukabuto (*C. maxima* x *C. moschata*), are being tested for their compatibility with melon scions, and their resistance to major soil-borne pathogens. The species *C. ficifolia* is also being used to control root rots in cucumber, providing resistance to salt and low temperatures (Pavlou et al., 2002).

7 Breeding Methods and Techniques

As stated before, screening of large germplasm collections of wild species and landraces has provided a genetic variability that has been the basis for winter squash breeding. Morphological traits have been mostly identified by visual selection in fields or greenhouses. Screenings for pest and disease resistance and for quality traits have been conducted under both controlled and open-air conditions.

Interspecific crosses have played a significant role in winter squash breeding for transferring these favourable attributes both from wild into cultivated and between cultivated species. Artificial crosses are easily performed in *Cucurbita*. Most *Cucurbita* types are monoecious. The blossoms open in the morning and are pollinated primarily by specially-adapted solitary bees (Lira-Saade, 1995). The large, connated corollas can be tied up both in male and female flowers to prevent pollinator entry the day before anthesis, when the colour of the petals begins to turn to yellowish-orange. Male flowers produce big pollen grains which can be directly deposited on the fleshy stigmas the next morning. Petals of the female flowers can then be tied up again until fruits are developed. Usually, manual pollinations are carried out with fresh pollen, although pollen from pre-anthesis flowers that are kept for a few days at low temperatures and high humidity can also be used (Robinson and Decker-Walters, 1997).

Many of the species of *Cucurbita* can be successfully crossed (all are diploid 2n=40). However, repeated pollination, bud pollination, and mixed pollen pollination are frequently used to get successful crosses. In some cases embryo culture is also necessary (Wall, 1954; Metwally et al., 1996; Sisko et al., 2003;). F1 and later generations of the interspecific crosses are sometimes sterile or exhibit reduced fertility which makes the introgression of desired traits difficult and time-consuming.

Among the wild mesophytic species, *C. lundelliana* and *C. okeechobeensis* are morphologically similar and cross-compatible. *C. okeechobeensis* can be easily crossed with *C. ecuadorensis*. Between wild and cultivated taxa, *C. lundelliana* crosses with *C. moschata*, *C. maxima*, and *C. ficifolia* to produce partially fertile hybrids, and with *C. pepo* and *C. argyrosperma* to produce hybrids only through embryo culture. Also *C. ecuadorensis* crosses with *C. maxima* to produce fertile F1 hybrids, and *C. okeechobeensis* subsp. *martinezii* crosses with *C. moschata* and *C. argyrosperma*, and with *C. pepo* via embryo culture.

The other wild xerophytic species are weakly compatible with the cultivated *Cucurbita* species. This group can be divided into two subgroups based on cross-compatibility: one consisting of *C. palmata* S. Wats., *C. digitata*, *C. cylindrata* L.H. Bailey, and *C. cordata* S. Wats., and the other only of *C. foetidissima*. Hybrids between *C. maxima* and *C. foetidissima* have been obtained thought embryo culture. Furthermore, interspecific trisomics have been developed in *C. moschata* with chromosomes of *C. palmata* (Graham and Bemis, 1979).

Interspecific crosses are also possible between domesticates. The combinations *C. maxima x C. moschata*; *C. pepo x C. moschata*; *C. pepo x C. argyrosperma* are possible, but do not occur readily. Recently, Wessel-Weaver *et al.* (2004) studied the compatibility between *C. argyrosperma* and *C. moschata*, indicating a high degree of compatibility when *C. argyrosperma* is used as the female parent, but incompatibility in the reciprocal cross. *C. moschata* is considered to be the extant species with the most ancestral-like genome because of its wide cross compatibility. For this reason it has been used frequently as a genetic bridge to transfer genes between other less-compatible domesticates (for example, between *C. pepo* and *C. maxima*). Differences in the crossing affinity between cultivars of different origins in each species has been reported. Interspecific hybrids can be determined by plant and fruit morphology, or by using protein- or DNA-based molecular markers. Furthermore, as the differences in genome size within the genus are large, the interspecific origin of hybrids can be confirmed by evaluating nuclear DNA content (Sisko et al., 2003).

Crosses between cultivated species can be used directly, even used as commercial hybrids (Whitaker and Robinson, 1986), especially the crosses between *C. maxima* and *C. moschata* (Murkovic et al., 2002). In addition to interspecific hybridization, intraspecific crosses between cultivars of different groups have been frequently used to produce new fruit types or to introduce specific traits from one to another.

However, usually interspecific or intraspecific crosses are followed using selection programs. Segregant populations are tested in field or greenhouses. If plant traits are selected, it is possible to control pollination in the same generation, if fruit traits are tested, the progeny will be used in controlled crosses. Backcrossing, combined or not with pedigree selection, is the most-used method for introgressing single alleles, mainly when the traits are transferred from wild species with undesirable characters. As we have seen before, many interesting traits are controlled by one or two major *loci* and are relatively easy to select (Paris and Brown, 2005). However, interesting genes are frequently recessive or have modifiers. Additionally, many other characters are highly complex, like fruit size, fruit shape, earliness, yield, adaptation to certain environmental conditions, quality, and other performance characteristics. In these cases, breeding may be achieved using large progeny populations, which is difficult when working with *Cucurbita* crops because of the large size of the plant and the long growing season some types need. Pedigree and mass selection could be useful methods for these traits (Robinson and Decker-Walters, 1997).

Another breeding scheme frequently used in *Cucurbita* has been the direct use of variation in open-pollinated cultivars. Although primarily outcrossers, individual plants of *Cucurbita* are self-compatible and there is little to no inbreeding depression. Selected variants become genetically uniform after several generations of self-fertilization and selection, and can then be used to make F1 hybrids.

The final product of all these breeding programs can be open-pollinated cultivars, inbreeding lines, and more recently F1 hybrids. The introduction of F1 hybrids in winter squash is increasing. Hybrids provide seed companies with cultivar protection, allow breeders to more easily combine complementary traits, and generally have greater uniformity than open pollinated cultivars. The possible expression of heterosis is another advantage of F1 hybrids. Diallel crossing schemes have been used to study heterosis in *Cucurbita*. Heterosis for earlier flowering, for fruit weight, for seed weight per fruit, and for seed yield per plant has been reported in *C. pepo* (Cui and Loy , 2002). Other species, such as *C. maxima* and *C. moschata* also exhibit heterosis for some traits (earlier flowering, earlier fruit maturity, fruit number per plant, weight of fruit, days to maturity, soluble solid levels, taste rating, etc.) (Loy, 2004).

8 Integration of New Biotechnologies in Breeding Programmes

New biotechnologies are being increasingly applied in *Cucurbita* breeding (Brown, 2001), although it is still in the early stages when compared to other major Cucurbits.

In vitro techniques are very helpful in breeding as they can assist propagation, interspecific hybridization, reduction in generation times, and production of haploids or dihaploids. *In vitro* propagation via shoot tip culture has been used successfully in several cucurbit genera (Rahman et al., 1993; Sarowar et al., 2003). Plant protoplasts are also very useful for genetic manipulation, as they allow the fusion of cells in crosses that are difficult or impossible to make by sexual methods. The hybridization between *Cucumis melo* and an interspecific hybrid *C. maxima x C. moschata* has been attempted using this technique (Yamaguchi and Shiga, 1993). However, hybridity disappeared at later growth stages, showing the difficulty of transferring genes between different genera. The regeneration of plants *in vitro* from cultured protoplasts, cells, or tissues can be accomplished by organogenesis or embryogenesis. Some attempts at organogenesis have been made in *C. pepo* and *C. maxima*, leading to the generation of shoots from mature cotyledons (Lee et al., 2003). Embryogenesis was also achieved in *C. pepo* (Leljak et al., 2004) and *C. ficifolia* (Urbanek et al., 2004). Finally, little work has been done to produce haploid or double haploid tissues from ovule, anther or pollen grain cultures in *Cucurbita*. Only in *C. pepo*, haploid plants were produced from unpollinated ovules and anthers (Metwally et al., 1998).

Molecular markers have been used in the genus *Cucurbita* since the 1960s for elucidating the taxonomical relationships among cultivated and wild species. Isozymes were the first molecular markers to be used for this purpose (Weeden and

Robinson, 1990). More recently, DNA-based polymorphisms have provided a high number of new molecular markers. Chloroplast DNA fragment length, and sequence comparison have been used (Sanjur et al., 2002). These studies have supported previous knowledge about the relationships among the domesticates, namely that *C. moschata, C. pepo* and *C. argyrosperma* are genetically closer than *C. maxima* and *C. ficifolia*. The latter seems to be the most genetically distant.

Molecular markers have also been used to describe intraespecific genetic diversity. In *C. pepo*, studies with isozymes allowed the grouping of wild and cultivated taxa in subspecies and varieties. Similar groupings where found with DNA markers, like RAPD, SSR, ISSR, AFLP and SRAP (Decker-Walters et al., 2002; Katzir et al., 2002; Ferriol et al., 2003a; Paris et al., 2003). In *C. moschata*, RAPDs, SRAPs, and AFLPs have been used to analyse the diversity of landraces in different centres of diversity (Gwanama et al., 2000; Ferriol et al., 2004a). Fewer studies have been performed in *C. maxima* (Baranek et al., 2000). Using enzymatic systems, Decker-Walters *et al.* (1990) and Junior (1999) did not observe any relationship between the morphological and molecular variations. More recently, RAPDs, AFLPs and SRAPs have been used to study the diversity of Spanish *C. maxima* landraces (Ferriol et al., 2003b and 2004b). Finally, little has been done in studying the genetic intraspecific diversity in *C. argyrosperma* and *C. ficifolia*. For the former, using enzymatic systems, the different subspecies could not be differentiated (Merrick, 1990; Montes-Hernández and Eguiarte, 2002). On the other hand, *C. ficifolia* appears as the least diverse species of the cultivated *Cucurbita* species by means of isozymes, but some differential types have been identified using DNA-based markers (Andres, 1990; Ferriol et al., 2005).

The first map of *Cucurbita* was constructed from 11 isozyme loci in five linkage groups based on the F2 of the cross *C. maxima x C. ecuadorensis*. Since then, RAPD markers have been the most used for developing *Cucurbita* molecular maps, with the inconvenience that these markers are dominant and population-specific. Lee *et al* (1995) mapped 28 RAPD markers into five linkage groups from a F2 of *C. pepo* x *C. moschata*. The largest map, covering about 75% of the *Cucurbita* genome, was constructed by Brown and Myers (2002). It was based on the BC1 generation *C. pepo* x (*C. pepo x C. moschata*) and included 148 RAPD markers in 28 linkage groups, in addition to five morphological traits. More recently, a map of the F2 progeny of *C. pepo* (oil-seed pumpkin) x *C. pepo* (zucchini) has been constructed by Zraidi and Lelley (2004). This map contains 36 linkage groups which include 254 RAPD markers and morphological traits. Furthermore, more reproducible and transferable markers have been added.

Marker-assisted selection can only be useful for a few genes responsible for the expression of disease resistances, such as isozymes for WMV in crosses between *C. maxima* and *C. ecuadorensis*. Markers linked to ZYMV are only at a preliminary stage. On the other hand, a high number of morphological traits have been linked to molecular markers, like isozymes, RAPDs, and SSRs (Paris and Brown, 2005).

Cucurbita species have been used as models to characterize genes at the molecular level. These are genes involved in general physiological pathways with little direct use to squash breeders (Guo et al., 2004; Hu et al., 2005; Hossain and Fujita, 2006). Molecular tools are also allowing the characterisation of economically significant *Cucurbita* mutations, such as the hull-less seed trait in *C. pepo*. A comparative study of molecular changes during the development of seed coats in the wild-type and a recessive hull-less mutant of pumpkin has been recently undertaken with the goal of identifying key genes involved in secondary cell wall development in the testa (Bezold et al., 2005). An understanding of these genes will facilitate the manipulation of seed coat development in *Cucurbita* and other species for diverse commercial applications.

The genus *Cucurbita* was one of those in which genetic engineering was first applied. In fact, one of the first transgenic crops released was *C. pepo* with resistance to ZYMV. However, genetic engineering has mostly been applied to breeding summer squashes against viruses using coat protein (see chapter on "Summer squash"), but not to winter squashes.

9 Seed Production

The method of seed production depends on the cultivar type. As stated before farmers save seeds of landraces for the following year's planting. These also can be done with open-pollinated commercial cultivars. However, if seeds are recovered from natural crossings performed by bees, genetic identity can be alternated by cross pollination with other domesticated or wild species in the vicinity. For commercial production is then necessary to separate production fields, or use greenhouses or isolation cages. Controlled hand pollination is also an alternative. As we have commented before hand pollinations are easily conducted in *Cucurbita.* Hand pollinations are recommended to be performed early in the morning, when female flowers are receptive, using a high amount of pollen, as incomplete pollination produces poorly developed fruit with a low number of seeds, and removing open pollinated fruits early to increase fruit set. If pollination by insects is used, several factors must be considered to improve the fruit set ratio and seed production. For example, native bee populations can be artificially increased in field or greenhouses, the application of insecticides during the early and midday hours must be avoided to prevent killing pollinating bees, and also periods of cloudy and rainy weather that minimize bee activity should be avoided.

Environmental factors alter the flower ratio, also affecting seed production. For example, low light levels may cause female flower abortion. Fewer female flowers reach anthesis during prolonged periods of cloudy weather, and under conditions of excessive leaf canopy cover. A delay in fruit set by high temperatures may in turn result in increased vegetative growth and shading of developing pistillate flowers, thereby contributing to suppression of female flower development. In acorn squash long days and high temperatures favoured staminate flowering, whereas short days and lower temperatures favoured the pistillate phase of flowering. The daylength sensitivity of cultivars of *C. moschata*, *C. argyrosperma* and *C. ficifolia* should also be considered during seed production (Loy, 2004).

The commercial production of F1 hybrid seeds is performed by interplanting selected inbred lines. The female line needs to be emasculated. Hand emasculation is feasible due to the size of flowers and the facility of removing male buds before anthesis. This is easier in bushy than in vine types, so the production of hybrids in winter squash where many cultivars are still viney is sometimes difficult. One method of circumventing emasculation for hybrid seed production is to convert the seed parent to the gynoecious condition for an extended period through the use of the chemical ethephon, an ethylene-releasing compound (Robinson et al., 1978) female conversion is more easily accomplished with a bush rather than a vining inbred seed parent. Other treatments, with GA or Ag, that stimulate male flower production may be necessary in other cases. Male sterility can also be used. Genic male sterility that causes male flowers abort has been described in *C. pepo*, *C. maxima* and *C. moschata* (genes *ms-1, ms-2* and *ms-3*) (Paris and Brown, 2005). However, it is difficult and expensive to maintain male sterile populations. When the fruits reach maturity, they are harvested. Seeds can be stored in the fruit after harvest. In *C.maxima* the maximum seed quality in terms of total seed germination was obtained from seeds of 60 days after harvest (Demir et al., 2003). Seeds are removed from the fruit, cleaned, dried and packaged.

10 Acknowledgements

The authors acknowledge INIA (projects RF03-003 and RF2004-00003-00-00) for funding this research.

References

Ahmed, E. A., Ibn Oaf, H. S., El Jack, A. E., and Abdelmohsin, M. E. 2001. Evaluation of the cross Eskandarany x Whitaker for powdery mildew resistance (PMR), zucchini yellow mosaic virus resistance (ZYMR) and some yield characters, Cucurbit Genet. Coop. Rep. 24:35-36.

Andrade-Cetto, A., and Heinrich, M. 2005. Mexican plants with hypoglycaemic effect used in the treatment of diabetes, J. Ethnopharmacol. 99(3):325-348.

Andres, T. C. 1990. Biosystematics, theories on the origin, and breeding potential of *Cucurbita ficifolia*, in: *Biology and Utilization of the Cucurbitaceae*, D. M. Bates, R. W. Robinson, and C. Jeffrey, ed., Cornell University Press, New York, pp. 102-119.

Andres, T. C. 2000. Searching for *Cucurbita* germplasm: collecting more than seeds, Acta Hort. 510:191-198.

Andres, T. C. 2004a. Diversity in tropical pumpkin (*Cucurbita moschata*): cultivar, origin and history, in: *Progress in Cucurbit Genetics and Breeding Research, Proceedings of Cucurbitaceae 2004, the 8th EUCARPIA Meeting on Cucurbit Genetics and Breeding*, A. Lebeda and H. S. Paris, ed., Palacky University, Olomouc, pp. 113-118.

Andres, T. C. 2004b. Diversity in tropical pumpkin (*Cucurbita moschata*): a review of infraspecific classifications, in: Progress in: *Cucurbit Genetics and Breeding Research, Proceedings of Cucurbitaceae 2004, the 8th EUCARPIA Meeting on Cucurbit Genetics and Breeding*, A. Lebeda and H. S. Paris, ed., Palacky University, Olomouc, pp. 107-112.

Baranek, M., Stift, G., Vollmann, J., and Lelley, T. 2000. Genetic diversity within and between the species *Cucurbita pepo*, *C. moschata* and *C. maxima* as revealed by RAPD markers. Cucurbit Genet. Coop. 23:73-77.

Bavec, F., Gril, L., Grobelnik-Mlakar, S., and Bavec, M. 2002. Production of pumpkin for oil, in: *Trends in New Crops and New Uses*, J. Janick and A. Whipkey, ed., ASHS Press, Alexandria, pp. 187-193.

Berenji, J. 2000. Breeding, production, and utilization of oil pumpkin in Yugoslavia. Cucurbit Genet. Coop. 23:105-109.

Bezold, T. N., Mathews, D., Loy, J. B., and Minocha, S. C. 2005. Molecular analysis of the hull-less seed trait in pumpkin: expression profiles of genes related to seed coat development. Seed Sci. Res., 15(3):205-217.

Brown, R. N. 2001. The use and development of molecular breeding tools in *Cucurbita*: a literature review. Cucurbit Genet. Coop. 24:87-90.

Brown, R. N., and Myers, J. R. 2002. A genetic map of squash (*Cucurbita* sp.) with randomly amplified polymorphic DNA markers and morphological markers, J. Am. Soc. Hort. Sci. 127(4):568-575.

Brown, R. N., Bolanos-Herrera, A., Myers, J. R., and Miller, J. M. 2003. Inheritance of resistance to four cucurbit viruses in *Cucurbita moschata*, Euphytica 129(3):253-258.

Castetter, E. F. 1925. Horticultural groups of cucurbits, Proc. Am. Soc. Hort. Sci. 22:338-340.

Castro, H., Galvez, M., Gonzalez, S., and Villamil, C. 2006. Protein composition of *Cucurbita maxima* and *C. moschata* seeds. Biol. Plantarum 50(2):251-256.

Cohen, R., Hanan, A., and Paris, H. S. 2003. Single gene resistance to powdery mildew in zucchini squash of *C. pepo*, Euphytica, 130:433-441.

Cui, H., and Loy, J. B. 2002. Heterosis for seed yield exhibited in hull-less seeded pumpkin, in: Cucurbitaceae 2002, D. N. Maynard, ed., ASHS Press, Alexandria, pp. 323-329.

Culpepper, C. W., and Moon, H. H. 1945. Differences in the composition of the fruits of *Cucurbita* varieties at different ages in relation to culinary use, J. Agr. Res. 71(3):111-136.

Cutler, H., and Whitaker, T. 1967. Cucurbits from the Tehuacan caves, in: The Prehistory of the Tehuacán Valley, Environment and Subsistence, D. S. Byers, ed., Univ. of Texas Press, Austin, Vol. 1, pp. 212-219.

Chávez-Servia, J. L., Arias-Reyes, L. M., Jarvis, D. I., Tuxill, J., Lope-Atzina, D., and Eyzaguirre, C. 2002. *Managing Crop Diversity in Traditional Agroecosystems. Proceedings of a Sympossium, 13-16 February, 2002, Mérida Mexico*, IPGRI, Rome, pp. 84.

Chesney, P., Wessel-Beaver, L., and Maynard, D. N. 2004. Both traditional and semi-bush tropical pumpkin can be intercropped with beans or cowpeas, HortScience 39:525-528.

Cho, M. C., Om, Y. H., Huh, Y. C., Mok, I. G., and Park, H. G. 2003. Two oriental squash varieties resistant to powdery mildew bred through interspecific Crosses, Cucurbit Genet. Coop. 26:40-41.

Daniel, A. L., Brecht, J. K., Sims, C. A., and Maynard, D. N. 2005. Sensory analysis of bush and vining types of tropical pumpkin, Proc. Fla. State Hort. Soc. 108:312-316.

Decker-Walters, D. S., Walters, T. W., Poluszny, U., and Kevan, P. G. 1990. Genealogy and gene flow among annual domesticated species of *Cucurbita*, Can. J. Bot. 68:782-789.

Decker-Walters, D. S., and Walters, T. W. 2000. Squash, in: *The Cambridge World History of Food*, K. F. Kipple and K. C. Ornelas, ed., Cambridge University Press, New York, pp. 335-351.

Decker-Walters, D. S., Staub, J. E., Chung, S. M., Nakata, E., and Quemada, H. D. 2002. Diversity in free-living populations of *Cucurbita pepo* (Cucurbitaceae) as assessed by random amplified polymorphic DNA, Syst. Bot. 27(1):19-28.

Demir, I., Eraslan, K., and Sariyildiz, Z. 2003. Seed quality in winter squash (*Cucurbita maxima* L.) seeds stored at high moisture content in the fruit after harvest, Eur. J. Hort. Sci. 68(5):201-203.

Denna, D. W., and Munger, H. M. 1963. Morphology of thc bush and vinc habits and thc allelism of the bush genes in *Cucurbita maxima* and *C. pepo* squash, Proc. Amer. Soc. Hort. Sci. 82:370-377.

Desbiez, C. A., Gal-On, A. B., Girard, M. A., Wipf-Scheibel, C. A., and Lecoq, H. 2003. Increase in zucchini yellow mosaic virus symptom severity in tolerant zucchini cultivars is related to a point mutation in P3 protein and is associated with a loss of relative fitness on susceptible plants, Pytopathology 93:1478-1484.

Diez, M. J., Pico, B., and Nuez, F. (Compilers), 2002. *Cucurbit Genetic Resources in Europe*, International Plant Genetic Resources Institute, Rome, pp. 58.

Diaz Obregon, D., Lloja Lozano, L., and Carbajal Zuniga, V. 2004. Preclinical studies of *Cucurbita maxima* (pumpkin seeds) a traditional intestinal antiparasitic in rural urban areas, Rev. Gastroenterol Peru 24(4):323-327.

Eldestein, M., Burger, Y., Horev, C., Porat, A., Meir, A., and Cohen, R. 2004. Assesing the effect of genetic and anatomic variation of *Cucurbita* rootstocks on vigour, survival and yield of grafted melons, J. Hort. Sci. Biotecnol. 79:370-374.

Esquinas-Alcázar, J. T., and Gülick, P. J. 1983. *Genetic Resources of Cucurbitaceae*. Int. Board for Plant Genet. Res., Rome, pp. 101.

FAOSTAT data, 2006, http://faostat.fao.org/faostat/collections?subset=agriculture.

Ferriol, M., Picó, B., and Nuez, F. 2003a. Genetic diversity of a germplasm collection of *Cucurbita pepo* using SRAP and AFLP markers, Theor. Appl. Gen. 107:271-282.

Ferriol, M., Picó, B., and Nuez, F. 2003b. Genetic diversity of some accessions of *Cucurbita maxima* from Spain using RAPD and SRAP markers, Gen. Resour. Crop Evol. 50:227-238.

Ferriol, M., Picó, B., Fernández de Córdova, P., and Nuez, F. 2004a. Molecular diversity of a germplasm collection of squash (*Cucurbita moschata*) determined by SRAP and AFLP markers, Crop Sci. 44(2):653-664.

Ferriol, M., Picó, B., and Nuez, F. 2004b. Morphological and molecular diversity of a collection of *Cucurbita maxima* landraces, J. Am. Soc. Hort. Sci. 121(1):60-69.

Ferriol, M., Picó, B., and Nuez, F. 2005. Genetic diversity of *Cucurbita* spp. in the Canary Islands: a bridge between America and Europe, in: *Plant Genetic Resources of Geographical and Other "Islands" (Conservation, Evaluation and Use for Plant Breeding), Book of Abstracts of the XVII EUCARPIA Genetic Resources Section Meeting 30 March-2 April*, S. Bullita, ed., CNR, Castelsardo, pp. 9.

Fritz, G. J. 1994. Precolumbian *Cucurbita argyrosperma* ssp. *argyrosperma* (Cucurbitaceae) in the Eastern woodlands of North America, Econ. Bot. 48(3):280-292.

Gathman, A. C., and Bemis, W. P. 1990. Domestication of buffalo gourd, *Cucurbita foetidissima*, in: p. 318-324. In: *Biology and Utilization of the Cucurbitaceae*, D. M. Bates, R. W. Robinson, and C. Jeffrey, ed., Cornell University Press, New York, pp. 335-348.

González-Román, M., and Wessel-Beaver, L. 2002. Resistance to silverleaf disorder is controlled by a single recessive gene in *Cucurbita moschata* Duchesne, Cucurbit Genet. Coop. Rep. 25:49-50.

Graephe, S. 2003. Crop and soil variability in traditional and modern mayan maize cultivation of Yucatan, Mexico, J. Agric. Rur. Dev. Trop. Subtrop. 75:1-72.

Graham, J. D., and Bemis, W. P. 1979. Six interspecific trisomics (2n *C. moschata* + 1 *C. palmata* chromosome) and one primary trisomic of *Cucurbita moschata*, Cucurbit Genet. Coop. Rep. 2:37.

Guo, H. N., Chen, X. Y., Zhang, H. L., Fang, R. X., Yuan, Z. Q., Zhang, Z. S., and Tian, Y. C. 2004. Characterization and activity enhancement of the phloem-specific pumpkin PP2 gene promoter, Transgen. Res. 13(6):559-566.

Gwanama, C., Labuschagne, M. T., and Botha, A. M. 2000. Analysis of genetic variation in *Cucurbita moschata* by random amplified polymorphic DNA (RAPD) markers, Euphytica 113:19-24.

Gwanama, C., Botha, A. M., and Labuschagne, M. T. 2001. Genetic effects and heterosis of flowering and fruit characteristics of tropical pumpkin, *Plant Breed.* 120(3):271-272.

Gwanama, C., Nichterlein, K., Lungu, D., and Simabwachi, W. 2002. Variation of fruit B-carotene content of tropical pumpkin (*C. moschata* (Duchesne) Poirot) landraces in Zambia, Plant Gen. Resour. Newslet. 129:44-46.

Harvey, W. J., Grant, D. G., and Lammerink, J. P. 1997. Physical and sensory changes during the development and storage of buttercup squash, New Zeal. J. Crop Hort. Sci. 25(4):341-351.

Hidaka, T., Anno, T., and Nakatsu, S. 1987. The composition and vitamin A value of the carotenoids of pumpkins of different colors, J. Food Biochem. 11:59-68.

Hossain, M., and Fujita, M. 2006. Induction of pumpkin glutathione *S*-transferases by different stresses and its possible mechanisms, Biol. Plantarum 50(2):210-218.

Hu, G., Zhang, S. L., Xu, C. J., and Lin, S. Q. 2005. Cloning and analysis of the phloem specific promoter from *Cucurbita maxima* and construction of new plant expression vector, Acta Agric. Univ. Jiangxiensis 27(4):481-485.

Júnior, A. T. A. 1999. Divergência genética entre acessos de moranga do banco de germoplasma de hortaliças de Universidade Federal de Viçosa, Hort. Bras. 17:3-6.

Katzir, N., Portnoy, V., Yonash, N., Mozes-Daube, N., Tzuri, G., and Paris, H. S. 2002. Use of AFLP, ISSR, and SSR marker systems to assess genetic diversity in *Cucurbita pepo. Plant, Animal and Microbe Genomes X Conference, San Diego, California* 10:121.

Kristkova, E., and Lebeda, A. 2000. Resistance in *Cucurbita pepo* and *Cucurbita maxima* germplasm to watermelon mosaic potyvirus, Plant Gen. Resour. Newslet. 121:47-52.

Koike, K., Li, W., Liu, L., Hata, E., and Nikaido, T. 2005. New phenolic glycosides from the seeds of *Cucurbita moschata,* Chem. Pharm. Bull. 53(2):225-228.

Kwack, S. N. 1995. Inheritance of determinate growth habit in *Cucurbita moschata* Poir., J. Kor. Soc. Hort. Sci. 36(6):780-784.

Kwack, S. N., and Fujieda, K. 1985. Breeding high female lines through interspecific hybridization of *Cucurbita*, Cucurbit Genet. Coop. 8:78-79.

Lee, Y. K., Chung, W. I., and Ezura, H. 2003. Efficient plant regeneration via organogenesis in winter squash (*Cucurbita maxima* Duch.), Plant Sci. 164(3):413-418.

Lee, S. H., and Chung, G. C. 2005. Sensitivity of root system to low temperature appears to be associated with the root hydraulic properties through aquaporin activity, Sci. Hortic. 105(1):1-11.

Lee, Y. H., Jeon, H. J., Hong, K. H., and Kim, B. D. 1995. Use of random amplified polymorphic DNAs for linkage group analysis in interspecific hybrid F2 generation of *Cucurbita*, J. Kor. Soc. Hort. Sci. 36(3):323-330.

Leljak-Levanik, D., Bauer, N., Mihaljevic, S., and Jelaska, S. 2004. Somatic embryogenesis in pumpkin (*Cucurbita pepo* L.): control of somatic embryo development by nitrogen compounds, J. Plant Physiol. 161(2):229-236.

Lebeda, A., and Krístkova, E., and Paris, H. S. 2000. Interactions between morphotypes of *C. pepo* and obligate biotrophs. (*Pseudoperonospora cubensis, Erysiphe cichoracearum* and *Sphaerotheca fuliginea*), Acta Hort. 510:219-225.

Lira-Saade, R. 1995. *Estudios Taxonómicos y Ecogeográficos de las Cucurbitaceae Latinoamericanas de Importancia Económica, Systematic and Ecogeographic Studies on Crop Genepools 9*, Internacional Plant Genetic Resources Institute, Rome, pp. 281.

Lira-Saade, R., and Montes-Hernández, S. 1994. Cucurbits (*Cucurbita* spp.), in: *Neglected Crops: 1492 from a Different Perspective*, J. E. Hernández Bermejo and J. León, ed., Plant Production and Protection Series No. 26. FAO, Rome, pp. 63-77.

Loy, J. B. 2004. Morpho-physiological aspects of productivity and quality in squash and pumpkins (*Cucurbita* spp.), Crit. Rev. Plant Sci. 23(4):337-363.

Maynard, D. N. 2001. Variation among tropical pumpkin (*Cucurbita moschata*) cultivars in susceptibility to silverleaf, Cucurbit Genet. Coop. 24:24.

Maynard, D. N., Hochmuth, G. J., Vavrina, C. S., Stall, W. M., Kucharek, T. A., Webb, S. E., Taylor, T. G., and Smith, S. A. 2001. Cucurbit production in Florida, in: *Vegetable Production Guide for Florida*, D. N. Maynard and S. M. Olson, ed., Univ. Florida, IFAS, Extension, Gainesville, pp. 151–178.

Maynard, D. N., Elmstrom, G. W., Talcott, S. T., and Carle, R. B. 2002. "El Dorado" and "La Estrella": compact plant tropical pumpkin hybrids, HortScience 37:831-833.

Merrick, L. C. 1990. Systematics and evolution of a domesticated squash, *Cucurbita argyrosperma*, and its wild and weedy relatives, in: *Biology and Utilization of the Cucurbitaceae*, D. M. Bates, R. W. Robinson, and C. Jeffrey, ed., Cornell University Press, New York, pp. 77-95.

Metwally, E. I., Moustafa, S. A., El Sawy, B. I., and Shalaby, T. A. 1998. Haploid plantlets derived by anther culture of *Cucurbita pepo*, Plant Cell Tiss. Organ Cult. 52(3):171-176.

Metwally, E. I., Haroun, S. A., and El Fadly, G. A. 1996. Interspecific cross between *Cucurbita pepo* L. and *Cucurbita martinezii* through in vitro embryo culture, Euphytica 90(1):1-7.

McAuslane, H. J., Webb, S. E., and Elmstrom, G. W. 1996. Resistance in breeding lines of *Cucurbita pepo* to squash silverleaf, a disorder associated with feeding by *Bemisia argentifolii* (Homoptera: Aleyrodidae), Fla. Entomol. 79:206-221.

Montes-Hernández, S., and Eguiarte L. E. 2002. Genetic structure and indirect estimates of gene flow in three taxa of *Cucurbita* in western Mexico, Amer. J. Bot. 89(7):1156-1163.

Montes-Hernández, S., Merrick, L. C., and Eguiarte, L. E. 2005. Maintenance of squash (*Cucurbita* spp.) landrace diversity by farmers' activities in Mexico, Genet. Resour. Crop Evol., 52(6):697-707.

Montes, R. C., Vallejo, C. F. A., and Baena, G. D. 2004. Diversidad genetica de germoplasma colombiano de zapallo (*Cucurbita moschata* Duchesne Exp. Prior). Acta Agron., Univ. Nac. Col. 53(3/4):43-50.

Morgan, W., and Midmore, D. 2003. Kabocha and Japanese Pumpkin in Australia. *Rural Industries Research and Development Corporation*, 67, Barton, pp. 64.

Murkovic, M., Mulleder, U., and Neunteuflw, H. 2002. Carotenoid content in different varieties of pumpkins, J. Food Comp. Anal.15:633-638.

Nee, M. 1990. The domestication of *Cucurbita* (Cucurbitaceae), Econ. Bot. 44(3, suppl.): 56-68.

Oliszewski, N. 2005. Archaeobotany of mound structures in Campo del Pucará, Catamarca, Argentina (1750-1450 b.p.): ceremonial use or rubbish dumps?. Veg. Hist. Archaeobot. 14 (4):465- 471.

Pangalo, K. I. 1930. A new species of cultivated pumpkin. Bull. Appl. Bot. Genet. and Pl. Breed. 23:253-265.

Phillips, K. M., Ruggio, D. M., Ashraf-Khorassani, M. 2005. Phytosterol composition of nuts and seeds commonly consumed in the United States. J Agric Food Chem. 53(24):9436-45.

Provvidenti, R. 1990. Viral diseases and genetic sources of resistance in *Cucurbita* species, in: *Biology and Utilization of the Cucurbitaceae*, D. M. Bates, R. W. Robinson, and C. Jeffrey, ed., Cornell University Press, New York, pp. 427-435.

Provvidenti, R. 1997. Resistance to viral diseases of cucurbits conferred by biotechnological and natural resistance genes (In Chinese), China Veget. 4:55-57.

Pandey, S., Singh, J., Upadhyay, A. K., Ram, D., and Mathura, R. 2003. Ascorbate and carotenoid content in an indian collection of pumpkin (*C.moschata* Duch.ex Poir.), Cucurbit Genet. Coop. 26:51-53.

Paris, H. S. 1989. Historical records, origins, and development of the edible cultivar groups of *Cucurbita pepo* (Cucurbitaceae), Econ Bot. 43:423-443.

Paris, H. S. 2001. Characterization of the *Cucurbita pepo* collection at the Newe Ya'ar Research Center, Israel, Plant Genet. Res. News., 126:41-45.

Paris, H. S., and Brown, R. N. 2005. The genes of pumpkin and squash, HortScience 40: 1620-1630.

Paris, H. S., Yonash, N., Portnoy, V., Mozes-Daube, N., Tzuri, G., and Katzir, N. 2003. Assessment of genetic relationships in *Cucurbita pepo* (Cucurbitaceae) using AFLP, ISSR, and SSR markers, Theor. Appl. Genet. 106:971–978.

Pavlou, G. C., Vakalounakis, D. J., and Ligoxigakis, E. K. 2002. Control of root and stem rot of cucumber, caused by *Fusarium oxysporum* f. sp. *radicis-cucumerinum*, by grafting onto resistant rootstocks, Plant Dis. 86 (4):379-382.

Piperno, D. R., and Stothert, K. E. 2003. Phytolith evidence for early Holocene *Cucurbita* domestication in southwest Ecuador, Science 299 (5609):1054-1057.

Piperno, D. R., Holst, I., Wessel-Beaver, L., and Andres, T. C. 2002. Evidence for the control of phytolith formation in *Cucurbita* fruits by the hard rind (Hr) genetic locus: Archaeological and ecological implications, Proc. Natl Acad. Sci. U.S.A. 99(16):10923-10928.

Rahman, S. M., Hossain, M., Islam, R. and Joarder, O. I. 1993. Plant regeneration from internode segments of *Cucurbita maxima* Duch. x *Cucurbita moschata* Duch. Curr. Sci., 65(7):562-564.

Robinson, R. W., and Decker-Walters D. S. 1997. *Cucurbits*. CAB International, Wallingford, pp. 226.

Robinson, R. W., and John, C. A. 1987. Downy Mildew Resistance in *Cucurbita*, Cucurbit Genet. Coop. 10:87

Robinson, R. W., and Provvidenti, R. 1997. Differential response of *Cucurbita pepo* cultivars to strains of Zucchini yellow mosaic virus, Cucurbit Genet. Coop. 20:58-59.

Robinson, R. W., Whitaker, T. W., and Bohn, G. W. 1978. Promotion of pistillate flowering in *Cucurbita* by 2- cloroethyl phosponic acid, Euphytica, 19:180-183.

Sanjur, O. I., Piperno, D. R., Andres, T. C., and Wessel-Beaver, L. 2002. Phylogenetic relationships among domesticated and wild species of *Cucurbita* (Cucurbitaceae) inferred from a mitochondrial gene: implications for crop plant evolution and areas of origin. Proc. Natl. Acad. Sci. U.S.A. 99:535-540.

Sarowar, S., Oh, H.Y., Hyung, N. I., Min, B. W., Harn, C. H., Yang, S. K., Ok, S. H., and Shin, J. S. 2003. In vitro micropropagation of a *Cucurbita* interspecific hybrid cultivar – a root stock plant, Plant Cell Tiss. Organ Cult. 75:179-182.

Sisko, M., Ivancic, A., and Bohanec, B. 2003. Genome size analysis in the genus *Cucurbita* and its use for determination of interspecific hybrids obtained using the embryo-rescue technique, Plant Sci. 165(3):663-669.

Smith, B. D. 2005. Reassessing Coxcatlan Cave and the early history of domesticated plants in Mesoamerica, Proc. Natl Acad. Sci. U.S.A. 102(27):9438-9445.

Soukkary, F. A. H. 2001. Evaluation of pumpkin seed products for bread fortification, Plant Foods Hum. Nutr. 56(4):365-384.

Stevenson, D. G., and Yoo, S. H. 2005. Structural and physicochemical characteristics of winter squash (*Cucurbita maxima* D.) fruit starches at harvest, Carbohydrate Polym. 59 (2):153-163.

Tadmor, Y., Paris, H. S., Meir, A., Schaffer, A. A., and Lewinsohn, E. 2005. Dual role of the pigmentation gene B in affecting carotenoid and vitamin E content in squash (*Cucurbita pepo*) mesocarp, J. Agric. Food Chem. 53 (25):9759-9763.

Teppner, H. 2004. Notes on *Lagenaria* and *Cucurbita* (Cucurbitaceae) – review and new contributions. Phyton, 44 (2):245-308.

Tobias, I., and Palkovics, L. 2003. Characterization of Hungarian isolates of zucchini yellow mosaic virus (ZYMV, potyvirus) transmitted by seeds of *Cucurbita pepo* var *Styriaca*, Pest Manag. Sci. 59(4):493-497.

Urbanek, A., Zechmann, B., and Muller, M. 2004. Plant regeneration via somatic embryogenesis in Styrian pumpkin: cytological and biochemical investigations, Plant Cell Tiss. Organ Cult. 79(3):329-340.

Vinter, V., Křístková A., Lebeda, A., and Křístková, E. 2004. Descriptor lists for genetic resources of the genus *Cucumis* and cultivated species of the genus *Cucurbita*, in: *Progress in Cucurbit Genetics and Breeding Research, Proceedings of Cucurbitaceae 2004, the 8th EUCARPIA Meeting on Cucurbit Genetics and Breeding*, A. Lebeda and H. S. Paris, ed., Palacky University, Olomouc, pp. 95-99.

Wall, J. R. 1954. Interspecific hybrids of *Cucurbita* obtained by embryo culture, Proc. Amer. Soc. Hort. Sci. 63:427-430.

Wang, R., Chen, G. L., Song, W., Lu G. Y., Liang, J., and Li, W. X. 2003. Effects of NaCl stress on cation contents in seedlings of two pumpkin varieties, J. Plant Physiol. Mol. Biol. 32(1):94-98.

Weeden, N.F., and Robinson, R.W. 1990. Isozyme studies in *Cucurbita*, in: *Biology and Utilization of the Cucurbitaceae*, D. M. Bates, R. W. Robinson, and C. Jeffrey, ed., Cornell University Press, New York, pp. 51-59.

Wessel-Beaver, L. 2000. Evidence for the center of diversity of *Cucurbita moschata* in Colombia, Cucurbit Genet. Coop. 23: 54-55.

Wessel-Beaver, L., Cuevas, H. E., Andres, T. C., and Piperno, D. R. 2004. Genetic compatibility between *Cucurbita moschata* and *C. argyrosperma*, in: *Progress in Cucurbit Genetics and Breeding Research, Proceedings of Cucurbitaceae 2004, the 8th EUCARPIA Meeting on Cucurbit Genetics and Breeding*, A. Lebeda and H. S. Paris, ed., Palacky University, Olomouc, pp. 393-400.

Whitaker, T. W., and Bemis W. P. 1976. Cucurbits, in: *Evolution of crop plants,* N.W. Simmonds, ed, Longman, London, pp. 64-69.

Whitaker, T. W., and Davis, G. N. 1962. *Cucurbits. Botany, Cultivation, and Utilization*, World crops Books, London, pp. 250.

Whitaker, T. W., and Robinson, R.W. 1986. Squash breeding, in: *Breeding Vegetables Crops*, M. J. Bassett, ed. AVI Publishing Co, pp. 209-242.

Winkler, J. 2000. Breeding of hull-less seeded pumpkins (*Cucurbita pepo*) for the use of the oil, Acta Hortic. 510:123-128.

Wright, P. J., and Grant, D. G. 1999. Effects of pre-shipping storage conditions on buttercup squash quality rots, New Zeal. J. Crop Hort. Sci. 27:337-343.

Xianglin, Z. 1987. A study of the breeding of naked kernel pumpkin and its genetic behaviour, Acta Hort. Sinica 14:115-118.

Yamaguchi, J. and Shiga, T. 1993. Characteristics of regenerated plants via protoplast electrofusion between melon (*Cucumis melo*) and pumpkin (interspecific hybrid, *Cucurbita maxima x C. moschata*), Japan. J. Breed. 43(2):173-182.

Zraidi, A., Pachner, M., Lelley, T., and Obermayer, R. 2003. On the genetics and histology of the Hull-less character of Styrian Oil-Pumpkin (*C.pepo* L.), Cucurbit Genet. Coop. 26:57-61.

Zraidi, A., and Lelley, T. 2004. Genetic map for pumpkin *Cucurbita pepo* L. using random amplified polymorphic DNA markers, in: *Progress in Cucurbit Genetics and Breeding Research, Proceedings of Cucurbitaceae 2004, the 8th EUCARPIA Meeting on Cucurbit Genetics and Breeding*, A. Lebeda and H. S. Paris, ed., Palacky University, Olomouc, pp. 507-514.

Summer Squash

Harry S. Paris[1]

[1] Agricultural Research Organization, Department of Vegetable Crops, hsparis@volcani.agri.gov.il

1 Introduction

Summer squash are the edible immature fruits of *Cucurbita pepo* L., a highly diverse species of the gourd family, Cucurbitaceae. Summer squash are an easy-to-grow, short-season crop best adapted to temperate and subtropical regions. In the New World tropics, some *Cucurbita moschata* Duchesne are grown for their edible immature fruits. Some *Cucurbita maxima* Duchesne are also grown for this purpose in South America.

According to the F.A.O. (2006), over 1,000,000 hectares of pumpkins, squash, and gourds are harvested annually with a total yield of nearly 19,000,000 tonnes. Only a small amount of this area and production is from summer squash. On the other hand, production and marketing of summer squash is concentrated in the economically advanced countries, and therefore the monetary value of this crop probably exceeds by far the remaining crops encompassed by this F.A.O. category. A conservative estimate of the worldwide value of this crop is several billion dollars annually, ranking summer squash relatively high among vegetable crops in economic value, and per capita consumption appears to be rising. Among the largest producers of summer squash are Turkey, Italy, Egypt, Spain, U.S.A., and Mexico (Paris, 1996).

Summer squash are harvested when they are shiny. Usually the size preferred ranges from 100 to 200 g. Most often this size is achieved between two and five days past anthesis, depending on growing conditions. If the fruits are not harvested on time, they continue to grow and begin to lose their shininess. Dull, oversize fruits are generally unsaleable.

2 Origin and Domestication

Archaeological, linguistic, and historical records concur that the genus *Cucurbita* L. (pumpkins, squash, and some gourds) is native to the Americas and was dispersed to other continents by transoceanic voyagers around the turn of the 16th century (Whitaker, 1947).

Archaeological evidence of domestication of *Cucurbita pepo* in southern Mexico dates back 10,000 years (Smith, 1997). Evidence has been found to indicate the presence of domesticates in northeastern Mexico and eastern U.S.A. over 4,000 years ago. Domestication is considered to have occurred at least twice, in southern Mexico and eastern United States (Decker, 1988).

Linguistic evidence is derived from reports by 16th-century European explorers of peculiar Native American names. Indeed, the word "squash" itself is a derivative of the Native American "asquash", the plural form of "asq", designating a fruit that is immature or incomplete (Gray and Trumbull, 1883). Other words for squash derived from Native American languages of what is now the United States are "cushaw" and "macocqwer" or "macock".

Historical evidence is based on the observation that the earliest-known records of *Cucurbita* in Europe are post-Columbian. These records are paintings that appeared in a prayer book prepared for Queen Anne of France in Touraine, Loire Valley, completed in 1508 (Paris et al., 2006), and in festoons on a ceiling in the Villa Farnesina in Rome, completed in 1518 (Janick and Paris, 2006). A quarter-century later, illustrations of *Cucurbita* began to appear in the great botanical herbals of the Renaissance, notably in the *De Historia Stirpium* of Fuchs (1542) and culminating with the *Stirpium Sciagraphia et Icones* of Chabrey (1666).

Cucurbita pepo is native to North America (Trumbull, 1876; Erwin, 1931; Whitaker, 1947) and can be found growing wild in northeastern Mexico and southern, southeastern, and central U.S.A. (Nee, 1990). As yet undiscovered wild populations might still exist in central or southern Mexico and the wild range might have extended to what is now the northeastern U.S.A. (Petersen and Sidell, 1996).

Wild *C. pepo* plants are spreading, viney, with slender, angular stems, spiculate petioles and pentalobate laminae approximately 25 cm long and wide (Lira Saade et al., 1993). The plants are monoecious, bearing separate staminate and pistillate flowers at the stem nodes. The corollas are orange-yellow, approximately 10 cm broad, opening at dawn and withering by noon. The fruits are small (35–100 mm diameter), striped, spherical, oblate, ovate, obovate or pyriform, with a lignified rind encasing thin, coarsely fibrous, usually bitter, light yellow-orange or greenish white flesh enclosing hundreds of small (8–11 mm long), more-or-less oval, beige seeds (Bailey, 1943). Domesticated *Cucurbita* plants generally have larger leaves and flowers, thicker stems, and fewer branches and bear larger and fewer, non-bitter fruits with thicker peduncles. The fruits of domesticates contain larger and fewer seeds and have a wider range of exterior colours and colour patterns, with relatively thin, less durable fruit rinds. Their flesh is thicker, not bitter, less coarsely fibrous and can be highly coloured (Whitaker and Bemis, 1964).

As wild gourds are usually round or nearly so, it is very likely that the first *Cucurbita* that were domesticated for culinary purposes were also round-fruited. As the pumpkins of Mexico and Guatemala have thin, tough, fibrous mature-fruit flesh and relatively thick, lignified rinds, they appear to represent an early stage of domestication. These pumpkins differ from the wild gourds by having larger plants, leaves, flowers, fruits, and seeds, and their lack of fruit bitterness.

Although the first use of *Cucurbita* by people as food appears to have been the consumption of its seeds, the use of immature *Cucurbita* fruits as a vegetable appears nonetheless to be thousands of years old (Cutler and Whitaker, 1961), probably antedating the use of the mature fruit flesh (Paris, 2000). Initially, immature fruits larger than those that are referred to today as summer squash were preferred. To the present day, primitive landrace pumpkins that appear to have been harvested between 10 days and two weeks past anthesis are marketed in Mexico and Guatemala. Fruits of this age have attained nearly their maximum size without having yet developed rind lignification (Schaffer et al., 1986). These landrace pumpkins appear to be grown for a dual purpose, consumption of the immature fruits and consumption of the seeds of their mature fruits, possibly echoing the initial culinary uses of *Cucurbita* by ancient peoples.

Even until modern times, the fruits of summer squash were not usually picked at an age of several days past anthesis (Tapley et al., 1937; Lorenz, 1949). Instead, they were mostly picked approximately a week past anthesis, after attaining greater size, which would result in greater yields (MacGillivray, 1960) and better shipping quality (Oemler, 1883). The first clear-cut records of commercialization of harvesting the high-quality young fruits are of cocozelle squash in 19th-century France (Mérault, 1827; Seringe, 1847). The flowers of cultivated *Cucurbita* are quite large, 10–20 cm in diameter, and both staminate and pistillate flowers are consumed in some regions, the open flowers being harvested early in the morning. Since the 16th century, some cocozelles have been grown for their extremely large, meaty pistillate flowers (Paris and Janick, 2005) and staminate flowers can also serve as a culinary item.

There is an association among cultivated cucurbits in general, and among cultivars of *Cucurbita pepo* in particular, between the stage of development at which the fruits are harvested for culinary use and the length-to-maximum-width ratio of the fruit (Paris, 1989). Cultivars intended for the culinary use of the mature fruits deviate little from a 1:1 ratio and those that are intended to be consumed when young diverge markedly from this ratio. The divergence results from the desire for a narrower placental cavity and proportionally more of the coloured exocarp and firm mesocarp tissue. Narrowing of the seed cavity is associated with shortening of the seeds (Paris and Nerson, 2003) and lower seed yield per fruit (Nerson, 2005). Even though the fruits of the primitive cultivars from Mexico and Guatemala are consumed when young, they are globular or oblate, reflecting the culinary importance of their long, large seeds.

A number of *Cucurbita pepo* cultigens were developed by Native Americans in pre-Columbian times. Seeds or fruits of cultigens originating from different parts of North America reached Europe within a few decades of European contact with American horticulture. Many ended up being grown in close proximity of one another for the first time, in gardens of royalty, doctors, apothecaries, botanists, and others. Cross-pollination was inevitable. On one hand, this diluted the outstanding qualities of each of the donor parents but, on the other hand, resulted in genetic recombination and new phenotypes. Apparently, a small minority of these recombinants were selected that proved to be quite valuable in the long term, leading to the development of new cultivar-groups that were to eventually contain a larger number of cultivars, more phenotypic variation, and far more economic importance than their North American ancestors (Paris, 2000).

3 Varietal Groups

Much confusion has afflicted the terminology of the Cucurbitaceae since ancient times. On one hand, the family is extremely polymorphic for size, shape, and colour of the fruits; on the other, the fruits of some cultigens of one species can exhibit great similarity to those of other species. Often, the result has been different names for the same species and the same name for different species. Confusion is enhanced by translation of names to different languages. Cucurbitaceous fruits have been valued by humans in both, the New World and the Old World, for thousands of years, for food and a multitude of various other uses. People the world over have been fascinated by the fast growth of cucurbits, from seed to a rampant vine bearing prominent, attractive fruits within two or three months. Cucurbits are frequent subjects of art, literature, and myth. Metaphorically, they are associated with warmth, sunshine, health, vitality, fertility, sexuality, and abundance, leading to mirth and laughter (Norrman and Haarberg, 1980).

Of all the species of the Cucurbitaceae, *Cucurbita pepo* is perhaps the most polymorphic for fruit characteristics, indeed one the most polymorphic of the plant kingdom (Duchesne, 1786; Naudin, 1856). The introduction of this species together with *C. maxima* and *C. moschata*, both of which are also highly polymorphic, into Europe 500 years ago further confused cucurbit terminology and compounded the myths concerning crossing among cucurbits. This is an easily understood consequence, as the size of the mature fruits of *C. pepo* can range from several centimetres in diameter to over 20 kg in weight. Fruit shape can vary from round to disc to very long. The fruits can be smooth or warted, with or without longitudinal ribs (10 rounded protrusions), longitudinal grooves (shallow, narrow depressions, usually 20 in number), furrows (usually 10 deep, broad, angular depressions alternating with angular ridges of the same number), or wavy lobes ("scalloping", usually 10 broad, rounded projections alternating with the same number of indentations). Exterior colour can be green, orange, or yellow, but range in shading and intensity from almost black to almost white and can appear in patterns of longitudinal striping, which can be broad and contiguous, narrow and non-contiguous, irregular, and/or in latitudinal bicolour patterns, all superimposed on barely discernable to obvious light-coloured speckling. The positioning of the longitudinal phenomena is in accordance with the subsurface placement of the (usually) 10 main carpellary vein tracts that pass through the ovary which connect the (usually) five sepals of the calyx and same number of petals of the corolla with the peduncle. The colour of the mature fruit flesh is most often light yellow-orange, but can range from greenish white to intense orange; it can be relatively thick or thin, and coarsely fibrous and tough to finely fibrous and tender. *C. pepo* has considerable polymorphy for vegetative and flowering characteristics as well, including colour of stem, length of stem internodes, amount of branching, length and angle of petioles and the amount and harshness of their spicules, depth of lobing of the laminae, earliness of flowering, sexuality, size of the flowers, and more.

The word "gourd", in the broadest sense, can be synonymous with "cucurbit". In the narrow sense, "gourd" applies to any inedible *Cucurbita*. The term "pumpkin" applies to any more-or-less round edible *Cucurbita* and the term "squash" applies to any edible non-round *Cucurbita*. The term "summer" of "summer squash" is derived from the harvest of immature fruits during the summer rather than of mature, 6-week-old or more fruits a month or two later, in the autumn of temperate climates. Besides the landrace pumpkins of Mexico and Guatemala, the fruits of a few round-fruited cultigens are harvested immature, and hence these can be referred to as "summer pumpkins".

Antoine Nicolas Duchesne (1747—1827) was the first to apply modern scientific inquiry to understand relationships among squash, pumpkins, and *Cucurbita* gourds. By studying crossability among them, he identified three species, *C. maxima*, *C. moschata*, and *C. polymorpha*. This last species encompassed within it four species of *Cucurbita* that had been designated by the systematist Carl von Linné, as Duchesne showed that these four freely intercrossed with one another. Although the name *C. polymorpha*, designated for this species by Duchesne, is highly appropriate, the rules of botanical nomenclature dictate use of the Linnean name, *C. pepo*. Duchesne also proposed, based mostly on fruit characteristics, that the 93 cultigens of *C. pepo* in his possession be divided into five "races". Two of these races, the orange gourds and the oviform/pyriform gourds, he considered to be ancestral to the races of warted gourds, pumpkins, and patissons (scallop squash).

By definition, all members of the same species should be interfertile if crossed with one another. Thus, infraspecific classification does have its limitations as, theoretically, characteristics can be freely exchanged among the various genotypes of a species and, within the course of a generation, result in intermediates. Nonetheless, for any domesticated plant species, desirable phenotypic combinations tend to be selected by people. These combinations can be rare and their desirability a variable of an individual preference or a regional culture but, if heritable, they can become preserved and reproduced. Eventually, slight variations might be selected, all of which still have in common the genes conferring the desired combination. Specifically, the original cultivar possessing the unique genetic combination might spawn the development of a number of similar cultivars, and thus the concept of "cultivar-group", formerly "convariety", used by horticultural plant taxonomists (Brickell et al., 2004).

Infraspecific groupings of cultivars of a crop species are scientifically meaningful only if they reflect genetic relationships. However, the degree of usefulness of these groupings depends on the ease with which they can be recognized by all concerned with the crop, including growers, wholesalers, retailers, and consumers, as well as breeders. While isozyme and DNA-sequence polymorphisms have proven to be useful tools for observing genetic relationships and have revealed a fundamental division within *Cucurbita pepo* (Decker, 1985), they cannot be readily observed or recognized outside of the laboratory. On the other hand, capricious groupings of cultivars not based on genetic relationships may be fine for marketing purposes, but market demands are ephemeral, tending to change with time, and are useless for scientific evaluation. Furthermore, groupings should be worldwide in scope and rather than founded entirely on the market types present in one country or region.

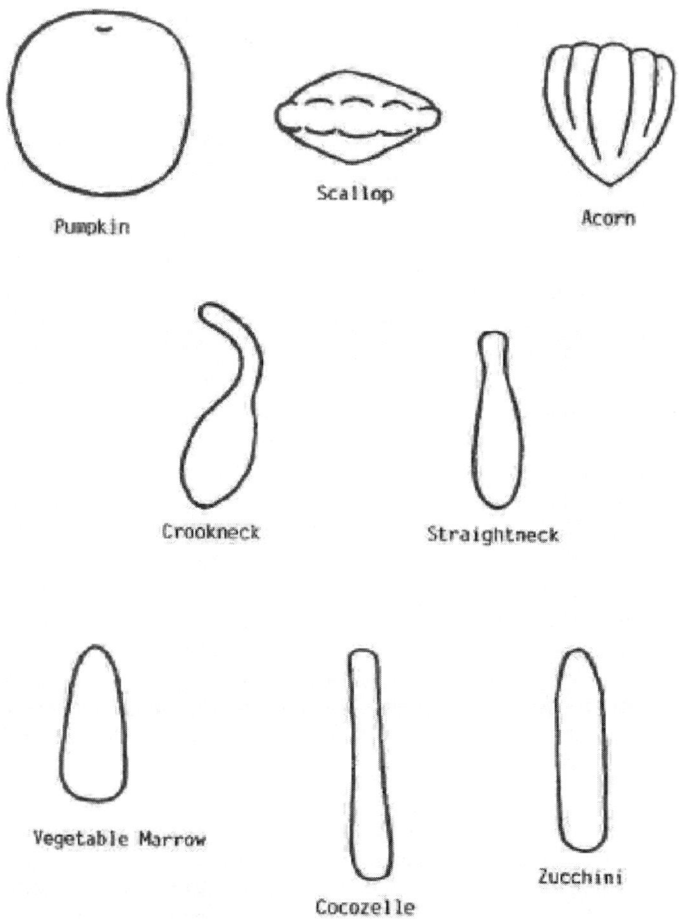

Fig. 1. Schematic representation of fruit shape of the eight edible cultivar-groups of *Cucurbita pepo* (after Paris, 1986).

Fortunately, for *Cucurbita pepo*, there is a proposed grouping of cultivars that seems to meet all of the above criteria. Fruit shape is easily recognized by all concerned with summer squash, from the scientist or breeder through the growers, buyers, produce personnel and consumers. Fruit shape is also a highly polygenic characteristic (Emerson, 1910; Sinnott, 1935; Sinnott and Kaiser, 1934) and therefore can be expected to be reflective, at to least to some extent, of genetic relationships (Paris, 1986), a supposition that later received substantial support from evidence obtained in investigations using several DNA marker systems (Paris et al., 2003; Ferriol et al., 2003). Accordingly, six cultivar-groups of summer squash were designated (Paris, 1986), these being the Scallop Group, the Crookneck Group, the Straightneck Group, the Vegetable Marrow Group, the Cocozelle Group, and the Zucchini Group. In addition, there are two cultivar-groups of *C. pepo* which are usually grown for consumption of their mature fruits, the Acorn Group and the Pumpkin Group (Figure 1 and Figure 2). Finally, there are three groups that are not grown for culinary purposes but instead for ornament: the Round, Smooth-Rinded Gourd Group, the Oviform, Smooth-Rinded Gourd Group, and the Warted Gourd Group (Table 1). Altogether, *Cucurbita pepo* is currently considered to contain 11 cultivar-groups (Paris, 2000). All cultivar-groups except for the Straightneck and Zucchini have been in existence for centuries (Paris, 2000; Paris and Janick, 2005).

Fig. 2. Young (summer squash, 2–5 days past anthesis) fruits of eight edible-fruited cultivars, one from each of the edible-fruited cultivar-groups of *Cucurbita pepo*. Left to right: subspecies *pepo*, Zucchini Group 'True French', Cocozelle Group 'Striato d'Italia', Vegetable Marrow Group 'Verte Petite d'Alger', Pumpkin Group 'Tondo Verde Scuro di Piacenza'; subspecies *texana*, Scallop Group 'Yellow Bush Scallop', Acorn Group 'Table Queen', Straightneck Group 'Early Prolific Straightneck', Crookneck Group 'Yellow Summer Crookneck'.

Table 1. Infraspecific classification of cultivated *Cucurbita pepo* (after Paris, 1986 and Paris, 2000).

Cultivar-group name, synonym(s)	Fruit shape	Common fruit usage	Subspecies
Acorn, Table Queen	Turbinate, broad at peduncle end, convex at stylar end, furrowed	Esculent when mature	*texana*
Cocozelle, Italian Marrow	Long to extremely long, length-to-broadest width ratio at least 3.5, cylindrical but bulbous at stylar end	Esculent when immature	*pepo*
Crookneck	Long, peduncular half with narrow, slightly to very curved neck, broad stylar half, convex at stylar end	Esculent when immature	*texana*
Pumpkin	Round: spherical, globular, oblate, ovate, obovate	Esculent, mostly when mature	*pepo*
Scallop, Patty Pan, Patisson, Button, Custard Marrow	Flattened, with scalloped margins	Esculent when immature	*texana*
Straightneck	Cylindrical with short neck or constriction near the peduncle, broad stylar half	Esculent when immature	*texana*
Vegetable Marrow, Middle Eastern	Short, length-to-broadest width ratio 1.5—3.0, tapered cylindrical, narrow at peduncle end, broad at stylar end	Esculent, mostly when immature	*pepo*
Zucchini, Courgette	Uniformly cylindrical, length-to-width ratio 3.5—4.5	Esculent when immature	*pepo*
Round, Smooth-rinded Gourd	Small, spherical to oblate with flattened or depressed ends, smooth	Ornament	*pepo*
Oviform, Smooth-rinded	Small, oval to pear-shaped, smooth	Ornament	*texana*
Warted	Small, spherical, oblate, and other	Ornament	*pepo*

As could be expected, a number of phenotypic characteristics accompany the fruit-shape groupings. Among the most striking of these are the sizes and shapes of the seeds. The cultivar-groups of gourds, which have the smallest fruits, also have the smallest seeds. The crooknecks, straightnecks, scallops, and acorns have smaller fruits and seeds than do the cocozelles, vegetable marrows, zucchinis, and pumpkins. The pumpkins, with their round fruits, have the largest, flattest seeds whilst the zucchinis, with their uniformly cylindrical fruits, have the shortest, thickest seeds (Paris and Nerson, 2003).

Studies of isozyme polymorphisms revealed a fundamental division within *Cucurbita pepo* that is above the level of cultivar-groups (Decker, 1985; Andres, 1987). This division is expressed botanically with three subspecific designations, subsp. *pepo*, subsp. *texana* (Scheele) Filov (syn. subsp. *ovifera* (L.) Decker) and subsp. *fraterna* (Bailey) Lira, Andres & Nee. Subsp. *pepo* includes the Cocozelle, Vegetable Marrow, Zucchini, Pumpkin, Warted Gourd, and Round, Smooth-Rinded Gourd Groups; no wild plants of this subspecies have as yet been described. Subsp. *texana* includes the Crookneck, Straightneck, Scallop, Acorn, and Oviform, Smooth-Rinded Gourd Groups and wild plants found growing in the U.S.A. Wild plants of subsp. *fraterna* have been seen in northeastern Mexico (Andres, 1987) but no cultigens have been found as yet that are closely related to them (Paris et al., 2003). The collected samples of subsp. *fraterna* are more closely related to subsp. *texana* than to subsp. *pepo* (Sanjur et al., 2002). DNA-sequence polymorphisms suggest that 'Miniature Ball', a gourd cultivated for ornament, is more closely related to each of the subspecies than any of the subspecies are to one another, suggesting that this cultivar represents or is closely related to ancestral *C. pepo* (Paris et al., 2003).

4 Genetic Resources

Wild *C. pepo* has been found repeatedly, since the mid-19th century, in much of the southeast and central U.S.A., concentrated in Texas but occurring as far north as Kentucky and Illinois (Erwin, 1931; Nee, 1990; Decker-Walters et al., 1993). Only relatively recently have sightings of wild *C. pepo* been reported in Mexico (Bailey, 1943; Andres, 1987) and then only in the northeastern Mexican states of Tamaulipas and Nuevo Leon. It is thought that *C. pepo* subsp. *pepo* originated further south in Mexico (Decker, 1988), but wild plants have not as yet been reported from there.

Wild plants of *C. pepo* are viney, branched, and bear fruits that are small (4–7 cm in diameter), hard, thin-fleshed, fibrous, and bitter. 'Miniature Ball' shares the characteristics of the wild gourds but has even smaller fruits, approximately 3–4 cm in diameter. Conceivably, it was collected in the wild some time ago, having since been increased and grown in cultivation for its decorative value. Neither 'Miniature Ball' nor any of the wild or other cultivated gourds seem to contain genes for resistance to the major diseases afflicting summer squash and therefore their value as genetic resources for breeding appears to be quite limited. The genetic resources of potential value for breeding summer squash, then, would consist almost entirely of other edible-fruited cultigens. Primitive landrace pumpkins are widely grown in Mexico. These have larger plant parts, fruits, and seeds than the wild gourds and the fruits are non-bitter, and in some localities there is notable variability for fruit shape (Lira Saade and Montes Hernandez, 1994), but otherwise have few advances over the gourds and have not been reported as having any disease resistances. Therefore it would seem that the potential usefulness of genetic resources within *C. pepo* for breeding would need to be found in more advanced germplasm. Astonishingly, most of the germplasm variation is not found concentrated near the centre of origin or for that matter anywhere in North America, to the point that even Vavilov (Chester, 1951) tentatively assigned *C. pepo* to the near-eastern centre of origin of cultivated plants. By comparing the photographs presented by Zhiteneva (1930) of *C. pepo* fruits collected in Mexico with those collected in Asia Minor, it is understandable, given the considerably greater variation found in the latter, how *C. pepo* could have been thought to have originated in that region.

Numerous landraces or primitive cultivars of summer squash of *Cucurbita pepo* subsp. *pepo* have been found in Italy and Turkey. Many of these have undesirable traits of more ancestral *C. pepo*, including much branching, highly spiculate foliage, late maturity, and strongly male sexuality. However, most of them have foliage that is bushy rather than viney, a trait that is quite important for facilitating multiple harvesting. Moreover, they harbour a great deal of genetic variation. The variability among the Italian landraces is striking, as some of them are much less branching, less spiculate, early maturing, or more strongly female. Some of the landrace cultivars bear round fruits (summer pumpkins), others bear dumpy fruits (vegetable marrows), and yet others bear quite long, bulbous fruits (cocozelles). Fruit colour ranges from intense to light green or striped. All three groups contain a host of characteristics that could be exploited, including extreme earliness and nearly complete female sex expression. Old cultivars of *C. pepo* subsp *pepo*, particularly cocozelles, serve as germplasm resources for the enrichment of the cylindrical-fruited group of cultivars, the zucchini. The Zucchini Group is the most modern, most economically important, but has the least genetic variation of the groups of *C. pepo* subsp. *pepo* (Paris et al., 2003). This group has and can be expected to continue to benefit from introgression of desirable traits available in the old cultivars of the other groups. On the other hand, cultivars of *C. pepo* subsp. *texana* can hardly be found in Italy or Turkey. These were grown centuries ago by Indian tribes in what is now the U.S.A. A few of them, little changed from long ago, are still offered by some American seedsmen. Some unique patisson (scallop) cultivars can be found in more northerly parts of Europe.

The old cultivars of Italy are still preserved by individuals (Perrino et al., 1988), farmers' cooperatives, and can be found in local seed stores. These usually have been inbred somewhat with the goal of greater uniformity. They are becoming scarce, though, as modern, productive hybrids are replacing them in many areas. Some seed companies in other parts of Europe offer a few of the Italian cultivars as well as some patissons. There are also private organizations of enthusiasts dedicated to preserving genetic heritage. These organizations, such as the Seed Savers Exchange in the U.S.A., have members that maintain seed collections of old, "heirloom" cultivars.

There are also government organizations, mostly referred to as plant introduction stations or germplasm banks that are dedicated to preserving genetic variation. A large number of accessions of *Cucurbita pepo* are housed in facilities of such organizations in Europe (Diez et al., 2002), Mexico, the U.S.A., and Costa Rica (Lira Saade and Montes Hernandez, 1994).

In considering the genetic resources of *Cucurbita pepo*, it is necessary to add that this species is sparingly crossable with several other species of *Cucurbita*, among them *C. moschata* (Castetter, 1930; Whitaker and Bemis, 1964; Whitaker and Bemis, 1975; Merrick, 1995). A tropical species, *C. moschata* harbours a variety of disease resistances, several of which have been transferred to *C. pepo*. This species has also served as a bridge in transferring powdery mildew resistance from the wild taxon *C. okeechobeensis* Bailey subsp. *martinezii* (Bailey) Andres & Nabhan ex Walters & Decker-Walters to *C. pepo* (Munger, 1990; Jahn et al., 2002). Therefore, in a sense, specific oligogenically inherited characteristics from other *Cucurbita* could potentially serve as supplementary genetic resources to *C. pepo*.

5 Major Breeding Achievements

5.1 Non-Bitterness and Larger Fruit Size

Major achievements in breeding squash were accomplished by Native Americans thousands of years ago. These peoples domesticated wild *Cucurbita pepo* gourds, and selected for non-bitter fruits and larger fruits and seeds. As indicated by existing primitive Mexican and Guatemalan pumpkin landraces, they fixed these characteristics in their domesticates.

5.2 Fruit Shape and Colour Variation

Native Americans also found and selected fruits of *Cucurbita pepo* deviating from the round shape and the broad-striped colour pattern of wild-growing plants. However, there do not seem to be records of any particular Mexican landrace that is uniform for a fruit shape other than the one that can be characterized as the wild-type, round (oblate, globular, spherical, ovate, obovate). There is no evidence found so far that indicates the existence of more-or-less uniform cultivars of *C. pepo* subsp. *pepo* bearing non-round, elongate fruits developed in North America prior to the European contact. Intense selection for long-fruitedness in this subspecies apparently was first conducted in Italy in the 16th century. On the other hand, in *C. pepo* subsp. *texana*, selection for deviation from fruit roundness was fixed in several directions prior to the European contact. A number of cultivars having flat, scalloped fruit, that is the Scallop Group, differing from one another in fruit colour, were grown along the Atlantic seaboard of what is now the United States when Europeans first set foot in North America. The same can be concluded for turbinate, furrowed cultivars, that is, the Acorn Group. Cultivars having elongate fruits, the Crookneck Group, also are a breeding achievement of American Indians.

5.3 Bush Growth Habit

Summer squash, the very young fruits of *Cucurbita pepo* grown for culinary use, are harvested daily or at 2- or 3-day intervals over a period spanning several weeks or months. This requires many harvests over the season. Bush growth habit, a mutant characteristic conferred by a single gene, *Bu*, greatly facilitates multiple harvesting. This characteristic, which is due to shortened, thickened internodes, seems to be rare or absent from Mexican landraces. It can be recognized in some illustrations in botanical tomes published centuries ago, but was not common at that time. Squash of the Scallop Group (*Cucurbita pepo* subsp. *texana*) having bush growth habit were first selected centuries ago by Native Americans of what is now the eastern seaboard of the U.S.A. Selection and fixation of the gene for bush growth habit seems to have been intensively carried out over 200 years ago by breeders in Italy, in the Cocozelle Group (*C. pepo* subsp. *pepo*). On the other hand, in the case of the Vegetable Marrow Group, bush growth habit was not common to nearly all contemporary cultivars until well into the first half of the 20th century (Paris, 2000).

Some cultivars having bush growth habit have relatively erect-growing plants whilst most tend to be prostrate or become prostrate as they age. Erect growth further facilitates field management and harvesting.

5.4 Less Branching

As summer squash are harvested many times throughout the growing season, contained, orderly vegetative growth is important. The short, thick internodes of bush plants play a major role in containing growth. Other vegetative characteristics, such as lack of branching, are also important. It is difficult to find fruits that are borne on plants that are highly branched, as they are hidden in the foliage. Here too, Italian breeders led the way in developing less-branched, "open growth habit" summer squash, reaching its pinnacle just prior to the turn of the 20th century with the development of a new group of cultivars which was to become known as the Zucchini.

5.5 Femaleness

The landrace pumpkins of Mexico as well as most of the old summer squash cultivars bear very many more staminate flowers than pistillate flowers. In Italy, there are a number of old cultivars belonging to the Pumpkin, Vegetable Marrow, and Cocozelle Groups that are highly female in sex-expression. Two such cultivars have commercial names, 'Bolognese' and 'Striato Pugliese'. These bear only a few male flowers, and then only at the first few nodes of the main stem of the plant. Selection for femaleness is yet another achievement of pre-20th century Italian squash breeding.

5.6 Earliness

This characteristic is a function of plant vigour and female tendency. Two of the vegetative characteristics discussed above affect earliness. Bush plants begin to bear earlier than vine plants and non-branched plants bear earlier than branched plants. The 'Quarantina Vera Nana' (Tamaro, 1901), a zucchini developed in Italy at the end of the 19th century, was definitely one of the earliest of its time, its name indicating that it began to bear in 40 days.

5.7 The Zucchini

This new group of cultivars, the Zucchini, sporting uniformly cylindrical fruits of attractive, uniformly distributed intense green colour was developed just prior to the "discovery" of Mendelian genetics in 1900, which ushered in the era of modern plant breeding. Today the zucchini dominates the summer squash market and breeding efforts of seed companies. Another new group of cultivars, the Straightneck, is a development of importance for markets in the eastern U.S.A.

5.8 The Greatest Achievement of Modern Squash Breeding: F_1 Hybrids

The results of research on plant hybridization in the early years of the 20th century contributed greatly to the commercial vegetable seed industry. The considerable heterosis expressed in *Cucurbita pepo* (Passmore, 1930) led Curtis (1939) to propose widespread commercialization of hybrid summer squash, and indeed countless F_1 hybrid summer squash have been commercialized since then. Many of these hybrids far exceed their open-pollinated counterparts of the same cultivar-group and market type in yield, earliness, vigour, and uniformity. F_1 hybrids act as patents, being irreproducible by all except to whom the parents of the hybrid are available. They protect the years of time and effort invested in cultivar development, and therefore have provided a strong economic incentive to squash breeding.

5.9 Disease Resistance

The most recent major achievement of summer squash breeding is the development of F_1 hybrids with resistance to some diseases. As described below, success was recently achieved by the use of both, conventional breeding and biotechnology. The new hybrids have, in general, exhibited satisfactory resistance in the field (Fuchs et al., 1998). However, the hybrids that have been commercialized thus far tend not to equal their susceptible counterparts in all of the horticulturally desirable characteristics.

6 Current Goals of Breeding

Breeding summer squash appears to be more intensive now than ever before and is conducted in an increasing number of countries. There are strong regional preferences for cultivar-group and market type. For example, squash of the Crookneck Group are extremely popular in the southeast U.S.A. and of the Straightneck Group in the northeast. Outside of that country, both groups are hardly known. Squash of the Vegetable Marrow Group are popular in the Middle East and North Africa, with overall light green colour preferred in the former region and light green with non-contiguous dark green stripes preferred in the latter. Squash of the Cocozelle Group are popular in Italy and other countries bordering the Mediterranean Sea to the north; while in most areas striped fruits are more popular, in others entirely light-coloured fruits are preferred. Cultigens of the Pumpkin Group are grown as summer squash in Italy and France; like the vegetables marrows, the placenta is removed and the seed cavity stuffed with meat, rice, or cheese. Of all of the summer squash, the Zucchini Group is by far the most widely grown, commercially produced on a large scale on the six arable continents. Thus, by far the greatest international investment by seed companies has focused on breeding this group of cultivars.

As is true for many other crops, two extremely important characteristics for squash growers are earliness and productivity. Earlier and increased production means earlier and increased income. Many modern hybrids are extremely early and productive. There are a number of horticultural characteristics specific to summer squash that are important as well.

6.1 Horticultural Characteristics

Flavour, nutritive value, and texture are important characteristics determining sensory quality of summer squash, and have been ranked as best when the fruits are young, at 3 days past anthesis, decreasing rapidly thereafter (Culpepper, 1937). Hence, young fruits are preferred over larger, more mature fruits.

Fruit flavour differs among summer squash. In particular, yellow-coloured squash differ distinctly from green-coloured squash. But flavour can also differ greatly among squash of the same colour. Many summer squash are rather bland but others have a rich, distinct flavour. Obviously, for the consumer, flavour is extremely important, but this characteristic does not seem to have received much attention from seed companies.

As summer squash are immature fruits, their vitamin and mineral contents are not high. However, for most constituents, they rank markedly higher than cucumbers (*Cucumis sativus* L.) (MacGillivray et al., 1942; Gebhardt and Thomas, 2002). Moreover, summer squash groups differ from one another with regard to some constituents. Zucchini has considerably higher potassium than crookneck, straightneck, and scallop squash (U.S.D.A., 2006). Zucchini, crookneck and straightneck have more provitamin A than scallop squash, which can be attributed to their more intense external coloration. The intense green zucchinis have six-fold more chlorophyll than the light green vegetable marrows (Globerson, 1969).

Developing placenta and seeds adversely affect fruit texture. They occupy a greater proportion of the volume of round and short fruits and hence in the summer pumpkins and the vegetable marrows the placenta and seeds are removed for culinary preparation. For almost all other methods of cooking, there is a strong preference for long-fruited squash, viz. cocozelle, zucchini, crookneck, and straightneck.

A most important component of summer squash quality is fruit shape. The fruit shape (cultivar-group) has strong regional preferences. New cultivars need to conform as closely as possible to the ideal of the cultivar-group being bred. For example, the fruit of zucchini cultivars should be as uniformly cylindrical as possible.

Colour is another important aspect of fruit quality and chances of success of a new cultivar increase if the fruit colour conforms to the current demands of the particular target market. There are strong regional preferences of consumers in regards to fruit colour, although vibrant, intense colour is more often preferred. Intense colour is conferred by the complementary dominant *L-1* and *L-2* alleles (Paris and Nerson, 1986). Transfer from ornamental gourds to squash of the *B* gene, which confers yellow fruit hue, led to the development of cultivars having an especially vivid, yellow colour (Shifriss, 1988). Over a dozen gene loci affecting fruit colour have been identified, several of which are multiple-allelic (Paris and Brown, 2005).

Another important aspect of quality is fruit gloss. Young fruits of *C. pepo* are glossy, but lose their gloss as they grow. Glossiness is a sign of freshness and palatability and some cultivars retain their gloss considerably longer than others. Summer squash also have a short shelf-life, only a few days, but some cultivars have a considerably longer shelf life than others (Mencarelli et al., 1982; Sherman et al., 1985; Sherman et al., 1987). These two characteristics need to be checked for any prospective new cultivar. Cultivars also differ for size of the blossom scar, which should not be excessively large. Longer, narrower peduncles facilitate detachment of the fruit from the plant.

The cultivar-groups of subsp. *pepo* tend to have larger plant parts than those of subsp. *texana* and their fruits grow more quickly (Lorenz, 1949). Intersubspecific and intergroup crosses have been made to exploit heterosis, resulting in extremely vigorous hybrids, which would be advantageous to growers (Anido et al., 2004). However, the fruit of these intergroup F_1s also grow extremely fast, resulting in too many oversize fruit, which is a distinct disadvantage to the growers.

Parthenocarpy, the setting of fruits without pollination, has been reported in *Cucurbita pepo* (Globerson, 1971; Rylski, 1974; Nijs and Veldhuyzen van Zanten, 1982; Robinson and Reiners, 1999; Menezes et al., 2005). It is expressed in protected crops in winter and spring and parthenocarpic tendency varies among cultivars. Parthenocarpy of squash is not expressed nearly as strongly as in parthenocarpic greenhouse cucumbers. As summer squash is an increasingly high-value crop, the tendency to grow it under protected conditions is likely to increase. Maintaining the viability of pollinators, such as honeybees, under closed conditions, is problematic and therefore the ability to set fruits parthenocarpically can be expected to become an increasingly important trait in squash breeding.

A number of vegetative characteristics are important for growers, for example, bush growth habit as discussed earlier. Besides having short internodes, the plants should be erect rather than prostrate, for even easier management. The plants should be unbranched, to facilitate harvest and the petioles should be held horizontally rather than vertically. Lack of branching and horizontal petiole angle results in "open growth habit", which makes for fast, efficient picking of the fruits (Baggett, 1972). The petioles and abaxial leaf surfaces of *Cucurbita pepo* typically are studded with stiff, sharp spicules. The spicules can scratch the fruit at harvest, which detracts from their appearance. Spicule size and density varies among cultivars. Obviously, a desirable trait for any cultivar would be for the spicules to be few and small or absent entirely. The plants should be relatively small, as excessive foliage is unnecessary for high productivity and hides the fruits, thus slowing harvest. On such plants, moreover, the fruits often grow excessively fast, resulting in oversizing. This trait is undesirable because the economic value of the fruits is inversely proportional to their size, the smaller fruits commanding a higher price in nearly all markets.

6.2 Disease Resistance

Disease resistance has been considered to be the most important goal of breeding squash (Whitaker and Robinson, 1986). The diseases and disorders afflicting the Cucurbitaceae are very numerous. Many of them have been compiled, described, and presented pictorially by the U.S.D.A. (1969), Bernhardt et al. (undated), Zitter et al. (undated), and Blancard et al. (undated). Summer squash is subject to a number of these maladies, the more widespread and important ones include viruses, fungi, and a physiological disorder.

Overall, the most damaging diseases to summer squash are viruses. Among the seed companies, by far the number one goal of breeding has been for resistance to viral diseases. The focus of breeding for such resistances is entirely due to commercial reasons: virus diseases can completely destroy a crop and therefore virus resistance is the biggest sales initiator. Both, conventional and genetic engineering approaches have been successful recently and a number of new virus-resistant zucchini and crookneck hybrids are available. However, these may not be the best of their respective cultivar-groups for horticultural characteristics.

Of the viruses, the most widespread include three members of the *Potyviridae*: zucchini yellow mosaic virus (ZYMV), watermelon mosaic virus (WMV), and papaya ringspot virus W (PRSV-W). Another virus of great importance is cucumber mosaic virus (CMV) (*Cucumoviridae*). All of these are transmitted by aphids in a non-persistent manner. Regionally important viruses that can cause widespread damage to summer squash are squash mosaic virus (SqMV) (*Comoviridae*), which is transmitted by beetles, and the newly emerged and highly destructive squash leaf curl begomovirus (SLCV) (*Geminiviridae*), which is transmitted by whiteflies (Cohen et al., 1983). Chemical control of the transmitters of viruses is ecologically damaging and has proven to be not feasible because, for it to be effective, a high frequency of applications is required. Reflective, silver-colored plastic mulch is effective in delaying the spread of the aphid-borne viruses (Smith, 1970; Conway et al., 1989) and spun-bonded polyester row covers exclude whiteflies (Webb and Linda, 1992). However, economic viability of the use of both mulch and row covers is better achieved when they are combined with intensive cultivation practices and both must be disposed of when the season is over.

Breeding for virus resistance is a goal that has been shared by many squash breeders for a long time. Virus-resistant cucumber cultivars have been available for over 50 years. The lack of virus-resistant summer squash cultivars cannot be attributed to a lack of effort. Programs for breeding for resistance to CMV, WMV, PRSV, and SqMV have been underway for decades but no commercially viable summer squash resistant to these viruses have resulted from them. Instead, the lack of success can be attributed to the absence of good sources of resistance within *Cucurbita pepo*. This has necessitated interspecific hybridization. Resistance has most often been found in *C. moschata*. Many attempts have been made over several decades to transfer virus resistance to *C. pepo*, almost all of which met with failure. Moreover, resistance to one virus usually will not confer resistance to other viruses, and therefore resistance to each virus usually needs to be transferred separately. The interspecific-hybrid and initial-backcross plants often suffer from sterility, making introgression of resistance difficult and time consuming. Recovery of the desired horticultural characteristics of the recurrent parent together with the newly derived resistance, if successful, often requires much backcrossing and spans a decade or more. Often, resistance to a single virus is conferred by several recessive genes or by complementary genes. Therefore, accumulating resistances to different viruses in the same genetic background can require an enormous amount of effort. Furthermore, in any given region, the seriousness of infection of different viruses often varies from year to year and new, destructive viruses can suddenly appear (Nameth et al., 1986). Zucchini yellow mosaic virus, the most widespread and severe virus affecting summer squash today, was first described in the early 1980s and SLCV, another severe virus, was first described even more recently.

Over the past two decades, breeding squash for virus resistance has focused on ZYMV. These efforts have met with a fair amount of success. Not only were several sources of resistance in *Cucurbita moschata* identified, the genetic relationships among the resistances from different sources have been revealed (Pachner and Lelley, 2004; Paris and Brown, 2005). Resistance derived from two different sources has been transferred to *C. pepo* and introgressed into the Zucchini Group (Provvidenti, 1997; Paris and Cohen, 2000). These efforts have resulted in the development of several ZYMV-resistant zucchini cultivars.

Powdery mildew is probably the most devastating of the fungal diseases afflicting summer squash. The powdery mildew fungi afflicting cucurbits are *Podosphaera xanthii* (Castag.) U. Braun & N. Shish. (formerly known as *Sphaerotheca fuliginea*) and *Golovinomyces cucurbitacearum* (R. Y. Zheng & G. Q. Chen) Vakalounakis & Klironomou (formerly known as *Erysiphe cichoracearum*), with the former apparently much more widespread. Powdery mildew is disseminated by wind and difficult to control, particularly in dry areas. Timeliness of application of fungicides is an important component of successfully preventing an epidemic (McGrath and Staniszewska, 1996). However, repeated applications of fungicides have often resulted in selection for resistance of the pathogen (Schroeder and Provvidenti, 1969; O'Brien et al., 1988; McGrath, 1996). Severity of powdery mildew infection is also determined by various environmental factors (Aust and Hoyningen-Huene, 1986) such as temperature (Yarwood et al., 1954) and light intensity (Leibovich et al., 1996), and also by plant age (Palti, 1961). Some of the *C.*

pepo cultivar-groups are less susceptible than others to powdery mildew, with the acorns and scallops being generally less susceptible and the cocozelles and vegetable marrows being generally more susceptible (Cohen et al., 1993; Lebeda and Kristkova, 1994).

Many *Cucurbita pepo* accessions have been tested for resistance to powdery mildew, but none with a high level of resistance were found (Sowell and Corley, 1973). Sources of resistance to powdery mildew were identified, however, in two wild species, *Cucurbita lundelliana* Bailey (Rhodes, 1964) and *C. okeechobeensis* (Contin, 1978). Resistance introgressed from *C. okeechobeensis* is now available in *C. pepo* and has been used to develop new cultivars (Jahn et al., 2002; Paris and Cohen, 2002). This resistance to powdery mildew is conferred by a single incompletely dominant gene (Paris and Brown, 2005).

Powdery mildew-resistant breeding lines have been observed to suffer from reduced fruit yield as compared with their susceptible relatives (McGrath and Staniszewska, 1996). This reduction in productivity could be attributable to possible presence of a portion of the *C. okeechobeensis* genome in resistant *C. pepo*. On the other hand, disease resistance has been associated with reduced yield potential in cucumber, *Cucumis sativus* L., (Staub and Grumet, 1993) which, like summer squash, is a cucurbit that is grown for the production of its young fruits and harvested repeatedly.

A number of physiological races of *Podosphaera xanthii* have been identified in melon, *Cucumis melo* L., based on differential susceptibility of various melon cultivars (Cohen et al., 2004). No such race-specific resistance has been reported, as of yet, in *Cucurbita pepo*.

Another important fungal disease is downy mildew, *Pseudoperonospora cubensis* (Berk. et Curt.) Rost. This fungus is most often a problem in humid regions. There are several pathotypes of this fungus, not all of which attack squash (Cohen et al., 2003). Differences in the degree of susceptibility to downy mildew have been observed among cultivar-groups, with the Cocozelle, Vegetable Marrow, and Zucchini Groups being less susceptible (Lebeda and Kristkova, 2000).

Leaf silvering is an important physiological disorder of summer squash. This disorder is confused by some with silver mottling, that is, the appearance of patches of silver in the axils of the veins of the leaf laminae, a trait that is conferred by a single dominant gene, *M* (Paris and Brown, 2005). Leaf silvering, in mild cases, appears as silvering along the leaf veins. In severe cases, the entire upper surface of the leaf laminae takes on a silver, instead of green, appearance and petioles, stems and fruits are bleached. Photosynthesis is impaired (Burger et al., 1988) and the plants become unproductive. Leaf silvering is most often caused by the feeding of immature whiteflies (*Bemisia* sp.) (Costa et al., 1993; Yokomi et al., 1995) and is exacerbated by drought stress (Paris et al., 1993a). Some cultivars of the Cocozelle Group and of the Vegetable Marrow Group are resistant to silvering (Paris et al., 1993b) but it is not clear if such resistance merely allows the foliage to remain green or also permits unimpaired fruit production (Chen et al., 2004).

Other diseases as well as spider mites and various insects, including aphids, whiteflies, and beetles, attack summer squash. Damage by cucumber beetles to seedlings is positively associated with the concentration of cucurbitacins in the cotyledons and cucurbitacin content differs greatly among cultigens (Ferguson et al., 1982).

7 Breeding Methods and Techniques

As with other crops, breeding is best done under the same or similar conditions in which the crop will be grown. Summer squash are mostly grown in the field and therefore selection is often best conducted in the field. In an increasing number of countries, they are widely grown under protected conditions. Selection for adaptation to growing under greenhouse conditions is probably best conducted in a greenhouse. Wider adaptation, however, may be achieved by alternating selection between field and greenhouse. Summer squash is not, or at least not yet, as highly specialized for growing under field versus under protected conditions, as are the most important of the vegetable crops, such as tomato and cucumber. Nonetheless, growing conditions in the field or under protection vary greatly in different parts of the world. For example, there is much less natural light during the winter months for growing under protected conditions in Europe than in north Africa. Many seed companies have facilities in different parts of the world, allowing selection under widely different conditions.

Summer squash plants are relatively large, requiring considerable space for vegetative growth and a considerable volume of soil or medium for expansion of their shallow but intricate root system. For breeding, usually 50 or 60 cm is the distance between plants that is required to determine the potential of a given individual, with 180 cm between rows to allow easy passage. In segregating populations, promising individual plants can be marked with a ribbon or with a small flag. Successive plantings can be made in many areas in order to spread out the burden of selection over a longer period of time or allow selection from within larger populations. Another way of spreading out selection time is transplanting at the beginning of the season. Transplants are often prepared in plastic trays having inverse-pyramid 3.5 × 3.5 × 6.0 cm cells. Squash seedlings, like those other cucurbits, will become desiccated if bare-root transplanted, and therefore must be transplanted with the root medium intact.

In areas having long summers, progress by as many as three generations per year can be obtained, two in the open field and one in the greenhouse. Usually, in such areas, the second planting in the field is later in the summer, when pests and diseases are prevalent and can destroy the breeding plots. Considerable investment can be needed in order to allow successful growing and selection.

Summer squash plants are monoecious, bearing separate staminate and pistillate flowers at the stem nodes, in the axils of the leaves. Staminate flowers are differentiated first and occur at almost all of the initial nodes. Pistillate flowers differentiate later in ontogeny, but develop more quickly so that, depending on environmental conditions, anthesis of the two sexes tends to be synchronized. Cultivars differ among one another in sexuality, some having a higher proportion of pistillate flowers. Usually a single flower, staminate or pistillate, develops at each node, but in the cultivar-groups of *Cucurbita pepo* subsp. *texana*, more than one flower can be formed at each node.

The flowers open and function at the break of dawn and wither before noon, not to open again. They are conspicuous and large, the orange-yellow corollas ranging from 10–20 cm in diameter. The corolla usually consists of five, but sometimes six or even seven petals that are fused at the base and free at the apex. The calyx is green, short, with the same number of awl-shaped sepals as there are petals. Pistillate flowers have a short pedicel and inferior ovary, the shape of which is similar to that of the fruit. The style usually has three stigmatic lobes and a large nectary surrounds its base. The staminate flowers have a long pedicel and the stamens have united filaments and anthers shaped as a short column. The proportion of staminate and pistillate flowers is influenced by the environment, with more staminate flowers tending to be produced under hot or stressful conditions (Free, 1970).

The pollen is clumped, sticky, and heavy. It cannot disperse by wind and, as the male and female flowers are borne separately, mechanical pollination is necessary. The most common agent of pollen dispersal is the domesticated honeybee, which is attracted to the copiously produced nectar (Free, 1970). There is no well-documented case of self-incompatibility in *Cucurbita pepo*. No inbreeding depression was reported upon self-pollinating cultigens (Haber, 1928; Scott, 1934). However, for derivatives of wild plants, inbreeding depression has been well-documented (Hayes et al., 2005). Heterosis is quite common and striking. This, as well as economic considerations of seed producers, has resulted in the great emphasis on breeding hybrid summer squash.

Controlled pollinations, be they self-pollinations or cross-pollinations, are relatively simple to conduct in squash because of the monoecism and the large size of the flowers. Flower buds closest to anthesis are larger than the other buds on the plant. On the day prior to anthesis, they begin to swell and the exterior of the corolla shows some yellowing. For controlled pollinations, bees and other possible pollinators must be excluded. This requires that the flower buds that will open on the following morning need to be secured on the previous day. This can be accomplished on both sexes by placing a strip of stiff paper tightly around the apical part of the corolla and attaching a clip or clasp. On the following morning, the male flower, with its pedicel, is detached and brought to the female flower. The clip and paper are released from the male flower, which is then stripped of its corolla and calyx. The clip and paper are then released from the female flower. While holding the male flower by its pedicel, much in the manner of using a fine paint brush, the stamens are then brought into contact with the pistil, the result being that the pollen clumps stick to the pistil. It is best to turn the pedicel back and forth so as to roll the anther column over much of the surface of the pistil, as this insures the greatest amount of pollen transfer. The female flower is then clasped shut again with a clip and paper to prevent subsequent insect visitation. A tag on which is written the parentage and date of pollination is then looped around the pedicel of the pollinated female flower. Development of the flower into a fruit is more likely if the pollination is done earlier in the day as the pollen quickly loses viability as the day progresses. Any previously set fruits on the plant must be removed as they will compete with the control-pollinated flower and quite likely prevent its development. The fruit is harvested no sooner than 40 days past anthesis unless conditions dictate otherwise. Seeds extracted from fruits that are not fully mature will germinate poorly if at all.

Detached fruits can be kept in a greenhouse or at room temperature for several weeks, usually with no detriment to the seeds. Often, the seeds from these detached fruits will attain fuller maturity, become plumper, and germinate better (Vining and Loy, 1998).

In order to save time and effort, pollinations are best conducted after selection for vegetative and fruit characteristics. However, the pollinations should be conducted promptly after selections are made as the plants may become weaker with age. This can be especially true for strongly female germplasm, in which the number of staminate flowers produced by the plant is limited and restricted to the first few days of flowering.

Selection for resistance to viruses is conducted at the seedling stage, usually of plants grown in pots in the greenhouse. Typically, seedlings are inoculated by dusting carborundum abrasive on them and then rubbing them with virus inoculum. Resistant plants can then be transplanted to larger pots or the field. On the other hand, testing for resistance to fungal pathogens, such as powdery mildew, is often possible only late in the season. Plants bearing maturing fruits are more susceptible to mildew than younger plants.

Plant breeding methods most often employed with squash begin with hybridization, followed by pedigree breeding, backcross breeding, or backcross-pedigree breeding (Allard, 1960). Pedigree breeding involves self-pollination and selection for particular traits in every generation. It is the most commonly used method when both parents of the hybrid have several horticulturally valuable characteristics. Backcross breeding involves selection of a plant carrying a particular, simply inherited, desirable characteristic and crossing it into germplasm which lacks the particular characteristic but is otherwise quite desirable. This is the method usually employed in the development of disease-resistant breeding material. Backcross-pedigree breeding is similar to backcross breeding and used in similar circumstances, except that the desirable characteristic is recessive. Identification and selection of plants carrying the desirable recessive characteristic therefore require cycles of backcrossing and selfing. Once true-breeding germplasm combining all of the desired features is obtained through one of these three methods, the germplasm can be bulk increased by selfing or sib-pollinations.

The breeding process is most often conducted under wide spacing in order to facilitate easy identification and pollination of desirable plants. However, testing of advanced breeding lines and new hybrids is best conducted under commercial conditions, in which the plants are more crowded. Generally, breeding lines having smaller, less-branched plants are less adversely affected by the increased plant density than breeding lines have larger, more highly branched plants.

8 Integration of New Biotechnologies in Breeding Programs

As far as is known, all members of the genus *Cucurbita* have 20 pairs of chromosomes ($2n = 2x = 40$). Given this high chromosome number, it is not surprising that, even given the over 100 genetic loci that have been identified, few cases of genetic linkage have been reported (Paris and Brown, 2005).

Cucurbita pepo is not ranked among the five economically most important vegetable crops worldwide and therefore budgeting for research with this species has not been forthcoming. As a result, gene mapping of *Cucurbita* lags far behind that of cucumber and melon. Two partial genetic maps, using random amplified polymorphic DNA (RAPD) markers, have been constructed (Brown and Myers, 2002; Zraidi and Lelley, 2004). However, RAPD markers are population-specific and therefore cannot be easily integrated.

The readily available, easily manipulated genetic diversity within *Cucurbita pepo* and within another species with which it is sparingly crossable, *C. moschata*, combined with the large capital investment needed for development, has not favoured the widespread application of biotechnologies to the genetic improvement of summer squash. On the other hand, one of the first applications of biotechnology to plant breeding was accomplished for this crop.

Plant viruses can be important limiting factors to summer squash production, from significantly decreasing yields to totally destroying the crop. *Cucurbita pepo* is highly susceptible to viruses from several families and totally lacks sources of resistance to them. The genetic barriers to crossing, the rather low degree of success, the large amount of time required to introgress various resistances, and the ever-changing relative importance of the different viruses affecting summer squash have resulted in the application of biotechnology to the production of transgenic virus resistance in *C. pepo* (Gaba et al., 2004). Efforts to produce transgenic virus resistance in *C. pepo* have focused on ZYMV, WMV, and CMV. The potential to engineer transgenic virus resistance depends on the ability to transform a given crop (Gaba et al., 2004) and the availability of sequence data for cloning a fragment of a viral gene (Tricoli et al., 2002). Resistance is usually conferred as a single dominant gene and multiple resistances for cucurbits have been produced in a single operation by using several genes in a single construct. Transgenic virus resistance has proven quite successful, as crookneck squash have been developed which carry resistance to two or all three of these viruses (Clough and Hamm, 1995; Fuchs et al., 1998).

9 Seed Production

The reproduction of seeds of "open-pollinated" or non-hybrid cultivars of summer squash is relatively easy and inexpensive, merely requiring growing plants of the cultivar in an isolated field and allowing them to pollinate one another freely. However, 100% purity requires isolation by a considerable distance, several kilometres, from the nearest *C. pepo*, be it of a summer or winter squash, pumpkin, or cultivated gourd or wild gourd. Adequate isolation from gourds is vital, as gourds often carry bitter and poisonous alkaloid compounds known as cucurbitacins. Bitterness is conferred by a single dominant gene, meaning that bitter, poisonous fruits could develop on "rogue" plants grown from contaminated seed stock (Herrington, 1983).

The cost of hybrid seed production is greater than for open-pollinated seed production. Nonetheless, hybrids dominate the commercial summer squash market today first and foremost because heterosis is strongly expressed in very many instances (Anido et al., 2004). Hybrids are often far superior in yield to non-hybrid varieties of the same cultivar-group or of the same market type. Moreover, as hybrids do not breed true, they are a natural patent, safeguarding the investment of effort by the breeder in developing new cultivars. Initially, hybrid seeds were produced by controlled hand-pollination or by removing staminate flowers, prior to their anthesis, from the plants of the female parent of the hybrid (Curtis, 1939). Both methods are rather costly. There is no reported cytoplasmic male sterility in *Cucurbita.* Genetic male sterility conferred by a single recessive gene has been found (Eisa and Munger, 1968) but has not, at least not yet, been widely applied for hybrid seed production because use of this genetic male sterility requires the identification and removal of male fertile plants prior to the anthesis of the pistillate flowers. Even under the best of circumstances, 50% of the plants in the population carrying genetic male sterility will be male fertile and would have to be removed from the seed field. Ethylene, a plant hormone, increases femaleness and suppresses maleness in several cucurbits, including squash (Robinson et al., 1970). The ethylene-releasing compound known commercially as ethephon is widely used today for hybrid seed production. Ethephon is applied by spraying on plants when they have two true leaves expanded. At this stage of seedling growth, the sexuality of the flower buds through node 5 or 6 has usually been differentiated as male. Upon spraying of ethephon, these male flower buds abort and flower buds that differentiate during the week after application develop as female (Hume and Lovell, 1983). This allows an interval of one to two weeks in which only female flowers are produced on the female parent of the hybrid. Untreated plants of the male parent grown in the same isolated field provide the only source of pollen (Shannon and Robinson, 1979) and cross-pollination can be quite efficient if a hive of honeybees is placed in the field. Ethephon, however, has its limitations, as 100% purity is seldom achieved given the occasional staminate flower that develops, reaching anthesis during the otherwise male-free interval. Ethephon can also lower seed number per fruit and decrease the size of the seeds (Edelstein et al., 1987).

When the fruits reach maturity, at least 40 days past anthesis, they are harvested, cut, and the seeds removed. The seeds, together with the surrounding placental tissue, can be placed in a receptacle to ferment for two or three days, allowing for easier extraction and cleaning of the seeds. The seeds then are dried quickly, and packaged and stored.

References

Allard, R. W. 1960. *Principles of Plant Breeding*, John Wiley & Sons, New York, pp. 43–49.

Andres, T. C. 1987. *Cucurbita fraterna*, the closest wild relative and progenitor of *C. pepo*. Cucurbit Genet. Coop. Rep. 10: 69–71.

Anido, F. L., Cravero, V., Asprelli, P., Firpo, T., Garcia, S. M., and Cointry, E. 2004. Heterotic patterns in hybrids involving cultivar-groups of summer squash, *Cucurbita pepo* L. Euphytica 135: 355–360.

Aust, H.-J., and Hoyningen-Huene, J. v. 1986. Microclimate in relation to epidemics of powdery mildew. Ann. Rev. Phytopathol. 24: 491–510.

Baggett, J. R. 1972. Open growth habit in summer squash, HortScience 7: 288.

Bailey, L. H. 1943. Species of *Cucurbita*, Gent. Herb. 6: 266–322.

Bernhardt, E., Dodson, J., and Watterson, J., undated. *Cucurbit Diseases, a Practical Guide for Seedsmen, Growers & Agricultural Advisors*. Petoseed, Saticoy, California.

Blancard, D., Lecoq, H., and Pitrat, M., undated. *A Colour Atlas of Cucurbit Diseases*. Manson, London.

Brickell, C. D., Baum, B. R., Hetterscheid, W. L. A., Leslie, A. C., McNeill, J., Trehane, P., Vrugtman, F., and Wiersema, J. H. (eds.) 2004. *International Code of Nomenclature for Cultivated Plants*, *Acta Horticulture* 647, Leuven, Belgium, 144 pp.

Brown, R. N., and Myers, J. R. 2002. A genetic map of squash (*Cucurbita* sp.) with randomly amplified polymorphic DNA markers and morphological markers. J. Amer. Soc. Hort. Sci. 127: 568–575.

Burger, Y., Schwartz, A., and Paris, H.S. 1988. Physiological and anatomical features of the silvering disorder of *Cucurbita*. J. Hort. Sci. 63: 635–640.

Castetter, E. F. 1930. Species crosses in the genus *Cucurbita*, Amer. J. Bot. 17: 41–57.

Chabrey, D. 1666. *Stirpium Sciagraphia et Icones*, P. Gamoneti and J. de la Pierre, Geneva, Switzerland, pp. 129–135.

Chen, J., McAuslane, H. J., Carle, R. B., and Webb, S. E. 2004. Impact of *Bemisia argentifolii* (Homoptera: Auchenorryhyncha: Aleyrodidae) infestation and squash silverleaf disorder on zucchini yield and quality. J. Econ. Entomol. 97: 2083–2094.

Chester, K. S. 1951. *Selected writings of N. I. Vavilov. The Origin, Variation, Immunity and Breeding of Cultivated Plants*, Chronica Botanica, Waltham, MA U.S.A.

Clough, G. H., and Hamm, P. B. 1995. Coat protein transgenic resistance to watermelon mosaic and zucchini yellows mosaic virus in squash and cantaloupe. Plant Disease 79: 1107–1109.

Cohen, R., Leibovich, G., Shtienberg, D., and Paris, H. S. 1993. Variability in the reaction of squash (*Cucurbita pepo*) to inoculation with *Sphaerotheca fuliginea* and methodology of breeding for resistance. Plant Pathology 42: 510–516.

Cohen, R., Burger, Y., and Katzir, N. 2004. Monitoring physiological races of *Podosphaera xanthii* (syn. *Sphaerotheca fuliginea*), the causal agent of powdery mildew in cucurbits: factors affecting race identification and the importance for research and commerce. Phytoparasitica 32: 174–183.

Cohen, S., Duffus, J. E., Larsen, R. C., Liu, H. Y., and Flock, R. A. 1983. Purification, serology, and vector relationships of squash leaf curl virus, a whitefly-transmitted geminivirus. Phytopathology 73: 1669–1673.

Cohen, Y., Meron, I., Mor, N., and Zuriel, S. 2003. A new pathotype of *Pseudoperonospora cubensis* causing downy mildew in cucurbits in Israel. *Phytoparasitica* 31: 458–466.

Contin, M. E. 1978. *Interspecific Transfer of Powdery Mildew Resistance in the Genus* Cucurbita. Ph.D. thesis, Cornell University, Ithaca, New York.

Conway, K. E., McCraw, B. D., Motes, J. E., and Sherwood, J. L. 1989. Evalutions of mulches and row covers to delay virus diseases and their effects on yield of yellow squash. Appl. Agric. Res. 4: 201–207.

Costa, H. S., Ullman, D. E., Johnson, M. W., and Tabashnik, B. E. 1993. Squash silverleaf symptoms induced by immature, but not adult, *Bemisia tabaci*. Phytopathology 83: 763–766.

Culpepper, C. W. 1937. Composition of summer squash and its relationship to variety, stage of maturity, and use as a food product, Food Res. 2: 289–303.

Curtis, L. C. 1939. Heterosis in summer squash (*Cucurbita pepo*) and the possibilities of producing F_1 hybrid seed for commercial planting. Proc. Amer. Soc. Hort. Sci. 37: 827–828.

Cutler, H. C., and Whitaker, T. W. 1961. History and distribution of the cultivated cucurbits in the Americas, Amer. Antiq. 26: 469–485.

Decker, D. S. 1985. Numerical analysis of allozyme variation in *Cucurbita pepo*. Econ. Bot. 39: 300–309.

Decker, D. S. 1988. Origin(s), evolution, and systematics of *Cucurbita pepo* (Cucurbitaceae), Econ. Bot. 42: 4–15.

Decker-Walters, D. S., Walters, T. W., Cowan, C. W., and Smith, B. D. 1993. Isozymic characterization of wild populations of *Cucurbita pepo*. J. Ethnobiol. 13: 55–72.

Diez, M. J., Pico, B., and Nuez, F. (Compilers), 2002. *Cucurbit Genetic Resources in Europe*, International Plant Genetic Resources Institute, Rome, 58 pp.

Duchesne, A. N. 1786. *Essai sur l'histoire naturelle des courges*. Panckoucke, Paris, 46 pp.

Edelstein, M., Nerson, H., Paris, H. S., Karchi, Z., and Burger, Y. 1987. Early vegetative development of spaghetti squash is unaffected by seed size. Cucurbit Genet. Coop. Rep. 10: 78–79.

Eisa, H. M., and Munger, H. M. 1968. Male sterility in *Cucurbita pepo*. Proc. Amer. Soc. *Hort. Sci.* 92: 473–477.

Emerson, R. A. 1910. The inheritance of sizes and shapes in plants. Amer. Nat. 44: 739–746.

Erwin, A. T. 1931. Nativity of the cucurbits. Bot. Gaz. 91: 105–108.

FAO, 2006, Rome; http://faostat.fao.org/faostat/collections?subset=agriculture. 31 January 2006.

Ferguson, J. E., Metcalf, E. R., Metcalf, R. L., and Rhodes, A. M. 1982. Cucurbitacins of cotyledons of Cucurbitaceae cultivars as related to diabroticite beetle attack. Cucurbit Genet. Coop. Rep. 5: 42–43.

Ferriol, M., Pico, B., and Nuez, F. 2003. Genetic diversity of a germplasm collection of *Cucurbita pepo* using SRAP and AFLP markers. Theor. Appl. Genet. 107: 271–282.

Free, J. B. 1970. *Insect Pollination of Crops*, Academic Press, London, pp. 297–313.

Fuchs, L. 1542. *De Historia Stirpium*, Isingrin, Basel, pp. 698–699.

Fuchs, M., Tricoli, D. M., Carney, K. J., Schesser, M., McFerson, J. R., and Gonsalves, D. 1998. Comparative virus resistance and fruit yield of transgenic squash with single and multiple coat protein genes. Plant Disease 82: 1350–1356.

Gaba, V., Zelcer, A., and Gal-On, A. 2004. Cucurbit biotechnology – the importance of virus resistance. In Vitro Cell. Dev. Biol.—Plant 40: 346–358.

Gebhardt, S. E., and Thomas, R. G. 2002. Nutritive Value of Foods. *U.S.D.A., A.R.S.,* Home and Garden Bulletin 72.

Globerson, D. 1969. The inheritance of white fruit and stem colour in summer squash, *Cucurbita pepo* L., Euphytica 18: 249–255.

Globerson, D. 1971. Effects of pollination on set and growth of summer squash in Israel. *Expl. Agric.* 7: 183–188.

Gray, A., and Trumbull J. H. 1883. Review of DeCandolle's origin of cultivated plants, Amer. J. Sci. 25: 370–379.

Haber, E. S. 1928. Inbreeding the Table Queen (Des Moines) squash, Proc. Amer. Soc. Hort. Sci. 25: 111–114.

Hayes, C. N., Winsor, J. A., and Stephenson, A. G. 2005. Environmental variation influences the magnitude of inbreeding depression in *Cucurbita pepo* ssp. *texana* (Cucurbitaceae). J. Evol. Biol. 18: 147–155.

Herrington, M. E. 1983. Intense bitterness in commercial zucchini. Cucurbit Genet. Coop. Rep. 6: 75–76.

Hume, R. J., and Lovell, P. H. 1983. The control of sex expression in cucurbits by ethephon, Ann. Bot. 52: 689–695.

Jahn, M., Munger, H. M., and McCreight, J. D. 2002. Breeding cucurbit crops for powdery mildew resistance, pp. 239–248 in: Bélanger, R. R., Bushnell, W. R., Dik, A. J., and Carver, T. L. W., *The powdery mildews a comprehensive treatise*. APS Press, St. Paul, Minnesota.

Janick, J., and Paris, H. S. 2006. The cucurbit images (1515–1518) of the Villa Farnesina, Rome. Ann. Bot. 97: 165–176.

Lebeda, A., and Kristkova, E. 1994. Field resistance of *Cucurbita* species to powdery mildew (*Erysiphe cichoracearum*). Z. Pflanzenkrank. Pflanzenshutz 101: 598–603.

Lebeda, A., and Kristkova, E. 2000. Interactions between morphotypes of *Cucurbita pepo* and obligate biotrophs (*Pseudoperonospora cubensis, Erysiphe cichoracearum* and *Sphaerotheca fuliginea*). In: Katzir, N., and Paris, H. S., eds., *Proceedings of Cucurbitaceae 2000*, Acta Hort. 510: 219–225.

Leibovich, G., Cohen, R., and Paris, H. S. 1996. Shading of plant facilitates selection for powdery mildew resistance in squash. Euphytica 90: 289–292.

Lira Saade, R., Andres, T. C., and Nee, M. 1993. *Cucurbita* L. In: Lira Saade, R., ed., *Estudios Taxonomicos y ecogeograficos de las Cucurbitaceae Latinoamericanas de importancia economica*. International Plant Genetic Resources Institute, Rome.

Lira Saade, R., and Montes Hernandez, S. 1994. Cucurbits (*Cucurbita* spp.), In: *Neglected crops: 1492 from a different perspective*, Hernandez Bermejo, J. E., and Leon, J., eds. FAO Plant Production and Protection Series No. 26. FAO, Rome.

Lorenz, O. 1949. Growth rates and chemical composition of fruits of four varieties of summer squash. Proc. Am. Soc. Hort. Sci. 54: 385–390.

MacGillivray, J. H. 1960. Summer squash. Effect of size on picking time and yield. Veg. Crops Ser. 109, University of California, Davis.

MacGillivray, J. H., Hanna, G. C., and Minges, P. A. 1942. Vitamin, protein, calcium, iron, and calorie yield of vegetables per acre and per acre man-hour. Proc. Amer. Soc. Hort. Sci. 41: 293–297.

McGrath, M. T. 1996. Increased resistance to triadimefon and to benomyl in *Sphaerotheca fuliginea* populations following fungicide usage over one season. Plant Disease 80: 633–639.

McGrath, M. T., and Staniszewska, H. 1996. Management of powdery mildew in summer squash with host resistance, disease threshold-based fungicide programs, or an integrated program. Plant Disease 80: 1044–1052.

Mencarelli, F., Anelli, G., and Tesi, R. 1982. Idoneità alla conservazione di alcune cultivars di carciofo e di zucca da zucchini. Frutticoltura 44(8): 47–50.

Menezes, C. B. de, Maluf, W. R., Azevedo, S. M. de, Faria, M. V., Nascimento, I. R., Nogueira, D. W., Gomes, L. A. A., and Bearzoti, E. 2005. Inheritance of parthenocarpy in summer squash (*Cucurbita pepo* L.). Genet. Mol. Res. 4: 39–46.

Mérault, A. J. 1827. *L'art du jardinier*. Paris, pp. 398–400.

Merrick, L. C. 1995. Squashes, pumpkins and gourds, in: Smartt, J., and Simmonds, N. W., eds., *Evolution of Crop Plants, 2nd ed.*, Longman Scientific & Technical, London, pp. 97–105.

Munger, H. M. 1990. Availability and use of interspecific populations involving *Cucurbita moschata* and *C. pepo*. Cucurbit Genet. Coop. Rep. 13: 49.

Nameth, S. T., Dodds, J. A., Paulus, A. O., and Laemmlen, F. F. 1986. Cucurbit viruses of California: an ever-changing problem. Plant Disease 70: 8–12.

Naudin, C. 1856. Nouvelles recherches sur les caractères spécifiques et les variétés des plantes du genre *Cucurbita*. Ann. Sci. Nat. Bot., ser. 4, 6: 5–73, 3 pl.

Nee, M. 1990. The domestication of *Cucurbita* (Cucurbitaceae). Econ. Bot. 44(3, supplement): 56–68.

Nerson, H. 2005. Effects of fruit shape and plant density on seed yield and quality of squash. Sci. Hort. 105: 293–304.
Nijs, A. P. M. den, and Veldhuyzen van Zanten, N. J. D. 1982. Parthenocarpic fruit set in glasshouse grown zucchini squash. Cucurbit Genet. Coop. Rep. 5: 44–45.
Norrmann, R., and Haarberg, J. 1980. Nature and language: a semiotic study of cucurbits in literature. Routledge & Kegan Paul, London.
O'Brien, R. G., Vawdrey, L. L., and Glass, R. J. 1988. Fungicide resistance in cucurbit powdery mildew (*Sphaerotheca fuliginea*) and its effect on field control. Aust. J. Exptl. Agric. 28: 417–423.
Oemler, A. 1883. Truck-farming at the south. New York, pp. 209–213.
Pachner, M., and Lelley, T. 2004. Different genes for resistance to zucchini yellow mosaic virus (ZYMV) in *Cucurbita moschata*, in: *Progress in cucurbit genetics and breeding research, Proceedings of Cucurbitaceae 2004*. Lebeda, A., and Paris, H.S., eds., pp. 237–243.
Palti, J. 1961. Prediction of powdery mildew outbreaks on cucurbits on the basis of seasonal factors and host age, Bull. Res. Counc. Israel 10D: 236–249.
Paris, H. S. 1986. A proposed subspecific classification for *Cucurbita pepo*. Phytologia 61: 133–138.
Paris, H. S. 1989. Historical records, origins, and development of the edible cultivar groups of *Cucurbita pepo* (Cucurbitaceae). Econ Bot. 43: 423–443.
Paris, H. S. 1996. Summer squash: history, diversity, and distribution, HortTechnology 6: 6–13.
Paris, H. S. 2000. History of the cultivar-groups of *Cucurbita pepo*, in: *Horticultural Reviews* 25(2001): 71–170, 4 pl., J. Janick, ed., Wiley, New York.
Paris, H. S., and Brown, R. N. 2005. The genes of pumpkin and squash. HortScience 40: 1620–1630.
Paris, H. S., and Cohen, S. 2000. Oligogenic inheritance for resistance to zucchini yellow mosaic virus in *Cucurbita pepo*. Ann. Appl. Biol. 136: 209–214.
Paris, H. S., and Cohen, R. 2002. Powdery mildew-resistant summer squash hybrids having higher yields than their susceptible, commercial counterparts. Euphytica 124: 121–128.
Paris, H. S., and Janick, J. 2005. Early evidence for the culinary use of squash flowers in Italy. Chron. Hort. 45(2): 20–21.
Paris, H. S., and Nerson, H. 1986. Genes for intense fruit pigmentation of squash. J. Hered. 77: 403–409.
Paris, H. S., and Nerson, H. 2003. Seed dimensions in the subspecies and cultivar-groups of *Cucurbita pepo*, Genet. Resources Crop Evol. 50: 615–625.
Paris, H. S., Stoffella, P. J., and Powell, C. A. 1993a. Sweetpotato whitefly, drought stress, and leaf silvering of squash. HortScience 28: 157–158.
Paris, H. S., Stoffella, P. J., and Powell, C. A. 1993b. Susceptibility to leaf silvering in the cultivar groups of summer squash. Euphytica 69: 69–72.
Paris, H. S., Yonash, N., Portnoy, V., Mozes-Daube, N., Tzuri, G., and Katzir, N. 2003. Assessment of genetic relationships in *Cucurbita pepo* (Cucurbitaceae) using AFLP, ISSR, and SSR markers. Theor. Appl. Genet. 106: 971–978.
Paris, H. S., Daunay, M. C., Pitrat, M., and Janick, J. 2006. First known image of *Cucurbita* in Europe, 1503–1508. Ann. Bot. 98: 41–47.
Passmore, S. F. 1930. Hybrid vigour in reciprocal crosses in *Cucurbita pepo*. Ann. Bot. 48: 1029–1030.
Perrino, P., Laghetti, G., and Hammer, K. 1988. Collection of plant genetic resources in Italy, 1987. Kulturpflanze 36: 377–390.
Petersen, J. B., and Sidell, N. A. 1996. Mid-Holocene evidence of *Cucurbita* sp. from central Maine, Amer. Antiq. 61: 685–698.

Provvidenti, R. 1997. New American summer squash cultivars possessing a high level of resistance to a strain of zucchini yellow mosaic virus from China. Cucurbit Genet. Coop. Rep. 20: 57–58.

Rhodes, A. M. 1964. Inheritance of powdery mildew resistance in the genus *Cucurbita*. Plant Disease Rptr. 48: 54–55.

Robinson, R. W., and Reiners, S. 1999. Parthenocarpy in summer squash. HortScience 34: 715–717.

Robinson, R. W., Whitaker, T. W., and Bohn, G. W. 1970. Promotion of pistillate flowering in *Cucurbita* by 2-chloroethylphosphonic acid. Euphytica 19: 180–183.

Rylski, I. 1974. Effects of season on parthenocarpic and fertilized summer squash. Expl. Agric. 10: 39–44.

Sanjur, O. I., Piperno, D. R., Andres, T. C., and Wessel-Beaver, L. 2002. Phylogenetic relationships among domesticated and wild species of *Cucurbita* (Cucurbitaceae) inferred from a mitochondrial gene: implications for crop plant evolution and areas of origin. *Proc. Natl. Acad. Sci. U.S.A.* 99: 535–540.

Schaffer, A. A., Boyer, C. D., and Paris, H. S. 1986. Inheritance of rind lignification and warts in *Cucurbita pepo* L. and a role for phenylalanine ammonia lyase in their control. Z. Pflanzenzücht. 96: 147–153.

Schroeder, W. T., and Provvidenti, R. 1969. Resistance to benomyl in powdery mildew of cucurbits. Plant Disease Rptr. 53: 271–275.

Scott, G. W. 1934. Observations on some inbred lines of bush types of *Cucurbita pepo*. Proc. Amer. Soc. Hort. Sci. 32 : 480.

Seringe, N. C. 1847. *Flore des jardins et des grandes cultures* 2: 531–541. C. S. Jeune, ed., Lyon.

Shannon, S., and Robinson, R. W. 1979. The use of ethephon to regulate sex expression of summer squash for hybrid seed production. J. Amer. Soc. Hort. Sci. 104: 674–677.

Sherman, M., Elmstrom, G. W., and Allen, J. J. 1985. Storage characteristics of three cultivars of yellow summer squash (*Cucurbita pepo* L.). Proc. Fla. State Hort. Soc. 98: 216–218.

Sherman, M., Paris, H. S., and Allen, J. J. 1987. Storability of summer squash as affected by gene *B* and genetic background. HortScience 22: 920–922.

Shifriss, O. 1988. On the emergence of *B* cultivares in squash. *HortScience* 23: 237–238, 431.

Sinnott, E. W. 1935. Evidence for the existence of genes controlling shape. Genetics 20: 12–21.

Sinnott, E. W., and Kaiser, S. 1934. Two types of genetic control over the development of shape. Bull. Torrey Bot. Club 61: 1–7.

Smith, B. D. 1997. The initial domestication of *Cucurbita pepo* in the Americas 10,000 years ago, Science 276: 932–934.

Smith, N. J. 1970. Yields increased fivefold with aluminum mulch. Amer. Veg. Grower, Nov., pp. 18–21.

Sowell, G. Jr., and Corley, W. L. 1973. Resistance of *Cucurbita* plant introductions to powdery mildew. HortScience 8: 492–493.

Staub, J. E., and Grumet, R. 1993. Selection for multiple disease resistance reduces cucumber yield potential. Euphytica 67: 205–213.

Tamaro, D. 1901. *Orticoltura*, 2nd ed., Milan, pp. 467–470.

Tapley, W. T., Enzie, W. D., and Eseltine, G. P. van, 1937. *The Vegetables of New York*. Vol. 1, Part 4, Albany, New York.

Tricoli, D.M., Carney, K. J., Russell, P. F., Quemada, H. D., McMaster, R. J., Reynolds, J. F., and Deng, R. Z. 2002. Transgenic plants expressing DNA constructs containing a plurality of genes to impart virus resistance. US Patent 6,337,431.

Trumbull, J. H. 1876. Vegetables cultivated by the American Indians, Bull. Torrey Bot. Club 6: 69–71.

U.S.D.A. 1969. Growing pumpkins and squashes. Farmers' Bulletin No. 2086. U. S. Government Printing Office, 21 pp.

U.S.D.A. 2006. A. R. S. Nutrient data laboratory. http://www.ars.usda.gov/nutrientdata. Last accessed 15 March 2006.

Vining, K., and Loy, J. B. 1998. Seed fill occurs in stored fruit of *Cucurbita pepo* L. harvested prematurely. Cucurbit Genet. Coop. Rep. 21: 57–58.

Webb, S. E., and Linda, S. B. 1992. Evaluation of spunbonded polyethylene row covers as a method of excluding insects and viruses affecting fall-grown squash in Florida. Hort. Entomology 85: 2344–2352.

Whitaker, T. W. 1947. American origin of the cultivated cucurbits, Ann. Missouri Bot. Gard. 34: 101–111.

Whitaker, T. W., and Bemis, W. P. 1964. Evolution in the genus *Cucurbita*. Evolution 18: 553–559.

Whitaker, T. W., and Bemis, W. P. 1975. Origin and evolution of the cultivated *Cucurbita*. Bull. Torrey Bot. Club 102: 362–368.

Whitaker, T. W., and Robinson, R. W. 1986. Squash breeding, in: Bassett, M. D., ed., *Breeding Vegetable Crops*, Avi, Westport, Connecticut, pp. 209–242.

Yarwood, C. E., Sidky, S., Cohen, M., and Santilli, V. 1954. Temperature relations of powdery mildews. Hildgardia 22: 603–622.

Yokomi, R. K., Jimenez, D. R., Osborne, L. S., and Shapiro, J. P. 1995. Comparison of silverleaf whitefly-induced and chlormequat chloride-induced leaf silvering in *Cucurbita pepo*. Plant Disease 79: 950–955.

Zhiteneva, N. E. 1930. The world's assortment of pumpkins. Trudy Prikl. Bot. Genet. Selek. 23: 157–207.

Zitter, T. A., Hopkins, D. L., and Thomas, C. E., undated. *Compendium of Cucurbit Diseases*. APS Press, St. Paul, Minnesota.

Zraidi, A., and Lelley, T. 2004. Genetic map for pumpkin *Cucurbita pepo* using random amplified polymorphic DNA markers, in: Lebeda, A. and Paris, H. S., eds., *Proceedings of Cucurbitaceae 2004*, Palacky Univ., Olomouc, Czech Republic, pp. 507–514.

Watermelon

Todd C. Wehner[1]

[1] North Carolina State University, Department of Horticultural Science, todd_wehner@ncsu.edu

1 Introduction

Watermelon (*Citrullus lanatus*) is a member of the cucurbit family (Cucurbitaceae). The crop is grown commercially in areas with long frost-free warm periods. Plants must be grown at a wide spacing because of their long, trailing vines. The exception is for dwarf cultivars where the plants can be grown at a tighter spacing. The crop may be established in the field by planting seeds or using containerized transplants. Management of plant pests (weeds, insects, and diseases, including nematodes) is essential during the production period. Three-fourths of the world production is grown in Asia, with China the leading country in production.

Watermelons are grown in most states of the United States, but the major producers are in the South and West (Florida, Georgia, California, and Texas) where the warm production season lasts longer. The fruit are harvested by hand, with the most experienced workers doing the cutting (removal of the fruit from the vine) and the others loading the bins or trucks. The fruit are shipped to markets throughout the United States, with some exported to Canada.

Watermelon fruit will keep for two to three weeks after harvest if they are stored properly at 10 to 15°C and 90% humidity. Besides whole watermelons, it is becoming popular to sell watermelon in pre-cut halves, quarters, slices, and chunks. Whole fruit usually are cut in the store under cold, aseptic conditions since the cut product does not ship or store well. Seedless watermelons are especially popular for pre-cut sales, since that shows their seedless quality.

In the 1800s, watermelon was grown mostly for local sales. However, with the development in the last few decades of rapid shipping in refrigerated railroad cars and trucks has led to distribution of watermelon throughout the United States from major production areas. Southern production areas begin shipping early in the year, and the harvest continues throughout the summer by moving to more northern areas.

Depending on the cultivar, watermelon fruit are produced in different sizes: ice box, small, medium, large, or giant; different shapes: round, oval, blocky, or elongate; different rind patterns: gray, narrow stripe, medium stripe, wide stripe, light solid, or dark solid; different flesh colours: white, yellow, orange, or red; and different types: seeded or seedless. Commercially, the most popular seeded cultivars are red flesh, blocky shape, and large sized (8–11 kg), like the cultivar Allsweet. For seedless watermelons, the popular cultivars are red flesh, oval shape, and medium sized (5–8 kg), like the cultivar Tri-X-313. Per capita consumption of watermelons in the United States is 7.2 kg.

Watermelon is served fresh as slices, as chunks (often in fruit salad), as juice, pickled rind, glacé candy, and as edible seeds (harvested from confectionary type cultivars). It is no longer just a summer fruit and is becoming an everyday fruit like apples, bananas, and oranges. The watermelon fruit is 93% water, with small amounts of protein, fat, minerals, and vitamins. In some arid regions, watermelon is used as a valuable source of water. The major nutritional components of the fruit are carbohydrates (6.4 g/100 g), vitamin A (590 IU), and lycopene (4,100 μg/100g, range 2,300–7,200), an anticarcinogenic compound found in red flesh watermelon. Lycopene may help reduce the risk of certain cancers, such as prostate, pancreas, and stomach. The lycopene content of the new dark red watermelon cultivars is higher than in tomato, pink grapefruit, or guava. Orange flesh types have only small amounts of lycopene, and the beta carotene content is similar to that of red flesh types. Canary yellow types do not contain lycopene, but do have a small amount of beta carotene. Watermelon seeds are rich in fat and protein.

Watermelon flowering and fruit development are promoted by high light intensity and high temperature. Watermelon is the only economically important cucurbit with pinnatifid (lobed) leaves; all of the other species have whole (non-lobed) leaves. The leaves are pinnately divided into three or four pairs of lobes, except for a non-lobed (sinuate) gene mutant controlled by the *nl* gene. Watermelon growth habit is a trailing vine. The stems are thin, hairy, angular, grooved, and have branched tendrils at each node. The stems are highly branched and up to 30 feet long, although there are dwarf types (*dw-1* and *dw-2* genes) with shorter, less-branched stems. Roots are extensive but shallow, with a taproot and many lateral roots.

Watermelon has small flowers that are less showy than those of other cucurbits. Flowering begins 4 to 8 weeks after seeding. Flowers of watermelon are staminate (male), perfect (hermaphroditic), or pistillate (female), usually borne in that order on the plant as it grows. Monoecious types are most common, but there are andromonoecious (staminate and perfect) types, mainly the older cultivars or accessions collected from the wild. The pistillate flowers have an inferior ovary, and the size and shape of the ovary is correlated with final fruit size and shape. In many cultivars, the pistillate or perfect flowers are borne at every seventh node, with staminate flowers at the intervening nodes. The flower ratio of typical watermelon cultivars is 7:1 staminate:pistillate, but the ratio ranges from 4:1 to 15:1.

The fruit of watermelon are round to cylindrical, up to 600 mm long and have a rind 10 to 40 mm thick. The edible part of the fruit is the endocarp (placenta). That contrasts with melon (*Cucumis melo*), where the edible part of the fruit is the mesocarp. Fruit as large as 120 kg have been recorded, but usually they weigh 4 to

16 kg. In Asia, even smaller watermelon fruit in the range of 1 to 4 kg are popular. That size is now becoming popular in the U.S. Fruit rind varies from thin to thick, and brittle to tough.

Seeds continue to mature as the fruit ripens and the rind lightens in colour. Seeds will be easier to extract from the fruit if the fruit is held in storage (in the shade or in a seed processing room) for a few days after removing them from the vine. If seeds are left too long in the fruit they will germinate *in situ*. There is no dormancy in watermelon seeds, so they can be harvested on one day, cleaned, dried, and planted on the next day. Seeds germinate in 2 days to 2 weeks depending on temperature and moisture conditions. Seeds will not germinate below 60°F. The optimum germination temperature is 85 to 90°F, especially for triploid seeds. For germination of triploid hybrid seeds, temperature and moisture are more critical, and it is especially important to avoid excess moisture.

2 Origin and Domestication

Watermelon has been cultivated in Africa and the Middle East for thousands of years, and in China since at least 900 AD. Watermelon was brought to the New World in the 1500s. In the United States, watermelon is a major vegetable crop that is grown primarily in the southern states. The major watermelon producing states are Florida, California, Texas, Georgia, and Arizona.

Through history, watermelon was distributed throughout the world as trade and knowledge of central Africa developed. The crop was grown in India by at least 800 AD, and in China by 1100 AD. The Moorish conquerors of Spain introduced watermelon into Europe, where it was noted in Cordoba in 961 AD and Seville in 1158 AD. The spread of watermelon into northern Europe was relatively slow, and it was not noted in the British Isles until late in the 16th century, perhaps because of the generally unfavourable climate for watermelon culture in much of Europe. About this time, watermelons were introduced into the New World, with culture of the plants noted in the Massachusetts colony in 1629. The introduction of watermelon into other parts of the world has followed established trade routes.

Watermelon has been improved by domestication and formal plant breeding from a late maturing vine with small fruit having hard, white flesh and bland or bitter taste, into an early maturing, more compact plant with large fruit having edible, sweet flesh. In the last century, plant breeders working in public or private programs in the United States and around the world have released cultivars having disease resistance, dwarf vines, larger fruit, higher sugar content, higher lycopene content, seedlessness, and new flesh colours, such as scarlet red, dark orange, and canary yellow. Recent advances in the breeding of seedless triploid hybrids have resulted in renewed popularity of watermelons, and per capita consumption has increased 37% since 1980.

2.1 Centres of Origin

Watermelon is thought to have originated in southern Africa because it is found growing wild throughout the area, and reaches maximum diversity there. It has been cultivated in Africa for over 4,000 years. The citron (*Citrullus lanatus* var. *citroides*) grows wild there, and is thought to be related to the wild ancestor of watermelon. In 1857, David Livingstone reported watermelon growing profusely after unusually heavy rainfall in the Kalahari Desert (the current nations of Namibia and Botswana). The natives there knew of sweet as well as bitter forms growing throughout southern Africa. De Candolle, in 1882, considered the evidence sufficient to prove that watermelon was indigenous to tropical Africa, more specifically the southern parts of Africa.

Citrullus colocynthis is considered to be a wild ancestor of watermelon, and is now found native in north and west Africa. Fruit of colocynth are small, with a maximum diameter of 75 mm. The flesh is bitter and the seeds are small and brown. Crosses of *C. lanatus* with *C. colocynthis* produced F_1 hybrids with nearly regular meiosis. The pollen was 30 to 40% fertile, and 35% of the seeds were fertile. The original wild watermelons probably had hard, non-sweet, sometimes bitter, white flesh, similar to the citron and colocynth.

2.2 Centres of Diversity

The primary centre of diversity for watermelon is southern Africa, with wild relatives also found in west Africa. The secondary centre is China, and related species can be found in India. Areas of the middle east as well as countries near the Mediterranean Sea may also be good places to collect old land races and wild accessions of *Citrullus*.

T. W. Whitaker considered *Citrullus colocynthis* to be the likely ancestor of watermelon. It is morphologically similar to *C. lanatus,* but with bitter fruit and small seeds. However, the bitter forms of *C. lanatus* were considered the probable ancestor of watermelon by others. That theory was supported based on the fact that they had the same number of chromosomes as *C. lanatus*, were freely intercrossable, and were found in the same areas of Africa and Asia. Citron was considered to be an intermediate stage between the primitive, bitter form of *C. lanatus* and the cultivated form of today.

Although *Citrullus* species grow wild in southern and central Africa, *C. colocynthis* also grows wild in India. India and China may be considered secondary centres of diversity for the genus. Cultivation of watermelon began in ancient Egypt and India, and is thought to have spread from those countries through the Mediterranean area, Near East, and Asia. The crop has been grown in the United States since 1629.

Germplasm is the foundation of breeding programs, so germplasm collection and evaluation are important aspects of breeding. Priorities for collection of *Citrullus* germplasm include India, especially the Indo-Gangetic plains and areas in the northwest parts of the country; Africa including the south and southwest (Kalahari Region); southern areas of the former USSR and Iran; and tropical Africa.

Recent work in germplasm collection and exchange has provided the USDA germplasm system with a total of 51 *Citrullus* accessions that were collected during a scientist exchange visit with the People's Republic of China led by Wehner in 1993. Later, in 1996, a team of four researchers led by Wehner collected germplasm of *Citrullus* in the Republic of South Africa.

3 Varietal Groups

Watermelon *(Citrullus lanatus)* has 22 chromosomes ($2n$=22, x=11). The genus *Citrullus* belongs to the subtribe Benincasinae. In 1930, L.H. Bailey proposed dividing cultivated watermelon *C. vulgaris*, into botanical variety *lanatus* and botanical variety *citroides*. The genus *Citrullus* has been studied taxonomically, and recently has been divided into four species: *C. lanatus* (syn. *C. vulgaris*), *C. ecirrhosus*, *C. colocynthis*, and *C. rehmii*. *C. ecirrhosus* is more closely related to *C. lanatus* than either is to *C. colocynthis*. There are two other closely related species: *Praecitrullus fistulosus* from India and Pakistan, and *Acanthosicyos naudinianus* from southern Africa. Other members of the Cucurbitaceae with 22 chromosomes include *Gymnopetalum, Lagenaria, Momordica, Trichosanthes,* and *Melothria*. None appear to be closely related to watermelon.

3.1 Citron

Watermelon has a close relative, citron or preserving melon, which is *C. lanatus* var. *citroides*. Its rind is used to make pickles, and the fruit are fed to livestock. The flesh of the citron is white or green, and may vary from bland to bitter tasting. Citron grows wild in the United States where it causes problems as a weed in crop production areas of the south, especially in Florida, Georgia, and Texas. Watermelon seed production fields should be isolated from weedy areas of citron since these two botanical varieties cross readily.

3.2 Egusi

Some watermelon accessions in the USDA-ARS germplasm collection show a particular phenotype usually described by breeders as Egusi seed type. These accessions have been misclassified on occasion. The Egusi watermelon is commonly known in Nigeria and the Congo as wild watermelon, Egusi melon, or Ibara. The Egusi watermelon is widely cultivated in Nigeria, where the protein- and carbohydrate-rich seeds are used as a regular part of the diet. The fruit are not edible because of their bitter, hard, white flesh. The origin of the Egusi phenotype is uncertain, and the developmental genetics of this seed phenotype are not known. Its seeds are coated by an adherent layer of tissues that may be remnants of nucellar tissues. The tissues are visible only after the second to third week of seed development, and can be removed at maturity for commercial use of the seeds. Egusi type watermelons are used to feed cattle in Africa. Egusi has sometimes been confused with *Citrullus colocynthis* and as a result, the Egusi watermelon has been sometimes considered a common name for *Citrullus colocynthis*. *Citrullus lanatus* (Thunb.) Matsum. & Nakai var. *lanatus* [=*Colocynthis citrullus* L.] is the cultivated watermelon, and can have Egusi phenotype, but accessions of the Egusi type are not colocynths (*Citrullus colocynthis*).

4 Genetic Resources

4.1 Germplasm Repositories

Several germplasm collections, along with current cultivars marketed by seed companies, represent the major sources of germplasm for watermelon breeders interested in the United States market. The USDA collection is stored at the Regional Plant Introduction Station, Griffin, Georgia with the backup collection at the National Seed Storage Laboratory, Fort Collins, Colorado. There are 1644 accessions in the collection, with about 90% currently available to researchers, and the rest needing to be regenerated to increase seed quantity or germination percentage. The collection includes representatives of all *Citrullus* species and botanical varieties. In addition, approximately 300 heirloom cultivars are kept at the National Seed Storage Laboratory.

The Cucurbit Genetics Cooperative has curators who volunteer to collect and maintain seeds of gene mutants published for many of the cultivated cucurbit species. Some gene mutants are no longer available, but small amounts of seeds of some of the gene mutants can be obtained from the curators for that species, T. C. Wehner and S. R. King.

Additional collections are kept by seed savers and other groups interested in heirloom cultivars, and by watermelon breeders around the United States. There are also watermelon germplasm collections in other countries that are being kept for use by the local research community.

4.2 Important cultivars

Watermelon cultivars have been described in the vegetable cultivar lists maintained by the American Society for Horticultural Science. A complete set of descriptions for all vegetable crops, including watermelon, from lists 1 through 27 are available on the world wide web. Seeds are available for many of the open pollinated and inbred cultivars on the list, but a large number of cultivars that are no longer available. Watermelon breeders should obtain and evaluate a sample of the cultivars available to become familiar with the diversity of germplasm. It is also useful to observe the improvement in horticultural traits that has been made in cultivars developed over time.

A breeding program usually is started by intercrossing the best cultivars currently available, or by crossing the best cultivars with accessions having one or more useful traits missing from the elite cultivars. Thus, in the beginning a watermelon breeder will need to obtain seeds of the best cultivars, a set of cultivars developed at different times in the past, a set of accessions from germplasm repositories, and lines with useful or interesting gene mutants.

A survey of popular cultivars in the ten major watermelon-producing states in the United States by D.N. Maynard in 2000 indicated that popular cultivars for commercial production were almost all hybrids, with few open-pollinated cultivars being used commercially. Popular diploid (seeded) open-pollinated cultivars ('Allsweet', 'Black Diamond', 'Calsweet', 'Crimson Sweet', 'Jubilee II', and 'Legacy') were grown mostly in one state each, suggesting regional adaptation or local demand. Hybrids generally were grown in several states, suggesting they have wider adaptation. The 'Allsweet' type, generally considered to be of high quality, was represented by more than half of the listed cultivars (three of the open-pollinated and 11 of the hybrids). The most popular diploid (seeded) cultivars were 'Sangria' and 'Royal Sweet' (seven states), 'Fiesta' (six states), and 'Mardi Gras' and 'Regency' (five states). For triploid (seedless) cultivars, almost half of the cultivars were 'Tri-X-313' type. The most popular triploid cultivars were 'Tri-X-313' (ten states), 'Summer Sweet 5244' (nine states), 'Millionaire' (eight states), 'Genesis' (five states), and 'Tri-X-Shadow' (four states).

In order to develop improved cultivars for an industry in a particular region of the world, the watermelon breeder will need to have seeds of cultivars, breeding lines, populations, plant introduction accessions, and gene mutants that express the traits of interest at a high level. The breeder should identify a source that has the highest level of expression. That would be true whether the trait is quantitatively inherited (fruit yield, earliness, size, sweetness) or qualitatively inherited (dwarfness, anthracnose resistance, flesh colour). If there is a choice of accession for a particular trait (for example, white flesh), it is better to use an adapted accession with the best genetic background. Thus, 'Cream of Saskatchewan' would be a better choice to use in the development of white flesh cultivars for use in the United States, than a wild-type, white-fleshed citron having large vines, late maturity, hard flesh, bitter flavour, large green seeds, and seed dormancy.

There were no defined cultivars of watermelon before the 1820s. Early cultivars include 'Black Spanish' (imported to United States from Portugal in 1827), 'Carolina' (available at least since 1827), and 'Imperial', 'Mountain Sprout', 'Seminole', and 'Mountain Sweet' (introduced by southern growers from 1840 to 1850). Other heirloom cultivars include 'Bradford', 'Clarendon', 'Odell', 'Ravenscroft', and 'Souter' (originating in South Carolina before 1850). Classic watermelon cultivars include 'Peerless' or 'Ice Cream' (1860), 'Phinney Early' (1870), and 'Georgia Rattlesnake' developed by M.W. Johnson in Atlanta, Georgia about 1870.

Planned cultivar development programs began in the United States in 1880 to 1900. Important cultivars developed for the southern United States included 'Cuban Queen' developed and marketed by Burpee in 1881, 'Round Light Icing' (1885), 'Kolb Gem' developed by Reuben Kolb of Alabama in 1885 and marketed by D.M. Ferry, 'Florida Favourite' selected from the cross of 'Pierson' x 'Georgia Rattlesnake' by Girardeau in Monticello, Florida in 1887, 'Dark Icing' developed in 1888 by D.M. Ferry, and 'Dixie' selected from the cross of 'Kolb Gem' x 'Cuban Queen' or 'Mountain Sweet' by George Collins in North Carolina and marketed by Johnson and Stokes. Important cultivars developed for the western United States included

'Chilean' (black or white seeded) brought from the west coast of South America and introduced to California in 1900, 'Angeleno' developed by Johnson and Musser in Los Angeles, California in 1908, and 'Klondike Solid' and 'Klondike Striped' of unknown origin developed about 1900. Important cultivars developed for shipping include 'Tom Watson' developed by Alexander Seed Co. in Augusta, Georgia in 1906, and 'Stone Mountain' developed by Hastings Co. in Atlanta, Georgia in 1924.

Important cultivars developed in the latter part of last century have built on past accomplishments. 'Charleston Gray' (USDA, Charleston, 1954), 'Crimson Sweet' (Kansas State University, 1963), 'Calhoun Gray' (Louisiana State University, 1965), and 'Dixielee' (1979), 'Jubilee' (1963), and 'Smokylee' (1971) (all from the University of Florida) have high resistance to Fusarium wilt. 'Dixielee' (University of Florida, 1979) and 'Sangria' F_1 (Syngenta - Rogers Brand, 1985) have dark red flesh. 'Millionaire' F1, 3*x* (Harris Moran, 1992) and 'Royal Jubilee' F_1 (Seminis) have consistently high yields. 'Crimson Sweet' (Kansas State University, 1963) and 'Sugarlee' (University of Florida, 1981) have high soluble solids. 'Kengarden' (University of Kentucky, 1975) has dwarf vines. 'Tri-X-313' F1 3*x* (Syngenta - American Seedless, 1962) is seedless. 'Minilee' (University of Florida, 1986), 'Mickylee' (University of Florida, 1986), 'New Hampshire Midget' (University of New Hampshire, 1951), 'Sugar Baby' (M. Hardin, Oklahoma, 1955), and 'Tiger Baby' (Seminis) are icebox size. 'Yellow Doll' (Seminis, 1977) has canary yellow flesh.

5 Major Breeding Achievements

5.1 Qualitative Traits

The inheritance of watermelon traits has been studied extensively, and single genes have been identified that are of value to plant breeding programs. Examples include *A* for monoecious vs. andromonoecious sex expression, *Ar-1* and *Ar-2* for resistance to anthracnose races 1 and 2, *C* for canary yellow flesh colour, *dw-1* and *dw-2* for dwarf vines, *E* for non-explosive rind, *F* for non-furrowed fruit surface, *Fo-1* for Fusarium wilt resistance, g^S for striped green rind pattern, *Go* for non-golden rind at maturity, *M* for non-mottled fruit skin, *o* for oval rather than elongate fruit shape, *Pm* for resistance to powdery mildew, *s* and *l* for short seeds, *Scr* for scarlet red flesh, y^o for orange flesh, and *Y* for coral red flesh.

A non-lobed leaf is a mutant expressed beginning in the seedling stage that is controlled by a single recessive gene. The single-gene trait can be useful for indication of hybrid plants. Hybrid seeds can be produced on one inbred line used as the female parent and having non-lobed leaves. If it is pollinated using bee pollination in an isolation block, and the male parent has normal, lobed leaves, then it will be possible to distinguish hybrid from non-hybrid at the seedling stage in the commercial seed lot. The hybrid seeds can then be planted in excess in grower fields and the non-lobed seedlings (produced by self- or sib-pollination) can be removed to leave just hybrid plants. Alternatively, non-hybrid seedlings can be removed from the flats during transplant production.

5.2 Inbreeding Depression and Heterosis

Watermelon is monoecious, and is naturally cross-pollinated like maize. However, there is not as much inbreeding depression or heterosis as one might expect. This is similar to other cucurbits such as cucumber (*Cucumis sativus*) and melon (*Cucumis melo*). It has been suggested that the lack of inbreeding depression is due to the small population size used by farmers during the domestication of the species. Watermelon plants are large, so only a few plants probably were grown in each area. Therefore, even with monoecious sex expression and insect-pollinated flowers, there would have been considerable inbreeding among the few plants representing the population. Since there is little inbreeding depression in watermelon, inbred lines are developed using self-pollination with little loss of vigour from the parental population.

In studies of heterosis in watermelon, some estimates have shown a 10% advantage of the hybrid over the high parent, but only for some parental combinations. The small amount of heterosis observed in watermelon hybrids makes hybrids unnecessary for high yielding commercial cultivars since inbreds should perform as well. However, hybrid cultivars are useful for combining traits inherited in a dominant fashion from the two parents. Examples of such traits include red or canary yellow flesh, resistance to Fusarium wilt and anthracnose, and resistance (actually lack of susceptibility) to powdery mildew. Hybrids also permit the protection of proprietary inbred lines from unauthorized use. However, one of the most important uses of hybrids is the production of seedless cultivars. The primary method for production of seedless watermelons involves the cross of a tetraploid female parent with a diploid male parent to produce a triploid, which will be sterile, and therefore, seedless. Currently, triploid hybrids are the most practical method for production of seedless watermelons.

6 Current Goals of Breeding

In watermelon breeding, it is important to have proper expression for many traits. Furthermore, lack of one key trait (such as scarlet red flesh colour) can make the cultivar unattractive for use in a particular market. This is the common situation for breeding most horticultural crops. With so many traits to work on, it is difficult to make improvements. However, by starting with a leading cultivar, and making crosses with other elite cultivars, it is possible to maintain the expression level for most traits while making gains for one or two traits. Important traits for watermelon are described below.

6.1 Vines

Vine length of watermelon varies from dwarf to long. For example, 'Charleston Gray' and 'Jubilee', large-fruited cultivars, have vines up to 30 feet long. Short or medium length vines are well suited to cultivars with small or medium sized fruit. For example, 'Sugar Baby', 'New Hampshire Midget', and 'Petite Sweet' are short vined, and 'Crimson Sweet' has intermediate vine length.

Dwarf mutants have been discovered in watermelon. Two genes cause dwarfing when they are in homozygous recessive condition: *dw-1* and *dw-2*. 'Kengarden' has the genotype *dw-1 dw-1*. Another gene mutant (Japanese Dwarf, *dw-2 dw-2*) has increased branching from the crown. Dwarf plants having both sets of genes (*dw-1 dw-1* and *dw-2 dw-2*) have hypocotyls 50% the length of normal vining plants, so can be selected in the seedling stage.

6.2 Sex Expression

Most current cultivars are monoecious, and that appears to be the preferred type of sex expression for commercial seed production of inbred lines and hybrid cultivars. Andromonoecy (*aa*) is recessive to monoecy.

Most cultivars have a ratio of 7 staminate to 1 perfect or pistillate flower. There are some cultivars with a ratio of 4 staminate to 1 pistillate flower. It may be possible to breed for gynoecious sex expression by selecting for increased proportion of pistillate nodes in a segregating population. There is no advantage to andromonoecious sex expression, since the perfect flowers must be pollinated by bees in order to set fruit. Thus, they are no more likely to set without bees or to be self-pollinated, than monoecious cultivars.

Male sterility is useful for the production of hybrid seeds without the requirement for expensive hand pollination. The glabrous male sterile (*gms*) mutant provides male sterility, but the plants are less vigorous, have poor seed set, and are susceptible to cucumber beetles because they lack hairs. A second male sterile mutant, the Chinese male sterile (*cms*), has been more useful for hybrid production.

Fruit can be set parthenocarpically. Although there are no gene mutants that make plants parthenocarpic, fruit set may be achieved without pollination by applying growth regulators to the plants. Thus, commercial production of seedless watermelon may be possible in areas where bees have been excluded by applying growth regulators at a particular growth stage to diploid pistillate flowers that would otherwise produce seeded fruit.

6.3 Yield

Yield varies among watermelon accessions and current cultivars. Growers want high weight per acre of marketable size fruit, with a low percentage of culls. The yield goal expressed by many growers is at least one load (45,000 lb.) per acre.

In the production of triploid hybrids, up to one third of the field must be planted to a diploid seeded cultivar. Therefore, higher yield of seedless watermelon per acre could be obtained by using a more efficient pollenizer that would allow more than two thirds of the field to be planted to the triploid cultivar. Alternatively, parthenocarpic fruit set (genetic or hormone-induced) to stimulate fruit set would permit the entire field to be planted to the triploid cultivar.

6.4 Earliness

Early maturity is desirable because prices for watermelon usually are best at the beginning of the local season. However, late maturity is associated with cultivars that have large fruit size and high yield. Thus, it may be necessary to sacrifice some earliness to obtain high yield or large fruit. Time from pollination to fruit harvest ranges from 26 days for early maturing, small-fruited cultivars such as 'Petite Sweet' to 45 days for large-fruited cultivars such as 'Super Sweet'.

The selection process for early maturity should involve both days from seeding or transplanting to first fruit set, and days from first fruit set to fruit harvest. Days to fruit harvest should be based on fruit having fully developed sugars as verified by a hand-held refractometer or by taste evaluation.

6.5 Fruit Type

Fruit size is an important consideration in a breeding program since there are different market requirements for particular groups of shippers and consumers. The general categories are: mini (<4.0 kg), icebox (4.0-5.5 kg), small, sometimes called pee-wee (5.5-8.0 kg), medium (8.0-11 kg), large (11-14.5 kg), and giant (>14.5 kg). Fruit size is inherited in polygenic fashion, with an estimated 25 genes involved. Shippers in the United States work with particular weight categories, such as 8.0-11 kg for seeded and 6.5-8.0 kg for seedless.

Old cultivars tend to have larger fruit size than current cultivars, because one of the things growers were interested in was winning competitions for fruit weight. Competitions are still being held to grow the largest fruit, but commercial production concentrates on high quality. Another reason for larger fruit in the past is that they are more efficient for hand harvest and shipping; large fruit handled individually permit more weight to be moved per unit. Also, there was demand for large fruit to be sold or served by the slice for restaurants and cafeterias. Today, most supermarkets request seedless fruit that weigh 6.5-8.0 kg for standard and 2.0-4.0 kg for mini types.

Small- or medium-fruited types were the result of adapting watermelon to the northern areas of the United States. Cultivars developed for the northern United States were bred from early maturing Asian cultivars brought from Japan and Russia. A. F. Yeager produced the early cultivars 'White Mountain' and 'New Hampshire Midget' from sources, which have 1.0-2.0 kg fruit with a 65-day maturity. The early cultivar 'Petite Sweet' has 2.0-4.5 kg fruit.

Even though icebox cultivars with 4.0-5.5 kg fruit have been developed to fit easily in a small refrigerator, most of the demand in the marketplace for small fruit has been met using sections cut from a large fruit. A large watermelon fruit cut into quarters has the same weight as an icebox melon, but it has a different shape, and consumers can see what they are buying. 'Sugar Baby', a small-fruited cultivar popular in some parts of the world, was selected in Oklahoma by M. Hardin in 1956.

Fruit shape is also an important part of market type. The general categories are round, oval, blocky, or elongate. There is one gene involved in round vs. elongate, with the F_1 being intermediate (blocky). In some cases, fruit shape is related to cotyledon shape at the seedling stage. Plants with elongate fruit have elongate cotyledons, and plants with round fruit have round cotyledons. However, others have concluded that selection for fruit shape at the seedling stage is ineffective. Among old cultivars with elongate-shaped fruit, there was greater susceptibility to production of gourd-neck or bottle-neck fruit, which are culls. Old cultivars with round fruit were more susceptible to hollowheart. Thus, some of the first hybrids were made between elongate and round inbreds to reduce the incidence of these defects. Recently, genetic resistance to those defects has been incorporated into new cultivars, and has made fruit shape less important to consider.

The third area of importance in market type is rind pattern, which can be gray, striped, or solid. Stripes on the rind can be narrow, medium, or wide where the stripes are the dark green areas. The striped pattern can be on light green or medium green background. Solid rind colour can be light or dark green. Solid dark green is dominant to gray rind pattern. Solid dark green is dominant to striped, and striped is dominant to solid light green rind pattern. However, the striped pattern can be seen on a solid dark green fruit after the colour has been bleached by the sun. The stripes can be over a light or medium green background. For example, 'Dixielee' has narrow stripes on a light green background, whereas 'Florida Favourite' has narrow stripes on a medium green background.

In addition to the common rind patterns, there is furrowed vs. smooth rind, controlled by the recessive gene, *f*. Most current cultivars have smooth rind. Another interesting mutant is golden rind, which is controlled by the recessive gene, *go*. Its usefulness as an indicator of fruit ripeness is limited because the change in fruit colour at fruit maturity is accompanied by chlorosis of the leaves. Furthermore, it does not appear to be a reliable indicator of ripeness, and may be disadvantageous for yield, especially if the grower is using a multiple harvest system.

6.6 External Fruit Quality

Rind durability is important on cultivars that are to be shipped to market. On large-fruited cultivars, the rind should be thick and tough; whereas on small-fruited cultivars, the rind should be thin and tough. Rind thickness should be a small percentage of flesh diameter to keep it in a balanced proportion for best appearance. Large-fruited cultivars look better with a thicker rind, and need the extra protection for postharvest handling and shipping. The rind can be tough and hard as in 'Peacock' or tough and soft as in 'Calhoun Gray'. Brittle rind as in 'New Hampshire Midget' is not useful for cultivars that are to be shipped to market.

Rind flexibility can be tested by cutting a 1/16 to 1/8 inch x 3 inch piece of rind from a fruit and bending the rind into an arc. If the rind bends into a tight arc, it is flexible and tough. If it breaks early in the attempt, it is tender and explosive.

Rind toughness can be measured by driving a spring-loaded punch into the rind. A tough rind would require more force to punch through, whereas a tender or brittle rind requires less force. Watermelon breeders often use faster methods to test for rind toughness, however. One method is to drop the fruit onto the ground from a particular height (for example, knee height) to see whether it breaks open or not. The drop height would depend on the soil type of the field being used. Another method is the "thumb" test, where the breeder presses on the rind at a particular location on each fruit. If the rind breaks when only a small amount of force is applied, then it has a tender rind; otherwise it should be resistant to shipping damage.

6.7 Internal Fruit Quality

Flesh colour is one of the primary traits consumers look for in a watermelon fruit. Colour can be scarlet red, coral red, orange, canary yellow, salmon yellow (golden), or white. Coral red (*YY*) is dominant to orange ($y^o y^o$), which is dominant to salmon yellow (*yy*). Canary yellow (*CC*) is dominant to non-canary yellow (*cc*), and epistatic to (overcomes) the *y* locus for red-orange-salmon yellow. Coral red is recessive to the white flesh colour, which is found in citron. Scarlet red colour (*Scr*) from 'Peacock' has been used to develop many new cultivars because of its attractive colour. Cultivars with dark red flesh include 'Dixielee', 'AU-Sweet Scarlet', 'Red-N-Sweet', and 'Sangria'.

Sugar content is a major component of flavour. Breeders select for high sugar content as indicated by taste and refractometer readings. Refractometer readings are easily made in the field using a handheld unit, and provide data on percentage of soluble solids (°Brix). These translate to sugar content, which should be a minimum of 10%. Newer cultivars have Brix as high as 14%. Some cultivars have higher levels of fructose, which tastes sweeter than sucrose. The difference in taste is not measured by a refractometer.

Selection should be made for good watermelon flavour, independent of sweetness (sugar content). Flavour should include freedom from bitterness, which is controlled by a single dominant gene, and may be introduced in crosses with *C. colocynthis* accessions. Another component is caramel flavour as in 'Sugar Baby' fruit, which some taste testers find unpleasant. The flavour is sometimes associated with dark red flesh colour. Its inheritance is not known, but caramel flavour does respond to selection. Thus, breeders should select lines with mild (not bitter) taste, high sugar content (°Brix), freedom from caramel flavour, and excellent "watermelon" taste. It is important that cultivars with excellent taste be included as checks in all selection blocks to provide a comparison for the plant breeder. Examples of cultivars with good quality that are commonly used include 'Allsweet', 'Crimson Sweet', and 'Sweet Princess'.

Flesh texture is an important part of internal quality. Watermelon fruit can have flesh that is soft or firm, and fibrous or crisp. The objectives for plant breeders should be to develop cultivars with flesh that is firm and crisp. The genes controlling those traits are not known, but they are heritable.

6.8 Watermelon Rind Pickles

The use of watermelon to make rind pickles is a small but interesting segment of the home gardening sector. Homeowners and small industries use the leftover watermelon crop, especially from cultivars having thick and crisp rind, to produce pickles. The watermelon fruit consists of the exocarp, mesocarp, and endocarp. The endocarp is the seed-containing part that is consumed as food, and the mesocarp and exocarp are usually referred to as the rind. The rind is used for making pickles after removing the thin exocarp, leaving the crisp, white mesocarp.

Many obsolete cultivars were discontinued from use in the market because of their thick rind, so they would be obvious candidates for use in making watermelon pickles. Some of those old cultivars are still used by home gardeners and heirloom collectors, and seeds are available from seed companies. 'Tom Watson', 'Georgia Rattlesnake', and 'Black Diamond' are three heirloom cultivars with good flavour, attractive rind pattern and colour, and thick rind. In addition, many hybrids currently cultivated for fruit production by commercial growers have rind that is thick enough for pickle production. Cubes of 10 mm per side can be cut from most of the hybrids tested, thus allowing the pickling of the rind of many modern cultivars, including seedless watermelons.

6.9 Seeds and Seedlessness

Seed colour can be white, tan, brown, black, red, green, or mottled. White seed colour usually is not preferred since it suggests that the fruit is immature, and can make it difficult to distinguish mature from immature seeds. On the other hand, white seeds may be a useful objective for the development of near-seedless cultivars that have few, small, and inconspicuous seeds. Black seed colour is attractive with scarlet red or canary yellow flesh colour. Black, brown, or tan seeds look good with orange flesh colour.

Seed size should be large for confectionery (edible seeded) type, and small or medium sized for the standard (edible flesh) type. A new seed size mutant discovered recently is called tomato seed. The seed size is about half that of the small watermelon seed size, and is controlled by a single recessive gene, *ts*.

Seed number should be high for the confectionery type, but should be low or medium for the edible flesh type. Seed number should be lower in small-fruited cultivars so that the seeds will not appear to include more than the usual percentage of the fruit volume. Seed number should be high enough to make seed production economical, but low enough to make the flesh easy to eat.

In theory, seedless triploid hybrids should provide higher yield than diploid hybrids because no energy is used in seed production. However, in practice this may not be the case. Fruit production in triploids is limited by the availability of viable pollen to induce fruit set.

During the development of tetraploid inbreds, seed yield is often low in early generations, so selection for fertility is essential. Some tetraploids are more fertile than others, and should be selected to keep seed costs low for triploid hybrid production, since the hybrid seeds are produced on the tetraploid parent line.

Triploid hybrids are generally seedless, but occasionally hard seed coats form in the fruit. The presence of objectionable seed coats is affected by environment, but can also be selected against in the development of the inbred parents of the hybrid. Inbred parents that do not develop objectionable seed coats in the fruit in different production environments should be selected for triploid hybrids.

6.10 Seedling Disease Tests

Disease resistance is an important objective of most breeding programs. Screening for resistance to several important diseases using greenhouse seedling tests is useful, and provides several advantages. Plants that are found to be resistant to the diseases being tested can be transplanted from the test flats to soil or other growth medium in bags or pots where they can be grown and self-pollinated, or crossed with other lines. Greenhouse tests can be run at a time when plants cannot be grown outside, permitting more generations of testing each year, and the disease testing greenhouses can be isolated from other watermelon research to keep the diseases from spreading. For seedling tests on gummy stem blight tests, plants should be isolated in one greenhouse, virus tests in another greenhouse, and breeding work in another greenhouse to prevent diseases from spreading from one to the other.

For some diseases such as anthracnose, it is useful to have a humidity chamber to incubate the disease after inoculation. A humidity chamber can be built on a greenhouse bench, usually one humidifier for each 3 m^2 of bench space. An air conditioner can be used to keep the temperature cool, since some diseases do best in cool and humid conditions. The greenhouse temperature is usually kept between 21° and 32°F for optimum plant growth, and the humidity chamber is usually kept between 47° and 57°C for optimum disease development. A less expensive option for disease chambers is to build a frame on a greenhouse bench and cover it with polyethylene film on the top and sides. Humidifiers placed inside the chamber several hours before disease inoculation should be able to raise the relative humidity above 95%.

6.11 Fusarium Wilt Resistance

Fusarium wilt is caused by *Fusarium oxysporu*m f. sp. *niveum*. The disease was first reported in 1889 in Mississippi, and was widespread throughout the southern parts of the United States by 1900. Three types of pathogen spores are commonly observed: small, colourless, oval, non septate microconidia; large, sickle shaped, septate macroconidia; and thick walled circular chlamydospores. There are three races known: 0, 1, and 2. Most current cultivars are resistant to race 0, and some also are resistant to race 1. Race 2 was discovered more recently, and occurs mainly in the south central production areas such as Texas and Oklahoma, but it also has been found in Florida.

Race 0 causes wilt in older, susceptible cultivars such as 'Florida Giant', 'Black Diamond', and 'Sugar Baby'. Race 1 is more virulent than race 0 and affects more plants within susceptible cultivars, but does not affect resistant 'Calhoun Gray'. Race 2 is highly virulent and can affect otherwise resistant cultivars such as 'Calhoun Gray', 'Summit', 'Smokylee', and 'Charleston Gray'. Races of Fusarium can be identified using differentials. 'Sugar Baby' and 'Black Diamond' are susceptible to all the three races; 'Quetzali', 'Mickylee', 'Charleston Gray', and 'Crimson Sweet' are susceptible to races 1 and 2, while 'Calhoun Gray' is susceptible to only race 2. Resistance to race 2 is available in PI 296341 and PI 271769 (Table 1).

Table 1. Reaction of cultivars or accessions of watermelon to Fusarium wilt races 0, 1 and 2 (S=susceptible, R=resistant).

Fusarium wilt race			
0	**1**	**2**	**Cultivar or accession**
S	S	S	Black Diamond (or Sugar Baby)
R	S	S	Quetzali (or Mickylee)
R	M	S	Charleston Gray (or Crimson Sweet)
R	R	S	Calhoun Gray
R	R	R	PI 296341 (or PI 271769)

Fusarium can survive in soil as a saprophyte. The pathogen is spread locally by moving soil, compost, manure, water, tools, and machinery from one field to another, as well as by humans and animals moving between fields. The pathogen can also persist on infested seeds for more than 2 years.

Fusarium enters plants through root tips and openings in roots where lateral roots emerge. Presence of root-knot nematodes is also thought to increase the incidence of the disease. After penetration, the fungus grows into the xylem where it accumulates materials that plug the xylem and cause wilting. Watermelon is attacked at all growth stages by the pathogen. At the seedling stage there is damping-off, and cotyledon wilt results in slower growth and stunting. The vascular tissue inside wilted stems may be discoloured. A white or pink coloured fungus growth usually appears on the surface of dead stems in wet weather conditions. The ideal temperature for infection and disease development is 80°F. However, seedling rot occurs at soil temperatures of 61° to 65°F, while seedling wilt is severe between 77° to 82°F. The disease is also promoted by high soil organic matter.

The first Fusarium wilt resistant cultivar 'Conqueror' was released in 1908. It was developed by W.A. Orton of the USDA using a wilt-resistant citron accession crossed with 'Eden'. 'Conqueror' did not have high fruit quality, so was not grown much after its release. However, cultivars developed using resistance from 'Conqueror' such as 'Iowa Belle' and 'Iowa King' had improved fruit quality, so were used commercially. More recent cultivars such as 'Calhoun Gray', 'Smokylee', and 'Dixielee' have resistance, as well as improved horticultural performance.

Two types of Fusarium wilt resistance are known, having different patterns of inheritance. Resistance to race 1 in 'Calhoun Gray' is controlled by a single dominant gene, with some modifier genes, and provides a high level of resistance that is easy to transfer into new breeding lines. There is also a source of resistance to race 1 which is controlled by several recessive genes. That source of resistance has been difficult to fix at a high level in stable, inbred lines. Cultivars resistant at high inoculum levels are 'Dixielee' and 'Smokylee'. In wild species, resistance to Fusarium has been reported to be polygenic. Resistance to race 2 has been reported in PI 296341, and the selection PI 296341-*FR* is resistant to all three races of Fusarium. Also, PI 271769 was reported to be highly resistant to race 2.

6.12 Anthracnose Resistance

Anthracnose caused by *Colletotrichum lagenarium* is an important disease of watermelon in the United States. Symptoms caused by this pathogen may occur on leaves, stems, and fruit. Lesions on leaves are irregular shaped, limited by the leaf vein, and brown to black in colour. Lesions on the stem are oval shaped and tan coloured with a brown margin. Lesions similar to those found on stems and leaves also appear on the fruit. Older fruit show small water-soaked lesions with greasy, yellowish centres that are somewhat elevated.

Seven races of the anthracnose pathogen have been reported. Races 4, 5, and 6 are virulent in watermelon, but races 1 and 3 are most important. Many cultivars are resistant to races 1 and 3, and resistance to race 2 will be needed in the near future.

The first source of resistance to anthracnose was identified in an accession, Africa 8, sent to D.V. Layton of the USDA by R.F. Wagner in Umtali, South Africa. Layton developed anthracnose resistant 'Congo', 'Fairfax', and 'Charleston Grāy' from that source. Resistance was later found to be inherited as a single dominant gene, *Ar-1*. The gene provides resistance to races 1 and 3, but not to race 2. 'Crimson Sweet' and many other current cultivars have that source of resistance. Several genes were found to be responsible for resistance to Race 2.

PI 189225, PI 271775, PI 299379, and PI 271778 have been reported to carry resistance to complex *Colletotrichum* species. Some of the other sources of resistance to anthracnose reported in the literature are PI 203551, PI 270550, PI 326515, PI 271775, PI 271779, and PI 203551. 'R 143' was reported to be resistant to race 2 of the pathogen. PI 512385 had the highest resistance to race 2 of the pathogen from a screening test involving 76 plant introductions.

6.13 Gummy Stem Blight Resistance

Watermelon is one of the most susceptible of the cucurbit species to gummy stem blight, caused by *Didymella bryoniae*. The disease occurs throughout the southern United States, particularly the southeast. Field and greenhouse tests are available, but the results are variable, and it can be difficult to get reproducible results.

The USDA collection of plant introduction accessions has been screened for gummy stem blight resistance by several teams of researchers. Some accessions have resistance to the disease, including PI 189225 and PI 271778.

6.14 Powdery Mildew Resistance

Watermelon is one of the most resistant cucurbit species to powdery mildew (*Sphaerotheca fuliginea*). However, there are a few regions of the world where powdery mildew is a problem on watermelon. For example, watermelons grown in southern India are affected with the disease, but not in northern India. In southern India, 'Arka Manik' is resistant to powdery mildew. The *pm* gene causes susceptibility to the disease, but most cultivars have the resistance allele. Powdery mildew is becoming more of a problem in the United States, especially in the western states, and has been reported in the southeastern states as well.

6.15 Yellow Vine Resistance

Yellow vine is a relatively new disease of watermelon, caused by an unknown, phloem-limited bacterium. Evidence indicates that leafhoppers vector the disease. The disease was first observed in central Texas and Oklahoma in 1991 and has caused severe losses in early-planted watermelon in some years. In 1998, the disease was detected in watermelon and pumpkin in Tennessee. Production areas of Georgia, Florida, and other parts of southeastern United States may be at risk in the future. Low levels of resistance or tolerance have been identified in a few open-pollinated and hybrid cultivars, although the mechanism of resistance is unknown. Research is needed to identify good sources of resistance.

6.16 Bacterial Fruit Blotch Resistance

Bacterial fruit blotch of watermelon is a serious disease of seedlings and fruit caused by *Acidovorax avenae* subsp. *citrulli*. Disease incidence increases under high humidity or where overhead irrigation is used. The disease was first reported to occur in commercial watermelon production areas in the United States in 1989. Early-season outbreaks can result in total loss of fruit by harvest time. Bacterial fruit blotch is also reported to attack cantaloupe fruit in the field, as well as other cucurbits. Bacterial fruit blotch epidemics during 1994 in certain states in the United States resulted in litigation, and had a devastating effect on the watermelon industry. Currently, most seed companies require growers to sign waiver forms to reduce the possibility of litigation. Some companies have restricted seed sales in certain states where the risk of disease is high. Seed costs have increased due to the changes in the seed handling, packaging and testing required for reducing the incidence of disease.

The characteristic symptoms of bacterial fruit blotch are the appearance of a dark olive green stain, or blotch, on the upper surface of infected fruit. Apart from attacking the fruit, the pathogen is also reported to attack the leaves and seedlings, and can be seed transmitted. D.L. Hopkins and co-workers reported that fermentation of seeds for 24 to 48 hours followed by 1% hydrochloric acid or 1% calcium hypochlorite treatment for 15 minutes prior to washing and drying were the most effective treatments for bacterial contaminated watermelon seeds. This treatment is for diploids; triploid seed germination is drastically reduced by fermentation. However, an effective, cost efficient, and environmentally safe method for disease control would be development of resistant cultivars.

A seedling test for early screening of watermelon fruit blotch was developed in 1992, and research on a few watermelon lines using this test has been reported. There has been some research to identify genetic resistance in the watermelon germplasm collection. Based on seedling tests, PI 295843 and PI 299378 were reported to be resistant to the pathogen. In 1993, D.L. Hopkins and co-workers conducted a study of 22 cultivars and 2 PI accessions for resistance to fruit blotch of watermelon and reported that none were immune to the pathogen. Research is underway to find sources of resistance in the germplasm collection.

Fruit resistance to the pathogen appears to be related to rind colour and ploidy, with diploid cultivars having light rind colour being most susceptible and triploid cultivars with dark rind colour being less susceptible. Fruit with stripes appeared to be intermediate in their resistance. Detached leaf tests have been developed that are effective in screening plants for resistance in a breeding program.

6.17 Bacterial Rind Necrosis Resistance

Bacterial rind necrosis is caused by *Erwinia* species. However, some other bacterial species (*Pseudomonas*, *Enterobacter*, and *Bacillus*) are also known to cause similar symptoms. Typical symptoms of bacterial rind necrosis on watermelon fruit are characterized by a light brown, dry, hard area of discoloration interspersed with light areas generally limited to the rind. The disease was first reported in Texas in 1968. The most resistant cultivars in studies conducted in Florida over a 3-year period were 'Sweet Princess' and 'Jubilee', while the most susceptible were 'Klondike Blue Ribbon' and 'Louisiana Queen'.

6.18 Root-Knot Nematode Resistance

Watermelon is susceptible to root-knot nematodes caused by *Meloidogyne* spp. The USDA collection of plant introduction accessions is being screened for resistance. Root-knot resistance may be an important future breeding objective if resistant accessions are identified.

6.19 Virus Resistance

The main virus problems in watermelon production in the United States are papaya ringspot virus-watermelon strain (PRSV-W, formerly watermelon mosaic virus-1), watermelon mosaic virus-2 (WMV-2), and zucchini yellow mosaic virus (ZYMV). The watermelon germplasm collection has been screened for resistance to some virus diseases. Accessions reported to be resistant to WMV-2 are PI 244018 and PI 244019. Resistance to ZYMV is found in PI 482299, PI 482261, PI 595203, and PI 255137. Research is in progress to identify sources of resistance to PRSV-W as well. Multiple virus resistance will be an important breeding objective for new cultivars in a few years.

6.20 Other Disease Resistance

Verticillium wilt is an increasing problem in the western United States, but little is known about sources of resistance. Resistance to Alternaria leaf spot has been identified in cultivars such as 'Sugar Baby', 'Fairfax', and 'Calhoun Gray'.

6.21 Physiological Disease Resistance

Many of the watermelon fruit defects have a genetic component. Breeders should select lines to be free of defects under conditions conducive to the problem. Fruit defects include hollowheart, rind necrosis, blossom-end rot, and cross stitch. Hollowheart is a separation of the tissue within the endocarp caused by rapid fruit growth and weak tissue. More research is needed to identify sources of defect resistance, and environmental conditions that help reduce their frequency.

6.22 Insect Resistance

Little research has been done on insect resistance in watermelon. This may be due to the fact that most insect pests can be controlled with insecticides. The major arthropod (insect and arachnid) pests of watermelon are aphids, pickleworm, spider mite, and spotted, striped, and banded cucumber beetles.

PI 299563 is resistant to melon aphid (*Aphis gossypii*). 'Congo' and 'Giza 1' were the most resistant of five accessions evaluated for resistance to spider mite. Several genes were found to control non-preference type resistance to spotted cucumber beetle in 'Hawkesbury' x a resistant accession. Resistance to spotted and banded cucumber beetles was due a single recessive gene.

A single dominant gene, *Fwr*, was responsible for resistance to the melon fruit fly (*Dacus cucurbitae*) in the watermelon line JI8-1. 'Afghan' is reported to have resistance to red pumpkin beetle (*Aulacophora foveicollis*), and 'Blue Ribbon' and 'Crimson Sweet' are resistant to pickleworm.

6.23 Stress Resistance

Little research has been done on stress resistance in watermelon. Water stress is an important cause of reduced yield in watermelon. It may be that some genotypes are more efficient in water use than others, but it probably will be difficult to develop highly efficient cultivars since watermelon fruit have very high water content. In Israel, deep-rooted cultivars are used in unirrigated desert areas.

Pollination problems are responsible for improper fruit development. It is necessary for all three lobes of the stigma to be fully pollinated if the fruit is to develop fully, and without curvature. Proper fruit development requires adequate numbers of honeybees or bumblebees during flowering, along with weather that is conducive to pollination. Bumblebees can be more effective pollinators than honeybees. Cold, rainy weather leads to poor pollen shed, and hot weather often leads to reduced bee activity. In the case of triploid hybrids, it is necessary to have up to one third of the field planted to a diploid pollenizer to assure adequate fruit development in the triploids which are male sterile.

Growers plant early in the season, often using transplants and plastic mulch (with row covers in some cases) when there is a danger of frost. Cucurbits are susceptible to chilling injury at air temperatures below 42°F. Chilling injury is a concern in watermelon because of the value of early harvested fruit. There might be chilling resistance in the watermelon germplasm collection that could be incorporated into new cultivars as has been done in other cucurbits. Watermelon appears to be more chilling resistant than melon and cucumber. Symptoms of chilling are white areas on the cotyledons and white or light brown margins on the fully expanded leaves. Chilling injury is increased by a longer duration of chilling, lower temperature, high intensity of light during chilling, high wind speed during chilling, or a higher growth temperature before chilling occurs. Watermelon is thermophilic, meaning that plants have a high optimum growth temperature. Although the optimum is probably 80-90°F, temperatures above 90°F reduce growth rate, and can reduce fruit yield. Above 105°F, plants can be injured, and young leaves will be light green with yellow margins.

Measles is a condition where green-brown spots develop on the fruit surface, covering a small area or even the entire surface, and starting out as minute water-soaked areas. The spots become tan, slightly raised areas with necrotic centres. The symptoms occur when excessive guttation is encouraged by periods of high humidity or during the early fall production season when the humidity is high and the nights are cool. The fruit symptoms become evident 21-25 days after the conducive environmental conditions occur. There is usually no economic loss from the stress, and it might be controlled by reducing the amount of irrigation in the fall production season.

7 Breeding Methods and Techniques

Major objectives for watermelon breeding include proper fruit type, early maturity, high fruit yield, high sugar content, tough flexible rind, and proper seed type. It is important to determine breeding objectives carefully before starting cultivar development. For example, seed type changes significantly for different market classes. Parental lines for seedless hybrids should have small seeds, whereas confectionery seed types should have large seeds. For commercial cultivars, black seeds are preferred because of their contrast with red, yellow, or orange flesh. Also, white seeds indicate immaturity to buyers, so white mature seed colour can be a confusing trait for them. Most of the old cultivars are diploid, open-pollinated or inbred lines, but hybrid diploid and hybrid triploid cultivars are taking over the commercial market in the United States.

After determining the breeding objectives, methods for measurement of the traits of interest should be developed, selection methods should be determined (specifying the operations to be carried out for each generation), and parents with high expression of the traits of interest should be chosen. Vine type should be long for commercial production and dwarf (bush) for home garden. It may also be possible to use the dwarf plant type for once-over harvest in commercial production. Sex expression should be monoecious, with a ratio of 7 staminate:1 pistillate flowers, or better (preferably 4:1). Andromonoecious sex expression and ratios of 15:1 are more typical of older cultivars.

For production in most areas of the United States, watermelon must have resistance to Fusarium wilt. Races 0 and 1 are common, and race 2 is becoming important, especially in Texas and Oklahoma where plastic mulch culture and fumigation are less common. Production areas in the southern United States usually have anthracnose race 1 and may also have problems with race 2. Gummy stem blight is a disease for which resistance is needed in most southern production areas. Powdery mildew is becoming a problem, especially in the western United States (possibly because of a new race), and should be a breeding objective for new cultivars. Bacterial fruit blotch was a problem in the 1990s, and resistant accessions have been identified. The disease can be effectively controlled by genetic resistance and by large-scale seed testing followed by destruction of contaminated seed lots. Protection from viruses in the United States production areas should include resistance to papaya ringspot virus-watermelon strain (formerly watermelon mosaic virus-1), watermelon mosaic virus (formerly watermelon mosaic virus-2), and zucchini yellow mosaic virus.

Finally, breeding objectives should emphasize early maturity, high fruit yield, durability for shipping, high internal quality, freedom from internal defects (hollowheart and rind necrosis), and proper seed type in a diploid (seeded) or triploid (seedless) hybrid. Internal quality traits include dark red flesh, high sugar content, proper sugar to acid ratio, excellent flavour, high nutritional value (vitamins and lycopene), firm (not soft) and non-fibrous texture. Seeds should be black colour, medium size (or small for inbreds to be made into tetraploids), and few to medium quantity per fruit (few for consumers, but medium to keep seed costs down). Flesh colour should be dark red (*Y* gene with modifier genes) with uniform colour throughout the fruit. For specialty types, flesh colour could be bright orange (y^o gene), canary yellow (*C* gene), or white (*Wf* gene). Other colours such as salmon yellow (*y* gene) exist, but are not preferred because the flesh looks overmature. Older cultivars have light red flesh, but dark red is becoming the preferred type. Diploid inbreds should be made into tetraploid inbreds and tested for fertility, seed yield, and ability to set fruit using controlled pollination. Tetraploid lines for use in triploid seedless hybrid production can be induced with colchicine. Finally, triploid hybrids should be tested for absence of seed coats in the fruit within a range of production environments.

7.1 Pollination Methods

Watermelon is a cross-pollinated species with monoecious or andromonoecious flowering habit. There is a popular myth that watermelon should not be grown close to other cucurbits such as cucumber, cantaloupe, or squash because of an adverse effect on horticultural traits such as flavour. However, watermelon will not cross with any other cucurbits except for species within the genus *Citrullus*. Furthermore, there is no effect of foreign pollen on fruit development (xenia) in watermelon.

7.2 Greenhouse Pollinations

Controlled pollinations can be made easily in a greenhouse or screenhouse since that eliminates the need to cover individual flowers the previous afternoon to protect them from pollinating insects such as bees. The greenhouse or screenhouse should be well sealed to prevent insects from getting in. In those structures, pollinations should be made in the morning, and plant maintenance work should be left for the afternoon. Computer controlled heating and cooling, and automated irrigation and fertilization make it possible to operate the greenhouse with fewer labour inputs.

Greenhouse plants can be grown in ground beds, plastic bags or pots containing the growth medium, or in various liquid media such as ebb and flow benches or nutrient film technique. If pots or bags are used, different container sizes should be evaluated to obtain the proper plant size. A good pot size for proper growth of watermelon plants is 200 mm diameter. Plants grown in larger pots will have longer vines that are more difficult to train and prune, larger fruit, and more seeds per pollination.

In the greenhouse, plants are usually trained vertically onto supports such as strings held by overhead wires. This saves floor space and makes better use of available light. The overhead wire should be 2 m above the walkway to permit most workers to reach the trellis without standing on a ladder, while being able to walk under it without ducking. Plants should be pruned to one main stem, usually with no branches. Because of their weight, fruit must be supported in a sling. Stem length of most watermelons usually requires that plants be trained up the string to the trellis wire, and back down again. Plants should be given sufficient floor space in the greenhouse to grow and flower. For elite cultivars and breeding lines, each plant should have 0.18 m^2 or more. It may be necessary to give wild accessions more space, perhaps 0.36 m^2 per plant or more.

In some latitudes, it may be necessary to provide supplemental lighting for plant growth. We find it difficult to grow plants in Raleigh, North Carolina in the winter without extra lighting. However, plants grow well and produce flowers, fruit, and seeds properly when grown in the spring (February through June) and fall (July through November) seasons.

7.3 Field Pollinations

Natural pollination of watermelons in the field is usually by honeybees that visit the flower to collect pollen and nectar. Bumblebees also are effective pollinators. Hand pollination of watermelon flowers is usually less effective than bee pollination. It is necessary to protect flowers from bee visits before and after making controlled pollinations. Flowers open shortly after sunrise and remain open for 1 day. Usually a pistillate flower and the staminate flower below it (proximal to it) open on the same day, making self pollination possible. Many breeders have found that hand pollination is more effective between 6 and 9 am than later in the day.

The two main methods for protecting controlled pollinations from insect pollination in the field are to begin pollinating before bees become active in the morning, or to cover the flowers the previous afternoon. For the first method, pollinations can be made on newly-opened flowers, which are then covered to keep bees away. This method requires less time per pollination, but care must be taken to stop pollinating when bees are observed in the field. Staminate and pistillate flowers can be covered with gelatin capsules (size OO), cotton wool, plastics caps, or paper rolled into a cylinder (often, holding a pencil inside as the paper is rolled) and closed at one end by folding. It is also possible to use inverted styrofoam or plastic cups (6-12 oz. size) held over the flower (and onto the soil surface) with a J-shaped wire (about 10 gauge thickness) stuck through the cup, or by a wooden stake glued to the cup. Breeders have also made flower covers using mesh or cloth bags, which in some cases are supported by a wire frame that can be stuck into the ground over the flowers to be protected.

The second method requires that flowers predicted to open the next morning be capped the previous afternoon. These flowers will be one or two nodes above the flowers (toward the shoot apex) that are newly opened, and should have some yellow colour in the petals. Flowers more than three nodes above the newly opened ones that are completely green will probably not open the next day. Capping of flowers is most useful if done on sunny days, since the pollen does not shed freely after rainy or cloudy days. The following morning, the caps are removed, flowers pollinated, and the caps replaced to keep bees away. This method permits the pollination crew to keep working longer as bees begin to work the field.

In a large field pollination nursery, workers often prefer to mark the flowers that have been capped in the afternoon with a flag (for example, white), which is then exchanged with a flag of a different colour (for example, blue) after the pollination has been made. Thus, it is easy to go to the white flags in the morning to make the pollinations, and to go to the blue flags in the afternoon to check whether the pollinations from previous mornings are developing properly. The setting of one fruit inhibits other fruit on the same plant from setting, so it is useful to remove pistillate flowers that have not been used for controlled pollinations as the pollinating crew moves through the field in the afternoon.

Andromonoecious plants have perfect flowers as well as staminate ones. Unfortunately, perfect flowers will not set fruit without being hand pollinated, or visited by a pollinating insect, so they are no more likely to be self-pollinated than pistillate flowers. After pollinating a pistillate flower, a tag is placed on the peduncle or on the stem just below the peduncle. Placing the tag on the stem causes less damage to the pollinated flower and developing fruit. The tag usually has the plot number of the female and male parents and the date the pollination was made. It can also have the initials of the person making the pollination, and the name of the study involved.

Controlled pollinations are made by removing a recently opened staminate flower from the plant to be used as the male parent. The petals of the staminate flower are bent back until they break. The flower can then be used like a paintbrush to pollinate a recently-opened pistillate flower on the plant to be used as the female parent.

A nursery for field pollination should be designed to make it easy to make controlled pollinations, and care for the plants. Direct seeding or transplants can be used. For direct seeding, the seeds should be treated with a registered fungicide before planting. Use of herbicides will significantly reduce the need for hand weeding. For transplants, plastic mulch and drip irrigation will help with weed control. Drip irrigation, or other low-level system (furrow, sub-irrigation) is superior to overhead irrigation to keep the plants dry, so hand pollinations can be made without having to wait for the watering to be completed, and to avoid having pollination caps washed off the flowers.

Pollinations are made easier by planting the lines to be crossed together in one area. Lines to be self-pollinated can be planted together in a second area. It is useful to plant each pair of lines to be crossed in adjacent rows or tiers.

If it is difficult to make self-pollinations in the field on a particular set of lines (perhaps selections from a trial), one or more cuttings can be taken from each of the plants to be selected. The cuttings can be rooted in moist sand in a greenhouse by burying the bottom (proximal) internode, with two to five nodes of leaves above. The resulting plants can be transplanted from the rooting bench to the greenhouse for trellising and self- or cross-pollination of the selections to produce seeds for the next generation.

7.4 Breeding Plans

Once the breeder has determined the objectives of the program, the choice of parental materials is one of the most important aspects of a breeding program. Using knowledge of the crop and predicting the traits consumers will be interested in having in future cultivars, the breeder gathers parental lines for crossing. The breeder should know which parent will contribute the traits of interest, and which methods will be used to evaluate the progeny for those traits. Thus, it is often necessary to collect and evaluate large numbers of PI accessions, cultivars, and breeding lines for the traits of interest to identify appropriate parents to use in the program. This work often continues in parallel with the main part of the breeding program.

The next step is to determine the breeding method to use for each part of the program. It is important for the breeder to consider the advantages and disadvantages of particular breeding methods, and how they can be incorporated into the overall breeding plan. Also, it is common to use more than one breeding method at a time in order to accomplish several sets of objectives. For example, one part of the program might be to use recurrent selection to develop a base population with general adaptation and the proper fruit type that also has high yield and early maturity. A second part of the program might be to use pedigree selection on the cross of two lines to develop inbred lines with the high yield, early maturity, and proper fruit type of one parent, and the dark red flesh colour, high sugar content, and firm crisp flesh texture of the other parent. A third part of the program might be to use backcross breeding to make a canary yellow flesh version of an elite red-fleshed hybrid with top performance.

7.5 Recurrent Selection

Although watermelon is a cross-pollinated crop, population improvement methods popular in some cross-pollinated crops have not been used. The main reason for that appears to be the large size of the plants, and the low rate of natural outcrossing that occurs. Also, because there are few plant breeders working on watermelon, and because of the requirement for many qualitative traits to be present in the new cultivars being tested for release, it is expensive to spend additional years in population improvement for quantitative traits.

It may be possible to improve quantitative traits such as yield in watermelon using recurrent selection i.e. repeated selection and massing of selected plants, but the populations should probably be developed initially to have the necessary qualitative genes in them. Those would include proper flesh colour, fruit size, and disease resistance. Due to large plant size and a 5-month generation time, recurrent selection methods should be those that have few generations per cycle, and few plants per family (or single-plant selection).

One approach would be to develop an elite population by intercrossing two to four of the best red fleshed hybrids available, trying to choose a set that was genetically unrelated. A population with a wide genetic base could also be developed by intercrossing 20 or more elite cultivars by hand for two or more generations, and using bees in an isolation block for two or more generations before beginning a mild selection pressure for important quantitative traits such as yield. Simple recurrent selection could be used for selection among single-plant hills for a set of highly heritable traits. A more complex method such as reciprocal recurrent selection would permit simultaneous improvement of two populations for combining ability for yield. This would be an expensive program to run, but would produce two populations that could be used to develop inbreds to be used as the female and male parents (respectively) of elite hybrids.

During population development, it would be necessary to identify methods for yield testing that were efficient for use in large yield trials. The usual guidelines for recurrent selection are to test at least 200 individuals (or progenies of individuals) per population, and to select at least 20 to intercross for the next cycle of selection. A yield trial involving 200 replicated families would require more resources than many breeding programs could afford if the trial were done using current methods.

Recurrent selection could be used to improve quantitative traits, such as yield, which are difficult to improve using qualitative methods such as pedigree and backcross breeding. Each year, the improved population would be used to begin the development of inbred lines to feed into other parts of the breeding program.

7.6 Pedigree Breeding

Probably the most common method for watermelon breeding is pedigree. In pedigree breeding, the breeder begins by choosing two or more adapted parents, which complement each other in their traits. For example, one parent might be generally good (yield, earliness, type) except for disease resistance and the other might be generally good (yield, earliness, type) except for fruit quality. The objective would be to produce new lines with high yield, early maturity, proper type, high fruit quality, and good disease resistance. The cultivars or breeding lines are crossed to form the hybrid (F_1) generation, which is then self- or sib-pollinated to form a segregating (F_2) population. The F_2 is self- or sib-pollinated while selecting for traits having high heritability to form the F_3 generation. If multiple plants are tested from each selected F_2 plant, then the breeder concentrates on selecting the best plants in each of the best F_3 families. This might include selection in the seedling stage in the greenhouse in the F_2 and F_3 generations for disease resistance such as Fusarium wilt races 0, 1, and 2 and anthracnose races 1 and 2.

Beginning at the F_4 generation, selection would begin to emphasize family-row performance for quantitative traits. Plants within family-rows that have excellent performance for qualitative traits should be selected for the next generation. As the families reach six generations of self-pollination (S_6 or F_5), they become more uniform, and can then be handled as inbred lines. This could include selection using eight-plant plots for early flowering, number of pistillate flowers, and fruit number.

The number handled might decrease from 54 F_2 plants of a cross to 36 F_3 families, 24 F_4 families, and 18 F_5 lines.

Single-seed-descent is a modification of pedigree breeding in which inbred lines are developed rapidly by self-pollination in greenhouses and winter nurseries, and selection is not practiced until later generations, such as S_3 to S_6. This method requires less record keeping and works better where the main objective is to improve quantitative traits such as yield and earliness, rather than qualitative traits such as flesh colour and disease resistance. However, traditional pedigree breeding is probably the more useful method for watermelon since there are many qualitative traits that can be selected in early generations. In that way, plants or families having unsuitable traits that are simply inherited (such as poor fruit flesh colour) can be eliminated in early generations. Otherwise, they would be carried along until the S_3 to S_6 generation when field-testing would be practiced in the single-seed-descent breeding method.

7.7 Backcross Breeding

Backcross breeding is used to transfer one qualitative (highly-heritable) trait into an otherwise superior inbred. The superior inbred is referred to as the recurrent parent. Often, six generations of selection and backcrossing to the recurrent parent are used to recover the genotype of the recurrent parent (except for the addition of the new trait) without the other undesirable traits from the non-recurrent (donor) parent. Two versions of the backcross method are used depending on whether the gene of interest is recessive or dominant.

For the transfer of a trait controlled by a recessive gene, the recurrent parent is crossed with the donor parent, and the F_1 backcrossed to the recurrent parent. In one scheme, the F_1 is self-pollinated to produce the F_2, which will segregate for the trait of interest. Individuals having the trait can then be backcrossed to the recurrent parent to produce the BC_1. The BC_1 generation is then tested for the trait, and individuals having it are self-pollinated once again to produce a segregating generation for selection and backcrossing to the recurrent parent. The process is repeated until the BC_6 generation when the best individuals are self-pollinated and selected for the trait to produce the improved inbred. The inbred does not need to be tested extensively in trials, because it will be identical to the original inbred, but with one new trait.

For the transfer of a trait controlled by a dominant gene, the recurrent parent is crossed with the donor parent, and the F_1 backcrossed to the recurrent parent. The BC_1 generation is then tested for the trait, and individuals having it are backcrossed to the recurrent parent. The process is repeated until the BC_6 generation when the best individuals are self-pollinated and selected for homozygous expression of the trait using progeny testing.

7.8 Inbred Development

The best selections from the recurrent selection program should be self-pollinated each cycle to begin inbred development. Pedigree selection, and backcross breeding result in the production of elite inbred lines. Each year, those inbred lines that are produced from the different parts of the breeding program should be increased by self-pollination, tested for useful horticultural traits, and used in the production of tetraploid inbred lines, as well as directly for the production of diploid hybrids based on the traits they have, and what is needed by the market.

Isolation blocks or screen cages can be used to make large seed increases of the inbreds if that is needed. Isolation blocks should be away from other watermelon fields, requiring a separation of at least 1 mile. Bees should be provided in the isolation block or cage by bringing in one strong hive, unless there are sufficient numbers of wild bees.

7.9 Hybrid Testing

The final stage of breeding is to produce hybrids for testing. Hybrids are usually made between two monoecious inbreds. For triploid hybrid production, the seed parent should have a distinctive rind pattern that has recessive inheritance. For hybrid production with less labour input, the seed parent could be male sterile. The seed increase of the male sterile inbred would be accomplished by pollinating male sterile plants with the heterozygote (*Ms ms*) as the pollen parent. For seedless hybrid production, the seed parent would be a tetraploid inbred.

Once they have been developed, all inbreds can be crossed in all possible combinations. However, that might produce too many entries to evaluate properly. For example, 20 inbreds could produce (20 x 19)/2 = 190 different hybrids, without including reciprocals. Thus, it may make more sense to make hybrids only from pairs of inbreds having complementing traits of the proper type.

Testing of experimental hybrids should progress in stages, with fewer hybrids to test in later stages where more effort is spent on each hybrid. The first year trials might have two replications in each of two locations. In the second year, the best hybrids could be evaluated in 8 to 12 locations using the conditions available at each (grower fields, state university experiment stations). In the third year, the hybrids would be sent to grower trials throughout the production regions of interest for trials involving 0.25 to 1.0 acre using a total of 5-10 lb. of seeds for all trials. Seeds should be screened for bacterial fruit blotch before sending to growers. One can usually get good data from at least 10 of the 50 trials. Information from the 3 years of trialling should lead to the release of the best one or two hybrids in the fourth year.

Although there is not much advantage of hybrids over open-pollinated cultivars for most traits, it is thought that the former are more uniform. Thus, it may be possible to get the same yield in fewer harvests because of more uniform growth and a more concentrated fruit set. Hybrids offer several advantages over open-pollinated cultivars. A major advantage is the production of seedless triploids, which are produced by crossing a tetraploid female inbred with a diploid male inbred. Hybrids also can express heterosis, with the hybrid performing slightly better than the best parent in some cases. The amount of heterosis in watermelon is around 10%. Another advantage is the ability to get an intermediate fruit shape by crossing an elongate-fruited inbred with a round-fruited one. Inbreds can be used to combine dominant genes for resistance from each parent into a hybrid that has more dominant genes expressed than either parent. A hybrid that has large seeds for the grower to plant and small seeds in the fruit sold to the consumer can be produced by crossing a large-seeded female inbred with a small-seeded male inbred. Finally, hybrids provide a way for the seed company to protect their proprietary inbreds from theft.

The disadvantages of hybrids are that they add an extra step to the breeding process, and increase the cost of seeds since they are produced by hand pollination rather than by bee pollination. Use of male sterile inbreds for seed production should help reduce the cost of hybrid seeds in the future.

7.10 Tetraploid Production

Use of triploid hybrids has provided a method for production of seedless fruit. The tetraploid method for seedless watermelon production was invented by H. Kihara. He began development of tetraploids in 1939, and had commercial triploid hybrids available 12 years later. The development of triploid cultivars adds several problems to the process of watermelon breeding: extra time for the development of tetraploids; additional selection against sterility and fruit abnormalities in tetraploid lines; choice of parents for low incidence of hard seed coats in the hybrids; the reduction in seed yield per acre; reduced seed vigour for the grower; and the necessity for the diploid pollenizer to use up to one-third of the grower's production field.

Seedless cultivars are produced by crossing a tetraploid ($2n=4x=44$) inbred line as the female parent with a diploid ($2n=2x=22$) inbred line as the male parent of the hybrid. The reciprocal cross (diploid female parent) does not produce seeds. The resulting hybrid is a triploid ($2n=3x=33$). Triploid plants have three sets of chromosomes, and three sets cannot be divided evenly during meiosis (the cell division process that produces the gametes). This results in non-functional female and male gametes although the flowers appear normal. Since the triploid hybrid is female sterile, the fruit induced by pollination tend to be seedless. Unfortunately, the triploid has no viable pollen, so it is necessary to plant a diploid cultivar in the production field to provide the pollen that stimulates fruit to form. Usually, one third of the plants in the field are diploid and two thirds are triploid, although successful production has been observed with as little as 20% diploids. Cultivars should be chosen that can be distinguished easily so the seeded diploid fruit can be separated from the seedless triploid fruit for harvesting and marketing.

Breeders interested in the production of seedless triploid hybrids need to develop tetraploid inbred lines to be used as the female parent in a cross with a diploid male parent. One of the major limiting steps in breeding seedless watermelons is the small number of tetraploid inbreds available. Development of seedless hybrids will be discussed in the following stages: (1) choice of diploid lines, (2) production of tetraploid plants, (3) tetraploid line development, and (4) hybrid production and testing.

Stage 1 involves choice of diploid lines to use in tetraploid production. Most of the tetraploid lines being used by the seed industry have gray rind so that, when crossed with a diploid line with striped rind, it will be easy to separate self-pollinated progeny (which will be seeded fruit from the female parent line) from cross-pollinated progeny (which will be seedless fruit from the triploid hybrid). The grower should discard the gray fruit so they are not marketed as seedless watermelons by mistake.

Stage 2 is the production of tetraploid plants. Many methods have been used effectively in other crops to produce polyploids, including tissue culture regeneration, temperature shock, and X-rays. In watermelon, tetraploids can be produced routinely using plants regenerated from tissue culture or using the herbicide oryzalin. Colchicine ($C_{22}H_{25}O_6N$), a poisonous alkaloid used in the treatment of gout, from the seeds and bulbs of *Colchicum autumnale* is a widely used method in watermelon for tetraploid production. Colchicine inhibits spindle formation, and prevents separation of chromosomes at anaphase. Of all the methods of colchicine application, shoot apex treatment at the seedling stage was found most effective.

For the seedling treatment method, the diploid line of interest is planted in the greenhouse in flats (8x16 cells is a popular size) on heating pads that keep the soil medium at 85°F for rapid and uniform germination. When the cotyledons first emerge from the soil, the growing point is treated with colchicine to stop chromosome division and produce a tetraploid shoot with four sets of chromosomes rather than two. The colchicine solution is used at a concentration of 0.1% for small-seed size cultivars ('Minilee', 'Mickylee', 'Sweet Princess'), 0.15-0.2% for medium-seed size cultivars ('Allsweet', 'Crimson Sweet', 'Peacock Striped', 'Sugar Baby'), and 0.2-0.5% for large-seed size cultivars ('Black Diamond', 'Charleston Gray', 'Congo', 'Dixielee', 'Klondike Striped Blue Ribbon', 'Northern Sweet'). Colchicine is applied to the seedling growing point in the morning and evening for 3 consecutive days, using 1 drop on small- or medium-seed size plants and 2 drops on large-seed size cultivars. The treatment produces plants that are diploid, tetraploid, or aneuploid, so it is necessary to identify and select the tetraploids in later stages. Treatment of the T_o diploids with colchicine results in about 1% of the seedlings (referred to as T_1 generation tetraploids) being tetraploids. Some diploid cultivars and breeding lines produce a higher percentage of tetraploids than others. For example, 'Early Canada' produces many tetraploids and 'Sweet Princess' does not.

Tetraploids can be detected by the direct method of counting chromosomes of cells under the microscope, or by comparing stem, leaf, flower, and pollen size with diploid controls. A popular method involves counting the number of chloroplasts in stomatal guard cells using a leaf peel under the microscope. Tetraploids have approximately 10-14 chloroplasts in each guard cell (20-28 total on both sides of the stomate), whereas diploids have only 5-6 in each guard cell (10-12 total). The method is useful for screening many plants for ploidy level in the seedling stage before transplanting to the main part of the greenhouse or field nursery for self-pollination. Usually, multiple methods are used, identifying tetraploid seedlings using their phenotype in flats before transplanting, the chloroplast number in the stomatal guard cells of the true leaves in seedling flats and greenhouse pots, and by the appearance of the fruit and seeds at harvest after self-pollination in the greenhouse. Tetraploids usually have thicker leaves, slower growth, and shorter stems than diploids.

Stage 3 involves tetraploid line development. Tetraploid plants are selected (using methods such as leaf guard cell chloroplast number) in the T_0 generation (plants from colchicine treated diploids) from the greenhouse flats where they were treated with colchicine. It is then necessary to plant the T_1 generation in flats to verify that the plants are tetraploids in that next generation, and transplant the selections to greenhouse pots for self-pollination. Seeds from those selections (T_2) can then be increased in larger plantings such as field isolation blocks to get sufficient numbers of seeds per tetraploid line to use in triploid hybrid production.

The fertility and seed yield of tetraploid lines will increase over generations of self- or sib-pollination, probably because plants with chromosome anomalies are eliminated, resulting in a tetraploid line with balanced chromosome number and regular formation of 11 quadrivalents. Seed yield of tetraploid lines in early generations is often only 50-100 seeds per fruit and sometimes as low as 0-5 seeds compared to 200-800 seeds for diploids. Another problem with early generation tetraploids is poor seed germination, making it difficult to establish uniform field plantings. It may require as much as 10 years of self-pollination before sufficient seeds of tetraploid lines can be produced for commercial production of triploid hybrids. Advanced generations of tetraploid lines usually have improved fertility, seed yield, and germination rate compared to the original lines. Some companies require more than 100 lbs. of seed of a tetraploid inbred to be available before beginning commercial production of the triploid hybrid cultivar. Approximately 110 tetraploid plants are required for production of each pound of triploid seeds.

Stage 4 is the evaluation of tetraploids (usually T_3 generation or later) as parents of triploid hybrids. The tetraploids should be evaluated directly for rind pattern, high seed yield, and other traits such as male sterility for reduced hand labour in hybrid seed production. The major test for tetraploids however, is as female parents in triploid hybrid seed production after making controlled crosses using diploid male parents. The resulting hybrids are tested in yield trials with two rows of triploid plots alternating with one row of diploid plots to assure adequate pollen for fruit set in the triploid hybrids. Useful tetraploid inbreds should produce triploid hybrids with excellent yield and quality for the market type and production area of interest.

7.11 Triploid Evaluation

Evaluation of triploid hybrids is similar to evaluation of diploid cultivars already discussed. There are a few special considerations, however. Triploids are not inherently superior to diploids, so triploid hybrids can be better or worse than their diploid parental lines. Therefore, as in the case of diploid hybrids, many combinations of parental lines should be evaluated in triploid yield trials to identify the ones producing hybrids with the best performance. In general, diploid inbred parents that have poor horticultural performance will produce triploid hybrids having poor performance.

One problem affecting triploid hybrids is empty seed coats (coloured or white) in the fruit. Under some environmental conditions, fruit are produced with large obvious seed coats that are objectionable to consumers. Triploid fruit should be evaluated for seed coat problems during trialling. Some selection should also be done on the parents before triploid production. Seed coats will be large in the hybrids if the parents have large seeds. Seed size is genetically controlled, with at least three genes involved: *l, s,* and *ts*. Use of tetraploid lines with small or tomato-size seeds may help solve the problem. Besides genetic effects, certain unknown environmental conditions seem to increase the number of hard seed coats in poor performing triploid hybrids.

Commercial production of elite triploid hybrid seed is done by hand in locations where labour is inexpensive, or by bee pollination in isolation blocks. The tetraploid and diploid inbreds are planted together in alternating rows, or in alternating hills within each row. Where labour is abundant, the staminate flowers can be collected from the male (diploid) parent and used to pollinate the pistillate flowers on the female (tetraploid) parent. Pollinated flowers should be capped the previous day to keep bees out, then covered after pollination to prevent self or sib-pollination after the cross has been made. The flowers should be tagged with the date so that the fruit can be harvested 35-50 days later.

A method that requires less hand labour is to plant the pollen and seed parents in alternating rows, and to remove all pistillate flowers from the seed parent rows during flowering time, usually a period lasting several weeks. Pistillate flowers on the female parent are tagged on the day they open with the date to assure that the fruit are mature when harvested, and to harvest only fruit that were pollinated during the time staminate flowers were removed from the female parent. Seeds that are harvested can also be sorted mechanically for size, weight or density to separate triploid seeds (resulting from cross pollination) from tetraploid seeds (resulting from self- and sib-pollination).

When seed production is by bee pollination in isolation blocks, the tetraploid flowers are sib- or cross-pollinated 84% of the time, producing 3x and 4x seeds (progeny). If the 2x and 4x parents of the 3x hybrid have different rind patterns, each of the three-ploidy levels can be distinguished at harvest. For safety, the pollen parent plants should be destroyed after fruit are set on the seed parent plants. A useful combination is for the tetraploid parent to have fruit with a gray rind pattern, and the diploid parent to have fruit with wide stripes, so the resulting triploid hybrid will have striped fruit, easily distinguished from the gray fruited tetraploids that result from self- or sib-pollination of the female parent.

7.12 Seed Harvest Mechanization

The job of watermelon breeding can be made easier and more efficient if mechanization is used for as many steps in the process as possible. Small plot equipment can be used for fieldwork to permit more germplasm to be tested with fewer workers and at a lower cost. Small-plot seeders can be used to plant seeds in the field with optimum seed spacing and planting depth using fewer workers than if seeds are planted by hand. If transplants are used to plant the test plots, machine transplanters can be used to punch the hole before the workers on the machine set the seedling into the hole, and follow up with water and fertilizer after the worker has pressed soil around the seedling, all while riding down the field row. Seeds can be packeted using a seed counter, and plot size can be optimized to gain the maximum information for the lowest cost. Research indicates that optimum plot shape is rectangular and block (replication) shape is square. It is difficult to mechanize harvest since it is done by hand, and each fruit is counted and weighed. However, some efficiency can be gained by using portable computers to collect and analyze data. In the advanced trials, it is useful to estimate flesh sweetness (fruit soluble solids content) using a refractometer, and rind toughness using a spring-loaded punch or penetrometer.

If a greenhouse generation is used to expedite inbred development or hybridization, automation systems are useful for handling the many plants to be grown for self- or cross-pollination. Such systems include automatic heating and cooling, drip irrigation with fertilizer and/or other chemicals injected into the water, trellis support for easy vertical training of the plants, automatic overhead curtains to keep the greenhouse from overheating during the day in the summer, and to keep the greenhouse warmer at night in the winter. Computer systems can provide efficient control of the greenhouse equipment and help provide optimum conditions for plant growth.

For seed harvesting and handling, it is useful to have a bulk seed extractor, washing screens, a seed sluice, and seed dryers. Seed companies have used such machines for years, and it is useful for the plant breeder to build smaller versions that match the size of the plant breeding program. Watermelon breeding is a labour intensive job, but mechanization can help make the most of the available workers, funds, and time.

8 Integration of New Biotechnologies into Breeding

Biotechnologies use an organism or a biological product to manipulate living cells and their molecules. In watermelon, biotechnology is being used to propagate plants in tissue culture, to study genes at the molecular level, to develop molecular markers for selecting useful genes, to isolate the DNA of useful genes, and to incorporate useful genes into cultivars using genetic transformation.

8.1 Tissue Culture

Tissue culture involves the production of cells or plants from plant parts. It offers a method for propagation of valuable plants such as tetraploid parental inbreds, or triploid seedless hybrids. With the increased demand for seedless cultivars, breeders are interested in producing tetraploids in quantities large enough for a hybrid seed production block. Since tetraploids watermelons have slow growth and low seed production, tissue culture can be useful for multiplying new tetraploid lines. Protocols have been developed for propagation of tetraploid plants.

Methods have been developed for the production of tetraploid plants by the regeneration of cotyledons of seedlings cultured in vitro. It was possible to produce tetraploid plants from different watermelon cultivars, and was an efficient alternative to the standard method of using colchicines to double the chromosome number of diploid plants.

8.2 Marker Assisted Selection

Molecular markers (usually sequences of DNA) can be used as reference points in mapping genes on a chromosome. The information is useful in the selection of plants that carry a marked gene of interest. Plant breeders can discard plants from a segregating population that are missing the gene of interest. In that way, field testing can be done using only the plants having a particular set of traits. Markers can also be used to identify cultivars (DNA fingerprinting), and to estimate the genetic relatedness of a set of cultivars or individuals in a population.

Over 40 genes have been described in watermelon. The genes are involved in disease resistance, flower type, fruit shape, and fruit quality. If molecular markers can be identified that are closely linked to those genes, then selection might be performed more rapidly, or earlier. For example, selection might be carried out in a seedling test instead of waiting for the plants to produce fruit.

Molecular markers have been used to estimate genetic relatedness of watermelon cultivars, and can be used to evaluate inbred lines for purity. More than 60 DNA sequences have been published, and some have been used to construct linkage maps. Finally, DNA markers have been used to detect the presence of pathogens on watermelon seeds.

8.3. Genetic Transformation

Genetic transformation provides methods for inserting single genes into plants while overcoming barriers to interspecific crossing. The soilborne bacterium *Agrobacterium tumefaciens* is often used as the vehicle for transferring genes into plants. Plant cells can also be transformed using microparticle bombardment, where DNA-coated particles are shot into plant cells. Several studies have reported transformation of watermelon using Agrobacterium or microparticle bombardment. Transformed cells must by regenerated from sterile culture to produce plants containing the new gene to be used in a breeding program.

Transformation of watermelon plants has been used to confer virus resistance. Several virus species cause disease in watermelon. These include Squash mosaic virus (SQMV), Cucumber mosaic virus (CMV), Papaya ringspot virus-watermelon strain (PRSV-W), Zucchini yellow mosaic virus (ZYMV), and Watermelon mosaic virus (WMV). Transfer of virus coat protein genes into watermelon plants may confer resistance to the virus disease, and provide plant breeders with new resistance genes.

9 Seed Production

Early watermelon cultivars were mostly inbred lines produced commercially by open pollination of bulk-increased or hand-pollinated breeder seeds. In the 1970s, large-scale production of diploid hybrid seed began. Diploid hybrids have now taken over most of the commercial production in North America, Western Europe, and Japan.

9.1 Hybrid Production

Hybrid seeds are produced in the seed parent by pollination from staminate flowers in the male parent. Hybrid production can be by hand pollination using inbred lines grown in adjacent rows in the field, or by planting the two parental lines in an insect-proof cage. Pollinations are marked for later seed harvest using tags or bags after pollination. Each fruit will have 200 to 800 seeds, and fewer than 4000 seeds are needed per acre of commercial production.

A less expensive alternative to hand pollination is to plant the two parental inbreds in an isolation block. Staminate flowers are then removed daily from the plants in the seed parent rows to avoid self- and sib-pollination. All pistillate flowers in the seed parent row that are pollinated during the days of staminate flower removal are tagged for hybrid seed harvest. Another solution would be to incorporate a recessive seedling marker such as non-lobed leaf or the glabrous gene into the seed parent. Seedlings resulting from self- or sib-pollinations would have the marker and could be removed from the planted field or removed from the transplant flats to get 100% hybrid seedlings. Conversion of the seed parent to a near-isogenic male sterile line offers the possibility of hybrid seed production without the work associated with the above three methods. However, genetic male sterility requires that male fertile plants be rogued out of the seed parent rows in the hybrid production block.

Seeds can be sorted after the seed cleaning operation by size, weight, or density to increase the proportion of hybrid seed in the lot. Diploid open-pollinated seed yields should be higher than 251 lb./acre (average for United States in 1976-1977). Very good seed yields would be 400 lbs./acre. Triploid seed yields average about 20-40 lbs./acre (about 10% what diploids would produce).

9.2 Commercial Systems

Most commercial watermelon seed production is located in arid or semi-arid areas of the world such as western China, Chile, Mexico, Thailand, and the United States (California and Colorado). Arid conditions favour the production of high quality, disease-free seeds. With the outbreak of bacterial fruit blotch of watermelon in the late 1980s, seed production in areas of low humidity and no rainfall has become even more desirable in order to produce disease-free seed.

Sanitation is important at all stages of production. Workers should wash their hands with antibacterial soap or rinse them with 70% isopropyl alcohol before handling plants or fruit and between seed lots. All equipment should be cleaned and all soil and plant material removed before use in production areas. Clean and disinfect harvesting tools and equipment with alcohol or 0.5% $NaOCl$ or $Ca(OCl)_2$ between seed lots. Sanitation, harvest, and control procedures for production of foundation and stock (parent) seed should be at least as stringent as that for commercial seed.

The process of growing watermelon seed crops is similar to that for growing market crops except that site selection is more critical. Choose a field that has not had any cucurbits (watermelon, cantaloupe, honeydew, cucumber, summer or winter squash, pumpkin, or gourd) in it for at least 2, but preferably 4 years. A field that has a history of Fusarium wilt or anthracnose should be avoided. Fields for open-pollinated watermelon seed production should be isolated by at least 1 mile from other watermelon fields to prevent contamination by outcrossing. Isolation also prevents disease spread from fields containing watermelon and cantaloupe crops of unknown origin or planted with seeds that have not been tested for seed-borne disease. The production site should be as far as possible from fields where bacterial fruit blotch occurred the previous year to reduce contamination from leftover debris. Wild cucurbits, such as citron and volunteer watermelons, must be removed from a 1-mile radius surrounding the production field to eliminate outcrossing and disease contamination.

Selection of parental seed from elite or foundation seed is the first critical element of seed production. Use seed that was produced in dry climates and has been tested to be free of the pathogens causing gummy stem blight, watermelon fruit blotch, anthracnose, and squash mosaic. Direct-seeded plantings reduce the risk of seedling contamination in greenhouses. If transplants are used, they should be produced in a greenhouse that does not contain other cucurbits. Irrigation of transplants in the greenhouse preferably should be from an ebb and flow or a float system. Overhead irrigation of seedlings in the greenhouse should be avoided. Greenhouses for transplant production should have good air circulation and low relative humidity.

Drip or furrow irrigation should be used in the production field instead of overhead irrigation to reduce leaf wetting and disease spread. Roguing of off-type and diseased plants within the field should be done throughout the growing season. There are four useful stages for roguing. The first is before flowering when vegetative characters are checked. The second stage is at early flowering when morphology of undeveloped fruit is checked. The third stage is when the developing fruit are checked for trueness to type, and the final roguing is confirming the external morphological characters of the fruit to be harvested. Roguing for off-types is not effective after pollination in a field for open-pollinated seed production. It is only effective when fruit have been self or cross-pollinated and the male has no off-types. Inspectors should be trained to recognize variations in watermelon fruit blotch symptoms.

Preventative applications of copper fungicide can also help in reducing fruit blotch contamination of seed. The first spray should be 2 weeks before flowering. Application of registered fungicides will reduce gummy stem blight seed contamination. Seed should not be harvested from fields where there is confirmation of fruit blotch or until the possibility of fruit blotch is eliminated. Seeds harvested from fields in which fruit blotch is confirmed or which were adjacent to contaminated fields should not be used.

All fruit should be inspected by trained technicians for symptoms that are suspected to be fruit blotch. All fruit suspected of having fruit blotch must be discarded. No fruit should be harvested from vines that have anthracnose or gummy stem blight symptoms. When seeds of open-pollinated fruit, and in some cases, hybrid fruit, are mature the fruit are windrowed by machine. Windrowed fruit are picked up by self-propelled vine seed harvesters that crush the fruit and separate the seeds and pulp from the rind. For some hybrid seed production, fruit are harvested by hand and various sized seed extractors are used. In either case, the diploid seed slurry is transferred to bins where it is allowed to ferment for 24 to 48 hours. During this time the sugars and gelatinous material surrounding the seeds are degraded.

Fermentation plus acid washing (1% hydrochloric acid) can reduce the chance of seed transmission of fruit blotch. Fermentation and acid treatment of triploid seed reduces seed viability, so is not recommended. Seeds extracted from tetraploid fruit for triploid seed production should be washed immediately. Seeds are separated from pulp and juice by washing in a rotary washer or flume system. Some seed lots are dried by heat from the sun. However, higher quality seeds are produced using forced air warmed by propane heaters. Seeds are placed on flat drying beds or in large rotary dryers. Dry seeds are run through a mill containing sizing screens that separates large seeds from trash and small seeds.

All seed lots should be assayed for the presence of the fruit blotch bacterium, squash mosaic virus, and gummy stem blight pathogen by the best methods available. In Asia, cucumber green mottle virus is a problem and is seed transmitted. For fruit blotch, seedling grow-outs of at least 10,000 seeds per lot are currently used, but polymerase chain reaction (PCR) techniques may provide more efficient and sensitive methods. Coupling seedling grow-outs with PCR may be necessary for some situations. Squash mosaic virus can be screened with grow-outs. For gummy stem blight, seedling grow-outs or blotter tests using a minimum of 1,000 seeds per lot are recommended. However, PCR techniques may provide better methods in the future. Commercial seeds should be treated with a registered protectant such as Captan and Thiram before sealing them into cans, bags, or packets. Seeds should be stored in hermetically sealed containers at 6.5% (no greater than 10%) moisture content. Under favourable storage conditions, seeds should last 4 years. To be salable, germination of the seed lot must be at least 70%.

References

Crall, J. 1981. Fifty years of watermelon breeding at ARC Leesburg. Proc. Fla. State Hort. Soc. 94: 156-158.

Eigsti, O. J. and Dustin, P. 1955. Colchicine in agriculture, medicine, biology, and chemistry. Iowa State College Press. Ames.

Fehner, T. 1993. Watermelon, *Citrullus lanatus* (Thunb.) Matsum. & Nakai. p. 295-314. In: G. Kalloo and B. O. Bergh (eds.). Genetic Improvement of Vegetable Crops. Oxford, Pergamon Press. New York.

Kihara, H. 1951. Triploid watermelons. Proc. Amer. Soc. Hort. Sci. 58: 217-230.

Mohr, H. C. 1986. Watermelon breeding. p. 37-66 In: M.J. Bassett (ed.). Breeding vegetable crops. AVI Publishing Co. Westport, Conn.

Orton, W. A. 1907. A study of disease resistance in watermelons Science 25: 288.

Parris, G. K. 1949. Watermelon breeding. Econ. Bot. 3: 193-212.

Porter, D. R. 1933. Watermelon breeding. Hilgardia 7: 533-552.

Rhodes, B. and Dane, F. 1999. Gene list for watermelon. Cucurbit Genetics Coop. Rpt. 22: 61-77.

Rhodes, B. and Zhang, X. 1999. Hybrid seed production in watermelon. p. 69-88 In: A. S. Basra (ed.). Food Products Press, New York.

Robinson, R. W. 2000. Rationale and methods for producing hybrid cucurbit seed. p. 1-47. In: A. S. Basra (ed.). Food Products Press, New York.

Robinson, R. W. and Decker-Walters, D. S. 1997. Cucurbits. CAB International, New York, NY; 226 pp.

Robinson, R. W., H. M. Whitaker, and Bohn, G. W. 1976. Genes of the Cucurbitaceae. HortScience 11: 554-568.

Wehner, T. C. 1999. Heterosis in vegetable crops. p. 387-397. In: J. G. Coors and S. Pandey (eds.). Genetics and exploitation of heterosis in crops. Amer. Soc. Agron., Madison, Wis.

Whitaker, T. W. and Davis, G. N. 1962. Cucurbits: botany, cultivation, and utilization. Interscience Publishers, Inc., New York.

Index

Acidovorax avenae, 398-399
Albugo occidentalis, 197
Arabidopsis, 26, 41, 105-106, 134, 137, 142, 160, 303
artichoke (*see* globe artichoke and cardoon)

Bacillus thuringiensis, 140
betalains, 233
Beta vulgaris, 219-220 (*see also* table beet)
Brassica oleracea, 119, 121, 123, 151-152, 158 (*see also* cabbage and kale, and cauliflower and
broccoli (*see* cauliflower and broccoli) broccoli)
Bremia lactucae, 89

cabbage (*see* cabbage and kale)
cabbage and kale, 119-142
- breeding achievements, 127-134
- breeding objectives, 135-139
- classification, 119
- disease resistance, 131-136
- diseases, 131
- domestication, 123-124
- economic importance, 121
- evolution, 123-124
- genetic maps, 122
- genetic resources, 125-127
- genetic transformation, 139
- germplasm collections, 125-127
- germplasm conservation, 126-127
- glucosinolates, 129, 132-134, 136-138, 141
- haploids, 138-139
- heterosis, 138
- hybrids, 127, 138, 142
- male sterility, 141
- molecular markers, 136, 141-142
- morphology, 121
- nutritional quality, 132-134
- origin, 122-124
- pest resistance, 129-131, 136-137, 139-140
- pests, 129-131
- QTLs, 136, 139, 141
- quality, 128
- reproductive biology, 122, 138
- seed production, 142
- selection, 138
- self-incompatibility, 122, 128, 142
- taxonomy, 119-120
- tissue culture, 138-140
- transgenics, 137, 139-140
- uses, 121

U triangle, 120
varietal groups, 121, 123-125
wild relatives, 123, 126
cardoon (*see* globe artichoke and cardoon)
cauliflower (*see* cauliflower and broccoli)
cauliflower and broccoli, 151-176
abiotic stress resistance, 164
antioxidants, 166
breeding achievements, 163-164
breeding objectives, 164-167
cultivars, 154-158
curd colour, 165-166
curd development, 154, 160-161
cytodemes, 161
cytogenetics, 158-160
disease resistance, 163
domestication, 152-154
economic importance, 151
evolution, 152
flowering, 175
gene pools, 159-160
genetic resources, 158-162
genomes, 158-160
germplasm collections, 161-162
glucosinolates, 164, 166, 170
haploids, 172-174
harvest index, 163
history, 152-154
hybrids, 157, 163, 168-170, 172, 175
in vitro culture, 171-174
interspecific hybridization, 170-171
male sterility, 168-169, 172
molecular markers, 152, 172
nutritional quality, 151
origin, 152-154
pest resistance, 163-164
pigments, 165-166
pollination, 168, 175
protoplasts fusion, 172
quality, 163, 165-168
reproductive biology, 168-169, 175-176
resistance genes, 164
seed production, 167-170, 174-176
self-incompatibility, 168-170, 176
taxonomy, 158
U triangle, 158
varietal groups, 154-158
wild relatives, 152, 159-160, 170, 174
Cercospora beticola, 232
chicory (*see* chicory and endive)
chicory and endive, 3-44
androgenesis, 37
backcrossing, 19
breeding objectives, 18-22
Cichorium spinosum, 5
cultivar groups, 3-4, 8-17, 43-44
curled endive, 3
cytoplasmic male sterility, 28-30, 38
diseases, 25-26
economic importance, 4
escarole, 3
flavonoids, 25
fructans, 22-24
fructose, 24-25
genetic maps, 31-32
genetic relationships, 33
genetic transformation, 37
genetic variation, 18
herbicide resistance, 26-27
heterosis, 17, 21
hybrids, 17, 19-22, 40, 44
industrial applications, 22-25
in vitro culture, 35-36
marker assisted selection, 30
molecular markers, 6, 20, 22, 30-35
morphology, 7
organogenesis, 35-36
origin, 6-7
pests, 26
phenolics, 25
pollination, 9
protoplasts fusion, 37-38

QTLs, 30-32
reproductive biology, 9, 17-18, 27-29, 39-41
resistance genes, 26
root chicory, 3
seed production, 38-43
selection, 19-22, 42
self-incompatibility, 27-28, 34
somatic embryogenesis, 36
synthetics, 21
taxonomy, 5-7
vernalization, 39-41
wild relatives, 5
Cichorium endivia, 5, 17, 19, 39, 43 (*see also* chicory and endive)
subsp. *endivia* var. *crispum*, 10
subsp. *endivia* var. *latifolium*, 10
Cichorium intybus, 5, 17, 20, 41, 43 (*see also* chicory and endive)
subsp. *intybus* var. *foliosum*, 10-17
subsp. *intybus* var. *sativum*, 10-17
Cichorium spinosum, 5
citron, 385
Citrullus lanatus, 381
colchicine, 410
Colletotrichum lagenarium, 497
cucumber, 241-273
architectural habit, 258-259
backcross breeding, 258
breeding achievements, 272-273
correlations, 249, 251-252
cultivars, 244-246, 259
diffusion, 242-243
disease resistance, 245-246, 254, 270
diseases, 245-246
domestication, 242-243
earliness, 251, 269
economic importance, 241
fruit size, 252, 270
gene pools, 246-247
genetic maps, 261-264
genetic markers, 265-267
genetic resources, 246-247
germplasm, 243, 247
germplasm collections, 247
gynoecy, 251
harvest index, 248
heritability, 248-249
history, 242-246
hormones, 250-252, 260
hybrids, 250-251, 258, 260
hybrid testing, 258
line extraction, 257-258, 272
marker assisted selection, 249, 260-272
molecular markers, 260-272
multiple lateral branching, 252, 269
nematode resistance, 259
origin, 242-244
parthenocarpy, 252-253, 270
pedigree breeding, 257
pests resistance, 254
plant habit, 245-246, 258
pollination, 260
population development, 256, 271
QTLs, 264-270
quality, 253-254, 270
recurrent selection, 256
reproductive biology, 245-246, 250-251
resistance genes, 254
seed production, 260
selection, 249, 255
sex expression, 245-246, 250-251, 268-269, 273
single-seed descent, 257
stress resistance, 255
taxonomy, 241-242
uses, 244-246
varietal groups, 244-246
wild relatives, 241-243
yield, 248-249, 268-270
yield components, 249, 268-270
cucurbitacin, 331
Cucurbita pepo, 351 (*see* pumpkin and winter squash, and summer squash)

Cucurbita sp., 317-318, 351 (*see* pumpkin and winter squash)
Cucumis hystrix, 241-244, 247
Cucumis melo, 283-284 (*see also* melon)
Cucumis sativus, 241 (*see also* cucumber)
curled endive, 3
Cynara cardunculus, 49 (*see also* globe artichoke and cardoon)
  var. *scolymus*, 49
  var *altilis*, 49-50

Daucus carota, 36
Didymella brioniae, 293, 397

egusi, 385-386
endive (*see* chicory and endive)
escarole, 3

flavonoids, 25
Fusarium oxysporum, 91, 292-293, 395-396
fructans, 22-24, 50
fructose, 24-25
globe artichoke and cardoon, 49-68
  antioxidants, 50-51
  breeding achievements, 59
  breeding objectives, 60-62
  cultivars, 53-55, 60-61
  diseases, 62
  domestication, 52-53
  economic importance, 49-50
  genetic diversity, 56-57, 58-59
  genetic maps, 62
  genetic resources, 55-58
  haploids, 64
  heritability, 59
  hybrids, 66-68
  inheritance of traits, 58-59
  inulin, 50
  in vitro culture, 64
  male sterility, 66-67
  micropropagation, 64
  molecular markers, 52, 54, 56-57, 62-63
  origin, 52-53
  pests, 62
  phenolics, 65
  QTLs, 62
  secondary metabolites, 65
  seed production, 65-68
  seed propagated cultivars, 66-68
  uses, 50-51
  varietal groups, 53-55
  wild relatives, 52
glucosinolates, 129, 132-134, 136-138, 141, 164, 166, 170
Golovinomyces cichoracearum, 291-292, 333, 367-368

kale (*see* cabbage and kale)

Lactuca saligna, 84
Lactuca sativa, 75 (*see also* lettuce)
Lactuca serriola, 83
Lactuca virosa, 84
lettuce, 75-108
  abiotic stresses, 104
  backcross, 99
  bacteria, 92-93
  breeding achievements, 86-89
  breeding objectives, 89-96
  composition, 96
  crossing technique, 97-98
  cultivars, 78, 81-82, 89, 99
  diseases, 83-85, 86-87, 89-93,103
  domestication, 77
  fungi, 89-91
  genetic transformation, 103-105
  genomics, 105-106
  germplasm collections, 85-86
  economic importance, 75-76
  herbicide resistance, 104
  history, 77-79
  hybridization, 97-98
  hybrids, 97
  markers assisted selection, 101
  molecular markers, 100-103, 105-106
  morphology, 77
  origin, 77-78

pedigree method, 98-99
pests, 83-85, 87, 93-94
physiological disorders, 94
QTLs, 102-103, 105-106
quality, 87-88, 95-96, 104-105
reproductive biology, 97-98, 107-108
resistance genes, 89, 91-93
seed conservation, 108
seed production, 106-108
selection, 99
uses, 75
taxonomy, 83
tissue culture, 100
transgenics, 104-105
varietal types, 79-82
viruses, 91-92
wild relatives, 77, 83, 95, 100, 102-103
yield, 88-89, 94-95

Meloidogyne, 259, 399
melon, 283-306
bitterness, 300
botanical groups, 284-290
center of diversification, 285
composition, 298-299
disease resistance, 291-296
diseases, 291-296, 306
domestication, 284
downy mildew, 292
economic importance, 283
flesh colour, 299
Fusarium wilt, 292-293
genetic maps, 302-305
genetic resources, 290
genetic transformation, 303
genome, 302
germplasm collections, 290
gummy stem blight, 293
haploids, 302
heterosis, 300-301
history, 284
hybrids, 300-301, 305-306
interspecific hybridization, 302
in vitro culture, 303
male sterility, 305
marker assisted selection, 303
molecular markers, 290, 302-303
morphology, 285-290
origin, 284
parthenocarpy, 300
pests, 295-296
plant growth, 296-297
plant habit, 296-297
pollination, 300-301
powdery mildew, 291-292
QTLs, 297, 300, 302-305
quality, 298-300
resistance genes, 291-294
reproductive biology, 297-298, 300-301
seed production, 305
sex expression, 285-290, 297-298
skin colour, 300
tetraploids, 302
transgenics, 303
tribes, 285
uses, 283-284
varietas, 284-290
vegetative propagation, 301, 305
volatile compounds, 300
viruses, 293-295, 306

phenolics, 25, 65, 189, 211
Peronospora farinosa, 196, 212
Plasmodiophora brassicae, 131-132
Podosphera xanthii, 291-292, 333, 367-368
Pseudoperonospora cubensis, 292, 333-334, 368
pumpkin (*see* pumpkin and winter squash)
pumpkin and winter squash, 317-343
adaptation, 332
breeding achievements, 330-335
bitterness, 331
carotenoids, 336-337
cucurbitacin, 331

cultivars, 320-327, 334-335
disease resistance, 333-334
diseases, 333-334, 338
domestication, 318-320, 331
downy mildew, 333-334
dry matter, 336-337
economic importance, 317-318
flowering, 332
fruit colour, 331, 336-337
fruit morphology, 320-327
fungi, 333-334
genetic maps, 341
genetic resources, 327-330
genetic transformation, 342
germplasm collections, 329-330
heterosis, 340
history, 318-320
hybrids, 339-340, 342-343
hull-less seeds, 334-335
interspecific hybridization, 338-339
in vitro culture, 340
male sterility, 343
marker assisted selection, 341
molecular markers, 340-341
nutritional value, 336
origin, 318-321
pest resistance, 333
pests, 333
plant architecture, 331-332
pollination, 338-339, 342-343
powdery mildew, 333
pumpkin oil, 334-335, 337
quality, 336-337
reproductive biology, 332, 338-339, 342-343
resistance genes, 333-334
rootstocks, 338
seed conservation, 343
seed production, 342-343
seed pumpkins, 334-335, 337
selection, 338-340
sex ratio, 332
spread, 320-324
uses, 317-318, 337
varietal groups, 317, 320-327
wild relatives, 318-320, 327-328
yield, 331-332
ZYMV, 334, 342

Rhizoctonia solani, 232-233
root chicory, 3

Sclerotinia sclerotiorum, 90
Sphaeroteca fuliginea, 397
spinach, 189-213
antioxidants, 189, 211
betalains, 233
breeding objectives, 196-202
composition, 189-190, 211
cultivars, 194, 198-202
disease resistance, 208-209
diseases, 196-202, 208-209
downy mildew, 196, 198-199
economic importance, 189, 191
genetic resources, 194-195
genetic transformation, 212
germplasm collections, 195
history, 194
hybrids, 194, 204-206
marker assisted selection, 212-213
molecular markers, 212-213
nutritional quality, 189-190, 211-212
origin, 191, 194
oxalate, 211-212
pests resistance, 209-211
pests, 209-211
pigments, 189, 211
phenolics, 189, 211
pollination, 205-206, 208
reproductive biology, 195-196, 202-204
resistance genes 198-202
root rot, 232-233
seed production, 202-208
selection, 194
sex determination, 195-196
sex reversion, 204
transgenics, 212

uses, 189
white rust, 197, 199-202
Spinacia oleracea, 190 (*see also* spinach)
summer squash, 351-373
bitterness, 361
branching, 362
breeding achievements, 361-363
CMV, 366
colour, 361, 365
composition, 364
cultivars, 358, 360-361
diseases, 366-368
disease resistance, 363
domestication, 352-354
downy mildew, 368
earliness, 363
economic importance, 351
etymology, 352
flavor, 364
flowering, 369-370
fruit shape, 356-357, 361, 365
fruit size, 361
genetic maps, 372
genetic resources, 359-361
germplasm conservation, 360-361
habit, 362, 365-366
history, 352-355
hybrids, 363, 372-373
landraces, 360
leaf silvering, 368
male sterility, 373
molecular markers, 359, 372
morphology, 354
origin, 352-354
parthenocarpy, 365
pests, 368
plant architecture, 362
pollination, 369-371
powdery mildew, 363-367
PRSW-W, 366-367
races, 355
reproductive biology, 353, 369-371
seed production, 372-373
selection, 371
sex expression, 362, 369, 373
taxonomy, 355
transgenics, 372
uses, 352-354, 357
varietal groups, 354-359, 362-363
viruses, 366-367, 372
wild relatives, 352-353, 359-360
WMV, 366-367
zucchini, 363
ZYMV, 366-367

table beet, 219-235
bolting, 227-228
breeding objectives, 225-227, 232-235
cultivars, 223-224
diffusion, 221-222
disease resistance, 225
diseases, 225, 232-233
domestication, 220-222
economic importance, 219-220
evolution, 222
flowering, 227-228
gene pools, 224
genetic resources, 224-225
history, 221-224
hybrids, 229-231
inbred lines, 229
inbreeding depression, 231
leaf blight, 232
male sterility, 226, 229, 231
molecular markers, 225, 227, 235
multigerm seeds, 227
origin, 220-222
pest resistance, 225
pigments, 233-234
pollination, 225, 229-231
quality, 234-235
reproductive biology, 225-228
seed production, 227-229
uses, 219-222
varietal groups, 223-224

selection, 222
self-incompatibility, 225-226
transgenics, 227, 235
vernalization, 227-228, 230
wild relatives, 220, 225

Verticillium dahliae, 91

watermelon, 381-417
anthracnose, 397
backcross, 407
bacterial fruit blotch, 398-399
bacterial rind necrosis, 399
breeding achievements, 388-389
centers of diversity, 384-385
citron, 385
commercial types, 382
composition, 382
cultivars, 386-389
disease resistance, 395-400
diseases, 395-400
domestication, 383
earliness, 391
economic importance, 381
egusi, 385-386
flavor, 393
flesh color, 393
flowering, 382, 402-405
fruit shape, 391-392
fruit size, 391
Fusarium wilt, 395-396, 401-402
genetic resources, 386-388
germplasm collections, 386
gummy stem blight, 397
inbred development, 407-408
inbreeding depression, 389
heterosis, 389, 408-409
history, 386-388
hybrids, 386-389, 394, 408-412, 415
male sterility, 390
marker assisted selection, 414
molecular markers, 414, 417
nematodes, 399
nutritional value, 382
origin, 383-384
pedigree breeding, 406
pests resistance, 400
pollination, 388, 402-405, 407, 412, 415
population development, 406
powdery mildew, 397
quality, 392-393
recurrent selection, 405-406
reproductive biology, 382, 390, 402-405
resistance genes, 388
rind pattern, 392
rind pickles, 393-394
seedlessness, 390, 394, 409
seed production, 383, 411, 412-413, 414-417
selection, 393, 405-407
sex expression, 390, 401
stress resistance, 400
tetraploids, 409-413
tissue culture, 413
triploids, 394, 409-412
uses, 381-382
varietal groups, 385
vine length, 389-390
viruses, 399
wild relatives, 384-385
yellow vine, 398
yield, 390
winter squash (*see* pumpkin and winter squash)

Xanthomonas campestris, 131

zucchini, 363 (*see also* summer squash)

Printed in the United States of America